Agricultural Soil Science: Properties and Management

Agricultural Soil Science: Properties and Management

Edited by Rick Thomas

SYRAWOOD
PUBLISHING HOUSE
New York

Published by Syrawood Publishing House,
750 Third Avenue, 9th Floor,
New York, NY 10017, USA
www.syrawoodpublishinghouse.com

Agricultural Soil Science: Properties and Management
Edited by Rick Thomas

International Standard Book Number: 978-1-68286-508-8 (Hardback)

Cataloging-in-Publication Data

Agricultural soil science : properties and management / edited by Rick Thomas.
p. cm.
Includes bibliographical references and index.
ISBN 978-1-68286-508-8
1. Soil science. 2. Agriculture. 3. Farm management. I. Thomas, Rick.
S591 .A37 2018
631.4--dc23

Printed in China.

TABLE OF CONTENTS

PREFACE

The purpose of the book is to provide a glimpse into the dynamics and to present opinions and studies of some of the scientists engaged in the development of new ideas in the field from very different standpoints. This book will prove useful to students and researchers owing to its high content quality.

Agricultural soil science is a sub discipline of soil science which studies the edaphic conditions related to the production of food. It provides a holistic understanding of soil and its relation to the ecosystem. Agricultural soil science helps in studying the chemical, biological and physical composition of soils, particularly in relation with agriculture. Agricultural scientists study methods of making soil more productive. Those in search of information to further their knowledge will be greatly assisted by this book. It is a valuable compilation of topics, ranging from the basic to the most complex advancements in the field of agricultural soil science. This book includes contributions of experts and scientists, which will provide innovative insights into this field.

At the end, I would like to appreciate all the efforts made by the authors in completing their chapters professionally. I express my deepest gratitude to all of them for contributing to this book by sharing their valuable works. A special thanks to my family and friends for their constant support in this journey.

Editor

Effect of Different Ammonia Concentrations on Community Succession of Ammonia-oxidizing Microorganisms in a Simulated Paddy Soil Column

Hu Baolan, Liu Shuai, Shen Lidong, Zheng Ping, Xu Xiangyang, Lou Liping*

Department of Environmental Engineering, Zhejiang University, Hangzhou, Zhejiang, China

Abstract

Ammonia oxidation is performed by both ammonia-oxidizing bacteria (AOB) and ammonia-oxidizing archaea (AOA). To explore the effect of ammonia concentration on the population dynamic changes of ammonia-oxidizing microorganisms, we examined changes in the abundance and community composition of AOA and AOB in different layers. Most of the archaeal *amoA* sequences were *Nitrosotalea*-related and the proportion that *Nitrosotalea* cluster occupied decreased in the surface layer and increased in the deep layer during the cultivation process. *Nitrosopumilus*-related sequences were only detected in the deep layer in the first stage and disappeared later. Both phylogenetic and quantitative analysis showed that there were increased *Nitrosomonas*-related sequences appeared in the surface layer where the ammonia concentration was the highest. Both AOA and AOB OTU numbers in different layers decreased under selective pressure and then recovered. The potential nitrification rates were 25.06 $\mu g \cdot N \cdot L^{-1} \cdot g^{-1}$ dry soil$\cdot h^{-1}$ in the mid layer which was higher than the other two layers. In general, obvious population dynamic changes were found for both AOA and AOB under the selective pressure of exogenous ammonia and the changes were different in three layers of the soil column.

Editor: Matthias Horn, University of Vienna, Austria

Funding: The authors wish to thank the National Natural Science Foundation of China (40801198), National Key Technologies Research and Development program of China (2012BAJ25B00) and the Natural Science Foundation of Zhejiang province (Y507227). The funders had no role in study design, data collection and analysis, decision to publish, or preparation of the manuscript.

Competing Interests: The authors have declared that no competing interests exist.

* E-mail: baolanhz@hotmail.com

Introduction

Nitrification is critical in the global nitrogen cycle and is a unique pathway linking reduced inorganic nitrogen with oxidised inorganic nitrogen in nature [1]. Ammonia oxidation is the first and rate-limiting step of nitrification. It was assumed that all ammonia oxidisers were bacteria until ammonia-oxidizing archaea were found. The findings of Venter *et al.* [2] and Treusch *et al.* [3] indicated the posibility of archaea involving in ammonia oxidation and the exist of ammonia-oxidation archaea was finally certified by the enrichments and pure cultures of several strains of AOA [4–10]. Recent research has demonstrated that ammonia-oxidizing archaea may form a separate and deep-branching phylum, the Thaumarchaeota [11–13].

Molecular biological research based on the functional gene *amoA* and the 16SrRNA gene has revealed a widespread distribution of AOA [14–17]. Quantitative PCR analyses found that AOA normally outnumbered their substrate competitors, such as AOB, in both soil [17] and water [18], but the relative contributions of AOA and AOB in the soil nitrogen cycle remained undetermined. Some researchers believed that AOB played a more important role than AOA in soil nitrification. Jia and Conrad [19] found that changes in potential nitrification rates were only correlated with the number of AOB *amoA* genes when they added substrate (NH_3) or inhibitor (C_2H_2) of ammonia oxidation to the soil. Subsequently, Di *et al.* [20] discovered the same phenomenon in New Zealand grassland. In contrast, other researchers regarded AOA as the main drivers of ammonia oxidation in soil. Gubry-Rangin *et al.* [21] reported that the potential nitrification rates were positively correlated with the number and transcriptional activity of AOA *amoA* genes. Zhang *et al.* [22] confirmed the dominant role of AOA in soil nitrification by means of stable isotope probe and the quantitative analysis of archaral *amoA* genes. Offre *et al.* [23] also demonstrated that the growth of only archaeal but not bacterial ammonia oxidizers occurred in microcosms with active nitrification. Differences in cell size, specific cell activity and other physiological characteristics may explain the different contributions of AOA and AOB to soil nitrification. In addition, AOA and AOB may compete mutually or exhibit functional redundancy under some conditions, whereas under other conditions the fundamental physiological differences between these organisms may lead to niche separation of AOA and AOB [24]. Therefore, environmental factors are important in determining the different nitrification activities and relative contributions of AOA and AOB.

Among all environmental factors, ammonia is the substrate of ammonia oxidation for which concentrations will directly affect nitrification activity. It was found that AOA were able to grow well and the growth of AOA was coupled with soil nitrification when the concentration of ammonia was relatively low or the supply of ammonia was through the mineralization of organic matter; however, AOB were more competitive in soil nitrification and the number of AOB *amoA* gene copies was greater than that of AOA when the concentration of ammonia was higher [19,20,23–26]. Adaptation to long-term energy stress is believed to be a crucial factor that distinguishes AOA from AOB [27]. Different specific affinities for substrate between AOA and AOB may explain their different growth patterns under low or high ammonia concentrations. *N. maritimus* exhibited a high affinity for ammonia and was able to grow and convert ammonia at an extremely low ammonia concentration (0.14 μg/L), while the ammonia-oxidizing activity was completely inhibited when the ammonia concentration reached 28 mg/L. The affinity of *N. maritimus* for ammonia can be more than 1,000-fold greater than that of *N. europaea* [28]. The affinity for ammonia of *N. viennensis* fell in between *N. maritimus* and *N. europaea*, and their growth was totally inhibited at ammonia concentration of 280 mg/L [7]. Other than ammonia concentration, other environmental traits such as pH [29], oxygen concentration [18] and organic carbon [30] could affect the abundance and diversity of ammonia-oxidizing microorganisms and consequently lead to niche separation of AOA and AOB.

Paddy soils are an important component of agricultural ecosystems in Asia, and these soils represent one of the major cultivated soil types in China. China has a long history of rice production, and the total area of the paddy soils in the country is 46 M ha [31]. Thus, research on nitrification in paddy soils is of great significance for improving our understanding of nitrogen fluxes in agricultural and soil ecosystems. The abundance and community structure of ammonia-oxidizing microorganisms in paddy soils have been described [32], but the mechanism underlying the response of these microorganisms to ammonia stress is unclear. Therefore, the main objective of this study was to verify the significant role that ammonia concentrations play in the population shifts of AOA and AOB in paddy soils.

Materials and Methods

Ethics Statement

No specific permits were required for the described field studies.

Simulated Soil Column and Sampling

The flooded paddy soil used in this study was collected from paddy land (30°26'N, 120°19'E) in Zhejiang Province, China. The upper 50 cm of paddy soil was collected and filled into the soil column. The soil column was divided into three layers due to the definition and classification of soil layer in pedology and the setting of sampling ports, the surface layer (2–5 cm), the mid layer (17–20 cm) and the deep layer (47–50 cm), to investigate the differences in ammonia-oxidizing microorganisms at different depths.

The simulated soil column (Fig. 1) was made of plexiglass and had a height of 70.0 cm and a diameter of 12.5 cm.It had a working volume of 2.5 L and a total volume of 3.0 L. Spherical fillings placed at the bottom of the column acted as a filter to avoid loss of biomass. Paddy soil in agricultural ecosystem were usually irrigated with water from river or rainfall, so natural river water were used in this study to simulate the in situ environment of paddy soil. Therefore, the experimental medium was prepared with natural river water (NH_4^+-N, 0.53±0.1 mg/L; NO_2-N, 0.1±0.03 mg/L; NO_3-N, 3.72±0.5 mg/L, filtrated by filter) adding ammonia, with final ammonia concentration of 42 mg/L. medium was introduced at the top of the column by spraying (5L per day), and then permeated to the deeper layer. Due to the consumption of microorganisms and adsorption by the soil, an ammonia concentration gradient formed in the soil column from the surface layer to the deep layer (surface layer, 143 μg NH_4^+-N g^{-1} dry soil; mid layer, 61.9 μg NH_4^+-N g^{-1} dry soil; deep layer, 39.9 μg NH_4^+-N g^{-1} dry soil). The soil column was placed in thermostatic chamber and the temperature was 30±1°C. From the top layer to the deep layer, pH of the soil were 7.24±0.12, 7.41±0.1, 7.09±0.15 and the humidity were 38.7±1%, 36.5±0.8% and 34.4±1.5%, respectively. The oxygen concentration in the top soil of the column was 2.48 mg/L.

Soil sampling was performed every six months, constituting the first stage, the second stage, the third stage and the last stage of the cultivation process. For each layer, triplicate samples were collected in each sampling port shown in Figure1 (from the point close to the sampling connection to the point off the sampling connection) using miniature soil sampler and mixed together. The samples were used to measure ammonium, nitrite and nitrate and a small subsample was stored at −80°C for DNA extraction.

Potential Nitrification Rates

The potential nitrification rates in the soil layers were determined via batch tests using 150 mL conical flasks containing 80 mL of liquid. The medium for determining the rates contained 3.5 mg/L NH_4^+-N. Medium with sterilized soil was incubated as a control. Tests were performed in triplicate for 8 hours at 30°C in the dark, and the concentrations of NH_4^+-N, NO_2^-N and NO_3^-N were periodically monitored per 1.5 hours during incubation. Potential nitrification rates were characterized on the basis of the consumption of NH_4^+-N. In consideration of some ammonia-oxidizing microorganisms which could accumulate substrates intracellularly [33], the results of potential nitrification rates we obtained may be to some extent higher than the real situation.

DNA Extraction and PCR Amplification

DNA was extracted using the Power Soil DNA Kit (Mo Bio Laboratories, California, USA) following the manufacturer's instructions. The extracted DNA was examined in a 1.0% agarose gel by electrophoresis.

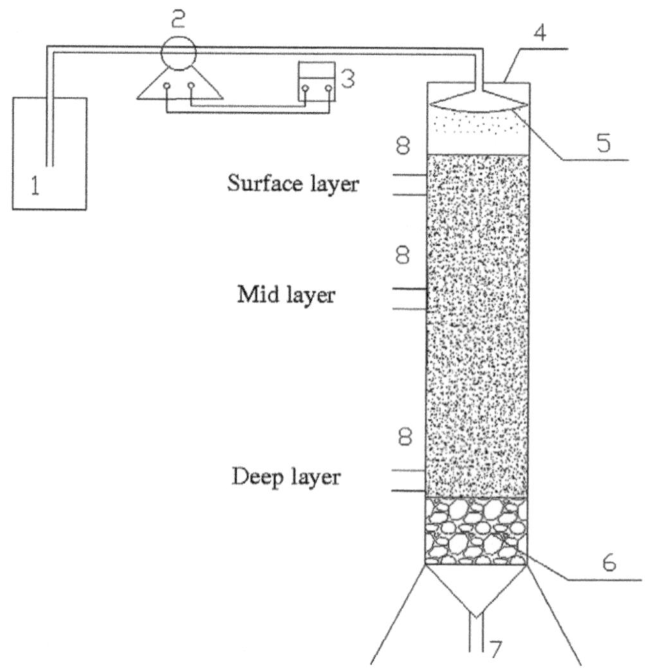

Figure 1. The simulated paddy soil column used in this study. (1) inlet pail, (2) peristaltic pump, (3) timer, (4) main body of soil column, (5) sprayer, (6) beaumontage, (7) outlet pipe, (8) sample connection.

Primers sets Arch-amoAF, Arch-amoAR [34] and amoA-1F, amoA-2R [35] were used for the amplification of archaeal and bacterial *amoA* genes, respectively. The PCR thermal cycling programs employed were described previously [36]. The amplified products were examined in a 1.0% agarose gel by electrophoresis.

Cloning and Sequencing

The obtained PCR products were cloned using the pMD19-T vector (TaKaRa, Bio Inc., Shiga, Japan) according to the manufacturer's instructions. Plasmid DNA was isolated using the Gene JET^{TM} Plasmid Miniprep kit (Fermentas Life Sciences, Burlington, Canada) and was digested with 5 U of *Eco*RI enzyme in *Eco*RI buffer for 1.5 h at 37°C. The digestion products were examined for an insert of the expected size using agarose (1.0%) gel electrophoresis. At least 20 positive clones from each library were randomly selected for sequencing using an ABI3100 automated sequencer (Applied Biosystems, California, USA).

Phylogenetic Analyses

All of the *amoA* gene sequences obtained were imported into MEGA4.1 to construct alignment files in combination with the most similar environmental sequences and sequences of known strains of AOA and AOB. Phylogenetic trees were constructed using the neighbour-joining method. Bootstrap analysis with 1000 replicates was applied to estimate the confidence values for the tree nodes.

Statistical Analyses

Operational taxonomic unit (OTUs) was used to determine the archaeal and bacterial *amoA* gene diversity in different soil samples. For both AOA and AOB, researchers found that 1%–3% 16S rRNA distance equate to a 15% *amoA* gene distance [37,38]. 85% sequence identity at the *amoA* gene level could be regarded as an approximate threshold below which ammonia-oxidizing microorganisms can be assigned to different species. Hence, OTUs were defined by 15% differences for both archaeal *amoA* gene and bacterial *amoA* gene in nucleotide sequences. OTU numbers were determined by the farthest neighbour algorithm in the DOTUR program [39]. The coverage of the clone libraries was calculated as $C = [1 - (n_1/N)] \times 100$, where n_1 is the number of unique OTUs and N is the total number of clones in a library.

Real-time Quantitative PCR (qPCR)

The primer used to quantify the copy numbers of *amoA* genes of AOA and AOB were the same with those used in the clone library construction. Primer pair P128r, P365r designed by Dang [40] was used to quantify the copy numbers of *Nitrosomonas amoA* gene. qPCR was performed using an iCycler iQ5 thermocycler and a real-time detection system (Bio-Rad, California, USA) according to the method described by Francis *et al.* [34], Mc Tavish *et al.* [41] and Dang [40]. Quantification standards consisted of 10^2–10^8, 10^1–10^7, 10^2–10^8 copy number dilution series of plasmid containing inserted *amoA* genes of AOA, AOB and *Nitrosomonas*, respectively. R^2 and the amplification efficiencies of the AOA, AOB and *Nitrosomonas* standard curves were 0.996, 0.998 and 0.996 (as shown in Figure S3, S4 and S5) and 96%, 89% and 95%, respectively.

Analytical Methods

Measurements of NO_3^--N, NO_2^--N and NH_4^+-N in the experimental medium were performed according to standard methods [42]. The total solid (TS) content was measured by the weighing method after drying at 105°C. The inorganic nitrogen contents in the soils were determined colorimetrically via flow injection analysis of soil-KCl extracts as previously described [43].

Accession Numbers

Sequences were deposited in GenBank under accession numbers JQ320498 to JQ320554.

Results

Diversity and Phylogenetic Analysis of AOA and AOB

185 derived archaeal *amoA* genes belonged to *Nitrososphaera* cluster, *Nitrososphaera* sister cluster, *Nitrosotalea* cluster and *Nitrosopumilus* cluster according to a new nomenclature system for archaeal *amoA* genes reported by Pester [37] (Figure 2), but no *Nitrosocaldus*-related sequences was detected. The sequences affiliated with *Nitrosotalea* cluster were the most and accounted for 69.2% of all the archaeal *amoA* genes obtained. Only two sequences were detected belonging to *Nitrosopumilus* cluster and the rest sequences belonged to *Nitrososphaera* cluster and *Nitrososphaera* sister cluster. A total of 9 OTUs were obtained using the DUTOR program based on a 15% *amoA* gene sequence distance cutoff [37] for all 185 AOA *amoA* sequences detected in this study. *Nitrosopumilus* cluster contained only one OTU and the other three clusters contained at least two OTUs. Sequences belonging to OTU1 were found in all three soil layers during the entire cultivation process, totalling 121 sequences, which accounted for 65.4% of the detected AOA *amoA* gene sequences.

As to AOB, all 201 sequences obtained for the AOB *amoA* gene in the clone libraries belonged to β-proteobacteria, as shown in Figure 3. Among these sequences, 91.0% were grouped into the *Nitrosospira* genus, and the remainders were grouped into *Nitrosomonas*. A total of 8 OTUs were obtained for the bacterial *amoA* genes using the DUTOR program based on a 15% *amoA* gene sequence distance cutoff [38]. *Nitrosomonas* cluster contained 2 OTUs, and the remainders were grouped into *Nitrosospira* cluster.

Community Succession of AOA and AOB

The community structure of the AOA and AOB in the soil column changed during the cultivation process and was closely related to soil depth. An obvious population shift was found for AOA in different layers when the proportions that each cluster occupied in the first stage and the last stage were compared. In the surface layer, the proportion of *Nitrosotalea* cluster decreased from 66.7% to 52.9% (Figure 4). Accordingly, an increase from 20% to 47.1% was found for *Nitrososphaera* cluster. The sequences affiliated with *Nitrososphaera* sister cluster disappeared during the 18-month cultivation process. Interestingly, in the mid layer, we found that cluster *Nitrosotalea*-related sequences represented more than 94.4% of the sequences detected, and this proportion did not change during the cultivation process. Accordingly, the proportion of *Nitrososphaera*-related sequences was less than 5.6% throughout the cultivation process. In the deep layer, the observed changes were dramatic. In the first stage, the diversity of the population was high and all the four AOA *amoA* clusters were found, the *Nitrosopumilus*-related sequences were only detected in the first stage of the deep layer and represented 9.1% of the sequences detected. However, 18 months later, the sequences belonging to *Nitrosopumilus* disappeared and the proportion that *Nitrosotalea* cluster occupied increased from 27.3% to 72.2%. In the 18-month cultivation process, the number of OTUs in all three layers decreased first and then increased. The number of OTUs in the deep layer was greater than in the other two layers in the first stage (Figure S1). In the second stage, the numbers of OTUs in all three layers were reduced dramatically. In the surface layer, the number of OTUs

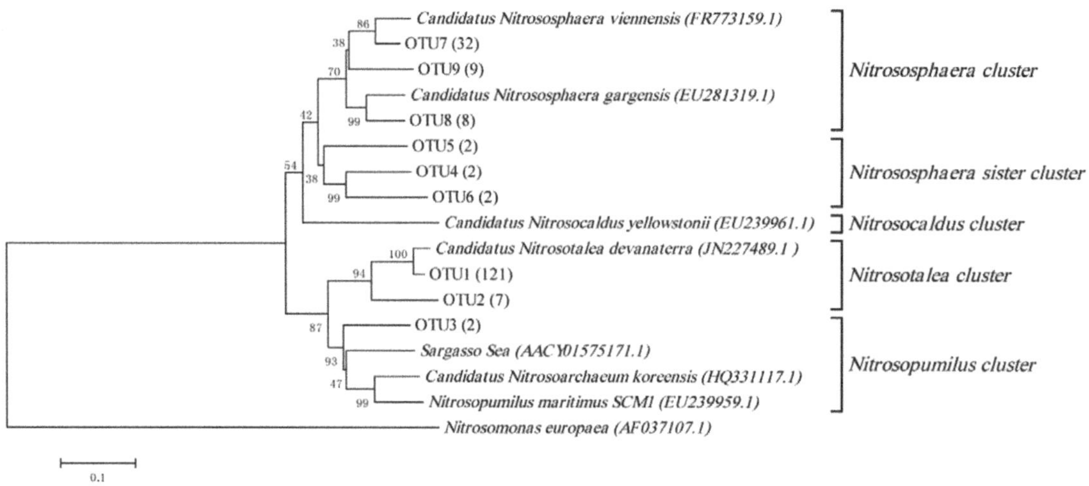

Figure 2. Neighbour-joining phylogenetic tree showing the relationships of the 185 sequences of archaeal *amoA* genes detected in the clone libraries constructed from samples from the soil column with their most similar sequences and sequences of known AOA. The numbers at the nodes are percentages that indicate the levels of bootstrap support from 1,000 replicates. The scale bar represents 0.1 nucleic acid substitutions per nucleotide position.

dropped from 4 to 1, while it decreased from 2 to 1 and from 7 to 3 in the mid and deep layer, respectively. The only difference was that the number of OTUs in the surface and mid layers recovered more rapidly than was observed for the deep layer. The number of OTUs in the surface and mid layers began to increase in the third stage; however, the number of OTUs in the deep layer did not increase until the last stage.

With respect to AOB, increasing sequences belonging to *Nitrosomonas* were detected in all three layers in the soil column, and the proportions that *Nitrosomonas*-related sequences occupied were 25%, 15.8% and 18.8% from the surface layer to the deep layer (Figure 5). The numbers of AOB OTUs varied in a similar way as was observed for the AOA OTUs. The AOB OTU numbers in the mid and deep layers decreased in the first three stages and increased in the last stage. However, in the surface layer, the number of OTUs remained stable at three after the third stage (Figure S2).

Real-time Quantitative PCR (qPCR)

The AOA and AOB *amoA* genes in the three soil layers were quantified every 6 months. AOA were more abundant than AOB in almost all soil samples, with the exception of the sample from the surface layer in the last stage, in which the AOB/AOA ratio was 16.95. The ratios of AOA/AOB ranged from 1.03–12.96 in the other 11 soil samples (Figure 6). The copy number of *Nitrosomonas amoA* genes increased substantially in the surface layer during the cultivation process, from 2.2×10^6 to 1.2×10^8. As for the other two layers, the change of was not obvious (Figure 7).

Nitrification Activity

Nitrification activity was tested after the 18-month cultivation period. Samples were collected from the three different layers of the soil column. Different nitrification activities were observed for the three layers of the soil column. The maximum specific nitrification activity in the mid layer was

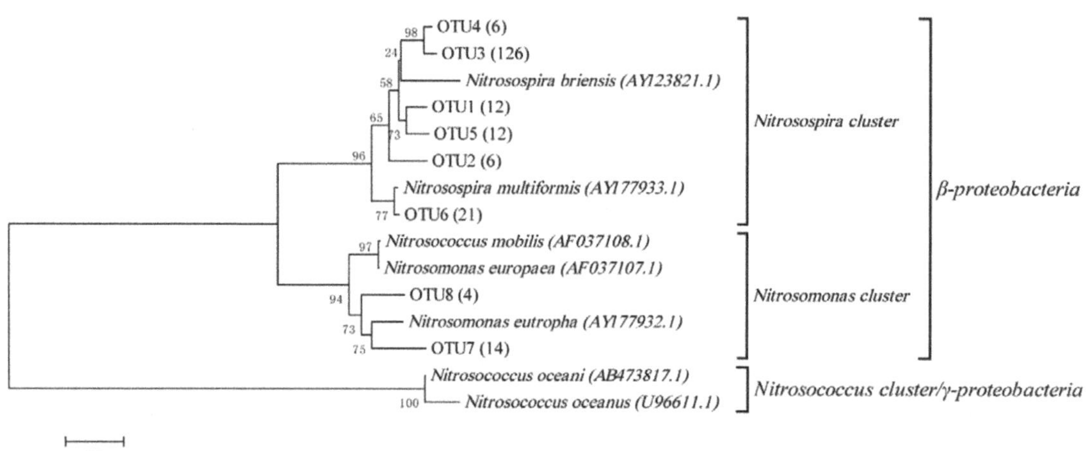

Figure 3. Neighbour-joining phylogenetic tree showing the relationships of the 201 sequences of bacterial *amoA* genes detected in the clone libraries constructed from samples from the soil column with their most similar sequences and the sequences of known AOB. The numbers at the nodes are percentages that indicate the levels of bootstrap support from 1000 replicates. The scale bar represents 0.05 nucleic acid substitutions per nucleotide position.

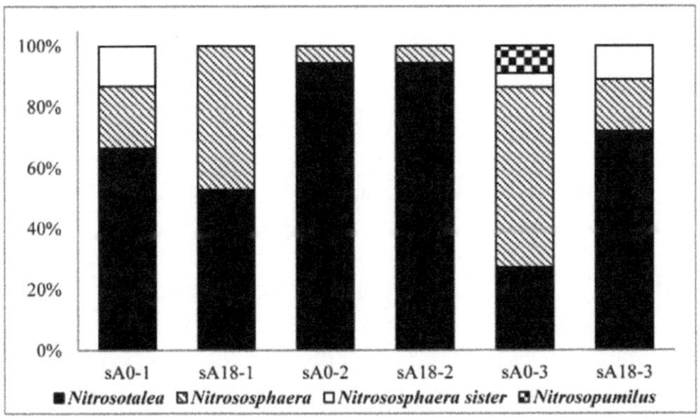

Figure 4. Population shifts of AOA in different layers. (sA0-1 represents the first stage in the surface layer; sA0-2, the first stage in the mid layer; sA0-3, the first stage in the deep layer; sA18-1, the last stage in the surface layer; sA18-2, the last stage in the mid layer; and sA18-3, the last stage in the deep layer).

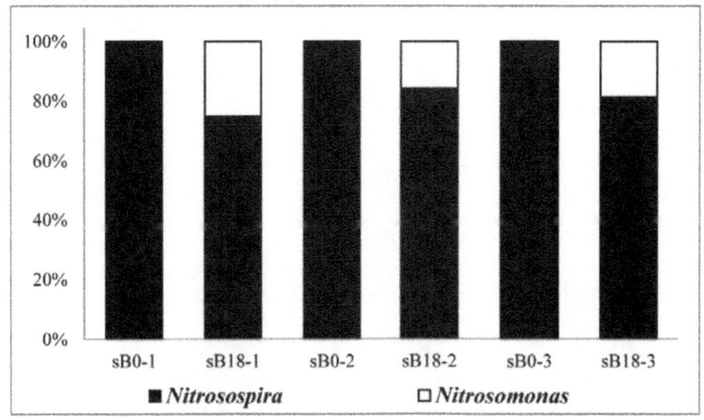

Figure 5. Population shifts of AOB in different layers. (sB0-1 represents the first stage in the surface layer; sB0-2, the first stage in the mid layer; sB0-3, the first stage in the deep layer; sB18-1, the last stage in the surface layer; sB18-2, the last stage in the mid layer; and sB18-3, the last stage in the deep layer).

25.06 $\mu g\cdot N\cdot L^{-1}\cdot g^{-1}\cdot h^{-1}$, while those in the surface and deep layers were 16.13 $\mu g\cdot N\cdot L^{-1}\cdot g^{-1}\cdot h^{-1}$ and 9.97 $\mu g\cdot N\cdot L^{-1}\cdot g^{-1}\cdot h^{-1}$, respectively. Therefore, ammonia-oxidizing microorganisms in the mid layer exhibited the highest nitrification activity among the three layers of the soil column.

Discussion

Community Succession of AOA and AOB

It was found that 69.2% sequences derived in our soil column were related to *Nitrosotalea devanaterra*-the only ammonia-oxidizing archaea from acid soil [6]. This was in accordance with the high-throughput sequencing results reported by Gubry-Rangina [44], which demonstrated that *Nitrosotalea devanaterra*-related sequences may be predominant in some agricultural soils. pH in our soil column were neutral or slightly alkaline and was not in accordance with the acid environment where *Nitrosotalea devanaterra* was found. The contradictory could be explained by that there may be existed AOA belonging to *Nitrosotalea devanaterra* cluster which could survive and oxidize ammonia to nitrite in neutral or slightly alkaline environment. The community composition of the AOA in the different layers of the soil column underwent great changes during the cultivation process. In the surface layer, the sequences belonging to *Nitrososphaera* sister cluster disappeared after 18-month cultivation. Since there was no isolated or enrichment strain of this AOA cluster [37], the physiochemical characteristics of *Nitrososphaera* sister cluster was unknown. Combined with the increased *Nitrososphaera*-related sequences in the surface and the decreased *Nitrososphaera* sister-related sequences in the deep layer, we deduced that the AOA affiliated with *Nitrososphaera* sister cluster had relative high affinity for ammonia and prefer to live in the environment with relative low ammonia concentration. *Nitrososphaera viennensis* [7] could grow well in the media containing ammonia concentration as high as 15 mM, indicating a low affinity of this strain of AOA. So, it was reasonable that the proportion *Nitrososphaera*-related sequences occupied increased in the last stage compared with the first stage in the surface layer, since it preferred conditions with relatively higher ammonia concentration. Correspondingly, in the deep layer, the proportion *Nitrososphaera*-related sequences occupied decreased from 59.1% to 16.7%. *Nitrosopumilus maritimus* showed extremely high affinity for ammonia and the ammonia inhibition concentration was as low as 2 mM to 3 mM [5], so it was not strange that the *Nitrosopumilus*-related sequences disappeared in the deep layer after exogenous ammonia was added during the cultivation process. No *Nitrosopumilus*-related sequence was detected in the surface and mid layer during the whole 18 months. *Nitrosotalea*-related sequences were the dominant species in our soil column throughout the cultivation process. In the surface layer, the proportion that the *Nitrosotalea*-related sequences occupied decreased from 66.7% to 52.9%; on the contrary, the proportion in the deep layer increased from 27.3% to 72.2%. It was reported that the acid AOA could grow well with the total NH_3& NH_4^+ concentration ranged from 10 µM to 10 mM (equivalent to 0.18 nM to 0.18 µM of NH_3) at pH 4.5 and no growth was detected at 50 mM and 100 mM (equivalent to 9 µM to 18 µM NH_3, pK_a for NH_3: NH_4^+, 9.25) [6]. The available NH_3 concentration in the medium was 44µM (pH, 7.24; 30°C), which was much higher than the ammonia

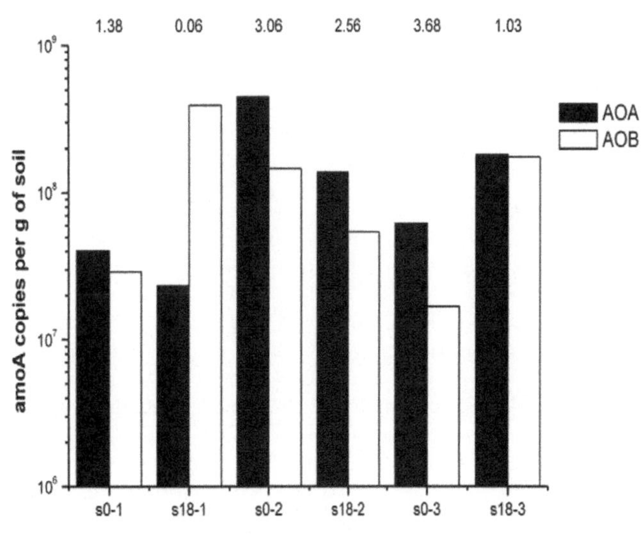

Figure 6. Archaeal and bacterial *amoA* gene copy numbers in different layers. (s0-1 represents the first stage in the surface layer; s0-2, the first stage in the mid layer; s0-3, the first stage in the deep layer; s18-1, the last stage in the surface layer; s18-2, the last stage in the mid layer; and s18-3, the last stage in the deep layer).

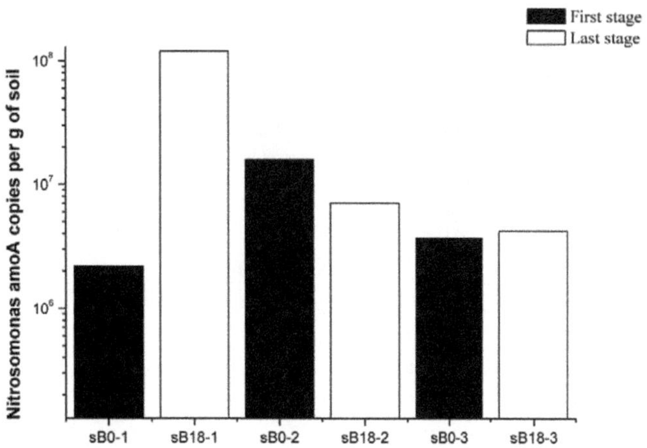

Figure 7. Nitrosomonas amoA gene copy numbers in different layers. (sB0-1 represents the first stage in the surface layer; sB0-2, the first stage in the mid layer; sB0-3, the first stage in the deep layer; sB18-1, the last stage in the surface layer; sB18-2, the last stage in the mid layer; and sB18-3, the last stage in the deep layer).

inhibition concentration (9 μM) of *Nitrosotalea devanaterra*. So, the sequences affiliated with *Nitrosotalea* cluster decreased in the surface layer where the ammonia concentration was relatively higher during 18 months cultivation process.

In our simulated soil column, we found that all three layers contained a certain amount of *Nitrosomonas*-like sequences after 18 months. In the surface layer the proportion occupied by *Nitrosomonas*-like sequences was 25.0%, which was higher than in the other two layers (mid layer, 15.8%; deep layer, 18.8%). Quantitative analysis also reached the similar conclusion that the copy number of *Nitrosomonas amoA* genes increased from 2.2×10^6 to 1.2×10^8 in the surface layer, and the copy numbers were much higher than that in the other two layers. This could be explained by differences in physiological characteristics between *Nitrosomonas* and *Nitrosospira*. Schramm *et al.* [45,46] have reported that the growth of *Nitrosomonas* species followed an r-strategy. The *Nitrosomonas* genus had a low ammonia affinity and was adaptable to conditions where the concentration of ammonia was high. In contrast, the growth of *Nitrosospira* species followed a k-strategy. The *Nitrosospira* genus had a high ammonia affinity and was adaptable to conditions where the concentration of ammonia was low. The appearance of and increases in *Nitrosomonas*-like sequences were reflection of the competitive advantage of *Nitrosomonas* species under higher ammonia concentrations when exogenous ammonia was added. The obtained *Nitrosomonas*-like sequences were primarily found in the surface layer because of the relatively higher ammonia concentration in this layer compared with the other two layers. Notably, no *Nitrosomonas*-related sequences was found in the first stage through construction of AOB clone libraries, but we indeed detected *Nitrosomonas amoA* gene using qPCR, this may be due to the different specificities of the primers used in phylogenetic and quantitative analysis.

The richness of AOA and AOB decreased under ammonia stress and then recovered after a period of adaptation, as reflected in the variation in OTU numbers. It seemed that the AOA in the deep layer were less sensitive to exogenous ammonia than those in the other two layers and had a relative longer adaptation period. Another explanation of the OTU numbers variation was the cultivation effect that led to the survival of only a few adapted species which take over the community. For instance, in the deep layer of the first stage, AOA population was constituted with OTU1-3 and OTU6-9, and 6 months later just OTU1, OTU7 and OTU9 were left. In the last stage, the population was composed with 4 OTUs and a new OTU was found even though it had never been derived in the first three stages, so finally the population was taken over by the survival species who could adapt to the selective pressure of the environment.

Abundance of AOA and AOB and Potential Nitrification Rates (PNR) in Different Layers

The ratio of AOA/AOB in soils was first reported by Leininger in 2006 [17]. He found that the ratio of AOA/AOB was correlated with soil depth in long-term fertilized soils and AOA *amoA* gene copy numbers could be three orders of magnitude greater than AOB *amoA* gene numbers at a depth of 30–40 cm. In the present study, the highest ratio of AOA/AOB was found to be 12.96 in the second stage in the surface layer. The growth and ammonia oxidation activity of *Nitrosopumilus maritimus* were inhibited at an ammonia concentration of 28 mg/L [28] which was lower than the inhibitory concentrations (7000–14000 mg/L) of ammonia found for AOB [47]. Therefore, it was expected that the numbers of bacterial *amoA* gene copies would exceed the archaeal *amoA* gene copy numbers in the surface layer during the last stage, where the ammonia concentration was the highest. The potential nitrification rates (PNR) were highest in the mid layer and were 1.6 times higher than that in the surface layer and 2.5 times higher than that in the deep layer. The lower potential nitrification rates in the deep layer may be the cause of the change in redox potential along with soil depth, as reported previously [48]. The redox potential in soils is determined by the relative concentrations of substances in an oxidized state versus a reduced state, which are influenced by oxygen concentrations. The deep layer exhibited the lowest redox potential, which resulted from the lowest oxygen concentration. The fact that the lowest pH (7.09 ± 0.15) existed in the deep layer compared with the other two layers (surface layer, 7.24 ± 0.12; mid layer, 7.41 ± 0.1) may also play an important role.

Conclusion

In general, obvious population dynamic changes were found for both AOA and AOB under the selective pressure of exogenous ammonia and the changes were different in three layers of the soil column.

Author Contributions

Conceived and designed the experiments: LLP XXY ZP. Performed the experiments: HBL LS SLD. Analyzed the data: HBL LS LLP. Contributed reagents/materials/analysis tools: LLP XXY ZP HBL. Wrote the paper: HBL LS.

References

1. Gruber N, Galloway JN (2008) An Earth-system perspective of the global nitrogen cycle. Nature 451: 293–296.
2. Venter JC, Remington K, Heidelberg JF, Halpern AL, Rusch D, et al. (2004) Environmental genome shotgun sequencing of the Sargasso Sea. Science 304: 66–74.
3. Treusch AH, Leininger S, Kletzin A, Schuster SC, Klenk H-P, et al. (2005) Novel genes for nitrite reductase and Amo-related proteins indicate a role of uncultivated mesophilic crenarchaeota in nitrogen cycling. Environ Microbiol 7: 1985–1995.
4. Blainey PC, Mosier AC, Potanina A, Francis CA & Quake SR (2011) Genome of a Low-Salinity Ammonia-Oxidizing Archaeon Determined by Single-Cell and Metagenomic Analysis. PLoS ONE 6.
5. Könneke M, Bernhard AE, de la Torre JR, Walker CB, Waterbury JB, et al. (2005) Isolation of an autotrophic ammonia-oxidizing marine archaeon. Nature 437: 543–546.
6. Lehtovirta-Morley LE, Stoecker K, Vilcinskas A, Prosser JI & Nicol GW (2011) Cultivation of an obligate acidophilic ammonia oxidizer from a nitrifying acid soil. Proceedings of the National Academy of Sciences of the United States of America 108: 15892–15897.
7. Tourna M, Stieglmeier M, Spang A, Könneke M, Schintlmeister A, et al. (2011) Nitrososphaera viennensis, an ammonia oxidizing archaeon from soil. Proceedings of the National Academy of Sciences of the United States of America 108: 8420–8425.
8. Jung MY, Park SJ, Min D, Kim JS, Rijpstra WIC, et al. (2011) Enrichment and Characterization of an Autotrophic Ammonia-Oxidizing Archaeon of Mesophilic Crenarchaeal Group I.1a from an Agricultural Soil. Applied and Environmental Microbiology 77: 8635–8647.
9. Hatzenpichler R, Lebedeva EV, Spieck E, Stoecker K, Richter A, et al. (2008) A moderately thermophilic ammonia-oxidizing crenarchaeote from a hot spring. Proceedings of the National Academy of Sciences 105: 2134–2139.
10. de la Torre JR, Walker CB, Ingalls AE, Könneke M & Stahl DA (2008) Cultivation of a thermophilic ammonia oxidizing archaeon synthesizing crenarchaeol. Environmental Microbiology 10: 810–818.
11. Brochier-Armanet C, Boussau B, Gribaldo S, Forterre P (2008) Mesophilic crenarchaeota: proposal for a third archaeal phylum, the Thaumarchaeota. Nat. Rev. Microbiol. 6: 245–252.
12. Spang A, Hatzenpichler R, Brochier-Armanet C, Rattei T, Tischler P, et al (2010) Distinct gene set in two different lineages of ammonia-oxidizing archaea supports the phylum Thaumarchaeota. Trends Microbiol 18: 331–340.
13. Pester M, Schleper C, Wagner M (2011) The Thaumarchaeota: an emerging view of their phylogeny and ecophysiology. Curr Opin Microbiol 14: 300–306.
14. Francis CA, Beman JM, Kuypers MMM (2007) New processes and players in the nitrogen cycle: the microbial ecology of anaerobic and archaeal ammonia oxidation. ISME J 1: 19–27.
15. Santoro AE, Francis CA, de Sieyes NR, Boehm AB (2008) Shifts in the relative abundance of ammonia-oxidizing bacteria and archaea across physicochemical gradients in a subterranean estuary. Environ Microbiol 10: 1068–1079.
16. Caffrey JM, Bano N, Kalanetra K, Hollibaugh JT (2007) Ammonia oxidation and ammonia-oxidizing bacteria and archaea from estuaries with differing histories of hypoxia. ISME J 1, 660–662.
17. Leininger S, Urich T, Schloter M, Schwark L, Qi J, et al. (2006) Archaea predominate among ammonia-oxidizing prokaryotes in soils. Nature 442: 806–809.
18. Wuchter C, Abbas B, Coolen MJL, Herfort L, Bleijswijk JV, et al. (2006) Archaeal nitrification in the ocean. Proc Natl Acad Sci 103: 12317–12322.
19. Jia Z, Conrad R (2009) Bacteria rather than archaea dominate microbial ammonia oxidation in an agricultural soil. Environ Microbiol 11: 1658–1671.
20. Di HJ, Cameron KC, Shen JP, Winefield CS, O'Callaghan M, et al. (2009) Nitrification driven by bacteria and not archaea in nitrogen-rich grassland soils. Nat. Geosci. 2: 621–624.
21. Gubry-Rangin C, Nicol GW, Prosser JI (2010) Archaea rather than bacteria control nitrification in two agricultural acidic soils. FEMS Microbiol Ecol 74: 566–574.
22. Zhang LM, Offre PR, He JZ, Verhanmme DT, Nicol GW, et al. (2010) Autotrophic ammonia oxidation by soil thaumarchaea. Proc Natl Acad Sci 107: 17240–17245.
23. Offre P, Prosser JI, Nicol GW (2009) Growth of ammonia-oxidizing archaea in soil microcosms is inhibited by acetylene. FEMS Microbiol Ecol 70: 99–108.
24. Verhamme DT, Prosser JI, Nicol GW (2011) Ammonia concentration determines differential growth of ammonia-oxidizing archaea and bacteria in soil microcosms. ISME J 5: 1067–1071.
25. Höfferle Š, Nicol GW, Pal L, Hacin J, Prosser JI, et al. (2010) Ammonium supply rate influences archaeal and bacterial ammonia oxidizers in a wetland soil vertical profile. FEMS Microbiol Ecol 74: 302–315.
26. Di HJ, Cameron KC, Shen JP, Winefield CS, O'Callaghan M, et al. (2010) Ammonia-oxidizing bacteria and archaea grow under contrasting soil nitrogen conditions. FEMS Microbiol Ecol 72: 386–394.
27. Valentine DL (2007) Adaptations to energy stress dictate the ecology and evolution of the archaea. Nat. Rev. Microbiol. 5: 316–323.
28. Martens-Habbena W, Berube PM, Urakawa H, de la Torre JR, Stahl DA (2009) Ammonia oxidation kinetics determine niche separation of nitrifying Archaea and Bacteria. Nature 461: 976–979.
29. Nicol GW, Leininger S, Schleper C, Prosser JI (2008) The influence of soil pH on the diversity, abundance and transcriptional activity of ammonia oxidizing archaea and bacteria. Environ Microbiol 10: 2966–2978.
30. Hommes NG, Sayavedra-Soto LA, Arp DJ (2003) Chemolithoorganotrophic growth of Nitrosomonas europaea on fructose. J Bacteriol 185: 6809–6814.
31. Liu QH, Shi XZ, Weindorf DC, Yu DS, Zhao YC, et al. (2006) Soil organic carbon storage of paddy soils in China using the 1: 1,000,000 soil database and their implications for C sequestration. Global Biogeochem Cy 20: GB3024.
32. Chen XP, Zhu YG, Xia Y, Shen JP, He JZ (2008) Ammonia-oxidizing archaea: important players in paddy rhizosphere soil? Environ Microbiol 10: 1978–1987.
33. Schmidt I (2004) Ammonium and hydroxylamine uptake and accumulation in Nitrosomonas. Microbiology 150: 1405–1412.
34. Francis CA, Roberts KJ, Beman JM, Santoro AE, Oakley BB (2005). Ubiquity and diversity of ammonia-oxidizing archaea in water columns and sediments of the ocean. Proc Natl Acad Sci 102: 14683–14688.
35. Rotthauwe JH, Witzel KP, Liesack W (1997) The ammonia monooxygenase structural gene amoA as a functional marker molecular fine-scale analysis of natural ammonia-oxidizing populations. Appl Environ Microbiol 63: 4704–4712.
36. Shen JP, Zhang LM, Zhu YG, Zhang JB, He JZ (2008) Abundance and composition of ammonia-oxidizing bacteria and ammonia-oxidizing archaea communities of an alkaline sandy loam. Environ Microbiol 10: 1601–1611.
37. Pester M, Rattei T, Flechl S, Gröngröft A, Richter A, et al. (2011) amoA-based consensus phylogeny ofammonia-oxidizing archaea and deep sequencing of amoA genes from soils of four different geographic regions. Environ Microbiol, 14: 525–539.
38. Purkhold U, Pommerening-Roser A, Juretschko S, Schmid MC, Koops HP, etal. (2000) Phylogeny of all recognized species of ammonia oxidizers based on comparative 16S and amoA sequence analysis: Implications for molecular diversity surveys. Appl. Environ. Microbiol. 66, 5368–5382.
39. Schloss PD, Handelsman J (2005) Introducing DOTUR, a computer program for defining operational taxonomic units and estimating species richness. Appl Environ Microbiol 71: 1501–1506.
40. Dang H, Li J, Chen R, Wang L, Guo L, et al. (2010) Diversity, Abundance, and Spatial Distribution of Sediment Ammonia-Oxidizing Betaproteobacteria in Response to Environmental Gradients and Coastal Eutrophication in Jiaozhou Bay, China. Applied and Environmental Microbiology 76: 4691–4702.
41. McTavish H, Fuchs JA, Hooper AB (1993) Sequence of the Gene Coding for Ammonia Monooxygenase in Nitrosomonas europaea. J Bacteriol 175: 2436–2444.
42. APHA AWWA, WEF (1998) Standard Methods for the Examination of Water and Wastewater, 20th ed. American Public Health Association, Washington, DC, USA.
43. Allen SE (1989) Chemical Analysis of Ecological Materials, second ed. Blackwell Scientific Publications, Boston, MA.
44. Gubry-Rangin C, Hai B, Quince C, Engel M, Thomson BC, et al. (2011) Niche specialization of terrestrial archaeal ammonia oxidizers. Proceedings of the National Academy of Sciences 108: 21206–21211.
45. Schramm A, de Beer D, Wagner M, Amann R (1998) Identification and activities in situ of nitrosospira and nitrospira spp.as dominant populations in a nitrifying fluidized bed reactor. Appl Environ Microbiol 64: 3480–3485.
46. Schramm A, de Beer D, van den Heuvel JC, Ottengraf S, Amann R (1999) Microscale distribution of populations and activities of Nitrosospira and Nitrospira spp. along a macroscale gradient in a nitrifying bioreactor quantification by in situ hybridization and the use of microelectrodes. Appl Environ Microbiol 65: 3690–3696.
47. Nicolaisen MH, Risgaard-Petersen N, Revsbech NP, Reichardt W, Ramsing NB (2004) Nitrificationa–denitrification dynamics and community structure of ammonia oxidizing bacteria in a high yield irrigated Philippine rice field. FEMS Microbiol Ecol 49: 359–369.
48. Bohrerova Z, Stralkova R., Podesvova J, Bohrer G, Pokorny E (2004) The relationship between redox potential and nitrification under different sequences of crop rotations. Soil Till Res 77: 25–33.

2

Competition Increases Sensitivity of Wheat (*Triticum aestivum*) to Biotic Plant-Soil Feedback

W. H. Gera Hol[1]*, Wietse de Boer[2,3], Freddy ten Hooven[1], Wim H. van der Putten[1,4]

1 Netherlands Institute of Ecology, Department of Terrestrial Ecology, Wageningen, The Netherlands, **2** Netherlands Institute of Ecology, Department of Microbial Ecology, Wageningen, The Netherlands, **3** Wageningen University, Department of Soil Quality, Wageningen, The Netherlands, **4** Wageningen University, Laboratory of Nematology, Wageningen, The Netherlands

Abstract

Plant-soil feedback (PSF) and plant competition play an important role in structuring vegetation composition, but their interaction remains unclear. Recent studies suggest that competing plants could dilute pathogenic effects, whereas the standing view is that competition may increase the sensitivity of the focal plant to PSF. In agro-ecosystems each of these two options would yield contrasting outcomes: reduced versus enhanced effects of weeds on crop biomass production. To test the effect of competition on sensitivity to PSF, we grew *Triticum aestivum* (Common wheat) with and without competition from a weed community composed of *Vicia villosa*, *Chenopodium album* and *Myosotis arvensis*. Plants were grown in sterilized soil, with or without living field inoculum from 4 farms in the UK. In the conditioning phase, field inocula had both positive and negative effects on *T. aestivum* shoot biomass, depending on farm. In the feedback phase the differences between shoot biomass in *T. aestivum* monoculture on non-inoculated and inoculated soils had mostly disappeared. However, *T. aestivum* plants growing in mixtures in the feedback phase were larger on non-inoculated soil than on inoculated soil. Hence, *T. aestivum* was more sensitive to competition when the field soil biota was present. This was supported by the statistically significant negative correlation between shoot biomass of weeds and *T. aestivum*, which was absent on sterilized soil. In conclusion, competition in cereal crop-weed systems appears to increase cereal crop sensitivity to soil biota.

Editor: James F. Cahill, University of Alberta, Canada

Funding: The work was funded by the EU project SOILSERVICE. SOILSERVICE is a collaborative, medium scale focussed research project, which is funded by EU's Seventh Framework Programme. (http://www4.lu.se/soil-ecology-group/research/soilservice.) The funders had no role in study design, data collection and analysis, decision to publish, or preparation of the manuscript.

Competing Interests: The authors have declared that no competing interests exist.

* E-mail: g.hol@nioo.knaw.nl

Introduction

Plants have selective effects on soil biota, with negative, neutral or positive consequences for future occupants of the same location [1,2]. These biotic plant-soil feedback (PSF) effects are caused by net effects of soil-borne mutualists and pathogens that develop in the rhizosphere during plant growth [3]. In addition, there are abiotic feedbacks such as changes in nutrient or water availability. If the performance of conspecific plants in the next growing cycle is stimulated, this is called a positive PSF. If conspecific plants perform worse in the next growing cycle, this is called a negative PSF. This process is important for plant succession [4,5] and diversity-productivity relationships [6,7], but it also plays an important role in applied areas, such as agriculture [8], restoration ecology [9,10] and invasion management [11]. The success of later successional plants traditionally has been attributed to their ability to constrain growth of predecessors [12], but more recently it has been acknowledged that species-specific PSF can influence predecessor-successor interactions [1,13,14].

In agriculture, crop species are being grown in rotation in order to avoid the development of negative PSF, as crop rotation prevents the accumulation of crop-specific soil pathogens [8]. Negative intraspecific PSF effects can be avoided by growing crops in successional cycles with other crops that are not sensitive to the pathogens from previous crops. The crop cycles can also be related to using positive interspecific PSF, for example green manures that increase nutrient availability and nitrogen-fixing plants that enhance the pools of available nitrogen in soil. Intercropping with legumes has been used to both reduce pathogen pressure (negative PSF) and simultaneously increase soil nitrogen (positive PSF) [15].

Empirical work on PSF has been done both with single plant species and with communities [2], mostly composed of wild plants. Shifts in relative abundance of plant species within a community indicate that PSF does affect competitive relations, e.g [5], but only a few studies have addressed the role of competition in PSF studies explicitly [1,13].

There are two mechanisms in relation to feedback in plant mixtures which would lead to opposite effects. Plants that are stressed due to competition can be more vulnerable to pathogens [16]. Successful competitors may need to invest in longer stems to compete for light or larger roots to compete for nutrients and these investments may come at the expenses of plant defense [17]. On the other hand, pathogen effects can be density-dependent and since the relative abundance of species in mixtures is lower than in monocultures less pathogenic effects can be expected and observed in mixtures [6,7].

Thus far there is limited evidence that intraspecific competition increases PSFs [2] or that interspecific competition decreased PSF

[18–20]; but see [1,21–23]. We analyze the effect of competition on plant soil feedback of wheat in the feedback phase. Competition can be studied with replacement (constant density) or additive designs (increasing density) [24]. We choose an additive design without the option to distinguish intra- from interspecific competition since this resembles the agricultural situation best. The outcome of the interaction between feedback and competition may depend on species identity [25], and thus it is important to use species which co-occur in order to provide relevant information for a particular system. All soils were conditioned with *T. aestivum* (common wheat), and nutrients were added to minimize the role of nutrient availability on plant growth responses and to focus on biotic feedback [23,26]. We determined how the presence of a weed community affects responses of common wheat (*Triticum aestivum*) to PSF, by performing a feedback experiment with monoculture and plant mixtures in the feedback phase. For our experiment soil was collected from 4 different locations to determine whether the effect of competition on PSF is general or rather depends on the specific soil biota. Given the high density of crop plants in agriculture, we expect that increased sensitivity to pathogens will overrule the reduction in pathogen density dependence effects by weeds. Thus we hypothesize that negative plant-soil feedback effects of *T. aestivum* plants will be enhanced when *T. aestivum* grows in plant mixtures. Our results show that plants with competition are more vulnerable to negative plant soil feedback effects.

Materials and Methods

The effect of plant mixtures on sensitivity to plant soil feedback was tested by growing *Triticum aestivum* L. plants for two growing cycles in the same soil, in the presence or absence of arable weeds in the second growing cycle. The first growing cycle consisted of wheat plants growing in sterilized soil with or without field soil inoculum. The aim of this phase was to standardize abiotic conditions and determine the variation in effects of soil biota from different farms on wheat growth. Since all soils were conditioned with *T. aestivum*, we minimized the role of soil nutrient feedback and focused on the biotic feedback.

Soils

Soil for the experiment was collected with approval from the owners of 4 farms from the Chilterns area (United Kingdom). The soil-type is chromic luvisol/leptosol with pH in the range of 7.2–7.6. At each farm, soil from three fields was collected. Soil was passed through a 1 cm sieve to remove stones and reduce the larger clumps of clay. For each farm 66 kg soil (a mix from the 3 fields) was sterilized by gamma radiation (25 kGray, Isotron, Ede, The Netherlands) in order to create a sterilized background soil. A small subsample from each field was kept apart and remained non-sterilized in order to serve as an inoculum for soil biota. The sterilized soil was used to fill square pots (11×11×12 cm) with 1 kg soil, including 90 g fresh living soil inoculum (9% inoculum). Eight pots per field were filled, resulting in 4 (farms) * 3 (fields) * 8 (pots) = 96 pots in total. In addition to the aforementioned treatments of sterilized soil with non-sterilised soil inoculum, a control treatment was included which consisted of 16 pots per farm filled with sterilized soil only. This added 4 (farms) * 16 (pots) = 64 pots to the experiment. The amount of replicates for the control pots was doubled from 8 to 16 to increase the balance when comparing inoculated vs non-inoculated pots.

Conditioning phase

For the first phase of the experiment, all pots were sown with 7 seeds of common wheat (*Triticum aestivum*) in a row in the middle of the pots. There were 24 inoculated (3 fields * 8 replicates) and 16 non-inoculated control pots per farm. Plants were kept under controlled conditions in the greenhouse (60% RH; 16 h L, 8 h D, 21°C/16°C) and placed in random order on a bench. Additional illumination was provided by 400W growing bulbs (Philips SON-T Agro, Philips, Eindhoven, The Netherlands). Light intensity at plant level was 225 µmol PAR. Pots were watered regularly with demineralized water and received 60 ml week^{-1} Hoagland solution (half strength, [27]). Nutrient addition may alter feedback effects [28], but was chosen to reduce possible nutrient feedback effects and focus on the biotic component. In addition, fertilization is common practice in agricultural systems. A thrips infestation was controlled by the predatory mite *Amblyseius cucumeris* (Koppert Biological Systems, Berkel en Rodenrijs, The Netherlands). Plants were harvested after 60 days; aboveground biomass was clipped and put in paper bags. Shoot dry weight was determined after 72 h drying at 70°C.

Feedback phase

For the second phase of the experiment, soil with roots was left in the pots and re-growing shoots were removed. After two weeks no more re-growth was observed and the pots were sown again with 7 seeds of *T. aestivum*. Seeds were sown perpendicular to the previous sown seeds, to be able to discriminate between potential re-growth and germination, but this precaution proved unnecessary. Half of the pots was planted with one seedling from each of the three weed species: hairy vetch (*Vicia villosa* Roth), lamb's quarters (*Chenopodium album* L.) and field forget-me-not (*Myosotis arvensis* (L.) Hill.). Seeds from *C. album* and *V. villosa* were collected in the wild in the Netherlands, while *M. arvensis* was ordered from a commercial supplier (Rieger-Hofmann, Blaufelden-Raboldshausen, Germany). Seeds were germinated on glass pearls by 16 h L, 8 h D, 23°C/15°C. *Chenopodium album* and *M. arvensis* are common arable weeds [29]; *Vicia villosa* is often applied as winter cover crop and might return as weed in follow-up crops. The weeds resulted in 2 treatments in the second phase: Mono and Mix (n = 12 for inoculated pots and n = 8 for non-inoculated control pots). The pots were kept in the same greenhouse as before, under the same light and temperature conditions. Pots were watered regularly with demineralized water and received 60 ml week^{-1} Hoagland solution (half strength, [27]). After 55 days the aboveground biomass was harvested as described above for the conditioning phase. Shoots from all pots were separated per species and dry weight per species per pot was determined after 72 h drying at 70°C. For a subset of plants (farm 3) the shoot material was ground in a Retsch mill and carbon and nitrogen was measured in three milligrams of the leaves by combustion with an elemental autoanalyzer Flash EA 1112 NC analyzer (Interscience, Breda, the Netherlands) to determine whether inoculation affected nitrogen and carbon content.

Statistical analyses

Differences in shoot biomass of *T. aestivum* on inoculated and non-inoculated soils in the conditioning phase were tested with a linear model: shoot biomass ~ Inoculum * Farm. Farm was included as fixed effect since we were interested in the generality of the tested responses. Normality of the residuals was tested with the non-parametric Kolmogorov-Smirnov test and Homogeneity of variances with the non-parametric Fligner test. Residuals were not normally distributed and this could not be improved by transformation or removal of outliers. However, non-parametric tests (Wilcoxon test per farm) gave the same results as the linear model and thus violation of the assumptions appeared not too influential. For consistency with the analyses in the feedback phase, we decided to present the linear model results. In the feedback phase we compared shoot biomass of *T. aestivum* between inoculated and non-inoculated soils with a linear model using Inoculation, Farm and Competition as fixed categorical factors. Model check plots were made to inspect behavior of the residuals and normality of errors; no transformations were necessary. We also run an additional model where weed biomass was included as a covariate to verify whether this could explain possible Inoculation effects. When covariates were significant, correlations were tested with Spearman's rank correlation tests. This was done separately for the two treatments in the feedback phase (inoculated and non- inoculated). Difference in nitrogen percentage in the shoots of plants growing on inoculated and non-inoculated soils was tested with a linear model nitrogen ~ Inoculum * Competition and verified non-parametrically with a Wilcoxon test. Since the interaction Inoculum * Competition was not significant ($P = 0.96$) we only present the main fixed effects for the model nitrogen ~ Inoculum + Competition. All statistical analyses were carried out with R 2.14.1 statistical package [30]. The data has been archived in the Data and Information Portal of the NIOO-KNAW http://data.nioo.knaw.nl/index.php and is available on request.

Results

Conditioning phase

The effect of inoculation with field soil on *T. aestivum* shoot biomass varied between farms (Fig. 1, Table 1). For two farms inoculated soil yielded more *T. aestivum* shoot biomass (10%±3) than non-inoculated soil, whereas the other two farms showed the opposite pattern (−12%±0.5).

Effect of crop-weed mixtures on vulnerability to PSF

In the feedback phase the effect of inoculation interacted with farm and competition (significant 3-way interaction, Table 2). The presence of weeds reduced *T. aestivum* shoot biomass in all cases (Fig. 2A, 2B), but the strength of the effect is modified by farm and inoculation. On farms 1, 2 and 3 the growth reduction by competition was smaller on sterilized soil (7–10%, Fig. 2A) than on

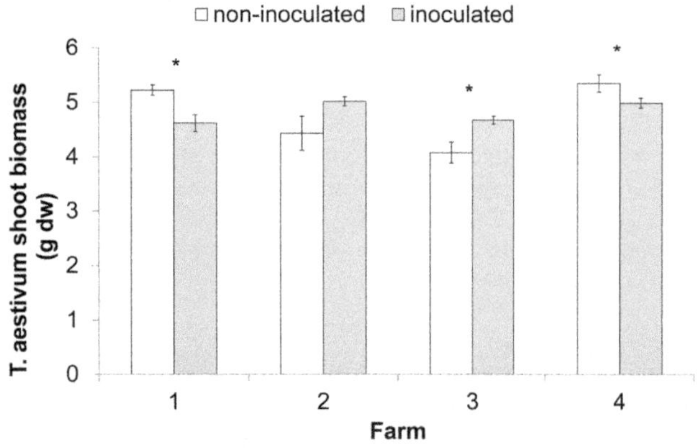

Figure 1. Soil inoculation and plant biomass. Shoot biomass of *Triticum aestivum* (mean ± SE) on sterilized soil without inoculum or with 9% field soil inoculum from 4 different farms in the conditioning phase. The asterisks indicate significant differences ($P<0.05$) between non-inoculated and inoculated soil.

Table 1. Linear model testing effect of inoculation and farm on shoot biomass of *Triticum aestivum* in the conditioning phase.

	Estimate	Std. Error	*t*-value	*P*-value
Intercept	4.61	0.13	34.45	<0.01
Inoculation	0.61	0.21	2.86	<0.01
Farm 2	0.40	0.19	2.10	0.04
Farm 3	0.06	0.19	0.31	0.76
Farm 4	0.37	0.19	1.95	0.05
Inoculation*Farm 2	−1.19	0.30	−3.97	<0.01
Inoculation*Farm 3	−1.20	0.30	−4.01	<0.01
Inoculation*Farm 4	−0.25	0.30	−0.82	0.41

inoculated soil (10–27%, Fig. 2B). Farm 4 is the exception, with stronger growth reduction by competition on the sterilized soil. Since total weed biomass was not significantly affected by inoculation, the difference in *T. aestivum* shoot biomass between sterilized and inoculated soils is probably not due to weeds being smaller on non-inoculated soil. When weed biomass was included as covariate, Inoculum is retained as a, albeit marginally, significant factor in the model. Apparently *T. aestivum* is more sensitive to competition when the field soil biota was present. Further support for this hypothesis can be found when considering the relation between shoot biomass of weeds and *T. aestivum*. For sterilized soil there is no correlation between those two (Fig. 3A), while for inoculated soil this correlation is significantly negative (Fig. 3B).

The presence of biota appeared to have increased either nutrient availability or uptake, since nitrogen percentage in the shoot during the feedback phase was slightly higher for inoculated soils (1.60±0.05% for inoculated soils, $n = 23$, 1.42±0.06% for non-inoculated soils, $n = 16$, lm Inoculum $P = 0.02$, Competition $P = 0.31$, no significant interaction). Legumes are known for increasing nitrogen, but in our experiment the presence of weeds including one legume species did not significantly affect nitrogen or carbon percentages in the shoot of *T. aestivum*.

Discussion

Effect of crop-weed mixtures on vulnerability to PSF

Plant mixtures in the feedback phase showed reduced wheat focal shoot biomass, which might be a combination of interspecific competition and enhanced sensitivity to plant soil feedback. In monoculture, however, there was very little evidence of plant soil feedback deviating from neutral. Negative effects of soil inoculation on shoot biomass in the conditioning phase may represent net pathogenic effects of soil biota, which might be even stronger in the feedback phase since the pathogens accumulated during conditioning. The same could be expected for the positive effects, unless positive biota need more time to establish, but in fact soils from only 1 out of 4 farms showed a repeated pattern of the conditioning phase in the feedback phase. Plants in soil from Farm 4 always grew better in sterilized soil compared to inoculated soil. Plants in soil from Farms 1, 2 and 3 did equally well on inoculated and non-inoculated soil in the feedback phase, despite the differences in seen in the conditioning phase. Therefore, regardless of the legacy effects in the conditioning phase, there was little difference between inoculated and non-inoculated soils for the

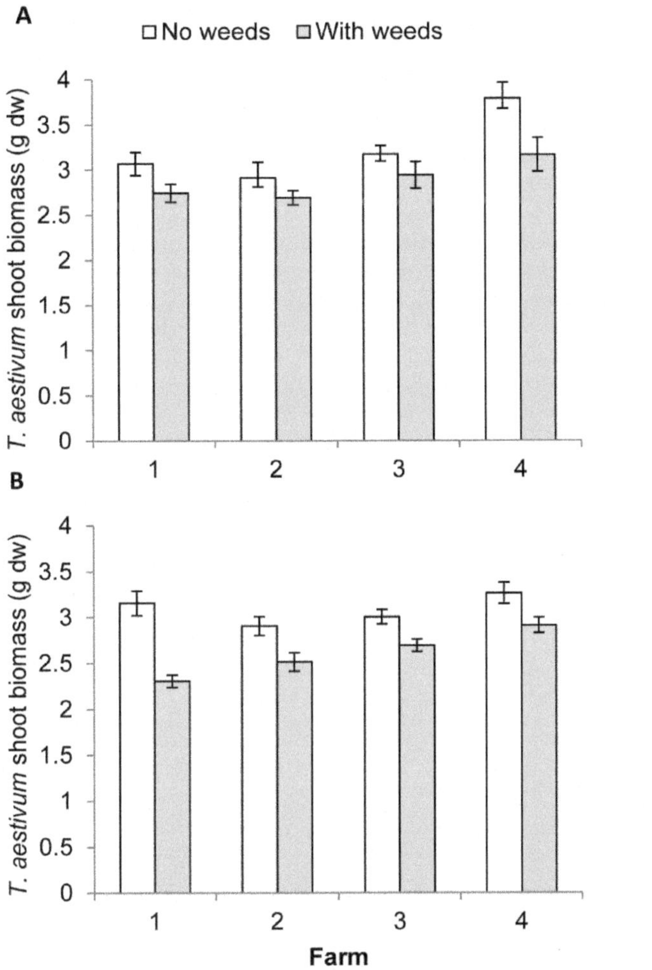

Figure 2. Competition interacts with inoculation effects. Shoot biomass of *Triticum aestivum* (mean ± SE) growing in soil with or without weeds. **A** growing in sterilized soil, **B** growing in inoculated soil.

Table 2. Linear model testing effect of inoculation, competition and farm on shoot biomass of *Triticum aestivum* in the feedback phase.

	Estimate	Std. Error	*t*-value	*P*-value
Intercept	3.15	0.10	30.32	<0.01
Inoculation	−0.08	0.16	−0.51	0.61
Farm 2	−0.25	0.15	−1.69	0.09
Farm 3	−0.15	0.15	−1.02	0.31
Farm 4	0.11	0.15	0.72	0.47
Competition	−0.85	0.15	−5.76	<0.01
Inoculation*Farm 2	0.09	0.23	0.38	0.70
Inoculation*Farm 3	0.25	0.23	1.08	0.28
Inoculation*Farm 4	0.61	0.23	2.63	<0.01
Competition*Farm2	0.46	0.21	2.20	0.03
Competition*Farm3	0.54	0.21	2.58	0.01
Competition*Farm4	0.50	0.21	2.40	0.02
Inoculation*Competition	0.52	0.23	2.22	0.03
Inoculation*Farm 2*Competition	−0.35	0.33	−1.06	0.29
Inoculation*Farm 3* Competition	−0.44	0.33	−1.33	0.19
Inoculation*Farm 4* Competition	−0.79	0.33	−2.40	0.02

monoculture treatment in terms of biomass and only the nitrogen data indicated that there is some biotic feedback. Feedback effects became more apparent under competition; addition of weed seedlings to half of the pots in the feedback phase led to consistently smaller *T. aestivum* plants on inoculated soils as compared to the non-inoculated ones. Likewise, Callaway et al. [22] only found effects of soil biota on *Centaurea maculosa* in greenhouse experiments when competitors were present. Stronger negative feedback effects under competition have been found before, in experiments with replacement designs [13,18,19], where total plant density is kept constant. Manipulation of soil communities independent of plant communities showed that

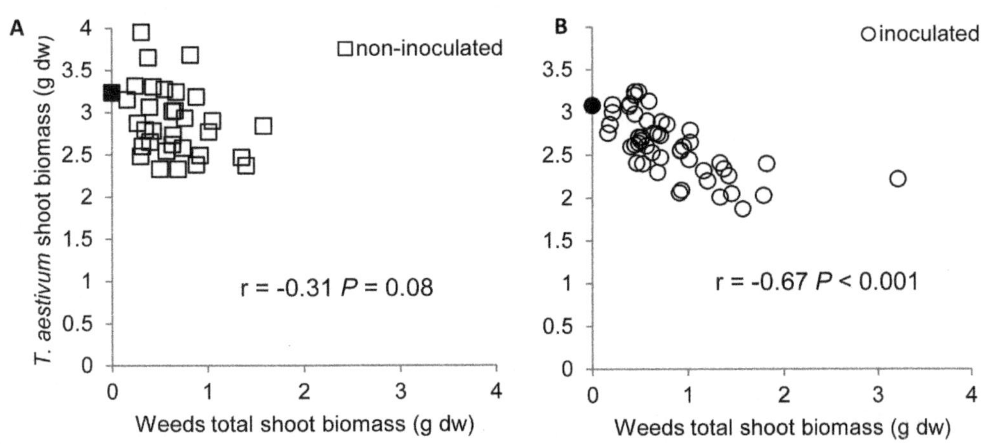

Figure 3. Correlation between crop and weeds depend on soil inoculation. Correlation between *T. aestivum* shoot biomass and weeds shoot biomass within the same pot on sterilized **A** non-inoculated soil and **B** inoculated soil in the feedback phase. Spearman rank correlation coefficients and *P*-values are shown in the graphs. Filled symbols indicate the average *T. aestivum* shoot biomass in the absence of weeds; n = 32 for non-inoculated soils and n = 48 for inoculated soils.

overyielding was due to plants inhibited by their own soil biota and unrelated to nutrient availability [31]. Shannon et al. [20] also used an additive design and found no evidence for feedback when plants were grown separately; only in competitive mixtures did feedback become apparent. Yet there is no consistency in the relationship between competition and feedback; feedback effects might also disappear under conditions of competition [22] and in some circumstances feedback effects might override competition effects [23]. The variety in effects might be related to soil origin and to plant species identity [22], although a study with 24 species showed that the majority of species suffered stronger negative feedback when grown in competition [19]. Even without competition plant soil feedbacks have been described as idiosyncratic [32,33] often meaning unpredictable [34] and the challenge is to find generalities in plant-soil feedback. The added value of the present study is the demonstration that biotic feedback effects were increased when co-occurring species compete and this was found across a wide range of farms and fields.

On non-inoculated soils *T. aestivum* plants had larger shoots and appeared to be less affected by competing weeds than on the inoculated soil. One possible explanation would be that the weeds did not grow well on the non-inoculated soil. However, weed biomass was not significantly different between non-inoculated and inoculated soils and including weed biomass as a covariate did not affect the significant difference between control soil and the inoculated soils. The degradation of wheat roots might also have differed between inoculated and non-inoculated soils, with consequences for plant nutrition. Since we made no measurements of plant growth during the experiment, we cannot exclude the possibility that there were initially differences in *T. aestivum* monocultures between sterilized non-inoculated and inoculated soil which disappeared in the later growth phase. This has been found before [13] and in plant mixtures the weeds could have taken advantage of the temporary growth delay of the wheat plants, thus effectively fixating the difference. This would also result in larger wheat plants in the feedback phase on sterilized control soils than on the inoculated soils. However, if weeds used the window of opportunity created in inoculated soil, then larger weed biomass would be expected for inoculated soil than for the sterilized control soils, but this was not found. Thus the most parsimonious interpretation is that *T. aestivum* plants growing in mixtures were more sensitive to their biotic PSF, e.g. due to a trade-off between resources allocated to either competitive or defense traits.

A possible mechanism for increased effects of PSF under competition could be the increased amount of roots in pots with plant mixtures. Root density is an important factor in development of fungal diseases [35]. However, the question is whether this would also work for specific pathogens. Generally, increasing plant diversity is thought to dilute species-specific pathogens [6,7]. The fact that the current study reveals that *T. aestivum* growing in plant mixtures was more sensitive to PSF might depend on the dominance of *T. aestivum*. This should be tested in a PSF experiment where the focal plant density is kept constant and the competitors are added in a large range to test whether there is a tipping point from increased negative effects of PSF in mixtures towards a dilution of plant-specific pathogens. Aguilera [36] modeled two-species competition under a range of competitive abilities and feedback scenario's and demonstrated that feedbacks can reverse the outcome of competition. The existence of such a tipping point will depend on the strength of the PSF versus the competitive effects. Knowledge of such a point could be used to determine optimal planting densities in mixed cropping systems.

Acknowledgments

We thank Simon Mortimer and Stephanie Harris for providing soil from farms in the UK. This is publication 5443 of the NIOO-KNAW.

Author Contributions

Conceived and designed the experiments: WHGH WdB WvdP. Performed the experiments: WHGH FtH. Analyzed the data: WHGH. Wrote the paper: WHGH WvdP WdB FtH.

References

1. Bever JD (1994) Feedback between plants and their soil communities in an old field community. Ecology 75: 1965–1977.
2. Kulmatiski A, Beard KH, Stevens JR, Cobbold SM (2008) Plant-soil feedbacks: a meta-analytical review. Ecol Lett 11: 980–992.
3. Bever JD, Westover KM, Antonovics J (1997) Incorporating the soil community into plant population dynamics: the utility of the feedback approach. J Ecol 85: 5261–573.
4. Reynolds HL, Packer A, Bever JD, Clay K (2003) Grassroots ecology: plant-microbe-soil interactions as drivers of plant community structure and dynamics. Ecology 84: 2281–2291.
5. Kardol P, Bezemer TM, Van der Putten WH (2006) Temporal variation in plant-soil feedback controls succession. Ecol Lett 9: 1080–1088.
6. Maron JL, Marler M, Klironomos JN, Cleveland CC (2011) Soil fungal pathogens and the relationship between plant diversity and productivity. Ecol Lett 14: 36–41.
7. Schnitzer SA, Klironomos JN, HilleRisLambers J, Kinkel LL, Reich PB, et al. (2011) Soil microbes drive the classic plant diversity-productivity pattern. Ecology 92: 296–303.
8. Bullock DG (1992) Crop rotation. Crit Rev Plant Sci 11: 309–326.
9. Eviner VT, Hawkes CV (2008) Embracing variability in the application of plant-soil interactions to the restoration of communities and ecosystems. Restor Ecol 16: 713–729.
10. Jordan NR, Larson DL, Huerd SC (2011) Evidence of qualitative differences between soil-occupancy effects of invasive vs. native grassland plant species. Invasive Plant Sci Management 4: 11–21.
11. Inderjit, Van der Putten WH (2010) Impacts of soil microbial communities on exotic plant invasions. TREE 25: 512–519.
12. Tilman D (1982) Resource competition and community structure. Princeton University Press, Princeton, N.J.
13. van der Putten WH, Peters BA (1997) How soil-borne pathogens may affect plant competition. Ecology 78: 1785–1795.
14. Harrison KA, Bardgett RD (2010) Influence of plant species and soil condition on plant-soil feedback in mixed grassland communities. J Ecol 98: 384–395.
15. Hauggaard-Nielsen H, Gooding M, Ambus P, Corre-Hellou G, Crozat Y, et al. (2009) Pea-barley intercropping and short-term subsequent crop effects across European organic cropping conditions. Nutr Cycl Agroecosyst 85: 141–155.
16. Gilbert GS (2002) Evolutionary ecology of plant diseases in natural ecosystems. Ann Rev Phytopathol 40: 13–43.
17. Fine PVA, Miller ZJ, Mesones I, Irazuzta S, Appel HM, et al. (2006) The growth-defense trade-off and habitat specialization by plants in Amazonian forests. Ecology 87: S150–S162.
18. Kardol P, Cornips NJ, van Kempen MML, Bakx-Schotman JMT, Van der Putten WH (2007) Microbe-mediated plant-soil feedback causes historical contingency effects in plant community assembly. Ecol Monogr 77: 147–162.
19. Petermann JS, Fergus AJF, Turnbull LA, Schmid B (2008) Janzen-Connell effects are widespread and strong enough to maintain diversity in grasslands. Ecology 89: 2399–2406.
20. Shannon S, Flory SL, Reynolds H (2012) Competitive context alters plant-soil feedback in an experimental woodland community. Oecologia 169: 235–243.
21. Callaway RM, Thelen GC, Barth S, Ramsey PW, Gannon JE (2004) Soil fungi alter interactions between the invader *Centaurea maculosa* and North American natives. Ecology 85: 1062–1071.
22. Callaway RM, Thelen GC, Rodriguez A, Holben WE (2004) Soil biota and exotic plant invasion. Nature 427: 731–733.
23. Casper BB, Castelli JP (2007) Evaluating plant-soil feedback together with competition in a serpentine grassland. Ecol Lett 10: 394–400.
24. Snaydon RW (1991) Replacement or additive designs for competition studies. J Appl Ecol 28: 930–946.
25. Callaway RM, Mahall BE, Wicks C, Pankey J, Zabinski C (2003) Soil fungi and the effects of an invasive form on grasses: neighbor identity matters. Ecology 84: 129–135.

26. Brinkman EP, van der Putten WH, Bakker E-J, Verhoeven KJF (2010) Plant-soil feedback: experimental approaches, statistical analysis and ecological interpretations. J Ecol 98: 1063–1073.
27. Hoagland DR, Arnon DI (1950) The water-culture method for growing plants without soil. Calif Agricultural Experiment Stn Circular 347:1–32.
28. Manning P, Morrison SA, Bonkowski M, Bardgett RD (2008) Nitrogen enrichment modifies plant community structure via changes to plant-soil feedback. Oecologia 157: 661–673.
29. Lundkvist A, Salomonsson L, Karlsson L, Gustavssone AMD (2008) Effect of organic farming on weed flora composition in a long term perspective. Eur J Agron 28: 570–578.
30. R development Core Team (2008) R: a Language and Environment for Statistical Computing. R website. Available: http://www.R-project.org Accessed 2012 Oct 8.
31. Hendriks M, Mommer L, de Caluwe H, Smit-Tiekstra AE, van der Putten WH, et al. (2013) Independent variation of plant and soil mixtures reveal soil feedback effects on plant community overyielding. J Ecol 101: 287–297.
32. Brandt AJ, Seabloom EW, Hosseini PR (2009) Phylogeny and provenance affect plant-soil feedbacks in invaded California grasslands. Ecology 90: 1063–1072.
33. Jiang LL, Han X, Zhang GM, Kardol P (2010) The role of plant-soil feedbacks and land-use legacies in restoration of a temperate steppe in northern China. Ecol Res 25: 1101–1111.
34. Hol WHG, Meyer KM, Van der Putten (2011) Idiosyncrasy in ecology – what's in a word? Front Ecol Environ 9: 431–432.
35. Otten W, Gilligan CA (2006) Soil structure and soil-borne diseases: using epidemiological concepts to scale from frungal spread to plant epidemics. Eur J Soil Sci 57: 26–37.
36. Aguilera AG (2011) The influence of soil community density on plant-soil feedbacks: an important unknown in plant invasion. Ecol Model 222: 3413–3420.

3

Understanding Dry Matter and Nitrogen Accumulation with Time-Course for High-Yielding Wheat Production in China

Qingfeng Meng[1], Shanchao Yue[1], Xinping Chen[1], Zhenling Cui[1]*, Youliang Ye[2], Wenqi Ma[3], Yanan Tong[4], Fusuo Zhang[1]

1 Center for Resources, Environment and Food Security, China Agricultural University, Beijing, China, **2** College of Resources and Environmental Sciences, Henan Agricultural University, Zhengzhou, China, **3** College of Resources and Environmental Sciences, Hebei Agricultural University, Baoding, China, **4** College of Resources and Environmental Sciences, Northwest Sci-Tech University of Agriculture and Forestry, Yangling, China

Abstract

Understanding the time-course of dry matter (DM) and nitrogen (N) accumulation in terms of yield–trait relationships is essential to simultaneously increase grain yield and synchronize N demand and N supply. We collected 413 data points from 11 field experiments to address patterns of DM and N accumulation with time in relation to grain yield and management of winter wheat in China. Detailed growth analysis was conducted at the Zadok growth stages (GS) 25 (regreening), GS30 (stem elongation), GS60 (anthesis), and GS100 (maturity) in all experiments, including DM and N accumulation. Grain yield averaged 7.3 Mg ha^{-1}, ranging from 2.1 to 11.2 Mg ha^{-1}. The percent N accumulation was consistent prior to DM accumulation, while both DM and N accumulation increased continuously with growing time. Both the highest and fastest DM and N accumulations were observed from stem elongation to the anthesis stage. Significant correlations between grain yield and DM and N accumulation were found at each of the four growth stages, although no positive relationship was observed between grain yield and harvest index or N harvest index. The yield increase from 7–9 Mg ha^{-1} to >9 Mg ha^{-1} was mainly attributed to increased DM and N accumulation from stem elongation to anthesis. Although applying more N fertilizer increased N accumulation during this stage, DM accumulation was not improved, indicating that N fertilizer management and related agronomic management should be intensified synchronously across the wheat growing season to simultaneously achieve high yields and match N demand and N supply.

Editor: Mingliang Xu, China Agricultural University, China

Funding: The authors thank the National Basic Research Program of China (973, Program: 2009CB118606), and Special Fund for Agriculture Profession (200803030). The funders had no role in study design, data collection and analysis, decision to publish, or preparation of the manuscript.

Competing Interests: The authors have declared that no competing interests exist.

* E-mail: cuizl@cau.edu.cn

Introduction

As the largest wheat producer and consumer in the world, China produced around 115 million t wheat with a planting area of about 24.3 million ha in 2010 [1]. To meet the projected demands of population growth and increasing consumption, cereal production must increase by 70–100% until 2050 [2,3]. However, as one of major cereal, yield growth of wheat in China has dropped from 9% in 1970s to <3% annually in 2000s [1]. Because of the great difficulty of improving yield potential over the short-term through genetic improvement [4], great efforts should be made to improve agronomical management to enhance average farm yields. Knowledge of DM and nutrient accumulation associated with yield–trait relationships in crops would supply an efficient tool to improve management efficiency and thus increase yield.

Grain yield has usually been positively correlated with total DM production and nutrient accumulation in crops [5]. Meanwhile, DM accumulation and nutrient accumulation vary with growth stage of crops [6]. Many previous studies indicated that DM and N accumulation for wheat occurred mainly pre-anthesis, and that grain yields greatly depended on the translocation of pre-anthesis assimilates and N accumulation [7–10]. However, in practice, a significant correlation has also been found between grain yield and DM production post-anthesis in high-yield systems in China [11,12]. Because few studies have focused on the relationship between grain yield and DM and N accumulation with time-course (e.g., [13]), the pattern of DM and N accumulation pre- and post-anthesis remains unclear.

Due to lack of proper knowledge, most Chinese farmers tried to increase wheat DM and N accumulation as high as possible at earlier stage to pursue high yield in practice. For example, farmers applied a large amount of irrigation and N fertilizer at early stage (*e.g.*, regreening stage, GS25) [14,15], which leaded to higher DM through more spring tillering. However, higher tiller population would lead to more non-reproductive tillers [16], and more risk of lodging, pests and diseases when it is more than optimum effective plant population. Meanwhile, the non-reproductive tillers would have a negative effect as competitors for assimilates and plant nutrients [17,18]. Thus, further studies on DM and N accumulation with time-course are needed, which could help to identify the critical stage of optimal spring tillering through management to improve grain yield.

N management also influences the dynamics of DM and N accumulation by crops [19]. For example, N overuse leads to higher plant N concentrations due to so-called "luxury accumulation" by crops, while N shortage decreases plant N concentrations and sacrifices some grain yield [20,21]. Recently, N recommendations in crop production systems emphasized the need for greater synchrony between crop N demand and the N supply from all sources throughout the growing season [22,23]. However, most of previous studies have been made to quantify total N accumulation at harvest through yield goals and crop models [20,24–26]. Due to the limited understanding dynamics of N accumulation, N supply in current N management practices often does not match crop demand. For instance, farmers typically apply a large amount of N fertilizer for wheat before sowing (often >50% total N fertilizer), which results in large N losses to the environment because the plant capacity for N accumulation is small during this period [27].

Thus, much effort should be made for dynamic DM and N accumulations associated with yield–trait relationship to improve management efficiency and simultaneously achieve high yields and synchronize N demand and N supply. In this study, we collected 413 field-year data at nine sites in major wheat area in China, to investigate the relationship between DM and N accumulation and grain yield, the pattern of DM and N accumulation for different yield levels with time, and the dynamics of DM and N accumulation with different N management practices.

Materials and Methods

Eleven field experiments at nine sites in four key winter wheat domains of China were conducted in Dongbeiwang (DBW) in Beijing; Xiangyun (XY), Zhaobao (ZB), and Lankao (LK) in Henan Province; Dingzhou (DZ) and Quzhou (QZ) in Hebei Province; and Yangling (YL) in Shannxi Province. A typical winter wheat–summer maize rotation system was adopted at all sites. No specific permissions were required for these locations. The field studies did not involve endangered or protected species.

The amount and distribution pattern of precipitation varied widely from year to year, as affected by the continental monsoon climate. Wheat was irrigated two to four times based on soil water content at about 100 mm at each time point. More details such as experimental year, soil texture, organic matter content, total N content, Olsen-P, and NH_4OAc-K in the different experimental locations were shown in Table S1.

Experimental Design

For the eleven field experiments, the details of treatments, including varieties and N application rates were listed in Table S2. At least three N treatments were included for all eight N level experiments: no N as a control (N-0), an optimal N rate (N-opt), and the farmer's N practice (FNP). At XY (2008–2009, 2009–2010), LK, DZ and QZ sites, N treatments at sub- or supra-optimal N rates were also applied in experiments with five or six N treatments, including 40%, 50%, 70% and 75% of N-opt, 125%, 130%, and 150% of N-opt according to the experimental design at each site.

The optimal N rate was based on either in-season root zone N management (DBW and QZ sites) or yield goal-based N recommendation (XY, ZB, LK and DZ sites). For the in-season root zone N management (IRNM), wheat growing season was divided into two periods: from sowing to the stem elongation stage, and from the stem elongation stage to the maturity. The amount of N fertilizer applied at the beginning of each growing period was determined by deducting the amount of soil N_{min} (NH_4^+–N+NO_3^-–N) in the root-zone from the target N value, which was estimated based on the yield target and crop N accumulation. Detailed descriptions of the target N value and the soil N_{min} measurement have been previously reported [27,28]. The yield goal-based N recommendation was derived from the mass balance approach [29]. In this method, optimal N rate was estimated as total N accumulation of the grain less all of the other sources of grain N and adjusted for inefficiencies in the ability of the crop to recover fertilizer N from the soil [30]:

$$\text{Optimal N rate} = \left(\begin{array}{l}\text{N accumulation in the harvested grain} \\ -\ \text{Soil N in the harvest grain}\end{array}\right) / \text{Fertilizer N use efficiency} \quad (1)$$

where optimal N rate was estimated for a selected yield goal, fertilizer N use efficiency was the proportion of fertilizer N applied to the soil that was recovered in the grain. For the FNP treatment, the N rate varied from 300 to 360 kg ha^{-1}, mostly with a split application (50%:50%) as a broadcast application at pre-sowing, and as a side-dress application near the stem elongation stage.

For other three cropping systems experiments (DZ, QZ and YL sites), details of the system design were in the Text S1. In this study, the current farmers' system (FP) and higher yield system from agronomists (HY) were considered N-over systems, while the optimized system (HYHR) and integrated soil–crop management system (ISSM) were considered N-opt systems.

All of the experiments consisted of a randomized complete block design with three to four replications. The plot area varied from 20 m^2 at XY (2007–2008) to 1800 m^2 at QZ (2007–2010). Urea as the main N fertilizer has been used for all field experiments. Phosphorus (P), potassium (K) and other nutrients were applied as needed according to soil tests. No organic manure was applied. At all sites, winter wheat was seeded in early to mid-October and harvested the following June. During the study year, no obvious water, weed, pest, or disease stresses were observed during the wheat growing season at any site. No obvious winter killing happened for all field experiments.

Sampling and Laboratory Procedures

Before sowing, soil samples from the 0–30 cm soil layer were air-dried and sieved to remove un-decomposed plant material. Then the sieved samples were used to measure soil organic matter, total N, Olsen-P, and ammonium acetate-extractable K. To determine DM and N content, plant samples were collected four times: at the regreening (GS25), stem elongation (GS30), anthesis (GS60), and maturity (GS100) stages. Plant samples were dried at 60°C to determine the DM, and sub-samples were taken to measure the N content using the Kjeldahl method. At harvest, areas ranging from 6 m^2 at XY (2007–2008) to 18 m^2 at QZ (2007–2010) were harvested manually and dried in an oven at 60°C to calculate grain yield. The grain yield value was adjusted to 14% moisture. The number of grains per square meter was counted manually for at least two rows within 1.0 m, and the number per ear was counted manually to estimate a mean value from more than 20 ears. For the thousand kernel weight (TGW), 1000 grains were randomly counted and weighed.

At XY (2007–2008, 2008–2009, 2009–2010), ZB (2007–2008), LK (2008–2009), QZ (2008–2009), DZ (2008–2010), and YL (2008–2009), the total tillers in the N-opt treatment were investigated before tillering (GS20), before winter (GS23), and at stem elongation, anthesis, and maturity.

Data Analysis

According to grain yield, all data from the nine sites were divided into three groups: <7, 7–9, and >9 Mg ha^{-1}. According this yield range, wheat DM and N accumulation were analyzed with time -course. According N application, all data were also divided into three groups: N-0, N-opt and N-over (125%, 130%, and 150% of N-opt, and FNP) for DM and N accumulation analysis. The data were then further categorized into nine groups according to the yield ranges (1) <7 Mg ha^{-1}, N-0; (2) <7 Mg ha^{-1}, N-opt; (3) <7 Mg ha^{-1}, N-over; (4) 7–9 Mg ha^{-1}, N-0; (5) 7–9 Mg ha^{-1}, N-opt; (6) 7–9 Mg ha^{-1}, N-over; (7) >9 Mg ha^{-1}, N-0; (8) >9 Mg ha^{-1}, N-opt, (9) >9 Mg ha^{-1}, N-over. The investigated stems dynamics in the N-opt treatment at six sites (XY (2007–2008, 2008–2009, 2009–2010), ZB (2007–2008), LK (2008–2009), QZ (2008–2009), DZ (2008–2010), and YL (2008–2009)) were also grouped according to grain yield (<7, 7–9, and >9 Mg ha^{-1}).

For all collected data, the correlation coefficients (*r*) between grain yield and average yield components, DM production, and N accumulation at different growth stages, were analyzed. At harvest, the amount of DM production translocation was calculated as the difference between dry DM in the stover at anthesis and at harvest. N accumulation translocation was calculated as the difference between N accumulation in the stover at anthesis and at harvest. The rate of fertile tillers from stem-elongation to anthesis was calculated according the result that the number of total ears at anthesis was divided by the tiller number at stem-elongation stage. N accumulation was calculated as:

$$\text{N accumulation} = \text{Dry matter} \times \text{N concentration} \qquad (2)$$

At regreening, stem elongation and anthesis stage, N accumulation was calculated as straw DM multiplying straw N concentration. At maturity, N accumulation was the sum of straw and grain N accumulation.

Results

Grain Yield, DM and N Accumulation, and N Concentration

Overall, wheat grain yield averaged 7.3 Mg ha^{-1}, ranging from 2.1 Mg ha^{-1} to 11.2 Mg ha^{-1} (Table 1). According to published statistical data, the average wheat yield was 4.7 Mg ha^{-1} for China and 3.0 Mg ha^{-1} for the whole world in 2009 [1]; thus, the mean yield observed in the present study is 155% and 243% higher than the national and world yields, respectively. However, this is still lower than that in high-yield wheat production areas of Europe, which produce more than 8.0 Mg ha^{-1} [1].

Sixty-four percent of total DM accumulated pre-anthesis, while the other 36% accumulated post-anthesis (Figure S1). Before anthesis, 10% and 20% of DM accumulated in the regreening and stem-elongation stages, respectively. From stem elongation to anthesis, DM accumulated the most and the fastest across all four stages. More than 6.3 Mg ha^{-1} DM accumulated from stem elongation to anthesis, with an average accumulation rate of 253 kg ha^{-1} day^{-1} (Figure 1A). A total of 1.5 Mg ha^{-1} DM accumulated from sowing to regreening, 1.4 Mg ha^{-1} DM from regreening to stem elongation, and 5.1 Mg ha^{-1} DM accumulated post-anthesis, while the associated daily accumulation rate averaged 21, 38, and 130 kg ha^{-1} day^{-1}, respectively.

The percent N accumulation was consistent prior to DM accumulation (Figure S1). Unlike DM accumulation, 80% of N accumulated pre-anthesis; 27%, 20%, and 33% of N accumulated from sowing to regreening, from regreening to stem elongation, and from stem elongation to anthesis, respectively. Similar to DM accumulation, the highest and fastest N accumulation was observed from stem elongation to anthesis. During this period, 2.44 kg N ha^{-1} accumulated per day and 61 kg N ha^{-1} accumulated in total (Figure 1B). Moreover, 50 kg ha^{-1} N, with 0.7 kg N ha^{-1} per day, accumulated from sowing to regreening; 38 kg ha^{-1} N, with 1.1 kg N ha^{-1} per day, accumulated from regreening to stem elongation; and 37 kg ha^{-1} N, with 1.0 kg N ha^{-1} per day, accumulated post-anthesis.

The relationships between grain yield and DM production, N accumulation, and N concentration for different growth stages and yield components were shown in Table 2. For yield components, grains per ear was significantly correlated with grain yield (r = 0.54). Significant correlations between DM and grain yield were observed in each stage (0.58–0.95). Grain yield was significantly correlated with DM accumulation from stem elongation to anthesis and post-anthesis. No positive correlation was found between grain yield and harvest index.

Similar to DM production, significant correlations between grain yield and N accumulation were found at each of stage (r = 0.54–0.81; Table 2). The amount of N translocation from stover to grain was important for yield increase (r = 0.47), although DM translocation was not significant. Between grain yield and N concentration, no significant relationship was found with an exception of grain N concentration at maturity (Table 2). Moreover, no significant correlation was found between grain yield and N harvest index.

Dynamics of DM and N Accumulation with Different Yield Levels

To further understand the relationship between yield and temporal dynamics of DM and N accumulation, the collected data were further categorized into three groups according to the yield range (Table 1). In total, 179, 112, and 122 data points were grouped according to the yield ranges <7 Mg ha^{-1}, 7–9 Mg ha^{-1}, and >9 Mg ha^{-1}, respectively. The grain yield averaged 5.3 Mg ha^{-1} (2.1–7.0 Mg ha^{-1}), 8.1 Mg ha^{-1} (7.0–9.0 Mg ha^{-1}), and 9.7 Mg ha^{-1} (9.0–11.2 Mg ha^{-1}), respectively. According to the yield components, the yield difference was mainly related to the grain number per square meter (Table S3), which was determined by ear number and grains per ear. The grain number per square meter averaged 24,000 m^{-2} for 7–9 Mg ha^{-1} and 28,000 m^{-2} for >9 Mg ha^{-1}, an increase of 33% and 56% compared to the yield range of <7 Mg ha^{-1} (18,000 m^{-2}). Similar grain weight per 1000 grains (40–42 g) was observed among all three yield levels.

Across the whole wheat season, wheat DM and N accumulation of the yield for <7 Mg ha^{-1} were consistently lower than those for 7–9 Mg ha^{-1} and >9 Mg ha^{-1} (Figure 2). This difference started from the beginning of the growth stage. From the sowing to the regreening stage, 2.2 Mg ha^{-1} DM accumulated in 7–9 Mg ha^{-1}, 314% higher than that in <7 Mg ha^{-1} (0.7 Mg ha^{-1}) (Figure 2A). Similarly, the 71 kg ha^{-1} of N that accumulated in 7–9 Mg ha^{-1} is substantially higher than the 27 kg ha^{-1} of N that accumulated in <7 Mg ha^{-1} during this stage (Figure 2B). At regreening stage, the percentage of DM (7%) and N accumulation (20%) to total in <7 Mg ha^{-1} was 47% and 43% lower compared with that in 7–9 Mg ha^{-1} yield range (14% and 34%), respectively. The lower DM and N accumulation starting at sowing to regreening and continuing through the subsequent growth stages in yield range

Table 1. Descriptive statistics of yield for total samples, three yield ranges, and nitrogen levels (Mg ha^{-1}).

	n[a]	Mean	SD[b]	Minimum	25%Q[c]	Median	75%Q	Maximum
Total	413	7.3	2.1	2.1	5.8	7.5	9.2	11.2
Yield category								
<7 Mg ha^{-1}	179	5.3	1.3	2.1	4.2	5.6	6.4	7
7–9 Mg ha^{-1}	112	8.1	0.6	7	7.5	8.1	8.6	9
>9 Mg ha^{-1}	122	9.7	0.5	9	9.4	9.6	9.8	11.2
N-level								
N-0	79	6.6	2.5	2.1	4.1	7.3	8.7	10.4
N-opt	126	7.6	2	3.1	6.1	7.5	9.4	11.2
N-over	154	7.5	1.8	3.1	6.2	7.7	9.3	10.6

[a]n = number of observation.
[b]SD = standard deviation.
[c]Q = quartile.

<7 Mg ha^{-1}, suggested poor agronomic management (e.g., sowing quality, late sowing) may be the main cause, as the yield potential and weather conditions were similar at all sites.

Compared to the yield of 7–9 Mg ha^{-1}, DM and N accumulation at a yield of >9 M ha^{-1} was similar before the stem-elongation stage (Figure 2). The greatest difference between these two yield levels for DM and N accumulation took place from stem elongation to anthesis. During this stage, DM accumulation increased from 6.3 Mg ha^{-1} in 7–9 Mg ha^{-1} to 8.7 Mg ha^{-1} in >9 Mg ha^{-1} and was associated with an increase in N accumulation from 55 kg ha^{-1} to 83 kg ha^{-1}. Meanwhile, the 46% of DM and 35% of N accumulation to total from stem elongation stage to anthesis in >9 Mg ha^{-1} was 15% and 30% higher compared with that in 7–9 Mg ha^{-1} (40% and 27%), respectively. These results showed that DM and N accumulation from stem elongation to anthesis was very important to achieve the >9 Mg ha^{-1} grain yield. Post-anthesis, 5.6 Mg ha^{-1} DM and 45 kg ha^{-1} N accumulated in 7–9 Mg ha^{-1} while DM and N accumulation averaged 6.8 Mg ha^{-1} and 50 kg ha^{-1} in >9 Mg ha^{-1}, respectively.

The yield range <7 Mg ha^{-1} group always had the highest N concentration for all growth stages. At regreening stage, the N concentration in <7 Mg ha^{-1} averaged 3.72%, 13–16% higher than that in both 7–9 Mg ha^{-1} and >9 Mg ha^{-1} yield ranges (3.21–3.29%). At stem elongation stage, the N concentration in <7 Mg ha^{-1} group was 3.32%, 6–16% higher than that in other two yield ranges (2.85–3.13%). At anthesis, 1.75% of N concentration in <7 Mg ha^{-1} was 7–13% higher than in other two yield ranges (1.55–1.63%). At maturity, N concentration in <7 Mg ha^{-1} averaged 1.36%, 4–8% higher than that in other two yield ranges (1.26–1.31%). This may be because of the lower DM accumulation in <7 Mg ha^{-1}. For yield ranges 7–9 Mg ha^{-1} and

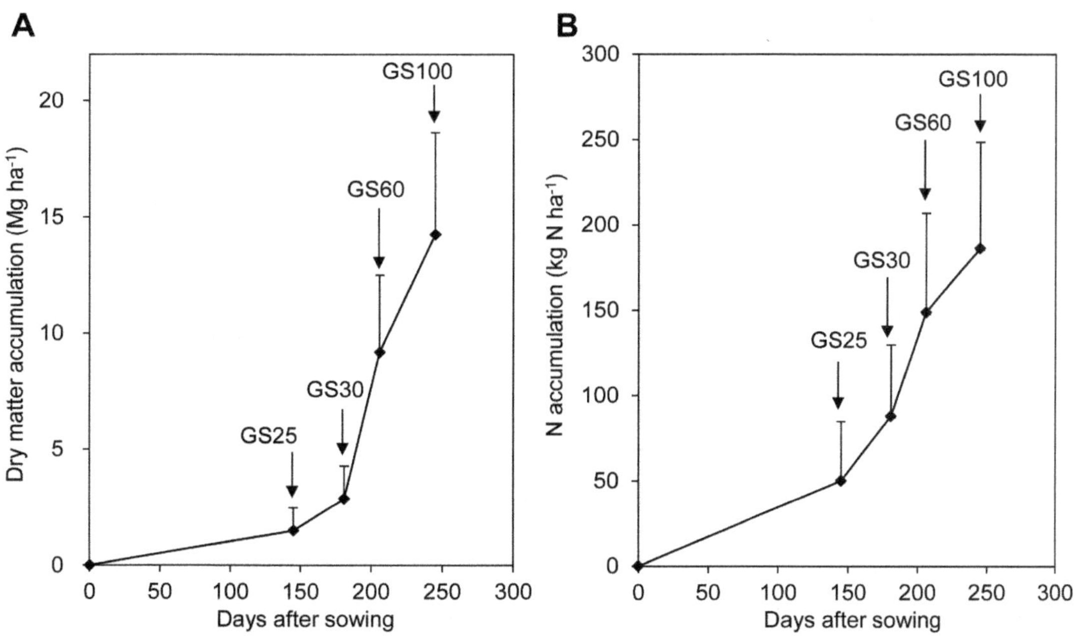

Figure 1. Dynamics of dry matter (A) and nitrogen (B) accumulation from sowing to maturity (n = 413). (GS25, GS30, GS60, and GS100 are the regreening, stem elongation, anthesis, and maturity stages, respectively.).

Table 2. Average yield components, dry matter and nitrogen accumulation in different growth stages, and correlative coefficient (r) between grain yield and these parameters for all samples (n = 413).

		Mean	r
Yield components	Ears number (10^4 ha^{-1})	600	0.36
	Grains per ear	39	0.54**
	Grain weight (g 1000^{-1})	41	0.10
Dry matter accumulation	Regreen (Mg ha^{-1})	1.5	0.60**
	Regreen to stem elongation (Mg ha^{-1})	1.4	0.29
	Stem elongation (Mg ha^{-1})	2.9	0.58**
	Stem elongation to anthesis (Mg ha^{-1})	6.3	0.65**
	Anthesis (Mg ha^{-1})	9.2	0.82**
	Post-anthesis (Mg ha^{-1})	5.1	0.70**
	Maturity (Mg ha^{-1})	14.2	0.95**
	Translocation (Mg ha^{-1})	1.3	0.29
N accumulation	Regreen (kg ha^{-1})	50	0.54**
	Regreen to stem elongation (kg ha^{-1})	38	0.24
	Stem elongation (kg ha^{-1})	88	0.59**
	Stem elongation to anthesis (kg ha^{-1})	61	0.37
	Anthesis (kg ha^{-1})	149	0.71**
	Post-anthesis (kg ha^{-1})	37	0.23
	Maturity (kg ha^{-1})	186	0.81**
	Translocation (kg ha^{-1})	97	0.47**
N concentration	Regreening-straw (%)	3.45	−0.29
	Stem elongation-straw (%)	3.14	0.00
	Anthesis-straw (%)	1.66	−0.14
	Maturity-straw (%)	0.63	0.24
	Maturity-grain (%)	2.15	−0.43**
HI and NHI	Harvest index	0.45	0.02
	N harvest index	0.74	−0.38

**$p<0.01$.

>9 Mg ha^{-1}, N concentration was similar for most stages but not at stem elongation stage. At stem elongation stage, N concentration in >9 Mg ha^{-1} was 3.13%, 10% higher than that in 7–9 Mg ha^{-1} (2.85%).

To further understand DM and N accumulation with time in the three yield levels, wheat stem dynamics was investigated at six sites with optimized N treatment (Figure 3). Although the basic seedling was similar among the three yield ranges, total stems differed before winter. Total stems in <7 Mg ha^{-1} was consistently lower compared to that in 7–9 Mg ha^{-1} and >9 Mg ha^{-1} from wintering to maturity. This further verified that sowing quality or other related agronomic management practices were poor and did not supply the optimal growth conditions for wheat. The final ear number was 0.77 million ha^{-1} higher in 9 Mg ha^{-1} than in 7–9 Mg ha^{-1}, although total tillers was 2.74 million ha^{-1} higher in 7–9 Mg ha^{-1} in the stem-elongation stage. As a result, earbering tiller percentage averaged 44% (from 16.69 to 7.36 million ha^{-1}) in >9 Mg ha^{-1} from the stem-elongation stage to maturity while it averaged only 34% (from 19.43 to 6.59 million ha^{-1}) in 7–9 Mg ha^{-1}. This suggested that the final ear number was determined by tiller quality, not total number, in earlier growth stages. In the stem-elongation stage, the N concentration in the stover for >9 Mg ha^{-1} averaged 3.27 g kg^{-1}, 7% higher than the 3.07 g kg^{-1} for 7–9 Mg ha^{-1} (Table 3). The higher N concentration in the plant in the stem-elongation stage may lead to higher stem quality (e.g., higher tiller weight), and thus more stems would survive from stem elongation to anthesis.

Dynamics of DM and N Accumulation with Different N Management Regimes

To further compare the partitioning of DM and N accumulation among different N application and yield levels, pooled data from all nine sites were grouped into three N application categories: N-0, N-opt, and N-over (Table 1). Accordingly, 79 samples with a mean yield of 6.6 Mg ha^{-1} (2.1–10.4 Mg ha^{-1}), 126 samples with a mean yield of 7.6 Mg ha^{-1} (3.1–11.2 Mg ha^{-1}), and 154 samples with a mean yield of 7.5 Mg ha^{-1} (3.1–10.6 Mg ha^{-1}) were grouped according to these three N application levels. Compared to the N-0 treatment, DM and N accumulation increased for the N-opt and N-over treatments in each growth stage for each yield range (Figure 4). Between N-opt and N-over treatments, DM accumulation was similar at each growth stage (Figure 4A). N accumulation was also similar between N-opt and N-over treatments in most growth stages but not at anthesis (Figure 4B). N accumulation averaged around 55 kg ha^{-1} at regreening stage and 100 kg ha^{-1} at stem elongation stage for both N-opt and N-over treatments. At anthesis, N accumulation in N-opt averaged 159 kg ha^{-1}, 6% lower compared with that in N-over (169 kg ha^{-1}). At maturity, N accumulation averaged about 200 kg ha^{-1} for both N-opt and N-over treatments. However, the N rate in all N-opt management averaged 179 kg ha^{-1} (n = 126), just 59% of the 304 kg ha^{-1} N rate in N-over (n = 154). This showed that more N fertilizer input over the rational level did not lead to more DM and N accumulation except N accumulation from stem elongation to anthesis, and not resulted in a higher grain yield.

For further comparison, the data for each N level were categorized according to the three yield ranges (<7, 7–9, and >9 Mg ha^{-1}). DM and N accumulation in the <7 Mg ha^{-1} yield was consistently lower than those in the 7–9 and >9 Mg ha^{-1} (Figure 5). When N was deficient, the difference in N accumulation occurred throughout the whole growing season in the 7–9 and >9 Mg ha^{-1} yield ranges, and the difference in DM accumulation was mainly post-anthesis (Figure 5A, D). Post-anthesis, DM accumulation in >9 Mg ha^{-1} averaged 6.5 Mg ha^{-1} (10.9–17.4 Mg ha^{-1}), 20% higher than that in 7–9 Mg ha^{-1} (9.9–15.3 Mg ha^{-1}) in the N-0 treatment (Figure 5A). When N was applied rationally, the differences in DM and N accumulation between the yield ranges of 7–9 and >9 Mg ha^{-1} occurred from stem elongation to maturity, particularly during the stem elongation to anthesis stage (Figure 5B, E). During this stage, 9.0 Mg ha^{-1} DM accumulation in >9 Mg ha^{-1} was 67% higher than that in 7–9 Mg ha^{-1} (5.4 Mg ha^{-1}), while 40 kg ha^{-1} more N accumulated in >9 Mg ha^{-1}. Post-anthesis, 0.7 Mg ha^{-1} more DM accumulated in >9 Mg ha^{-1} than in 7–9 Mg ha^{-1}, while the same 45 kg ha^{-1} N accumulated in both yield ranges. Unlike N-opt, the N accumulation was similar between yield ranges of 7–9 and >9 Mg ha^{-1} from sowing to anthesis when N was overused (Figure 5F). At anthesis, N accumulation averaged around 190 kg ha^{-1} for both 7–9 Mg ha^{-1} and >9 Mg ha^{-1}, which was similar to the N accumulation in >9 Mg ha^{-1} in the same stage in the N-

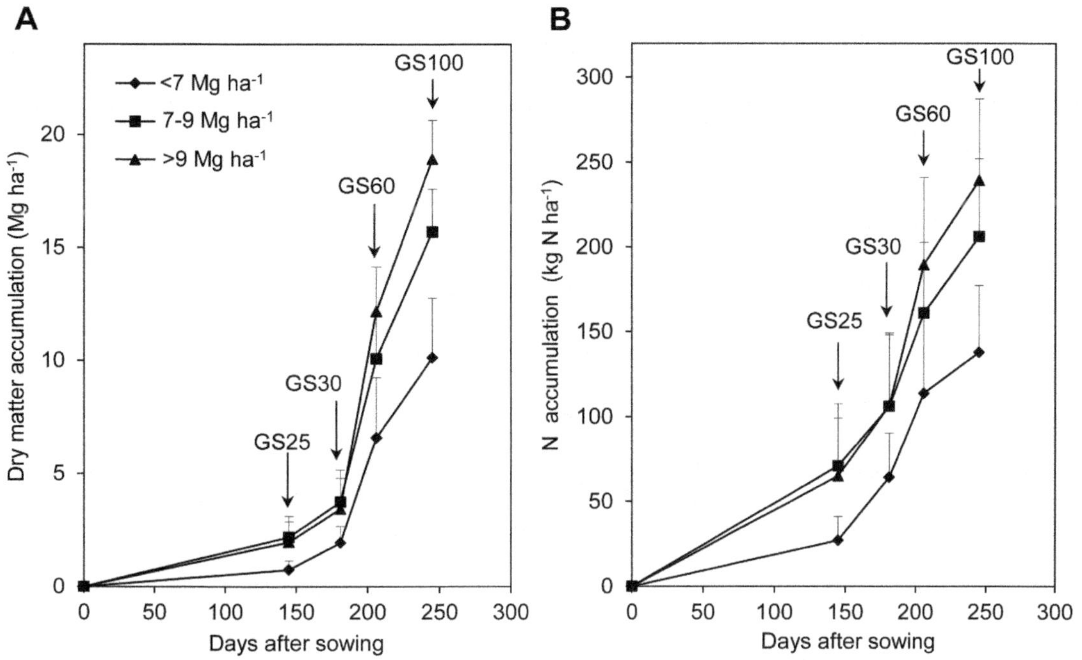

Figure 2. Total dry matter (A) and nitrogen (B) accumulation dynamics at yield ranges of <7 Mg ha^{-1} (n = 179), 7–9 Mg ha^{-1} (n = 112), and >9 Mg ha^{-1} (n = 122). (GS25, GS30, GS60, and GS100 are the regreening, stem elongation, anthesis, and maturity stages, respectively.).

opt treatment. Moreover, DM accumulation was similar at this stage in >9 Mg ha^{-1} for both N-opt and N-over (Figure 5B, C). However, DM in 7–9 Mg ha^{-1} in this stage was 21% lower than that in >9 Mg ha^{-1}, while DM for both of these was around 4.0 Mg ha^{-1} in stem-elongation stage in the N-over treatment (Figure 5C). These results indicated that higher N accumulation did not lead to more DM accumulation synchronously when N fertilizer was overused and thus N luxury accumulation occurred during growth, especially from the stem-elongation stage to anthesis.

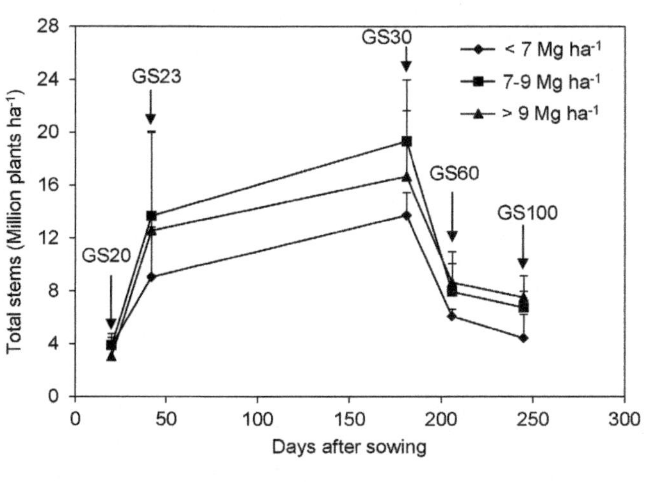

Figure 3. Wheat stem dynamics at grain yields of <7 Mg ha^{-1} (n = 19), 7–9 Mg ha^{-1} (n = 21), and >9 Mg ha^{-1} (n = 33) with optimized N treatment. (GS20, GS23, GS30, GS60, and GS100 are the pre-tillering, before winter, stem elongation, anthesis, and maturity stages, respectively.).

Discussion

We found that wheat yield improvement greatly depended on an increase in DM and N accumulation across the whole growing season. This finding contradicts some previous results, which showed that grain yield was mainly associated with the pre-anthesis assimilate contribution to grain filling and greater DM translocation efficiency [31,32]. This is because improving the harvest index (by introducing dwarfing genes into new varieties) has traditionally been considered the most important process for increasing the wheat yield potential of modern wheat varieties [33–35]. However, the utility of this strategy is limited because of the need to maintain sufficient leaf area and stem dry matter for interception of solar radiation, physical support, and storage of assimilates and N used in grain filling [34]. In the present study, the harvest index averaged 0.45, which was not significantly correlated with grain yield (Table 2). Given the relatively small possibility of further increasing yield by improving the harvest index, the greater yield potential must come from increases in net primary productivity (DM) [36]. The view that increasing DM and N accumulation over the entire growing season is important to achieve high grain yield, is also consistent with recent genetic gains in wheat breeding found in some studies [37,38].

We found a positive correlation between grain yield and DM accumulation post-anthesis (Table 2), in agreement with other research on modern wheat production in China [11,12]. Furthermore, yield increase was mainly attributable to DM and N accumulation from stem elongation to anthesis and post-anthesis (Figure 2). The stage from stem elongation to anthesis accumulated the most DM and N, and the accumulation rate was also highest among the four growth seasons (Figure 1). In plant physiology, the period a few weeks before anthesis (from terminal spikelet initiation to anthesis) is of paramount importance in determining the number of fertile florets at anthesis and final grain yield [39,40].

Table 3. Stover nitrogen concentration (%) from stem elongation to maturity and grain N concentration at maturity for the N-opt treatment for yield ranges of <7 Mg ha^{-1} (n = 19), 7–9 Mg ha^{-1} (n = 21), and >9 Mg ha^{-1} (n = 33), in which total stem dynamics was investigated.

	Stover			Grain
	Stem elongation (GS30)	Anthesis (GS60)	Maturity (GS100)	Maturity (GS100)
<7 Mg ha^{-1}	3.40±0.48[a]	1.74±0.18	0.50±0.07	2.30±0.14
7–9 Mg ha^{-1}	3.07±0.45	1.67±0.32	0.58±0.21	2.20±0.20
>9 Mg ha^{-1}	3.27±0.54	1.56±0.20	0.62±0.18	1.99±0.31

[a]average±SD.

The final ear number was similar or even higher in 9 Mg ha^{-1} than in 7–9 Mg ha^{-1}, although total stems was less in >9 Mg ha^{-1} than in 7–9 Mg ha^{-1} in the stem-elongation stage (Figure 3). This indicates that the higher ear number at maturity is likely attributable to population quality, not tiller number. In the stem-elongation stage, the N concentration in >9 Mg ha^{-1} was significantly higher than that in 7–9 Mg ha^{-1} (Table 3). The higher N concentration in plants the stem-elongation stage might lead to higher stem quality (e.g., higher tiller weight) and thus more stems would survive from stem elongation to anthesis. In practice, this suggests that tiller quality is of more importance than total tiller number.

DM accumulation was fastest (253 kg ha^{-1} day^{-1}) from stem elongation to anthesis, and then decreased significantly with time (Figure 1A). This differs from wheat production in most European countries. For example, a study in Germany showed that the DM accumulation rate reached 200 kg ha^{-1} day^{-1} during this stage, and then consistently increased to more than 250 kg ha^{-1} per day in the early grain-filling stage [41]. Meanwhile, the grain-filling stage for wheat is longer in Germany than in China because of cool temperatures during this stage. The dry and hot winds in May and June in China's wheat production area accelerate wheat maturation [42]. The higher DM accumulation rate during the earlier grain-filling stage and the longer grain-filling stage in Germany would lead to higher wheat yields.

In China, pursuing high grain yield has been the top priority in policy and in practice. The typical N rate applied by winter wheat farmers in the North China plain is around 369 kg N ha^{-1} [15], whereas results from region-wide experiments have demonstrated the optimal N rate to be 128 kg N ha^{-1} [27]. Most Chinese farmers apply large amounts of chemical fertilizer at a uniform rate as an "insurance" against low yields. Affected by the Green Revolution, most farmers still believe that more fertilizer and higher grain yield are synonymous. Our results showed applying more N fertilizer than the rational didn't result in higher wheat yield (Table 1).

On the other hand, being restricted by old knowledge and habits, farmers often apply large amounts of N fertilizer before sowing or during early growth stages (~50% of total N application). Our present study clearly demonstrated that this

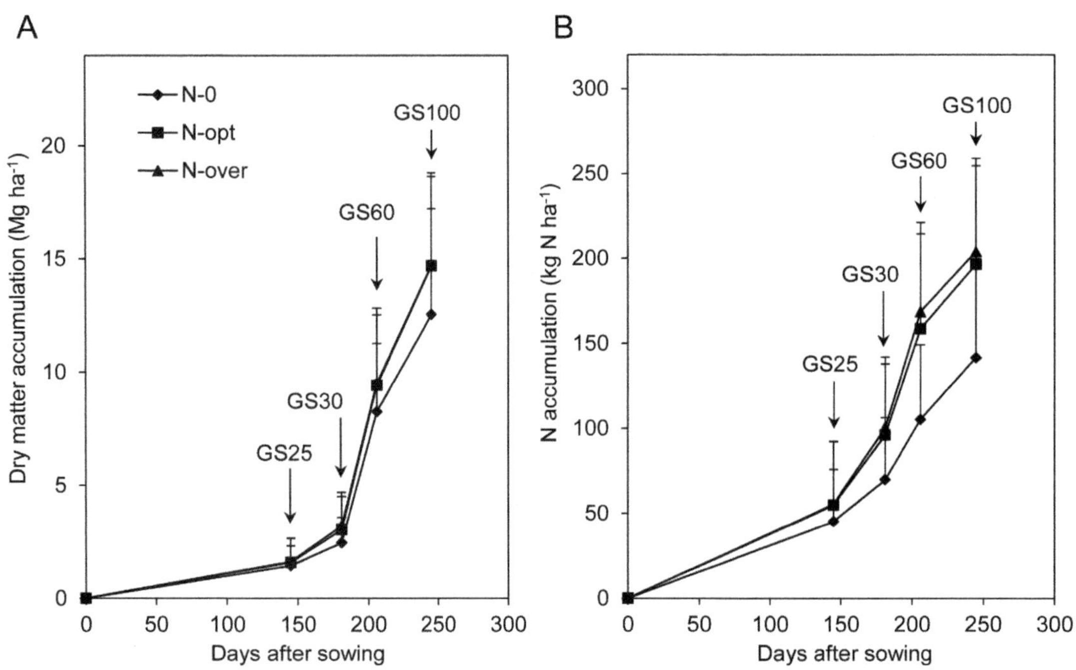

Figure 4. Total dry matter (A) and nitrogen (B) accumulation dynamics with N-0 (n = 79), N-opt (n = 126) and N-over (n = 154) treatments. (GS25, GS30, GS60, and GS100 are the regreening, stem elongation, anthesis, and maturity stages, respectively.).

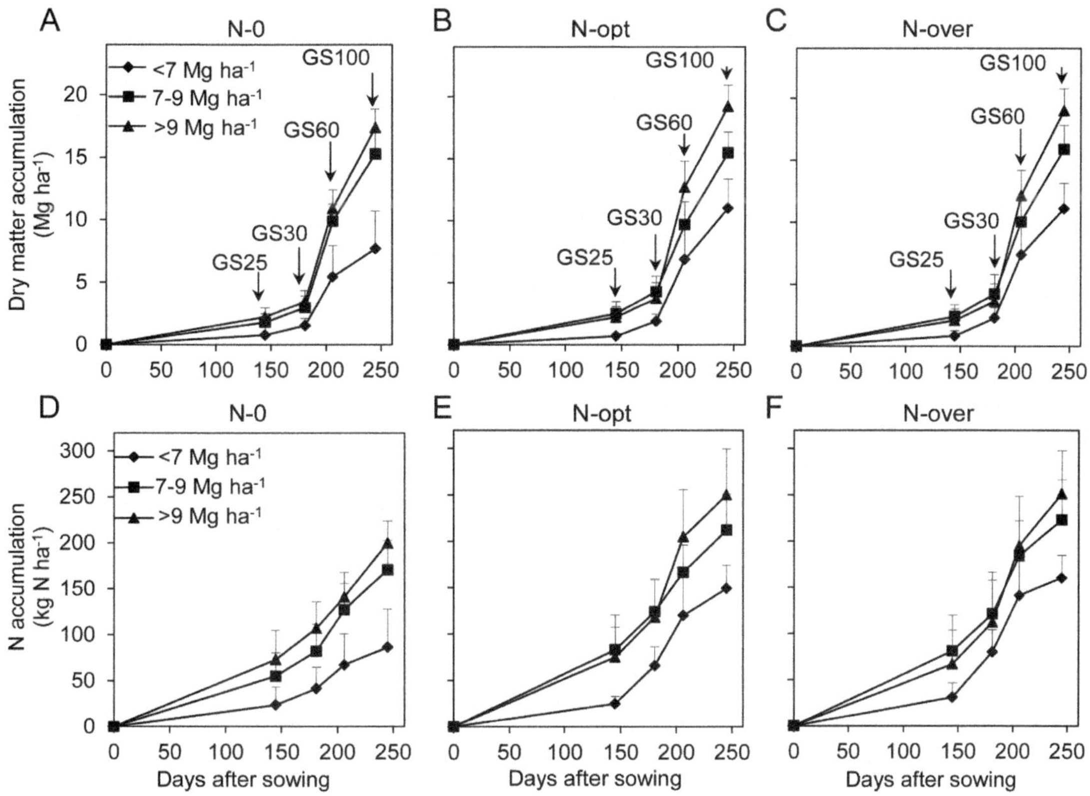

Figure 5. Total dry matter and nitrogen accumulation dynamics in the yield ranges <7, 7-9, and >9 Mg ha^{-1} with N-0 (n = 32, 34, and 13) (A and D), N-opt (n = 56, 29, and 41) (B and E), and N-over treatments (n = 67, 41, and 46) (C and F). (GS25, GS30, GS60, and GS100 are the regreening, stem elongation, anthesis, and maturity stages, respectively.).

large amount of N fertilization before wheat planting was completely unnecessary, and was likely to move beyond the root zone, particularly under irrigated conditions. In addition, surplus N at early stage would lead to more non-reproductive tillers [16], and more risk of lodging, pests and diseases when it is more than optimum effective plant population. To achieve high yield and N use efficiency at the same time, 60–70% of total N fertilizer application should be applied during the rapid crop-growth period, such as stem elongation stages for wheat.

Conclusions

A lack of information on DM and N accumulation dynamics with different yield ranges and N management regimes has resulted in N overuse and misuse and various environment problems in China and worldwide. We found significant correlations between grain yield and DM and N accumulation in each growth stage. Both the highest and fastest DM and N accumulations were observed from stem elongation to anthesis. During this stage, increased DM and N accumulation in >9 Mg ha^{-1} yield plots was key for a final higher yield compared with 7–9 Mg ha^{-1}. Applying more N could increase the N accumulation during this stage; however, DM accumulation was not further improved and thus final yield was not increased. This suggests that N management alone is insufficient to increase grain yield and that associated agronomic management practices should be intensified. Moreover, we found that higher final ear number at harvest was determined by higher tiller quality (e.g., higher tiller weight) in earlier stages, not by tiller number. Thus, more attention should be paid to increasing wheat tiller quality.

Supporting Information

Figure S1 Changes in dry matter and nitrogen accumulation expressed as a percentage of the levels at maturity (n = 413). (GS25, GS30, GS60, and GS100 are the regreening, stem elongation, anthesis, and maturity stages, respectively).

Table S1 Location, year, soil texture, and selected chemical properties in the top 30-cm soil layer at nine sites in intensive wheat production areas in China.

Table S2 Location, year, variety and treatment number, treatment and N rate for the eleven field experiments.

Table S3 Ear number, grains per ear, grain number per square meter and grain weight per thousand grains for grain yield ranges of <7 Mg ha^{-1} (n = 179), 7–9 Mg ha^{-1} (n = 112), and >9 Mg ha^{-1} (n = 122).

Text S1 Design of the system experiment.

Author Contributions

Conceived and designed the experiments: ZC XC FZ. Performed the experiments: QM SY YY WM YT. Analyzed the data: QM ZC. Contributed reagents/materials/analysis tools: ZC XC FZ YY WM YT. Wrote the paper: QM ZC.

References

1. FAO FAOSTAT–Agriculture Database. Available: http://faostat.fao.org/site/339/default.aspx. Accessed 2013 Jan 6.
2. Tilman D, Cassman KG, Matson PA, Naylor R, Polasky S (2002) Agricultural sustainability and intensive production practices. Nature 418: 671–677.
3. Tester M, Langridge P (2010) Breeding technologies to increase crop production in a changing world. Science 327: 818–822.
4. Tollenaar M, Lee EA (2002) Yield potential, yield stability and stress tolerance in maize. Field Crops Res 75: 161–169.
5. Rhoads FM, Stanley RL (1981) Fertilizer scheduling, yield, and nutrient uptake of irrigated corn. Agron J 73: 971–974.
6. Fageria NK, Baligar VC (2005) Enhancing nitrogen use efficiency in crop plants. Adv Agron 88: 97–185.
7. Cox MC, Qualset CO, Rains DW (1985) Genetic-variation for nitrogen assimilation and translocation in wheat.1. dry-matter and nitrogen accumulation. Crop Sci 25: 430–435.
8. Cox MC, Qualset CO, Rains DW (1985) Genetic-variation for nitrogen assimilation and translocation in wheat.2. nitrogen assimilation in relation to grain-yield and protein. Crop Sci 25: 435–440.
9. Heitholt JJ, Croy LI, Maness NO, Nguyen HT (1990) Nitrogen partitioning in genotypes of winter-wheat differing in grain N-concentration. Field Crops Res 23: 133–144.
10. Papakosta DK, Gagianas AA (1991) Nitrogen and dry-matter accumulation, remobilization, and losses for Mediterranean wheat during grain filling. Agron J 83: 864–870.
11. Pan QM, Yu ZW, Wang YF, Tian QZ (1999) Studies on uptake and distribution of nitrogen in wheat at the level of 9000 kg per hectare. (In Chinese). Acta Agron Sin 25: 541–547.
12. Ye Y, Wang G, Huang Y, Zhu Y, Meng Q, et al. (2011) Understanding physiological processes associated with yield–trait relationships in modern wheat varieties. Field Crops Res 124: 316–322.
13. Malhi SS, Johnston AM, Schoenau JJ, Wang ZH, Vera CL (2007) Seasonal biomass accumulation and nutrient uptake of pea and lentil on a black chernozem soil in Saskatchewan. J Plant Nutr 30: 721–737.
14. Wang ZM, Wang P, Lan LW, Zhou DX (2003) A water-saving and high-yielding cultivation system for bread wheat in Huang-Huai-Hai area of China. (In Chinese). Chinese Agric Sci Bullet 19: 22–43.
15. Cui ZL (2005) Optimization of the nitrogen fertilizer management for a winter wheat-summer maize rotation system in the North China Plain-from field to regional scale. China Agricultural University. Ph.D. Dissertation.
16. Weisz R, Crozier CR, Heiniger RW (2001) Optimizing nitrogen application timing in no-till soft red winter wheat. Agron J 93: 435–442.
17. Langer RHM, Dougherty CT (1976) Physiology of grain yield in wheat. Persp Exp Biol 2: 59–70.
18. Sharma RC (1995) Tiller mortality and its relationship to grain yield in spring wheat. Field Crops Res 41: 55–60.
19. Latiri-Souki K, Nortcliff S, Lawlor DW (1998) Nitrogen fertilizer can increase dry matter, grain production and radiation and water use efficiencies for durum wheat under semi-arid conditions. Eur J Agron 9: 21–34.
20. Greenwood DJ, Verstraeten LMJ, Draycott A (1987) Response of winter- wheat to N-fertilizer-quantitative relations for components of growth. Fert Res 12: 119–137.
21. Pask AJD, Sylvester-Bradley R, Jamieson PD, Foulkes MJ (2012) Quantifying how winter wheat crops accumulate and use nitrogen reserves during growth. Field Crops Res 126: 104–118.
22. Cassman KG, Dobermann A, Walters DT (2002) Agroecosystems, nitrogen-use efficiency, and nitrogen management. Ambio 31: 132–140.
23. Shanahan JF, Kitchen NR, Raun WR, Schepers JS (2008) Responsive in-season nitrogen management for cereals. Comput Electron Agric 61: 51–62.
24. Stanford G, Legg JO (1984) Nitrogen and yield potential. In Hauck RD, ed. Nitrogen in crop production. ASA, CSSA, and SSSA: Madison. 263 p.
25. Meisinger JJ, Randall GW (1991) Estimating nitrogen budgets for soil-crop systems. In Follett RF, Keeney DR, Cruse RM, eds. Managing nitrogen for groundwater quality and farm profitability. SSSA: Madison. 85 p.
26. Gastal F, Lemaire G (2002) N uptake and distribution in crops: an agronomical and ecophysiological perspective. J Exp Bot 53: 789–799.
27. Cui ZL, Zhang FS, Xinping CP, Miao YX, Li JL, et al. (2008) On-farm evaluation of an in-season nitrogen management strategy based on soil Nmin test. Field Crops Res 105: 48–55.
28. Zhao RF, Chen XP, Zhang FS, Zhang HL, Schroder J, et al. (2006) Fertilization and nitrogen balance in a wheat-maize rotation system in North China. Agron J 98: 938–945.
29. Stanford G (1973). Rational for optimum fertilization in corn produc tion. J Environ Qual 2: 159–166.
30. Lory JA, Scharf PC (2003) Yield goal versus delta yield for predicting fertilizer nitrogen need in corn. Agron J 95: 994–999.
31. Shearman VJ, Sylvester-Bradley R, Scott RK, Foulkes MJ (2005) Physiological processes associated with wheat yield progress in the UK. Crop Sci 45: 175–185.
32. Álvaro F, Isidro J, Villegas D, GarcÍa del Moral LF, Royo C (2008) Breeding effects on grain filling, biomass partitioning and remobilization in Mediterranean durum wheat. Agron J 100: 361–370.
33. Austin RB, Ford MA, Morgan CL (1989) Genetic-improvement in the yield of winter-wheat-a further evaluation. J Agric Sci 112: 295–301.
34. Donmez E, Sears RG, Shroyer JP, Paulsen GM (2001) Genetic gain in yield attributes of winter wheat in the great plains. Crop Sci 41: 1412–1419.
35. Acreche MM, Briceño-Félix G, Sánchez JAM, Slafer GA (2008) Physiological bases of genetic gains in Mediterranean bread wheat yield in Spain. Eur J Agron 28: 162–170.
36. Cassman KG (1999) Ecological intensification of cereal production systems: Yield potential, soil quality, and precision agriculture. Proc Natl Acad Sci USA 96: 5952–5959.
37. Singh RP, Huerta-Espino J, Rajaram S, Crossa J (1998) Agronomic effects from chromosome translocations 7DL.7Ag and 1BL.1RS in spring wheat. Crop Sci 38: 27–33.
38. Reynolds MP, Rajaram S, Sayre KD (1999) Physiological and genetic changes of irrigated wheat in the post-green revolution period and approaches for meeting projected global demand. Crop Sci 39: 1611–1621.
39. Fischer RA (1985) Number of kernels in wheat crops and the influence of solar-radiation and temperature. J Agric Sci 105: 447–461.
40. Slafer GA, Satorre EH, Andrade FH (1994) Increases in grain yield in bread wheat from breeding and associated physiological changes. In Slafer GA, ed. Genetic improvement of field crops. Marcel Dekker Inc: New York. 1 p.
41. Diekmann F, Fischbeck G (2005) Differences in wheat cultivar response to N supply. I: Differences in grain yield formation. J Agron Crop Sci 191: 351–361.
42. Sun HY, Liu CM, Zhang XY, Shen YJ, Zhang YQ (2006) Effects of irrigation on water balance, yield and WUE of winter wheat in the North China Plain. Agric Water Manage 85: 211–218.

4

Iron: The Forgotten Driver of Nitrous Oxide Production in Agricultural Soil

Xia Zhu[1,2,3], Lucas C. R. Silva[3], Timothy A. Doane[3]*, William R. Horwath[3]

1 Chengdu Institute of Biology, Chinese Academy of Sciences, Chengdu, China, **2** University of Chinese Academy of Sciences, Beijing, China, **3** Department of Land, Air, and Water Resources, University of California Davis, Davis, California, United States of America

Abstract

In response to rising interest over the years, many experiments and several models have been devised to understand emission of nitrous oxide (N_2O) from agricultural soils. Notably absent from almost all of this discussion is iron, even though its role in both chemical and biochemical reactions that generate N_2O was recognized well before research on N_2O emission began to accelerate. We revisited iron by exploring its importance alongside other soil properties commonly believed to control N_2O production in agricultural systems. A set of soils from California's main agricultural regions was used to observe N_2O emission under conditions representative of typical field scenarios. Results of multivariate analysis showed that in five of the twelve different conditions studied, iron ranked higher than any other intrinsic soil property in explaining observed emissions across soils. Upcoming studies stand to gain valuable information by considering iron among the drivers of N_2O emission, expanding the current framework to include coupling between biotic and abiotic reactions.

Editor: Ben Bond-Lamberty, DOE Pacific Northwest National Laboratory, United States of America

Funding: Support was provided by the California Department of Resources Recycling and Recovery (www.calrecycle.ca.gov), Agreement IWM09027; the J. G. Boswell Endowed Chair in Soil Science; and the University of California Agricultural Experiment Station (http://caes.ucdavis.edu/research/agexpstn). The funders had no role in study design, data collection and analysis, decision to publish, or preparation of the manuscript.

Competing Interests: The authors have declared that no competing interests exist.

* E-mail: tadoane@ucdavis.edu

Introduction

Emission of N_2O from soils is an extensively studied environmental process, given that N_2O is "at the heart of debates" [1] on several prevalent current issues. Approximately two-thirds of total global emission comes from soils; most of the emission from soils is in turn attributed to agriculture [2]. The intrinsic soil properties (as opposed to temporary changes) most commonly mentioned in research studies and models as controlling emission of N_2O are texture, pH, organic matter, and ability to supply inorganic nitrogen [3–12]. Production of N_2O in soil is generally attributed to microbiological processes [1,2,13–17], and therefore the factors that regulate the activity of N_2O-producing microorganisms should be the same factors that regulate N_2O production. These controlling factors are generally thought to be well recognized, but as research and related commentary on N_2O emission from agricultural soils continue to accumulate, the possible role of iron is rarely considered. This is in spite of its known involvement in enzymatic reactions [2,18,19] and non-enzymatic reactions [20–23] that generate N_2O. The connection between iron and N_2O may have been neglected because iron has never figured prominently in routine evaluations of soil for agronomic research or practical management decisions. Unlike the other soil properties cited above, iron does not have a direct and immediate bearing on the growth of most crops or on the agricultural suitability of a soil from either a physical or a chemical point of view. When it is considered, this is in instances of suspected plant deficiency or toxicity, not in the context of its potential connection with the nitrogen cycle. In addition, compared to other intrinsic properties, soil iron does not dramatically affect the short-term changes in microbiological activity generally associated with N_2O production. For these reasons, once interest in N_2O began to intensify, the previously reported connection with iron was already out of sight. The intent of our work was to reconsider the potential significance of iron in emission of N_2O from agricultural soils.

Materials and Methods

Ethics statement

The soils used in this study were collected under consent of the land owners, and the compost used was collected under consent of the compost facility management.

Soil characterization

Soils were collected from the top 15 cm in 10 agricultural fields throughout California, and were sieved to 2 mm following collection. Soil pH was measured in 1 M KCl (1:1 w:v). Percent clay, silt, and sand were determined by a modified pipet method [24]. Total carbon and nitrogen were determined on ball-milled samples by combustion-GC (Costech ECS 4010). Just prior to setting up the experiment, inorganic nitrogen (ammonium plus nitrate) was extracted by 0.5 M K_2SO_4 and determined colorimetrically [25,26]. Dissolved organic carbon (DOC) was determined in the same extract by UV-persulfate digestion (Teledyne-Tekmar Phoenix 8000).

We chose two commonly used, contrasting indices to characterize soil iron: that extractable by acid hydroxylamine (FeA), an index of reactive iron(III) minerals [27]; and that extractable by pyrophosphate (FeP), representing iron complexed with soil organic matter [28–30]. FeA was extracted by shaking 0.8 g soil for one hour with 40 ml 0.25 M hydroxylamine hydrochloride in 0.25 M HCl, followed by centrifugation for 30 minutes at 15600× G. FeP was extracted by shaking 1 g soil with 100 ml 0.1 M tetrasodium pyrophosphate for 16 hours, followed by centrifugation for 30 minutes at 15600× G; further centrifugation did not result in any difference in measured iron concentration, indicating that all fine iron colloids had been removed, an important consideration when using this extractant [29,30]. The concentration of iron in all extracts was determined colorimetrically [31]; pyrophosphate extracts were neutralized by a small addition of HCl prior to this determination. There was no interference from pyrophosphate in the colorimetric analysis. All analyses of soil properties were performed in duplicate. These properties are reported in Table 1.

Experimental treatments

As stated above, the properties most commonly believed to control emission of N_2O from agricultural soil include texture, pH, organic matter, and the inherent ability of the soil to release inorganic nitrogen. These are intrinsic properties which are not abruptly altered by environmental conditions; in contrast, our treatments were designed to manipulate the most common temporary extrinsic changes that influence N_2O production: water content, fertilization, and organic amendments. Since these can vary across a range of values, we necessarily limited our choice of treatments. Fertilizer and compost (as a model organic amendment) were either withheld or added at a rate typical of agriculture in California, and two water contents were chosen according to the range expected in agricultural soils. Field capacity, the amount of water a soil can retain against gravity, was chosen as the upper reference point. This is not uncommon, as soil moisture can temporarily exceed field capacity following irrigation or rainfall events [32,33]. In practice we used water holding capacity (WHC) to represent field capacity. As a contrasting treatment, we chose 50% WHC. This is near the permanent wilting point of most soils [34], and it is not likely that soil moisture will fall below this in the field except during unmanaged dry seasons. Although many intermediate values could have been selected as treatments, we chose to use both ends of a typical spectrum of values in order to present a broad yet concise study.

Experimental set-up

Prior to set-up, WHC was determined as follows: a soil sample was placed into a funnel lined with filter paper, which was then placed into a beaker of water such that just the tip of the funnel was always in contact with water; after the sample ceased to take up water, the sample was allowed to drain, and the moisture content measured. To begin the experiment, the equivalent of 50 g dry soil was placed into cups, which were themselves placed into larger jars containing a small amount of water to avoid desiccation. The larger jars were sealed with lids containing a small foam plug to allow gas exchange with the atmosphere. To imitate the timing typical of agricultural operations, 2 g finely ground finished green waste compost (corresponding approximately to a field application of 60 t ha^{-1} in the top 15 cm) were mixed with the soils and incubated at 40% WHC for seven days. Treatments not receiving compost were similarly incubated. Following this preincubation, each soil received a fertilizer addition according to treatment: none, ammonium sulfate, or

Table 1. Characterization of the soils used in this study.

Location	Classification[a]	FeA[b]	FeP[c]	DOC[d]	Inorganic N	Total N	Total C	Sand	Silt	Clay	pH
		mg kg^{-1}	mg kg^{-1}	mg kg^{-1}	mg kg^{-1}	%	%	%	%	%	
Sacramento Valley											
Davis	Fine, montmorillonitic, thermic Mollic Haploxeralf	1800	170	17	2	0.09	0.85	30	42	24	5.4
Dixon 1	Fine-silty, mixed, nonacid, thermic Typic Xerorthent	2150	290	30	11	0.14	1.60	23	49	28	5.6
Dixon 2	Fine-silty, mixed, nonacid, thermic Typic Xerorthent	1900	210	19	5	0.11	1.18	15	41	44	5.5
Salinas Valley											
Castroville	Fine, montmorillonitic, thermic Ultic Palexerol	710	550	88	32	0.08	0.75	72	15	13	6.4
Salinas 1	Fine, montmorillonitic, thermic Pachic Argixeroll	390	150	44	5	0.07	0.66	64	23	13	7.2
Salinas 2	Fine, montmorillonitic, thermic Typic Pelloxerert	1890	240	88	28	0.16	1.78	22	36	42	7.4
Spence	Fine-loamy, mixed, thermic, Typic Argixeroll	670	270	63	18	0.11	1.28	50	29	21	6.6
San Joaquin Valley											
Five Points	Fine-loamy, mixed, superactive, thermic Typic Haplocambid	850	60	57	4	0.08	0.67	36	32	32	6.8
Modesto	Fine-loamy, mixed, superactive, thermic Typic Argixeroll	410	240	164	130	0.11	0.97	72	18	10	6.9
Sanger	Coarse-loamy, mixed, nonacid, thermic Typic Xerorthent	390	260	28	4	0.03	0.30	61	32	7	4.2

[a]United States Department of Agriculture official soil series description, [b] acid hydroxylamine-extractable iron, [c] pyrophosphate-extractable iron, [d] dissolved organic carbon.

potassium nitrate. The amount of nitrogen added was 100 mg kg^{-1} soil, corresponding approximately to a field rate of 150 kg ha^{-1}. Fertilizer solution was sprayed onto the soils to reach a water content of 50% or 100% WHC, depending on the treatment. For each soil there were three replicates per treatment. Samples were incubated for 14 days at 22 degrees C.

Samples for N_2O analysis were taken on days 0, 1, 2, 3, 5, 9, and 14 following addition of fertilizer. The jars containing the soil cups were closed with lids containing septa and allowed to stand for one hour. Gas samples were taken at 0, 30, and 60 minutes after closure and transferred to evacuated gas sampling vials. N_2O concentration was determined by gas chromatography-ECD detection (Shimadzu GC-2014). At each sampling date, the rate of N_2O emission (flux) was determined by linear interpolation of the 0, 30, and 60 minute measurements. Cumulative N_2O emission over the course of the incubation was calculated using these data, taking the flux measured at a given date to be the average flux for the interval represented by that date.

Statistical analysis

To identify the soil properties that most strongly explained N_2O emission in each experimental treatment, we studied the data using partial least squares (PLS) multivariate analysis, a form of structural equation modeling. This tool is particularly suitable when the number of predicting variables is greater than the number of observed variables, when multicolinearity is expected among predicting variables, and when multivariate normality can not be assumed [35–37]. PLS ranks the predicting variables by importance based on linear regression models that project the predicting variables and the observed variables to a new, multivariate space. Prior to subjecting the data to PLS analysis, predicting variables (soil properties) and the observed response (cumulative N_2O emission) were standardized by centering and scaling the data to have a mean of zero and a standard deviation of one. This ensures that the predicting variables are ranked based on how much of the variation is explained when all variables have the same weight.

Although correlations among variables are possible, especially in studies that involve soil properties, this does not change the interpretation given by PLS, which depicts the relative importance of each variable separately, independently of intrinsic links between variables. Nevertheless, a correlation matrix is presented (Table 2) as an aid in understanding the relationships between the soil properties used in our study.

Following the exploratory PLS analysis, linear regressions between iron and N_2O emission were calculated using unweighted, untransformed data, and were considered significant enough to report at $P<0.1$. All statistical analyses were performed using JMP 10 software.

Results and Discussion

The results of the PLS analysis are shown in Figure 1, where each soil property is ranked according to its ability to explain cumulative N_2O emission across all soils. This ranking was performed for each of the 12 different treatments studied. In five of these treatments, iron (as either FeA or FeP) ranked higher than any other measured soil characteristic in explaining observed emissions. In four additional treatments, iron was among the top four predictors.

As a complementary approach to further investigate the relationship between iron and N_2O emission, simple linear regressions were calculated in which N_2O data were compared against FeP and FeA. Whereas PLS was used to arrange a suite of soil properties according to their ability to explain N_2O emission, regressions indicate, by the value of r^2, how much of the variability in N_2O emission can be explained by a single property; regressions also indicate the direction of the effect (positive or negative slope) and degree of importance of the effect (absolute value of the slope). In most cases, a significant relationship between N_2O emission and a given variable can be expected when that variable is ranked highly by PLS. In certain cases, however, a variable ranked highly by PLS may not necessarily yield a significant linear relationship when that variable is considered apart from the other variables; conversely, certain treatments in which a variable is not ranked highly by PLS may nonetheless yield a significant regression. The primary reason for this occasional discrepancy is the nature of the PLS procedure: by considering all predicting variables together, new predictors are generated which are composites of the original variables. Table 3 reports the results of the regressions for treatments that showed a significant relationship between N_2O emission and either iron index. Despite a dataset of values for N_2O emission which spanned more than three orders of magnitude across soils, several notable connections between iron and N_2O emission emerged.

FeP was significantly related to N_2O emission in four treatments, in which it explained between 16 and 62 percent of the variability, with a positive slope in all cases (i.e. greater

Table 2. Correlation matrix of the soil properties evaluated in this study.

	FeA[a]	FeP[b]	DOC[c]	Inorganic N	Total N	Total C	Sand	Silt	Clay	pH
FeA	–	−0.07	−0.41	−0.37	0.68	0.70	−0.91	0.84	0.79	−0.15
FeP	−0.07	–	0.25	0.05	0.04	0.08	0.38	−0.37	−0.31	−0.10
DOC	−0.41	0.25	–	0.93	0.25	0.12	0.53	−0.68	−0.29	0.59
Inorganic N	−0.37	0.05	0.93	–	0.18	0.02	0.45	−0.54	−0.28	0.43
Total N	0.68	0.04	0.25	0.18	–	0.98	−0.57	0.37	0.66	0.47
Total C	0.70	0.08	0.12	0.02	0.98	–	−0.61	0.46	0.66	0.36
Sand	−0.91	0.38	0.53	0.45	−0.57	−0.61	–	−0.89	−0.91	0.13
Silt	0.84	−0.37	−0.68	−0.54	0.37	0.46	−0.89	–	0.63	−0.44
Clay	0.79	−0.31	−0.29	−0.28	0.66	0.66	−0.91	0.63	–	0.17
pH	−0.15	−0.10	0.59	0.43	0.47	0.36	0.13	−0.44	0.17	–

[a]acid hydroxylamine-extractable iron, [b] pyrophosphate-extractable iron, [c] dissolved organic carbon.

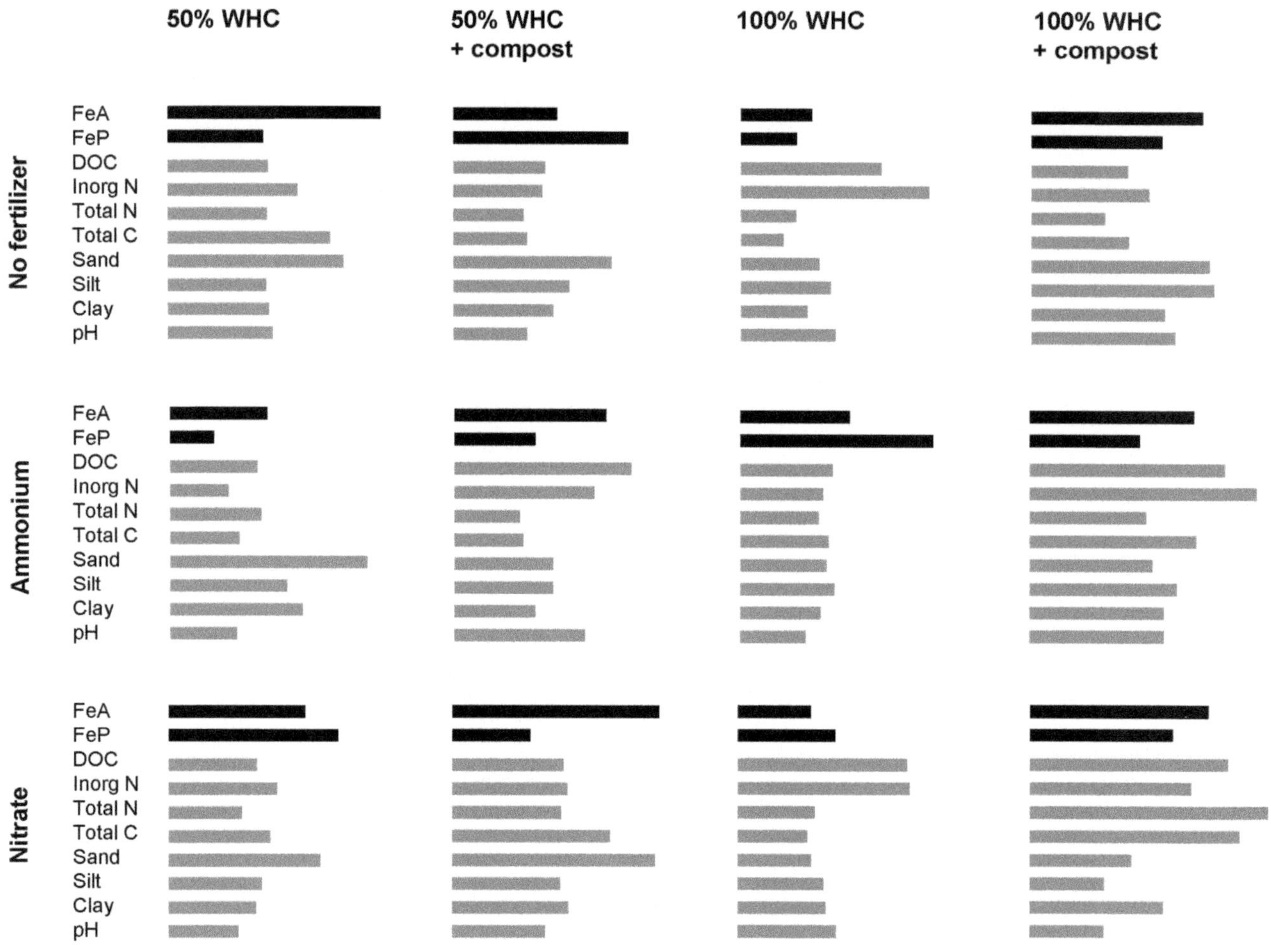

Figure 1. Relative importance of soil properties in explaining cumulative emission of N_2O under different conditions. Result of partial least squares multivariate analysis performed across ten soils for each of 12 different treatments. Two indices of soil iron (FeA: acid hydroxylamine-extractable iron and FeP: pyrophosphate-extractable iron) were ranked alongside other soil properties commonly considered to control soil N_2O emission. The size of each bar is given by the variable importance in the projection (VIP) value, and indicates the relative strength of each variable in explaining emission in that treatment. WHC = water holding capacity; DOC = dissolved organic carbon.

emission was associated with more FeP). This influence was greatest under 100% WHC when ammonium was present and compost was absent. Such a condition may be reasonably expected on occasion, since most fertilizers supply ammonium, and since this may occur close in time to irrigation or rainfall. In this treatment, an increase in FeP of 1 mg kg^{-1} corresponded to an increase in cumulative emission of 11.9 ng N_2O-N g^{-1} soil (averaged across all soils) during the course of the incubation (Table 3).

Like FeP, the connection between FeA and N_2O emission was also significant under several different conditions. Unlike FeP, however, which was positively related to N_2O emission, FeA was always negatively related to N_2O emission. There was no treatment in which both iron indices were significantly related to N_2O emission (Table 3). Considering that FeP and FeA bear almost no relationship to each other (Table 2), this difference in behavior suggests that these two indices indeed reflect two forms of iron that differ in reactivity. Also notable in Table 3 is the effect of

Table 3. Results of simple linear regression of cumulative N_2O emission (as ng N_2O-N g^{-1} soil) against iron, across ten soils and under 12 different conditions.

	50% WHC	50% WHC + compost	100% WHC	100% WHC + compost
No fertilizer	NS	FeP: 0.37, 0.38	FeA: 0.12, −0.09	NS
Ammonium	NS	FeA: 0.28, −0.20	FeP: 0.62, 11.9	FeA: 0.23, −0.62
Nitrate	FeP: 0.19, 0.46	NS	FeP: 0.16, 2.1	NS

The first value given is that of r^2, and the second value is the slope of the regression. NS = regression was not significant for either iron index. WHC = water holding capacity; FeA = acid hydroxylamine-extractable iron; FeP = pyrophosphate-extractable iron.

compost in fertilized treatments: the observed negative association between N_2O emission and FeA occurred only in the presence of compost, while the stimulating effect of FeP was observed only without compost.

The contrasting relationships of FeA and FeP with N_2O emission could be due to differences in the reaction of either form of iron with nitrogen compounds in the soil matrix. For example, hydroxylamine is produced from biological oxidation of ammonia, and is known to generate N_2O upon chemical reaction with iron(III) [20,38]. Reaction with FeP versus FeA, or locally high concentrations of either hydroxylamine or iron, could lead to more or less N_2O compared to other reaction products [38]. The ability of aerobic microorganisms to acquire iron can likewise depend on its chemical nature, consequently influencing the amount of reactive compounds produced or consumed through reactions that use iron-dependent enzymes. As soil water content increases, reducing conditions may develop, especially when the depletion of oxygen is accelerated by easily metabolized organic matter. The chemical nature of existing iron(III) may determine the ease with which it is reduced to iron(II) in anaerobic microsites. This will in turn control its participation in other reactions that produce N_2O, such as chemodenitrification, which includes the abiotic reduction of nitrite to N_2O by iron(II) [39,40]. Chemodenitrification can also produce other gases, and the relative amount of N_2O released may be affected by the form of iron present. A related anaerobic process is nitrate-dependent iron(II) oxidation [41]; a recent review [42] has highlighted, in the context of this process, how the simultaneous presence of nitrate-reducing and iron(III)-reducing areas can potentially be important to nitrogen cycling. Under anaerobic conditions, iron(III) can also be linked to ammonium oxidation [43,44]. If reactions that generate N_2O are active in any of the above processes, they may be stimulated or suppressed by different forms of iron, such as the two indices examined in this study. The degree of this influence under different conditions will then determine the importance of iron relative to other soil properties.

Our treatments consisted of two contrasting values for soil moisture and addition of amendments. This was done in order to explore the importance of iron across a wide range of conditions while at the same time avoiding a cumbersome dataset. It is clear from Figure 1 that the importance of iron can change between the two limits of each treatment variable. For example, between 50 and 100% WHC under ammonium fertilization, iron moves from a position of modest relevance to become the highest-ranked driver. Since our results show the importance of iron only at two distinct values, we do not know how its importance under intermediate conditions changes between the two end values. Even without such intermediate data, the differences between contrasting treatments can aid in understanding the mechanisms at work in generating N_2O. In the above example, the importance of iron rises markedly under ammonium fertilization as soil moisture increases from 50 to 100% WHC; FeP surpasses FeA in strength as well. As mentioned earlier, ammonia is oxidized to hydroxylamine, and this can react with iron(III) to produce N_2O. In a wetter soil, solutes are more mobile, which can lead to greater production of hydroxylamine as well as greater contact of hydroxylamine with iron. FeP is also likely to be more soluble than FeA. Any combination of these effects might elevate the importance of iron and change which form is more relevant in explaining the associated N_2O data.

The overall position of iron among other drivers of N_2O emission is determined by both its reactivity and the presence of processes subject to its influence. Ample opportunity for inquiry exists for defining the extent of the relationship between iron and N_2O in managed as well as unmanaged ecosystems, and this can provide useful practical and theoretical information. For example, including iron in current models of N_2O emission may strengthen their predictive ability. In addition, inasmuch as certain indices of iron can be related to its physical or chemical characteristics, observing the relationship between a given index and N_2O production, and how this changes under different conditions, may provide insight into the specific reactions at work. As stated earlier, production of N_2O is generally accepted to be a microbial affair, and it is logical to assume that the factors that regulate the activity of N_2O-producing microorganisms should be the same factors that regulate N_2O production. This is not incorrect, but is perhaps a somewhat restrictive rendering; a more accurate framework might include "biotic-abiotic reaction sequences" [39] that generate N_2O, such as those outlined above. Indeed, "the complex interactions that occur between microorganisms and other biotic and abiotic factors" have been suggested to be a key part of further understanding greenhouse gas production and improving predictions [17].

Conclusion

It has been recently emphasized [45] that solutions to environmental problems require explicit consideration of the couplings between element cycles. The environmental chemistry of iron has been well researched, as have many of the interrelated details of the nitrogen cycle. The specific connection between iron and N_2O in soil has also been recognized in both older and recent studies. However, iron and nitrogen have yet to be brought together in agricultural systems, the foremost source of soil N_2O emission. Our most important conclusion is simple: iron does indeed figure prominently among the soil properties controlling N_2O emission in contrasting conditions across diverse soils. Studies concerned with the potential of agricultural soil to emit N_2O will gain new momentum by remembering this "key biogeochemical engine" [46], building on a connection identified a long time ago but largely overlooked since then.

Acknowledgments

We gratefully acknowledge the assistance of those who collected the soil samples, as well as the constructive comments received during review, which were very helpful in improving our paper.

Author Contributions

Conceived and designed the experiments: XZ WRH. Performed the experiments: XZ. Analyzed the data: XZ LCRS. Wrote the paper: TAD.

References

1. Reay DS, Davidson EA, Smith KA, Smith P, Melillo JM, et al. (2012) Global agricultural and nitrous oxide emissions. Nat Clim Change 2: 410-416.
2. Thomson AJ, Giannopoulos G, Pretty J, Baggs EM, Richardson DJ (2012) Biological sources and sinks of nitrous oxide and strategies to mitigate emissions. Phil Trans R Soc B 367: 1157-1168.
3. Sahrawat KL, Keeney DR (1986) Nitrous oxide emissions from soils. Advances in Soil Sci 4: 103-148.
4. Eichner M (1990) Nitrous oxide emissions from fertilized soils: summary of available data. J Environ Qual 19: 272-280.
5. Bouwman AF, Fung I, Matthews E, John J (1993) Global analysis of the potential for N_2O production in natural soils. Global Biogeochem Cycles 7: 557-597.
6. Robertson K (1994) Nitrous oxide emission in relation to soil factors at low to intermediate moisture levels. J Environ Qual 23: 805-809.

7. Dobbie KE, McTaggart IP, Smith KA (1999) Nitrous oxide emissions from intensive agricultural systems: variations between crops and seasons, key driving variables, and mean emission factors. J Geophys Res 104: 26891-26899.
8. Li C, Aber J, Stange F, Butterbach-Bahl H (2000) A process-oriented model of N_2O and NO emissions from forest soils: 1. Model development. J Geophys Res 105: 4369-4384.
9. Skiba U, Smith KA (2000) The control of nitrous oxide emissions from agricultural and natural soils. Chemosphere Global Change Sci 2: 379-386.
10. Bouwman AF, Boumans LJM, Batjes NH (2002) Modeling global annual N_2O and NO emissions from fertilized fields. Global Biogeochem Cycles 16: 1080.
11. Freibauer A, Kaltschmitt M (2003) Controls and models for estimating direct nitrous oxide emissions from temperate and sub-boreal agricultural mineral soils in Europe. Biogeochemistry 63: 93-115.
12. Stehfest E, Bouwman L (2006) N_2O and NO emission from agricultural fields and soils under natural vegetation: summarizing available measurement data and modeling of global annual emissions. Nutr Cycl Agroecosys 74: 207-228.
13. Williams EJ, Hutchinson GL, Fehsenfeld FC (1992) NO_x and N_2O emissions from soil. Global Biogeochem Cycles 6: 351-388.
14. Bremner JM (1997) Sources of nitrous oxide in soils. Nutr Cycl Agroecosys 49: 7-16.
15. Freney JR (1997) Emission of nitrous oxide from soils used for agriculture. Nutr Cycl Agroecosys 49: 1-6.
16. Davidson EA, Keller M, Erickson HE, Verchot LV, Veldkamp E (2000) Testing a conceptual model of soil emissions of nitrous and nitric oxides. BioScience 50: 667-680.
17. Singh BK, Bardgett RD, Smith P, Reay DS (2010) Microorganisms and climate change: terrestrial feedbacks and mitigation options. Nat Rev Microbiol 8: 779-790.
18. Meiklejohn J (1953) Iron and the nitrifying bacteria. J Gen Microbiol 8: 58-65.
19. Glass JB, Orphan VJ (2012) Trace metal requirements for microbial enzymes involved in the production and consumption of methane and nitrous oxide. Front Microbiol doi: 10.3389/fmicb.2012.00061.
20. Chao TT, Kroontje W (1966) Inorganic nitrogen transformations through the oxidation and reduction of iron. Soil Sci Soc Am Proc 30: 193-196.
21. Buresh RJ, Moraghan JT (1976) Chemical reduction of nitrate by ferrous iron. J Environ Qual 5: 320-325.
22. Chalamet A, Bardin R (1976) Action of ferrous ions on reduction of nitrous acid in hydromorphous soils. Soil Biol Biochem 9: 281-285.
23. Bremner JM, Blackmer AM, Waring SA (1980) Formation of nitrous oxide and dinitrogen by chemical decomposition of hydroxylamine in soils. Soil Biol Biochem 12: 263-269.
24. Burt R, editor (1992) Soil Survey Laboratory Methods Manual. Soil Survey Investigations Report no. 42.WashingtonDC:USDA .
25. Verdouw H, van Echteld CJA, Dekkers EMJ (1978) Ammonia determination based on indophenol formation with sodium salicylate. Water Res 12: 399-402.
26. Doane TA, Horwath WR (2003) Spectrophotometric determination of nitrate with a single reagent. Anal Lett 36: 2713-2722.
27. Lovley DR, Phillips EJP (1987) Rapid assay for microbially reducible ferric iron in aquatic sediments. Appl Environ Microbiol 53: 1536-1540.
28. Bremner JM, Heintze SG, Mann PJG, Lees H (1946) Metallo-organic complexes in soil. Nature 158: 790-791.
29. Schuppli PA, Ross GJ, McKeague JA (1983) The effective removal of suspended materials from pyrophosphate extracts of soil from tropical and temperate regions. Soil Sci Soc Am J 47: 1026-1032.
30. Loveland PJ, Digby P (1984) The extraction of Fe and Al by 0.1 M pyrophosphate solutions: a comparison of some techniques. J Soil Sci 35: 243-250.
31. Dominik P, Kaupenjohann M (2000) Simple spectrophotometric determination of Fe in oxalate and HCl soil extracts. Talanta 51: 701-707.
32. Cassel DK, Nielsen DR (1986) Field capacity and available water capacity. In: Klute A, editor. Methods of Soil Analysis.Part 1.Physical and Mineralogical Methods.Madison,WI :Soil Science Society of America. pp. 901-926.
33. Veihmeyer FJ, Hendrickson AH (1931) The moisture equivalent as a measure of the field capacity of soils. Soil Sci 32: 181-193.
34. Hendrickson AH, Veihmeyer FJ (1945). Permanent wilting percentages of soils obtained from field and laboratory trials. Plant Physiol 20: 517-539.
35. Tenenhaus M, Vinzi VE, Chatelin YM, Lauro C (2005) PLS path modeling. Comput Statd Data Anal 48: 159-205.
36. Marcoulides G, Chin WW, Saunders C (2009) A critical look at partial least squares modeling. MIS Quart 33: 171-175.
37. Vinzi V, Chim WW, Henseler J, Wang H, editors (2010) Handbook of Partial Least Squares: Concepts, Methods, and Applications. Heidelberg: Springer.
38. Bengtsson G, Fronaeus S, Bengtsson-Kloo L (2002) The kinetics and mechanism of oxidation of hydroxylamine by iron(III). J Chem Soc Dalton Trans 12: 2548-2552.
39. Burger M, Venterea RT (2011) Effects of nitrogen fertilizer types on nitrous oxide emissions. In: Guo L, Gunasekara AS, McConnel LL, editors. Understanding Greenhouse Gas Emissions from Agricultural Management.Wa-Washington DC:American Chemical Society. pp. 179-202.
40. Hanse HCB, Borggaard OK, Sorensen J (1994) Evaluation of the free energy of formation of Fe(II)-Fe(III) hydroxide-sulphate (green rust) and its reduction of nitrite. Geochim Cosmochim Acta 58: 2599-2608.
41. Straub KL, Benz M, Schink B, Widdel F (1996) Anaerobic, nitrate-dependent microbial oxidation of ferrous iron. Appl Environ Microbiol 62: 1458-1470.
42. Weber KA, Achenbach LA, Coates JD (2006) Microorganisms pumping iron: anaerobic microbial iron oxidation and reduction. Nat Rev Microbiol 4: 752-764.
43. Clement JC, Shrestha J, Ehrenfeld JG, Jaffe PR (2005) Ammonium oxidation coupled to dissimilatory reduction of iron under anaerobic conditions in wetland soils. Soil Biol Biochem 37: 2323-2328.
44. Shrestha J, Rich JJ, Ehrenfeld JG, Jaffe PR (2009) Oxidation of ammonium to nitrite under iron-reducing conditions in wetland soils. Soil Sci 174: 156-164.
45. Finzi AC, Cole JJ, Doney SC, Holland EA, Jackson RB (2011) Research frontiers in the analysis of coupled biogeochemical cycles. Front Ecol Environ 9: 74-80.
46. Silver WL, Hall SJ, Liptzin D, Yang WH (2011) The iron redox engine drives carbon, nitrogen, and phosphorus cycling in terrestrial ecosystems. American Geophysical Union Abstracts, http://adsabs.harvard.edu/abs/2011AGUFM.B11F.01S, accessed Oct.28, 2012.

5

Eco-Stoichiometric Alterations in Paddy Soil Ecosystem Driven by Phosphorus Application

Xia Li[1], Hang Wang[1], ShaoHua Gan[1], DaQian Jiang[2], GuangMing Tian[1], ZhiJian Zhang[1,2,3]*

1 College of Environmental and Resource Science, Zhejiang University, Hangzhou, China, **2** Department of Earth and Environmental Engineering, Columbia University, New York, New York, United States of America, **3** China Academy of West Region Development, Zhejiang University, Hangzhou, China

Abstract

Agricultural fertilization may change processes of elemental biogeochemical cycles and alter the ecological function. Ecoenzymatic stoichiometric feature plays a critical role in global soil carbon (C) metabolism, driving element cycles, and mediating atmospheric composition in response to agricultural nutrient management. Despite the importance on crop growth, the role of phosphorous (P) in compliance with eco-stoichiometry on soil C and nitrogen (N) sequestration in the paddy field remains poorly understood in the context of climate change. Here, we collected soil samples from a field experiment after 6 years of chemical P application at a gradient of 0 (P-0), 30 (P-30), 60 (P-60), and 90 (P-90) kg ha^{-1} in order to evaluate the role of P on stoichiometric properties in terms of soil chemical, microbial biomass, and eco-enzyme activities as well as greenhouse gas (GHG: CO_2, N_2O and CH_4) emissions. Continuous P input increased soil total organic C and N by 1.3–9.2% and 3%–13%, respectively. P input induced C and N limitations as indicated by the decreased ratio of C:P and N:P in the soil and microbial biomass. A synergistic mechanism among the ecoenzymatic stoichiometry, which regulated the ecological function of microbial C and N acquisition and were stoichiometrically related to P input, stimulated soil C and N sequestration in the paddy field. The lower emissions of N_2O and CH_4 under the higher P application (P-60 and P-90) in July and the insignificant difference in N_2O emission in August compared to P-30; however, continuous P input enhanced CO_2 fluxes for both samplings. There is a technical conflict for simultaneously regulating three types of GHGs in terms of the eco-stoichiometry mechanism under P fertilization. Thus, it is recommended that the P input in paddy fields not exceed 60 kg ha^{-1} may maximize soil C sequestration, minimize P export, and guarantee grain yields.

Editor: Jose Luis Balcazar, Catalan Institute for Water Research (ICRA), Spain

Funding: This work is supported by Natural Science Foundation of China (40701162)(http://www.nsfc.gov.cn/Portal0/default152.htm). The funders had no role in study design, data collection and analysis, decision to publish, or preparation of the manuscript.

Competing Interests: The authors have declared that no competing interests exist.

* E-mail: zhangzhijian@zju.edu.cn

Introduction

The balance of elements has been a main focus of global change ecology and biogeochemical cycling research. Phosphorus (P) application remains an indispensable practice for agricultural crop production. However, P export from soil to surface waters may stimulate outbreaks of water eutrophication [1]. Meanwhile, carbon (C) storage in ecosystems is controlled by the mass conservation principle and the supply of other key nutrients, such as nitrogen (N) and P [2]. Therefore, maintaining a sustainable C-N-P balance in the soil ecosystem is necessary for coping with climate change, maximizing agricultural production, and optimizing P practice.

Ecological stoichiometry (Eco-stoichiometry) is based on stoichiometric theory and the metabolic theory of ecology, which involves the balance of energy and multiple chemical elements in ecological interactions at the subcellular to ecosystem scale [3]. Eco-stoichiometry, expressed as C:N:P stoichiometric ratio, can predict nutrient cycling and microbial biomass production in ecosystems [4,5,6] and plays an important role in element regulation during biosphere-scale processes, such as soil C storage and element balance in the soil biomass [7], and also governs greenhouse gas (GHG) emissions in terrestrial ecosystems [8]. Therefore, P fertilization coupled with element eco-stoichiometry may be a determining incentive in defining the dynamics that balance C-N-P and predicting GHG emissions in the soil ecosystem.

Microorganisms drive Earth's biogeochemical cycles [9] by a "consumer-driven nutrient recycling" (CDNR-like) mechanism that determines nutrient cycling, biomass stoichiometry, and community composition [10], and mediates the global C cycle during climatic changes [11]. In turn, this influences the ecological metabolic rate [4]. Measurements of the proportion of C, N, and P in the microbial biomass may thus be a practical tool for assessing the nutrient limitations of an ecosystem. For example, a low C-to-P ratio of microorganism biomass (MBC:MBP) may stimulate soil microorganisms to release nutrients and enhance the available P pool in the environment, while a high MBC:MBP ratio could cause the microorganisms to compete for available P and enhance soil P immobilization [12]. Conceptually, plasticity and homeostasis are the fundamental mechanisms by which organisms adjust the stoichiometric equilibrium to cope with environmental disturbances [5,13]. Exogenous P input would alter the primary stoichiometric balances among the soil-microorganisms complex, which could change soil C and N storage. However, the mechanisms on interaction between the exogenous P and soil organism stoichiometry as well as the ecological feedback to dynamics of soil C and N are still unknown.

Eco-enzyme activity represents an intersection of the ecological stoichiometry, wherein eco-enzyme activity (EEA) links environmental nutrient availability with microbial production [3]. Enzyme expression is regulated by environmental signals, while ecoenzymatic activity is determined by environmental interactions [11]. This in turn mediates nutrient cycling, sequestration from soil organic matter, and decomposition biochemistry [3]. The most widely assayed eco-enzymes, β-1,4-glucosidase (BG), β,4-N-acetylglucosaminidase (NAG), leucine aminopeptidase (LAP), and acid (alkaline) phosphatase (AP), hinge functional stoichiometries in relation to organic nutrient acquisition and are used as indicators of microbial nutrient demand [3,11]. These extracellular enzymes deconstruct plant and microbial cell walls into soluble substrates for microbial assimilation, and are a measure of microbial nutrient demand [14], which reveals the rate limitations of enzymatic catalysis in relation to soil carbon storage [15]. As such, the EEA should be sensitive to the effects of P application on microbial function and provide a mechanistic indicator of P for resource acquisition in the soil ecosystem. However, to date, few studies have focused on microbial function for resource acquisition under P input in soil, and thus the underlying mechanisms are largely unknown.

The interactions among C, N, and P cycling also determine the effect of GHG emissions on the Earth's climate through their influences on C and N sequestration in soil [16,17]. Agroecosystems contribute a large percentage of global emissions, including ~60% of N_2O, ~39% of CH_4, and ~1% of CO_2 [18]. As one of the important cereal crops, paddy fields (approximately 28.4 M ha) in China contribute to approximately 30% of the total global rough rice yield [17,19]. Combined with CO_2, N fertilization may generate intermediate nitrogenous gases (N_2O) [20]. Better paddy fertilization management strategies are thus needed to mitigate GHG emissions [21] and preserve soil productivity [22]. Relative to N, P is a static entity, which is strongly retained in the soil matrix [23] due to its lack of a significant gaseous phase. Therefore, P management coupled with stoichiometric methodology may offer an advantageous technology for controlling GHGs emitted from paddy fields, but remains inadequately studied.

To date, few investigations have probed the role of P fertilization on soil eco-stoichiometry in paddy fields, which hinders the optimization of C-N-P biogeochemical cycles. In 2005, a paddy field experiment with annual applications of chemical P fertilizer at rates of 0.0, 30, 60, and 90 kg P $ha^{-1}.y^{-1}$ was conducted in the Yangtze River delta in southeastern China to understand P driving soil eco-stoichiometry in a paddy field ecosystem. In 2011, we collected soil samples from these experimental paddy plots to investigate the stoichiometry of soil chemical, microbial biomass, and eco-enzyme activities in response to P fertilization. Data on GHG fluxes in July and August were also probed to understand the role of P on typical GHG emissions. We hypothesized that P application could change both soil C-N-P balance and eco-stoichiometric features in paddy soil and thus stimulate soil C sequestration.

Results

The Effect of Phosphorus on Soil Biochemical Features and GHG Emissions

Soils collected from four treatments varied noticeably in chemical, microbial, and eco-enzymatic properties (Table 1). Continuous P application significantly enhanced total soil P by 16–75% compared to P-0. In addition, total soil C was significantly ($p<0.05$) increased by 1.3% to 9.2% for P input compared to P-0, while soil N increased from 3–13% ($p<0.05$). Similarly, soil MBC was increased by 20% to 27% under P-60 and P-90; MBN was also increased under P-60 and P-90 by 28–50%, while an increase in MBP of 52–195% was observed. Due to P application, the ratios of soil MBC:MBP, MBC:MBN, and MBN:MBP were significantly increased with increased ratios of soil C:P ($p = 0.001$), soil C:N ($p = 0.039$), and soil N:P ($p = 0.002$), respectively (Fig. 1). The activities of soil eco-enzymes, such as BG, were enhanced from 39% to 75% compared to P-0, with AP decreasing by 14–33% ($p<0.01$) and NAG+LAP showing no significant changes (Table 1).

Net CO_2-C flux ranged from −88.7 to −204 mg·m^{-2}· h^{-1} and increased significantly ($p<0.05$) with increasing P application (Fig. 2). The net emission of CO_2 increased by 7–45% under P input compared to P-0 in July, with increases of 17% to 40% in August. CH_4-C flux, which ranged from 1.7–8.8 mg·m^{-2} h^{-1} in July, was significantly ($p<0.05$) lower by 14%–57% under P input compared to P-0. However, the highest CH_4 flux occurred with P-30 in August. The N_2O-N flux in our study ranged from −0.02 to 0.05 mg·m^{-2} h^{-1}. N_2O emission in July was significantly ($p<0.05$) reduced by 34–75% in the tested paddy field with increasing P fertilization compared with P-0; no significant difference was found in August (Fig. 2). Analysis of variance (ANOVA) showed that the soil CO_2 flux significantly responded to single factors of P application and sampling time; CH_4 flux significantly responded to single factors of P application and interactions of treatment and sampling date, while N_2O flux was significantly affected by P application and sampling date as well as interactions (Table 2).

The Effect of Phosphorus on Soil Eco-stoichiometry

The stoichiometric ratios of soil C:P declined (from 156 to 97) significantly ($p<0.01$) with increasing P applications, and the ratio of soil N:P was reduced from 11.4 to 7.3 (Fig. 3). However, no significant difference was found for the soil C:N ratio among the four treatments. Soil MBC:MBP and MBN:MBP ratios were both found to be significantly ($p<0.05$) decreased with increases in P application, while the MBC:MBN ratio remained unchanged, except for P-90 (Fig. 3). The acquisition of C relative to organic P indicated by the ratio of ln(BG):ln(AP) was significantly increased with increasing P application (Fig 3). The ln(BG):ln(NAG+LAP), which refers to the acquisition of C relative to organic N, showed no significant differences within P application treatments, but was significantly ($p<0.01$) increased compared to P-0. However, ln(NAG+LAP):ln(AP), which refers to the acquisition of N relative to organic P, showed no significant differences with P application (Fig. 3). Scatter plots of C:N:P stoichiometry for soil chemistry, microbial biomass, and eco-enzymatic activities for the paddy system under P application compared to P-0 indicated that P application caused a C-N co-limitation on ecological stoichiometric processes (Fig. 4). This result suggests that there was a higher C and N demand for soil microorganisms relative to P, integrated soil stoichiometric ratios with biochemical properties, and eco-enzymatic activities (Table 1).

The linear relationships between soil C:P and ln(BG):ln(AP) ($p = 0.001$), MBC:MBP and ln(BG):ln(AP) ($p = 0.000$), and MBN:MBP and ln(NAG+LAP):ln(AP) ($p = 0.050$) were significant and showed a negative trend (Fig. 1). However, there was no significant relationship between soil C:N and ln(BG):ln(NA-G+LAP), soil N:P and ln(NAG+LAP):ln(AP), or MBC:MBN and ln(BG):ln(NAG+LAP). The significant relationships showed that soil C:P, soil C:N, soil N:P, soil C:P, MBC:MBP, and MBN:MBP substantially influenced MBC:MBP, MBC:MBN, MBN:MBP, ln(BG):ln(AP), ln(BG):ln(AP), and ln(NAG+LAP):ln(AP), respectively, and could account for the respective 69.1%, 36.2%, 64.6%,

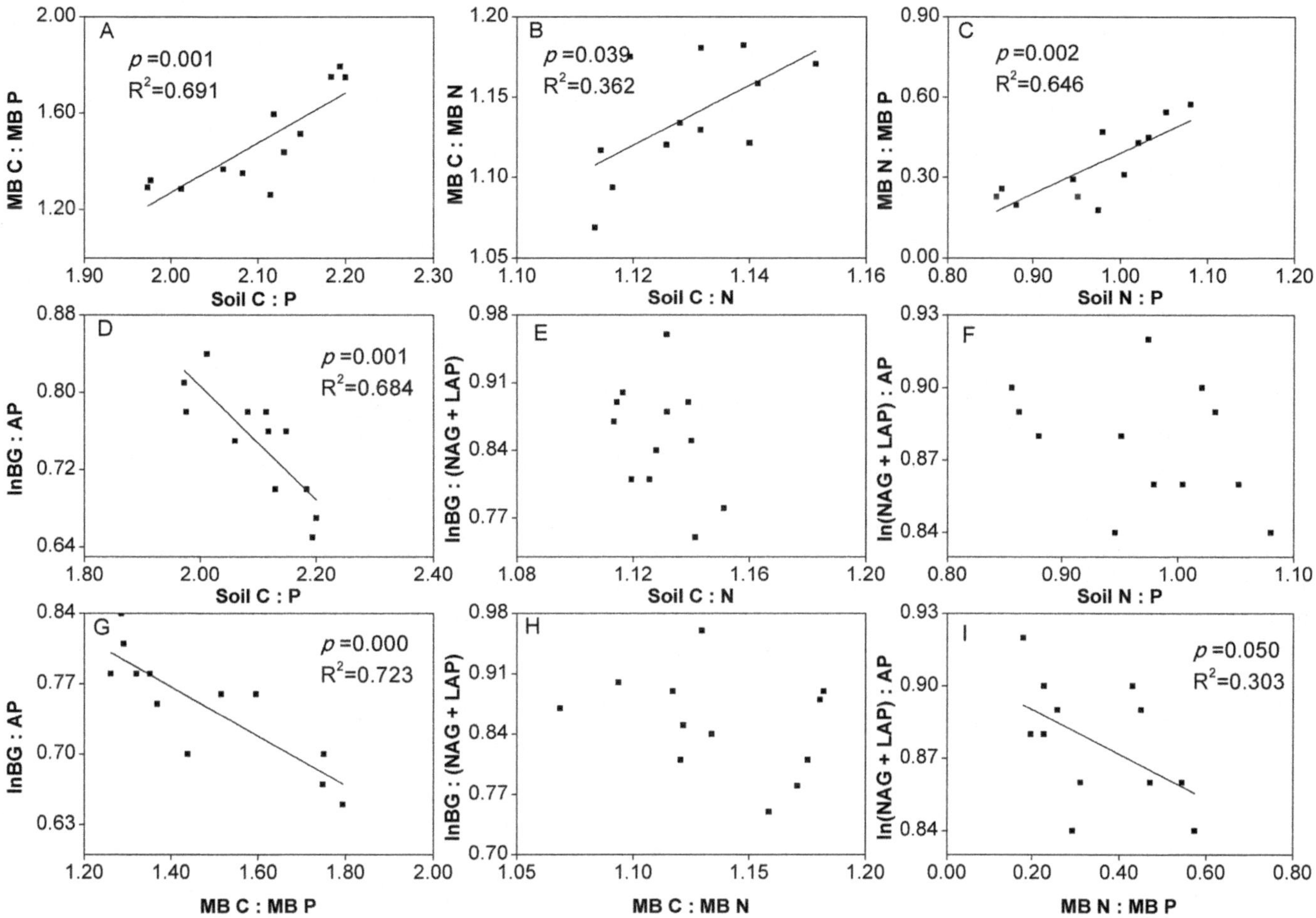

Figure 1. Linear regression of C:N:P stoichiometry for soil chemistry, microbial biomass, and eco-enzymatic activities. Summary of standardized major axis analysis of log10-transformed molar nutrient concentrations. Ratios of C:P, C:N, N:P acquisition activity, as indicated by ratios of ln(BG):ln(AP), ln(BG):ln(NAG+LAP), and ln(NAG+LAP):ln(AP), respectively.

68.4%, 72.3%, and 30.3% variances. Taken together, these results suggest that the microbial biomass stoichiometric ratios were positively influenced by the soil stoichiometric ratios, and both of them were negatively regulated by coenzyme activities under P application.

Discussion

Eco-stoichiometry upon Phosphorus Application and the Linkage to Soil Carbon and Nitrogen

Phosphorus amendment increased the decomposition of soil organic C through increased soil respiration [7]. However, the data in Table 1 indicate that paddy soil receiving P application had increased soil C and N pools by 1.3–9.2% and 3%–13%, respectively. Such percent change in soil organic C pools is consistent with a similar long-term field investigation [24], which verified that chemical fertilization (45 kg P ha^{-1}) could increase ~10% higher soil organic C.

The fact that the soil C:P declined (Fig. 1) while soil C and N pools increased (Table 1) upon P input indicated a higher P concentration relative to C in soils, which caused the organic matter pool to become less humic (protein-like) and provided potential energy for microbial utilization [25]. P input modified not only C:N:P ratios in the soil, but also the soil microbial biomass. The positively significant relationships between chemical and microbial stoichiometry (Fig. 1) clearly indicated that microorganisms adapted their acquisition ratios according to local resource ratios. In our study, the decreased soil MBC:MBP upon P input (Fig. 3) demonstrated that P application results in C-limitation in microbial biomass, despite the finding that soil C pools with P input were increased (Table 1). This also suggested that soil microorganisms were limited to soil C allocations and that P enrichment would improve the primary productivity, thus stimulating C sequestration in soil receiving P application, although this C-P synergistic effect [3,26] would plateau under higher P input rates at P-60 and P-90. Moreover, the distribution of ecoenzymatic C:N:P activity ratios may identify the boundaries of the microbial community response to fluctuations in nutrient availability [3,11]. The soil microbial metabolic pattern tended to be C limited (Fig. 4) with P input together with the increased ratio of lnBG:lnAP (Fig. 3), which closely responded to the decrease in soil MBC:MBP (Fig. 1). These results indicated that a greater C demand for microbial biomass due to P input enhanced the ecological function of organic C acquisition within the soil-microorganism complex. Hence, our hypothesis that soil C sequestration may be enhanced by P input was verified. Additionally, further analysis showed that MBC:MBN:MBP (49:3.3:1.0 to 21:1.6:1.0, respectively) with P application (Table 1) deviated significantly from the generally mean atomic C:N:P ratio (60:7:1) in the soil microbial biomass [27]. The C:N:P ratio in the

Table 1. Soil chemical, microbial, and eco-enzymatic properties in the tested paddy field after 5 years of phosphorus application.

Treatment	Soil chemical properties		
	Total C (mmol kg^{-1})	Total N (mmol kg^{-1})	Total P (mmol kg^{-1})
P-0	2070±46^{b}	151±3^{c}	13.3±0.5^{d}
P-30	2096±67^{b}	155±5^{bc}	15.5±1.0^{c}
P-60	2185±91^{ab}	162±3^{b}	17.9±0.4^{b}
P-90	2260±78^{a}	171±4^{a}	23.3±0.6^{a}
Treatment	**Soil microbial properties**		
	MBC (mmol kg^{-1})	MBN (mmol kg^{-1})	MBP (mmol kg^{-1})
P-0	40.7±2.3^{b}	2.76±0.14^{c}	0.83±0.08^{d}
P-30	43.9±1.4^{b}	3.14±0.22^{c}	1.26±0.30^{c}
P-60	48.8±2.3^{a}	3.54±0.16^{b}	2.07±0.22^{b}
P-90	51.8±1.9^{a}	4.14±0.27^{a}	2.45±0.07^{a}
Treatment	**Eco-enzymatic activity properties**		
	BG (nmol h^{-1} g^{-1})	NAG+LAP (nmol h^{-1} g^{-1})	AP (nmol h^{-1} g^{-1})
P-0	80.0±10.5^{c}	276±43^{a}	638±36^{a}
P-30	111.6±18.0^{b}	259±39^{a}	576±43^{b}
P-60	126.2±8.1^{ab}	256±46^{a}	538±24^{b}
P-90	140.3±15.2^{a}	231±35^{a}	451±51^{c}

The different letters listed beside the data represent significant differences at $p<0.05$ (Duncan test, one-way ANOVA).

biomass reflects the physiological and biochemical constraints on the elemental composition of primary production [6,27], but differences in soil organism habitats may preclude the emergence of constrained soil microbial element ratios [13]. Our data departed from the mean ratio, which suggested that the overall investment in structural cellular material in the microbial biomass was remarkably affected by P input, indicating that no rigorous homeostasis existed for the soil microbial community in the tested paddy soil.

Soil nutrient enrichment enhances root metabolic activity and the consequent excretion of organic C, which enters the mineral soil through rhizosphere deposition [28] and simulates N fixation by promoting nutrient equilibration [29]. P enrichment would exert a synergistic function of N fixation in the soil-microorganism-plant system [27], which may contribute to soil N accumulation

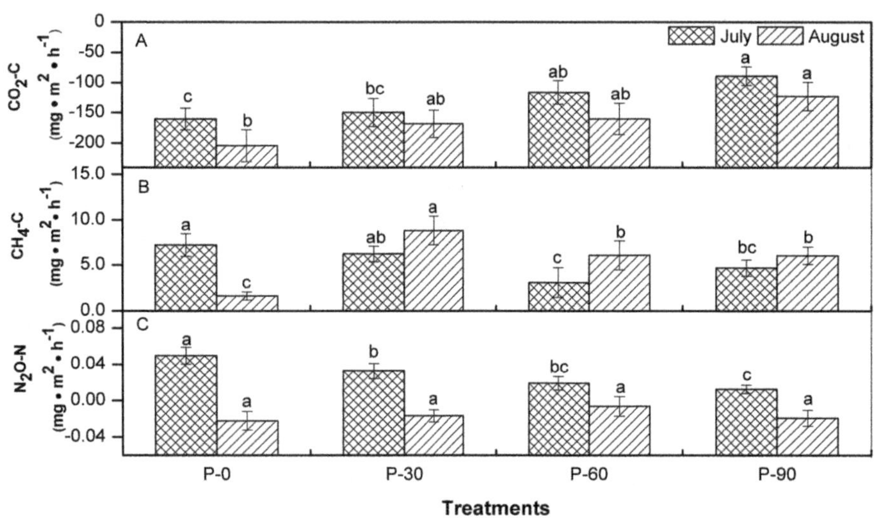

Figure 2. Emission intensities of greenhouse gases responding to phosphorus application for samplings in July and August. The different letters listed above bars represent significant differences at p<0.05 (Duncan LSD test).

Table 2. Results of two-way analysis of variance (repeated ANOVA) showing the *p* values for GHG emissions responding to P application and sampling time (July; August) for paddy field soil.

Factor	CO_2 flux	CH_4 flux	N_2O flux
Treatment	**<0.001**	**0.002**	**0.023**
Time	**0.001**	0.510	**<0.001**
Treatment×Time	0.732	**<0.001**	**0.001**

(Table 1), although the soil N:P stoichiometry somewhat decreased in response to P application (Fig. 1). Moreover, the newly-fixed N might be easily stored in the soil and/or biomass under P fertilization, because no significant difference on N-related enzymatic activities was found among the four treatments (Table 1). Based on element stoichiometry [30,31], soil microorganisms would utilize more plant residues and absorb more C and N for growth with the continuous P application as well as increased soil MBC and MBN (Table 1), which might be favorable for soil C and N sequestration [28,31] in the paddy ecosystem. However, no significant difference occurred in the lnBG:ln(NAG+LAP) (with the exception of P-0) and ln(NAG+LAP):ln(AP) ratios (Fig. 3) among the P treatments, indicating that a synergistic mechanism among the ecoenzymatic stoichiometry regulates the ecological function of microbial C, N, and P acquisition. Continuous P input caused a significantly ($p<0.05$) negative relationship between the N:P eco-enzymatic and microbial stoichiometries (Fig. 1), which suggests that an abundant source of P in soil microorganisms stimulates the ecological function as a sink for soil N cycle. Together, these results support our hypothesis that P application helps to define the dynamics of the multiple balance of C-N-P in paddy soil, where the synergism of ecoenzymatic stoichiometry enhances soil C and N sequestration.

P application clearly demonstrated a predominance of C and N co-limitations for microorganism metabolic activities and C-N acquisition in our study (Fig. 4). Mechanistically, these may arise through interactions among C, N, and P in bio-molecules such as mRNA [32] and linkage to soil bacterial communities [33,34]. A biogeochemical equilibrium model verified that ecological stoichiometry constraints microbial community metabolism [11]. Further investigations are needed to determine the inter-annual variations on relationship between molecular biology and eco-enzymatic stoichiometry as well as to illustrate alterations in ecological function caused by long-term P input in paddy ecosystem. It is also an important mechanism that 30–40% of sequestered C would be transferred into sub-soil (below 30 cm) through plant roots in grassland [35]. Although a plow pan layer exists following an arable horizon in a paddy field, the vertical distribution of C sequestration over the entire soil profile at the view of C-P eco-enzymatic stoichiometry is needed to further investigate for paddy plots receiving long-term P application.

The Role of Phosphorus on Greenhouse Gas Emission

The production of GHG emissions are regulated by soil physical properties and management practices, such as soil temperature, soil O_2 content, tillage, and alternations of wetting and drying

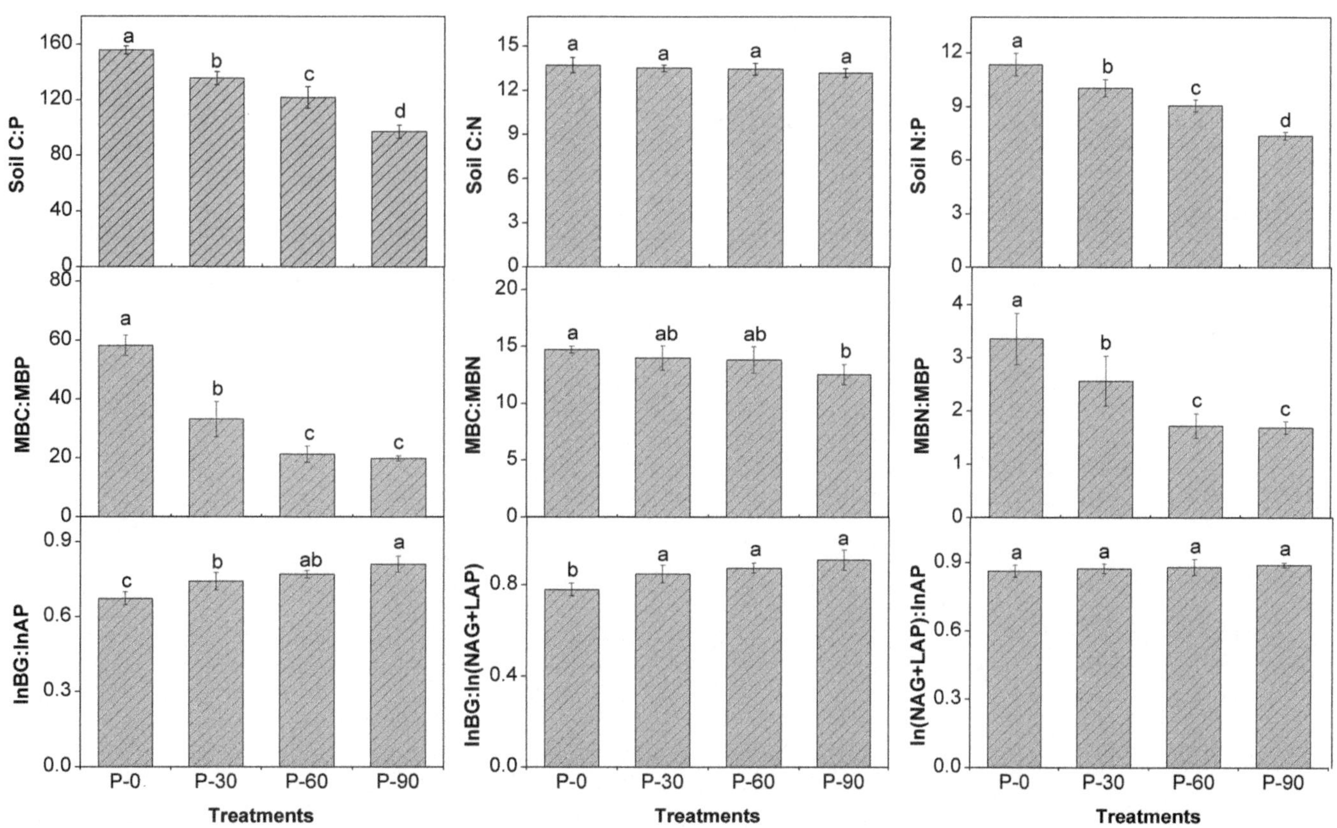

Figure 3. Stoichiometric ratios of soil elements, soil microorganisms, and eco-enzymatic activities in the tested paddy soil under phosphorus treatments. The different letters listed above bars represent significant differences at $p<0.05$ (Duncan test, one-way ANOVA).

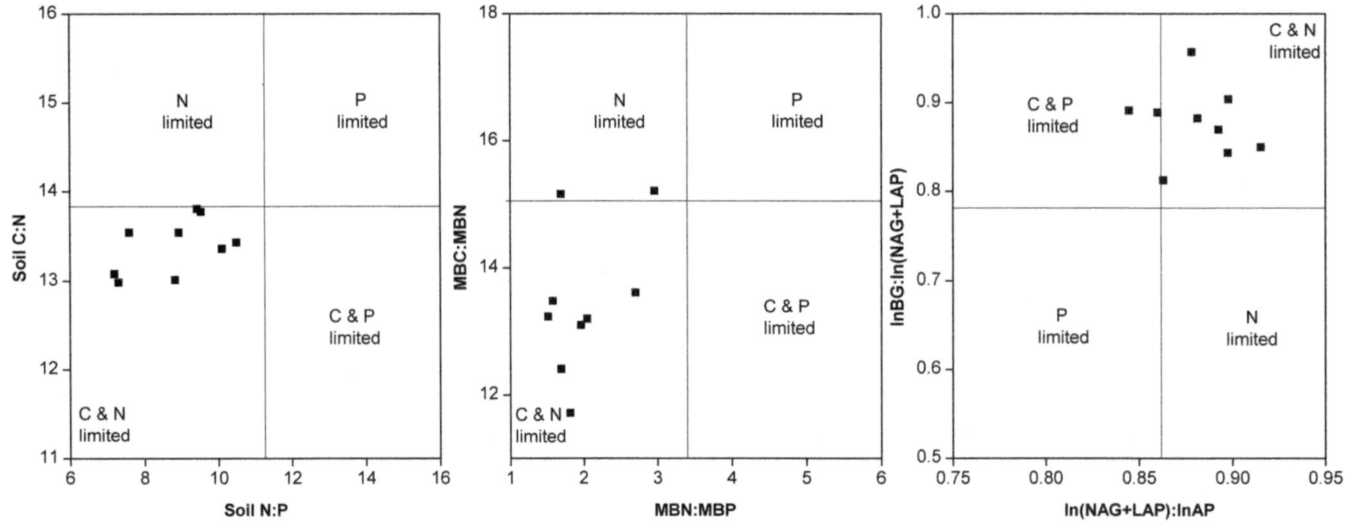

Figure 4. Scatter plots of C:N:P stoichiometry for soil chemistry, microbial biomass, and eco-enzymatic activities for the paddy system under P application compared to P-0.

[17,22,33]. In our study, the tested paddy soil varied only with P treatments, while the remaining environmental and agricultural factors were kept the same. Therefore, we hypothesized that the shifted GHG fluxes at two typical sampling seasons are mainly related to P treatment. Moreover, because the soil temperature is also a key factor for GHG emissions, July and August were chosen for estimating the GHG emissions subjected to P application due to the higher seasonal temperatures.

The net CO_2 flux is a balance of photosynthesis and respiration [17]. The negative value of the net CO_2 flux measured in this study was mainly due to the higher plant CO_2 uptake, rather than respiration emission. Increases in the soil C pool under continuous P input (Table 1) enhanced CO_2 fluxes over P application rates for both samplings (Fig. 2). A low P supply (fed by solution culture in 0.5 mg P L^{-1}) was found to stimulate root exudation and root aerenchyma development by more than two-fold compared to a high P supply (5–10 mg P L^{-1}) [36]. As might be expected, relatively higher CH_4 emissions were found for P-0 and P-30 sampling in July, while the highest emission occurred for P-30 sampling in August (Fig. 2). Shifting from continuous water-logging to midseason drainage [8,17,37] and increases of dissolved oxygen content of the rhizosphere [38,39] led to a drop in CH_4 flux and an increase in N_2O flux in paddy soil. CH_4 emissions (except P-0) were commonly found to be higher for sampling in July (water-logging) than in August (oxygen secretion through developed roots); however, an overwhelmingly opposite trend occurred for N_2O for the two samplings. The reason for the latter phenomenon is unclear and will require additional evaluation. The decrease ($p<0.05$) and mostly unchanged N_2O flux was found upon P application in July and August, respectively (Fig. 2), which is basically consistent with the trend found in the maize system [40]. These results suggest that P input may mitigate N_2O emission during the high temperature period in this study. Significant relationships between CO_2 flux and stoichiometric ratios (Table 3) indicated that CO_2 emission is not only influenced by soil chemical availability, which is basically in agreement with previous reports [7,17], but is also regulated by the co-limitation of the eco-enzymatic stoichiometric balance for paddy soil receiving P input. Although significant relationships related to stoichiometric ratios were found for both CH_4 and N_2O emissions in July, the effect of these stoichiometric ratios for predicting both CH_4 and N_2O is completely opposite to that of CO_2 (Table 3). Moreover, most of these significant relationships disappeared with increasing time (August) after P application. As such, a technical conflict would occur for simultaneously regulating three types of GHGs in terms of an eco-stoichiometry mechanism under conditions of P fertilization, and the associated strategies for such joint regulation would not be synchronous over time. GHG emissions are not only related to agricultural fertilization [38,39], but also the emission budget and associated net global warming potential, which are closely sensitive to the temporal and spatial traits [17,20,37]. Therefore, continuous probing of the eco-stoichiometric mechanisms for yearly GHG budgets responding to P input under the current field survey would be needed.

Implication: Phosphorus Management for Coping with Climate Change

Enhanced soil C sequestration induced by P input (Table 1) is worth considering as a means of addressing climate change. However, such a practice would require technical caution, because the environmental costs of nutrient pollution from agriculture practice have been substantial, including deteriorated downstream water quality and eutrophication [1,21]. A survey of P application indicated that paddy fields may also act as an ecological sink for P dynamic with no greater than 60 kg ha^{-1} of P input [23], provided that field water management (e.g., water-saving irrigation technology) [41] is optimized and takes into account physicochemical P soil absorption [29,42]. Supplemental cultivation of green manure during the fallow period (e.g., Chinese milk vetch, *Astragalus sinicus L)* or application of organic fertilizers are well-known means for enhancing soil C and N fixation [34,43]. Based on the element stoichiometric balance [3,27] and co-limitation of C and N for soil microbes (Fig. 4) in this study, excessive P in the paddy soil due to continuous P input might be effectively composited with C and N from milk vetch biomass with regard to the synergistic mechanisms among C-N-P in soil-microorganism complexes, leading to further soil C-N sequestration. Currently, rice grain yield (Table S1) among the experimental plots under P-30 (7580 kg ha^{-1}) was significantly higher than that of P-0 (6200 kg ha^{-1}), while no significant difference was

Table 3. Pearson correlation coefficients between CO_2, CH_4, and N_2O fluxes of paddy field and soil stoichiometric ratios.

	Month	GHG flux	Chemical ratio			Microbial ratio			Eco-enzyme ratio		
		C : P	C : N	N : P	MBC: MBP	MBC: MBN	MBN: MBP	ln (BG) : ln (AP)	ln (BG) : ln (NAG+LAP)	ln (NAG+LAP) : ln (AP)
July	CO_2 flux	−0.871**	−0.578*	−0.851**	−0.732**	−0.547	−0.675*	0.818**	0.786**	0.361
	CH_4 flux	0.572	0.194	0.589*	0.738**	0.203	0.814**	−0.618*	−0.616*	−0.232
	N_2O flux	0.841**	0.452	0.839**	0.862**	0.549	0.809**	−0.740**	−0.690*	−0.372
Aug	CO_2 flux	−0.809**	−0.446	−0.804**	−0.658*	−0.499	−0.555	0.778**	0.822**	0.155
	CH_4 flux	−0.348	−0.224	−0.341	−0.591*	−0.332	−0.475	0.520	0.414	0.451
	N_2O flux	−0.195	−0.692*	−0.099	−0.276	−0.216	−0.119	0.093	0.233	−0.308

Soil eco-enzymatic activities are presented as log10-transformed molar nutrient concentrations.

found between P-60 (9100 kg ha^{-1}) and P-90 (9070 kg ha^{-1}). Grain yield under P-60 is close to the highest level among a survey overview on rice yield [44] in ZheJiang, China. Thereafter, P input to paddy fields at a rate not to exceed 60 kg ha^{-1} is recommended for maximizing soil C sequestration, minimizing P export, and guaranteeing grain yields.

In summary, we found that continuous P application increased the soil C and N pools. P input directly modified soil C:N:P ratios, which shifted the microbial stoichiometry of the biomass to be both C-limited and N-limited. Moreover, a synergistic mechanism among the eco-enzymatic stoichiometry regulates the ecological function of microbial C, N, and P acquisition, thus stimulating C and N sequestration in soil receiving P application in paddy fields. The P application significantly mitigated N_2O and CH_4 emissions under P-60 and P-90 compared to P-30 or showed no difference for N_2O emissions in August, while enhanced CO_2 fluxes were observed in both July and August samplings. Therefore, these results suggest that technical conflicts would occur for the simultaneous regulation of three major GHGs through an eco-stoichiometry mechanism under the conditions of P fertilization. The integration of soil C sequestration, minimization of P export, and guarantee of the grain yield translate into a recommended P paddy field ecosystem input rate of no greater than 60 kg ha^{-1}.

Materials and Methods

(This work is unrelated to an ethics issues, and no specific permit was required for the described field study.).

Study Site

A plot experiment on P management for paddy field was established in April 2005 at the demonstration park of YuHang County Agricultural Research Station (30°18′51.84′′N, 119°54′13.37′′E) in ZheJiang, China. The experimental region has a subtropical monsoon climate with an average temperature of 17.8°C and an average annual rainfall of 1450 mm. The soil type is typical clay, blue-purple paddy soil. The dominant soil type in this region is a blue-purple paddy soil (*Mollic Endoaquoll*). The soil before experimental plots construction was composed of 3% sand, 47% silt, and 50% clay in the top 150 mm, while contents of soil total P, total C and total N were found 13.7, 2087, and 148 mmol kg^{-1}, respectively. Local farmers in this region routinely apply approximately 25–50 kg P ha^{-1} of inorganic P fertilizer or compound fertilizer in late July or early August in order to support one crop of rice and an over-wintering crop, such as wheat (*Triticum aestivum*) or rape (*Brassica Napus*).

Rice Field Plot Experiment and Soil and Greenhouse Gases Sampling

Rice field plot experiment. The construction of the field experimental plots, including specific designs on plot ridges, trenches, berms and inlets/outlets, was previously described by Zhang [23]. Briefly, twelve 4 m×5 m plots were constructed in two parallel rows in 2005. In order to keep each of these plots hydrologically isolated, a high-density polyethylene impermeable membrane of 0.75 mm (thickness)×105 cm depth was first inserted between two neighboring plots, and then concrete-brick walls of 12 cm (width)×105 cm (depth) were coupled on both sides of the membrane. The experiment was conducted using a completely randomized block design with three replicates for each treatment. P was applied at rates of 0 (P-0), 30 (P-30), 60 (P-60), and 90 (P-90) kg ha^{-1} in June, using superphosphate since 2005. These P treatments cover the routine rate of P application of local farmers and the excessive P rates for field experiments. All plots received 170 kg N ha^{-1} (urea) and 50 kg K ha^{-1} (KCl) each year. In 2011, 25-day-old rice seedlings (*Oryza sativa L.*) were transplanted at 150 mm×150 mm spacing, and rice was harvested on November 10, 2011. Details on rice grain yield and yield components are presented in Text S1.

Soil sampling. After application of P for 6 years from 2005 to 2011, the experimental paddy field enters a relatively stable state for soil biosphere. Therefore, soil sampling in 2011 was chosen for a comparison investigation on soil eco-stoichiometric characteristics under different P treatments. On May 21, 2011, one month before rice transplantation, each plot was divided into six subplots of grab-sampling to minimize edge effects. The arable soil samples for each plot consisted of six composited 2 cm diameter×10 cm deep cores (cultivated horizon) [45]. The samples were kept on ice and immediately transported to the laboratory. Air-dried soils were screened through 2 mm sieves and then stored in the dark at 4°C until analysis for total C, N, and P concentrations. The fresh soil samples were stored at 4°C for no longer than one week before microbial C, N, P, and enzyme activity analyses.

Gas collection and measurements. Sampling of GHGs was set in July and August 2011 following fertilization, representing the early-phase of rice tilling and middle-phase of rice jointing during the high temperature season. A series of lab-made gas collection apparatuses (Text S2) was constructed prior to field sampling, and then set in the middle of each plot. Gas sampling was performed between 9:00 am and 11:00 am using 10 ml vacuum glass tubes connected to the stomata sampling apparatus. For each flux determination, a series of gas samples was taken at

10 min intervals over one hour. At the same time, the air temperature within the chambers were measured, while ambient air pressure was recorded by an automatic weather station (FRT CS01A) assembled 95 m away from the experimental plots. The samples were analyzed for CO_2, N_2O, and CH_4 within 24 h using an Agilent 6890D GC (Agilent Technologies, Palo Alto, California, USA) equipped with an automated gas chromatographic system with an ^{63}Ni electron capture detector for N_2O analyses and flame ionization detector for CO_2 and CH_4.

The CO_2, N_2O, and CH_4 fluxes were calculated using linear regression of the change in gas concentration, based on the following equation:

$$F=\sigma\frac{P}{P_0}\frac{T_0}{T}\frac{dc}{dt}H \quad (1)$$

where F is the flux rate of $CO_2/CH_4/N_2O$ ($mg{\cdot}m^{-2}{\cdot}h^{-1}$), σ is the CO_2, CH_4 and N_2O gas density in standard state (1.96, 0.71, and 1.96 $kg{\cdot}m^{-3}$, respectively), P is the atmospheric pressure (kPa), T the air temperature (K), P_0 and T_0 is the atmospheric pressure (101.3 kPa) and the air temperature (273 K), dc/dt is the unit conversion factors for calculating CO_2 and N_2O flux rates, and H is the height of the chamber (m).

Soil fraction properties. Air-dried samples were analyzed for total organic C (TOC), total N, and P concentrations according to standard methods described by Bao [46]. The potassium dichromate oxidation method was used to determine total organic C contents. Total N was analyzed by the Kjeldahl method. Total P was digested by $HClO_4$-H_2SO_4 and measured spectrophotometrically using a continuous flow analyzer (Autoanalyzer III, BRAN+LUEBBE, Germany) set to 880 nm. The fresh soil samples used for determining microbial biomass C and N as well as the P measurements used the fumigation-extraction method as described by Wu et al. [47] and Brookes et al. [48]. Briefly, the fresh soil samples were split into two subsamples with one sample immediately extracted with 0.5 mol L^{-1} K_2SO_4 for microbial C and N or 0.5 mol L^{-1} $NaHCO_3$ for microbial P, while the other sample was fumigated with chloroform and then extracted. Following centrifugation, C, N, and P concentrations as well as microbial biomass element content were calculated from the difference between the fumigated and non-fumigated soil samples.

Extracellular Enzyme Assays

Four eco-enzymes (i.e., BG, AP, and NAG and LAP) were selected as indicators of microbial nutrient demand in the cycles of C, N, and P, respectively (Table S2). Fluorescence-based soil assays for β-1, 4-glucosidase (BG), β-1, 4-N-acetylglucosaminidase (NAG), L-leucine aminopeptidase (LAP), and acid phosphate (AP) used MUB-linked and AMC-linked artificial substrates [15,49] (Table S2). Briefly, Sample suspensions were prepared by homogenizing 1 g (wet weight) of soil with 125 ml of 50 mmol L^{-1} sodium acetate buffer using a vortex shaker for 1 min. The pH of sodium acetate buffer for the soil slurries is 6.0, which is the mean soil pH of the environmental samples [50]. Standard high throughput fluorometric enzyme assays were conducted in 96-well blank fluorescent plates (Corning Inc., costar 3603) after pipetting of buffer, slurries, references, and substrates followed a strict order and position on the well plate according to the order of Saiya-Cork [49] (Table S3). The microplates were covered and incubated in the dark at 20°C for 4 h. Then, 10 µl of 1.0 mol L^{-1} NaOH were added to each well to stop the reaction and increase the fluorescence of the substrates. Different buffers respond to NaOH addition in different ways. Thus, a time frame of 1 min between NaOH addition and the reading of plates in a fluorometer was used to reduce analytical variation [50]. Following the addition of NaOH, fluorescence was measured with a Bio-Tek Synergy HT microplate reader (Bio-Tek Inc., Winooski, USA) with 365 nm excitation and 460 nm emission filters. Enzyme activities were calculated and expressed as nmol h^{-1} g^{-1}.

Statistical Analysis

A completely randomized design was performed to examine significant differences in soil chemical composition, microbiological properties, eco-enzymes, and their stoichiometry among the various P fertilization treatments (P-0, P-30, P-60, P-90) using two-way analysis of variance (ANOVA) at $p<0.05$ levels. All assays were conducted in triplicate. A principal component analysis (PCA) was used to transform these variables to two factors. The acquisition ratios of ln(BG):ln(AP), ln(BG):ln(NAG+LAP), and ln(NAG+LAP):ln(AP) activities were also calculated and referred to the acquisition of organic C relative to organic P and N as well as organic N relative to organic P, respectively [3]. ANOVA and Pearson correlation statistics were also performed using the SPSS statistical software version 17.0. The clustering method and PCA were performed using the Minitab statistical software version 16.0. The Origin 8.0 (Origin Lab Corporation, USA) was also used for figure preparation.

Supporting Information

Table S1 Rice grain yield and yield components of experimental paddy field under P fertilization.

Table S2 Extracellular enzymes with corresponding substrate and the corresponding function.

Table S3 The order of pipetting buffer, slurries, references, and substrates in fluorometric enzyme assays.

Text S1 Rice grain yield and yield components.

Text S2 Gas collection apparatus fabrication.

Author Contributions

Conceived and designed the experiments: ZJZ DQJ. Performed the experiments: XL HW SHG. Analyzed the data: XL HW. Contributed reagents/materials/analysis tools: GMT. Wrote the paper: XL DQJ ZJZ.

References

1. Sims JT, Simard RR, Joern BC (1998) Phosphorus loss in agricultural drainage: Historical perspective and current research. Journal of Environmental Quality 27: 277–293.
2. Hobbie SE, Nadelhoffer KJ, Hogberg P (2002) A synthesis: The role of nutrients as constraints on carbon balances in boreal and arctic regions. Plant and Soil 242: 163–170.
3. Sinsabaugh RL, Hill BH, Shah JJF (2009) Ecoenzymatic stoichiometry of microbial organic nutrient acquisition in soil and sediment. Nature 462: 795-U117.
4. Allen AP, Gillooly JF (2009) Towards an integration of ecological stoichiometry and the metabolic theory of ecology to better understand nutrient cycling. Ecology Letters 12: 369–384.

5. Hall EK, Maixner F, Franklin O, Daims H, Richter A, et al. (2011) Linking microbial and ecosystem ecology using ecological stoichiometry: A synthesis of conceptual and empirical approaches. Ecosystems 14: 261–273.
6. Yu Q, Chen QS, Elser JJ, He NP, Wu HH, et al. (2010) Linking stoichiometric homoeostasis with ecosystem structure, functioning and stability. Ecology Letters 13: 1390–1399.
7. Bradford MA, Fierer N, Reynolds JF (2008) Soil carbon stocks in experimental mesocosms are dependent on the rate of labile carbon, nitrogen and phosphorus inputs to soils. Functional Ecology 22: 964–974.
8. Zhang Y, Su SL, Zhang F, Shi RH, Gao W (2012) Characterizing spatiotemporal dynamics of methane emissions from rice paddies in northeast China from 1990 to 2010. Plos One 7.
9. Falkowski PG, Fenchel T, Delong EF (2008) The microbial engines that drive Earth's biogeochemical cycles. Science 320: 1034–1039.
10. Cherif M, Loreau M (2009) When microbes and consumers determine the limiting nutrient of autotrophs: a theoretical analysis. Proceedings of the Royal Society B-Biological Sciences 276: 487–497.
11. Sinsabaugh RL, Shah JJF (2012) Ecoenzymatic stoichiometry and ecological theory. In: Futuyma DJ, editor. Annual Review of Ecology, Evolution, and Systematics, Vol 43. Palo Alto, CA: Annual Reviews. 313–343.
12. Peng PQ, Zhang WJ, Tong CL (2005) Soil C, N and P contents and their relationships with soil physical properties in wetlands of Dongting Lake fliood plain. Chinese Journal of Applied Ecology 16 (10): 1872–1878.
13. Sterner RW, Elser JJ (2002) Ecological Stoichiometry: The Biology of Elements From Molecules to the Biosphere. Princeton, NJ: Princeton University Press.
14. Moorhead DL, Sinsabaugh RL (2006) A theoretical model of litter decay and microbial interaction. Ecological Monographs 76: 151–174.
15. Sinsabaugh RL, Lauber CL, Weintraub MN, Ahmed B, Allison SD, et al. (2008) Stoichiometry of soil enzyme activity at global scale. Ecology Letters 11: 1252–1264.
16. Falkowski P, Scholes RJ, Boyle E, Canadell J, Canfield D, et al. (2000) The global carbon cycle: A test of our knowledge of earth as a system. Science 290: 291–296.
17. Cai ZC (2012) Greenhouse gas budget for terrestrial ecosystems in China. Science China-Earth Sciences 55: 173–182.
18. OECD (2000) Environmental indicators for agriculture methods and results. Executive summary Paris.
19. IRRI (2009) World rice statistics: Rough rice production by country and geographical region-FAO1961–2007. Available: http://beta.irri.org/solutions/index.php?option = com_content&task = view&id = 250. Accessed 2009 Jul 28.
20. Davidson EA (2009) The contribution of manure and fertilizer nitrogen to atmospheric nitrous oxide since 1860. Nature Geoscience 2: 659–662.
21. Vitousek PM, Naylor R, Crews T, David MB, Drinkwater LE, et al. (2009) Nutrient imbalances in agricultural development. Science 324: 1519–1520.
22. Snyder CS, Bruulsema TW, Jensen TL, Fixen PE (2009) Review of greenhouse gas emissions from crop production systems and fertilizer management effects. Agriculture Ecosystems & Environment 133: 247–266.
23. Zhang ZJ, Zhang JY, He R, Wang ZD, Zhu YM (2007) Phosphorus interception in floodwater of paddy field during the rice-growing season in TaiHu Lake Basin. Environmental Pollution 145: 425–433.
24. Huang QR, Hu F, Huang S, Li HX, Yuan YH, et al. (2009) Effect of Long-Term Fertilization on Organic Carbon and Nitrogen in a Subtropical Paddy Soil. Pedosphere 19: 727–734.
25. Marichal R, Mathieu J, Couteaux MM, Mora P, Roy J, et al. (2011) Earthworm and microbe response to litter and soils of tropical forest plantations with contrasting C:N:P stoichiometric ratios. Soil Biology & Biochemistry 43: 1528–1535.
26. Taylor PG, Townsend AR (2010) Stoichiometric control of organic carbon-nitrate relationships from soils to the sea. Nature 464: 1178–1181.
27. Cleveland CC, Liptzin D (2007) C : N : P stoichiometry in soil: is there a "Redfield ratio" for the microbial biomass? Biogeochemistry 85: 235–252.
28. Ciampitti IA, Garcia FO, Picone LI, Rubio G (2011) Soil carbon and phosphorus pools in field crop rotations in pampean soils of Argentina. Soil Science Society of America Journal 75: 616–625.
29. Vitousek PM, Porder S, Houlton BZ, Chadwick OA (2010) Terrestrial phosphorus limitation: mechanisms, implications, and nitrogen-phosphorus interactions. Ecological Applications 20: 5–15.
30. Hill BH, Elonen CM, Seifert LR, May AA, Tarquinio E (2012) Microbial enzyme stoichiometry and nutrient limitation in US streams and rivers. Ecological Indicators 18: 540–551.
31. Hessen DO, Agren GI, Anderson TR, Elser JJ, De Ruiter PC (2004) Carbon, sequestration in ecosystems: The role of stoichiometry. Ecology 85: 1179–1192.
32. Marklein AR, Houlton BZ (2012) Nitrogen inputs accelerate phosphorus cycling rates across a wide variety of terrestrial ecosystems. New Phytologist 193: 696–704.
33. Shange RS, Ankumah RO, Ibekwe AM, Zabawa R, Dowd SE (2012) Distinct soil bacterial communities revealed under a diversely managed agroecosystem. Plos One 7.
34. Reganold JP, Andrews PK, Reeve JR, Carpenter-Boggs L, Schadt CW, et al. (2010) Fruit and soil quality of organic and conventional strawberry agroecosystems. Plos One 5.
35. Bell LW, Sparling B, Tenuta M, Entz MH (2012) Soil profile carbon and nutrient stocks under long-term conventional and organic crop and alfalfa-crop rotations and re-established grassland. Agriculture Ecosystems & Environment 158: 156–163.
36. Liu DY, Zhang RF, Wu HS, Xu DB, Tang Z, et al. (2011) Changes in biochemical and microbiological parameters during the period of rapid composting of dairy manure with rice chaff. Bioresource Technology 102: 9040–9049.
37. Shang QY, Yang XX, Gao CM, Wu PP, Liu JJ, et al. (2011) Net annual global warming potential and greenhouse gas intensity in Chinese double rice-cropping systems: a 3-year field measurement in long-term fertilizer experiments. Global Change Biology 17: 2196–2210.
38. Shrestha M, Shrestha PM, Frenzel P, Conrad R (2010) Effect of nitrogen fertilization on methane oxidation, abundance, community structure, and gene expression of methanotrophs in the rice rhizosphere. Isme Journal 4: 1545–1556.
39. Ma K, Lu YH (2011) Regulation of microbial methane production and oxidation by intermittent drainage in rice field soil. Fems Microbiology Ecology 75: 446–456.
40. Adviento-Borbe MAA, Haddix ML, Binder DL, Walters DT, Dobermann A (2007) Soil greenhouse gas fluxes and global warming potential in four high-yielding maize systems. Global Change Biology 13: 1972–1988.
41. Zhang ZJ, Yao JX, Wang ZD, Xu X, Lin XY, et al. (2011) Improving water management practices to reduce nutrient export from rice paddy fields. Environmental Technology 32: 197–209.
42. Wang K, Zhang ZJ, Zhu YM, Wang GH, Shi DC, et al. (2001) Surface water phosphorus dynamics in rice fields receiving fertiliser and manure phosphorus. Chemosphere 42: 209–214.
43. Lee CH, Do Park K, Jung KY, Ali MA, Lee D, et al. (2010) Effect of Chinese milk vetch (Astragalus sinicus L.) as a green manure on rice productivity and methane emission in paddy soil. Agriculture Ecosystems & Environment 138: 343–347.
44. Wang DY, Zhand XF, Zhou CN, Zheng GS, Zhang GX, et al. (2010) Grain yield difference investigation and reasonable planting density analysis of rice production in Zhejing Province. Acta Agricuhurae Zhejiangens 22: 330–336.
45. Sander T, Gerke HH, Rogasik H (2008) Assessment of Chinese paddy-soil structure using X-ray computed tomography. Geoderma 145: 303–314.
46. Bao SD (2000) Agro-chemical analysis of soil. Beijing: China Agricultural Press. 78–290.
47. Wu J, Joergensen RG, Pommerening B, Chaussod R, Brookes PC (1990) Measurement of soil microbial biomass C by fumigation extraction - an automated procedure. Soil Biology & Biochemistry 22: 1167–1169.
48. Brookes PC, Powlson DS, Jenkinson DS (1982) Measurement of microbial biomass phosphorus in soil. Soil Biology & Biochemistry 14: 319–329.
49. Saiya-Cork KR, Sinsabaugh RL, Zak DR (2002) The effects of long term nitrogen deposition on extracellular enzyme activity in an Acer saccharum forest soil. Soil Biology & Biochemistry 34: 1309–1315.
50. German DP, Weintraub MN, Grandy AS, Lauber CL, Rinkes ZL, et al. (2011) Optimization of hydrolytic and oxidative enzyme methods for ecosystem studies. Soil Biology & Biochemistry 43: 1387–1397.

6

Microbial Detoxification of Bifenthrin by a Novel Yeast and Its Potential for Contaminated Soils Treatment

Shaohua Chen[9], Jianjun Luo[9], Meiying Hu*, Peng Geng, Yanbo Zhang

Laboratory of Insect Toxicology, Key Laboratory of Pesticide and Chemical Biology, Ministry of Education, South China Agricultural University, Guangzhou, People's Republic of China

Abstract

Bifenthrin is one the most widespread pollutants and has caused potential effect on aquatic life and human health, yet little is known about microbial degradation in contaminated regions. A novel yeast strain ZS-02, isolated from activated sludge and identified as *Candida pelliculosa* based on morphology, API test and 18S rDNA gene analysis, was found highly effective in degrading bifenthrin over a wide range of temperatures (20–40°C) and pH (5–9). On the basis of response surface methodology (RSM), the optimal degradation conditions were determined to be 32.3°C and pH 7.2. Under these conditions, the yeast completely metabolized bifenthrin (50 mg·L^{-1}) within 8 days. This strain utilized bifenthrin as the sole carbon source for growth as well as co-metabolized it in the presence of glucose, and tolerated concentrations as high as 600 mg·L^{-1} with a q_{max}, K_s and K_i of 1.7015 day^{-1}, 86.2259 mg·L^{-1} and 187.2340 mg·L^{-1}, respectively. The yeast first degraded bifenthrin by hydrolysis of the carboxylester linkage to produce cyclopropanecarboxylic acid and 2-methyl-3-biphenylyl methanol. Subsequently, 2-methyl-3-biphenylyl methanol was further transformed by biphenyl cleavage to form 4-trifluoromethoxy phenol, 2-chloro-6-fluoro benzylalcohol, and 3,5-dimethoxy phenol, resulting in its detoxification. Eventually, no persistent accumulative product was detected by gas chromatopraphy-mass spectrometry (GC-MS) analysis. This is the first report of a novel pathway of degradation of bifenthrin by hydrolysis of ester linkage and cleavage of biphenyl in a microorganism. Furthermore, strain ZS-02 degraded a variety of pyrethroids including bifenthrin, cyfluthrin, deltamethrin, fenvalerate, cypermethrin, and fenpropathrin. In different contaminated soils introduced with strain ZS-02, 65–75% of the 50 mg·kg^{-1} bifenthrin was eliminated within 10 days, suggesting the yeast could be a promising candidate for remediation of environments affected by bifenthrin. Finally, this is the first described yeast capable of degrading bifenthrin.

Editor: Arnold Driessen, University of Groningen, The Netherlands

Funding: The authors gratefully acknowledge the grants from the National Natural Science Foundation (No. 30871660) and Project of Scientific Technological Planning of Guangdong Province (No. 2009B020310005), People's Republic of China. The funders had no role in study design, data collection and analysis, decision to publish, or preparation of the manuscript.

Competing Interests: The authors have declared that no competing interests exist.

* E-mail: humy@scau.edu.cn

[9] These authors contributed equally to this work.

Introduction

Synthetic pyrethroid insecticides are frequently and widely used worldwide. These kinds of pesticides are synthetic analogs of the naturally occurring toxin, derived from the flowers of Chrysanthemum cinerariaefolium [1]. They have been classified as type I or type II based on the toxic symptoms and the absence or presence of a cyano group at the carboxyl alpha position. Recently, with the restriction on the use of organophosphates and carbamates, pyrethroids have generally been regarded as the replacements [2]. At present, they are the dominant insecticides among retail sales to consumers [3], and their usage is expected to further increase [4,5]. Due to these uses, pyrethroids have been reported in nearly all sediment samples tested from urban creeks [2,3,6,7,8].

The accumulation and widespread use of pyrethroids in agriculture have increased the public concern on potential human health risks that may result from chronic dietary exposure to pyrethroid residues on food [9,10]. For instance, numerous studies have shown that they may have neurotoxicity [11,12], immunotoxicity [13,14], reproductive toxicity [5,15], and genotoxicity [16,17] on non-target organisms. In addition, some current-use pyrethroids have recently been listed as potential endocrine disruptors by the Environmental Protection Agency (EPA) of the USA [18]. All these findings indicated that pyrethroids might be potentially harmful to human health. It is thus critically essential to develop an efficient and cost-effective remediation strategy to eliminate pyrethroids from the environments and agricultural products.

Bifenthrin[2-methylbiphenyl-3-ylmethyl-(Z)-(1RS)-3-(2-chloro-3, 3,3-trifluoroprop-1-enyl)-2,2-dimethylcyclopropane carboxylate] is one of the most popular pyrethroids effective against a broad spectrum of insect pests of economically important crops [5,16]. It is also extensively used for the control of residential pests such as termites in urban areas [19]. Its half-life in soil is usually between 65 and 125 days, but can range from 2 weeks to over 1 year, depending on the soil type, moisture, pH, temperature, and other conditions [20–22]. The World Health Organization (WHO) has classified bifenthrin as Toxicity Class II moderately hazardous [23]. In urban sediments, bifenthrin contributed the most to the observed toxicity among pyrethroids [2,6,8]. Therefore, bifenthrin was the pyrethroid of greatest concern with regards to aquatic life toxicity and widespread occurrence in water and soil [2,8].

Bifenthrin released to surface waters or sediments is subjected to hydrolysis, photodecomposition, volatilization, and aerobic degradation by microorganisms [20]. Microbial degradation is considered to be the most significant process determining the fate of bifenthrin and other pyrethroids in nature [24], and has received increasing attention as an effective, cheap, and safe approach to clean up contaminated environments [25]. To date, some pyrethroid-degrading microorganisms such as *Bacillus cereus* [26], *Pseudomonas fluorescens* [27], *Micrococcus* sp. CPN 1 [28], *Sphingobium* sp. JZ-2 [29], *Serratia* sp. JCN13 [30], *Ochrobactrum tritici* pyd-1 [31], and *Cladosporium* sp. HU [32] have been isolated. However, there was little information available on bifenthrin-degrading microorganism.

In the present study, a novel yeast *Candida pelliculosa* ZS-02 capable of degrading bifenthrin was isolated and characterized. The objective of this study was to optimize its degradation conditions, investigate its degradation pathway and evaluate its potential in bioremediation of bifenthrin-contaminated soils. Finally, obtained information illustrated that the isolated strain might have potential for use in bioremediation of bifenthrin-contaminated environments.

Materials and Methods

Chemicals and media

Technical grade bifenthrin (98% purity), cyfluthrin (95% purity), deltamethrin (98% purity), fenvalerate (91.2% purity), cypermethrin (92.9% purity), and fenpropathrin (93% purity) used in this study were obtained from Zhongshan Aestar Fine Chemical Inc., Ltd, China. Chromatographic grade acetonitrile were purchased from Sigma-Aldrich, USA. All other chemicals and solvents were purchased from Merck, Germany.

The mineral salt medium (MSM) containing (grams per liter) $(NH_4)_2SO_4$, 2; $MgSO_4 \cdot 7H_2O$, 0.2; $CaCl_2 \cdot 2H_2O$, 0.01; $FeSO_4 \cdot 7H_2O$, 0.001, $Na_2HPO_4 \cdot 12H_2O$, 1.5; and KH_2PO_4, 1.5; and yeast peptone dextrose (YPD) medium containing (grams per liter) yeast extract, 10; peptone, 20; and dextrose (or glucose), 20 were used for the isolation and cultivation of bifenthrin-degrading yeast, respectively. The final pH was adjusted to 7.2. Both media were autoclaved for 20 min at 121°C separately.

Isolation and screening of bifenthrin-degrading microorganisms

Activated sludge samples were collected as the inoculum from an aerobic pyrethroid-manufacturing wastewater treatment system located in Zhongshan (Guangdong, China), which had produced pyrethroids for many years. 30-mL of activated sludge was transferred into 250-mL Erlenmeyer flasks containing 50 mL sterilized MSM enrichment medium. Bifenthrin dissolved in acetone solution was added to a final concentration of 50 $mg \cdot L^{-1}$ as the sole carbon source. The enrichment culture was incubated for 7 days at 30±1°C with shaking at 150 rpm. A 5-mL from each enrichment culture was transferred into 50 mL of fresh enrichment medium containing 100 $mg \cdot L^{-1}$ of bifenthrin and incubated for another 7 days. Three additional successive transfers were made into media containing 200, 400, and 600 $mg \cdot L^{-1}$ of bifenthrin. The final cultures were serially diluted and spread on YPD agar plates. The plates were incubated for 5 days at 30°C, and colonies were picked and purified by re-streaking 3 times as described by Chen et al. [32,33]. The abilities of isolates to degrade bifenthrin were determined by high performance liquid chromatography (HPLC) (Agilent, USA) according to Chen et al. [34].

Characterization and identification of the bifenthrin-degrading isolates

One bifenthrin-degrading isolate that showed highest degradation efficiency was selected for further study. The isolate was grown on YPD agar plates at 30°C for 5 days and its morphology was investigated with a light microscope (Olympus, Japan) and scanning electron microscope (XL-30 ESEM, Philips Optoelectronics Co., Ltd, Holland). Colony morphology was observed on YPD agar plates incubated at 30°C at 1, 3, 5, and 7 days according to Barnett et al. [35]. The isolate was also subjected to sugar fermentation pattern analysis using API 20C AUX system (bioMérieux, France) according to the instructions of the manufacturer.

The isolate was confirmed by 18S rDNA sequence analysis. Total genomic DNA was prepared according to standard methods [36]. The 18S rDNA gene was amplified with the yeast universal primers EF4 (5′–GGAAGGGRTGTATTTATTAG–3′) and EF3 (5′–TCCTAAATGACCAAGTTTG–3′) [35]. Amplification was carried out in 50 μL reaction mixture containing 5 μL of 10×Ex *Tap* reaction buffer, 1 μL of 2.5 $mmol \cdot L^{-1}$ dNTP, 1 μL of 10 $\mu mol \cdot L^{-1}$ each primer, 1 μL of genomic DNA, 0.5 μL of 5 $U \cdot \mu L^{-1}$ Ex *Tap* DNA polymerase and 40.5 μL of ultrapure water. Reaction conditions consisted of initial denaturation at 94°C for 5 min, followed by 35 cycles of denaturation at 94°C for 1 min, annealing at 48°C for 1 min, and extension at 72°C for 2 min, with the last cycle followed by a ten-minute extension at 72°C. Polymerase chain reaction (PCR) product containing the amplified 18S rDNA gene fragment was purified with QIAquick Gel Extraction Kit spin column (Guangzhou Heda Technology Co. Ltd., China), ligated to the linear vector pMD20-T (TaKaRa Biotechnology Co. Ltd., China), and transformed into competent *Escherichia coli* DH5α cells. Positive clones were screened and sent to Shanghai Invitrogen Technology Co. Ltd., China, for sequencing. The 18S rDNA gene sequences with 1,452 bp were deposited at the GenBank under the accession number JN700989. The resulting sequence was compared with the genes available in the GenBank nucleotide library by a BLAST search through the National Center for Biotechnology Information (NCBI) internet site. Multiple alignments of 18S rDNA were carried out using CLUSTALX 1.8.1 software, and phylogenesis was analyzed using MEGA 4.0 software. An unrooted tree was built using the neighbor-joining method [37].

Inoculum preparation

To prepare the inoculum, the pure culture, obtained from individual colonies, was grown in YPD medium for 5 days, harvested by centrifugation at 4000×*g* for 5 min, washed twice with 0.9% sterile normal saline and re-suspended in MSM to set an OD_{600} of 0.3 by a spectrophotometer (Shimadzu, Japan). One percent of this suspension (1.0×10^7 $CFU \cdot mL^{-1}$) was used as the inoculum for studies.

Optimization of the bifenthrin-degrading conditions

The preliminary study indicated that temperature, pH and inoculum were significant variables for the degradation of bifenthrin by the yeast. In single-factor experiments, we determined the optimal ranges of the three factors, temperature: 20–40°C, pH: 5–9, and inoculum: $OD_{600} = 0.1–0.5$. These specified ranges were used as the reference levels. Furthermore, response surface methodology (RSM) based on the Box-Behnken design was used to optimize the main and interactive effects of the important parameters which significantly influenced the bifenthrin degradation by the yeast [38]. A three-variable Box-Behnken design consisting of 15 experimental runs with three replicates at the center point was generated by statistic analysis system (SAS) software (Version 9.0). The symbols and levels of three independent variables are shown in Table 1. The dependent variable was the degradation of 50 $mg{\cdot}L^{-1}$ bifenthrin in MSM by the yeast after 5 days of culture. Experiment was conducted according to Randomized block design and data were analyzed by response surface regression procedure of the SAS software to fit the following quadratic polynomial equation (Eq.(1)).

$$Y_i = b_0 + \Sigma b_i X_i + \Sigma b_{ij} X_i X_j + \Sigma b_{ii} X_i^2 \quad (1)$$

where Y_i is the predicted response, X_i and X_j are variables, b_0 is the constant, b_i is the linear coefficient, b_{ij} is the interaction coefficient, and b_{ii} is the quadratic coefficient.

Biodegradation of bifenthrin in liquid medium

Growth experiments with bifenthrin as the sole carbon source were performed in 250-mL Erlenmeyer flasks containing 50 mL sterile MSM with 50 $mg{\cdot}L^{-1}$ bifenthrin. Co-metabolism experiments were carried out in the same medium supplemented with 1% (*w/v*) glucose [39]. The seed suspension was aseptically inoculated into the MSM and incubated for 7 days at pH 7.2 at 32°C with shaking at 150 rpm. Each treatment was set in triplicate with non-inoculated samples as control. Samples (30-mL) were withdrawn periodically from the cultures to examine growth by recording the optical density (OD) value at 600 nm using spectrophotometer and to measure the bifenthrin concentration by HPLC as described previously [34,40].

Table 1. Box-Behnken experimental design and the response of dependent variable for bifenthrin degradation.

				Response
Run	X_1	X_2	X_3	**Bifenthrin degradation (%)** Y_{ZS-02}
1	−1	−1	0	55.5±1.5o
2	−1	1	0	65.2±0.6n
3	1	−1	0	67.6±0.3m
4	1	1	0	74.1±0.8i
5	0	−1	−1	72.3±0.5j
6	0	−1	1	75.5±0.4g
7	0	1	−1	75.2±0.4h
8	0	1	1	77.8±0.2f
9	−1	0	−1	68.4±1.2l
10	1	0	−1	79.5±0.4e
11	−1	0	1	70.6±0.5k
12	1	0	1	80.3±0.5d
13	0	0	0	86.5±0.4a
14	0	0	0	85.3±0.6b
15	0	0	0	83.1±1.0c

X_1: temperature, −1 (20°C), 0 (30°C), +1 (40°C); X_2: pH, −1 (5), 0 (7), +1 (9); X_3: inoculum, −1 ($OD_{600}=0.1$), 0 ($OD_{600}=0.3$), +1 ($OD_{600}=0.5$). The data presented are means of three replicates with standard deviation, which is within 5% of the mean. Different letters indicate significant differences ($p<0.05$, LSD test).

Biodegradation experiments by the yeast with different concentrations of bifenthrin (100–600 $mg{\cdot}L^{-1}$) and with various pyrethroids including bifenthrin, cyfluthrin, deltamethrin, fenvalerate, cypermethrin, and fenpropathrin (50 $mg{\cdot}L^{-1}$) were carried out in MSM supplemented with 1% glucose for 5 days at pH 7.2 at 32°C.

Identification of bifenthrin metabolites

The metabolites of bifenthrin in cell-free filtrates of the yeast cultures grown in MSM containing 50 $mg{\cdot}L^{-1}$ of bifenthrin were detected by gas chromatopraphy-mass spectrometry (GC-MS) (Agilent, USA). The cell-free filtrates were collected at 2, 4, 6, and 8 days, respectively. After acidification to pH 2 with 2 M HCl, the cultures were extracted with ethyl acetate and supernatant was dehydrated, dried and re-dissolved in methanol [28]. After filtration with 0.45 μm membrance (Millipore, USA), the samples were analyzed by GC-MS according to Zhang et al. [40].

Biodegradation of bifenthrin in soil

Soil samples were taken from a depth of 5–20 cm over a 25 m^2 area at grass-covered field, sieved to 2 mm and stored at 4°C in the dark for 14 days, before use [24]. The physicochemical properties of the soil were (grams per kilogram of dry weight): organic matter, 10.5; total N, 0.5; total P, 0.4; total K, 18.2; and pH, 6.9. The soil has a sandy loam texture (sand 65.0%, silt 28.0%, clay 7.0%). The soil intended for our bioremediation studies has not been used for agricultural purposes during the past 5 years, located in Guangzhou, Southern China. To investigate the removal potential of bifenthrin by the yeast in soils, 200 g of soil were placed in 500 mL-Erlenmeyer flask and their water contents were adjusted to 40% of water-holding capacity [39]. Soil moisture content was maintained at a constant level throughout the experiment by addition of distilled water when necessary. Bifenthrin was added to a final concentration of 50 $mg{\cdot}kg^{-1}$ of soil in acetone solution. After mixing and solvent evaporation the microbial suspensions were inoculated into soils (in triplicate) by drip irrigation to give the final concentration of 1.0×10^7 $CFU{\cdot}g^{-1}$ of soil and incubated at 32°C. The inoculum was thoroughly mixed under sterile conditions. In addition, sterilized soil (autoclaving at 121°C for 60 min) was also used to compare the removal efficiency of bifenthrin in soils [41,42]. Triplicate samples of non-inoculated soils were kept as control. After 2, 4, 6, 8, and 10 days, soil samples (20 g) were removed for analysis to determine the residual concentration of bifenthrin. The extracting and analytical methods were analogous to those used in the liquid medium.

Chemical analysis

The pyrethroid residues were determined on an Agilent 1100 HPLC system equipped with a Hypersil ODS2 C_{18} reversed phase column (4.6 nm×250 mm, 5 μm) with array detection from 190–400 nm (total scan) based on retention time and peak area of the pure standard [29]. In brief, 30-mL of cell-free medium was extracted using 60 mL of acetone/petroleum ether (1:1, *v/v*) in an ultrasonic bath. After partitioning, the supernatants were passed through a 0.22 mm polytetrafluoroethylene (PTFE) membrane filter (Millipore, USA), and the filtrates were concentrated using rotary evaporator (Heidolph, Germany). Residues were dissolved in acetone and analyzed by HPLC. A mixture of acetonitrile and water (85:15, *v/v*) was used as the mobile phase at a flow rate of 1.0 $mL{\cdot}min^{-1}$. The injection volume was 10 μL.

The metabolites of bifenthrin were identified on an Agilent 6890N/5975 GC-MS system equipped with auto-sampler, an on-column, split/splitless capillary injection system, and with HP-

5MS capillary column (30.0 m×250 μm×0.25 μm) with array detection from 30–500 nm (total scan). The operating conditions were as follows: the column was held at 160°C for 5 min, ramped at 10°C·min^{-1} to 200°C (first ramp), held at 200°C for 1 min, ramped at 10°C·min^{-1} to 280°C (second ramp), and then held at at 280°C for 8 min [40]. The temperatures corresponding to transfer line and the ion trap were 280°C and 230°C, respectively, and the ionization energy was 70 eV. The injection volume was 1.0 μL with splitless sampling at 250°C. Helium was used as a carrier gas at a flow rate of 1.5 mL·min^{-1}.

Statistical analysis

Results were assessed by analysis of variance (ANOVA) and statistical analyses were performed on three replicates of data obtained from each treatment. The significance ($p<0.05$) of differences was treated statistically by one-, two-, or three-way ANOVA and evaluated by post hoc comparison of means using lowest significant differences (LSD) using SAS software packages.

Ethics Statement

No specific permits were required for the described field studies. No specific permissions were required for these locations. We confirm that the location is not privately-owned or protected in any way. We confirm that the field studies did not involve endangered or protected species.

Results

Isolation and identification of the bifenthrin-degrading strain ZS-02

Isolates were able to grow with bifenthrin as the sole carbon source identified one strain, designated ZS-02 and completely metabolized 50 mg·L^{-1} of bifenthrin within 8 days. This strain was an obligate aerobe (oxidase- and catalase-positive). Colonies were small, ivory, round, and with the entire margin when grown on YPD agar plates for 5 days. Cells were oval, and 3–4 μm in length and 2–3 μm in width (Fig. 1). It was positive in utilization of glucose, sucrose, trehalose, maltose, glycerol, xylose, raffinose, sorbierite, methyl-α-D-glucopyranoside, and cellobiose. It was negative in utilization of arabinose, adonitol, xylitol, galactose, inositol, 2-keto-D-gluconate, lactose, melezitose, and N-acetyl-D-glucosamine (Table 2).

PCR amplification of 18S rDNA gene from strain ZS-02, a single fragment of 1,452 bp, GenBank accession number JN700989, was obtained and completely sequenced. Phylogenetic analysis (Fig. 2) based on the 18S rDNA sequences revealed that strain ZS-02 belonged to *Candida* group and was closely related to *Candida* sp. BG02-6-9-4 (99%). The isolate was further classified as *Candida pelliculosa* by API 20C AUX system (bioMérieux, France), with very good identification (99.9%). Thus, strain ZS-02 was identified as *C. pelliculosa* based on morphological, physio-biochemical characteristics, API 20C AUX test, and 18S rDNA gene analysis. This is the first report of pyrethroid-degrading species from the genus *Candida*.

Optimization of the bifenthrin-degrading conditions by strain ZS-02

The Box-Behnken design was applied to analyze the main and interactive effects of important variables including temperature (X_1), pH (X_2), and inoculum (X_3) based on earlier single-factor experiments. The experimental design matrix and the response of dependent variable for bifenthrin degradation are presented in Table 1. Subsequently, the data from Table 1 were processed by response surface regression procedure of SAS software and results were fitted with the second-order polynomial model equation (Eq.(2)):

$$\begin{aligned} Y_{ZS-02} = & 84.967 + 5.225X_1 + 2.675X_2 + 1.1X_3 - 9.933X_1^2 \\ & - 0.8X_1X_2 - 0.35X_1X_3 - 9.433X_2^2 \\ & - 0.15X_2X_3 - 0.333X_3^2 \end{aligned} \quad (2)$$

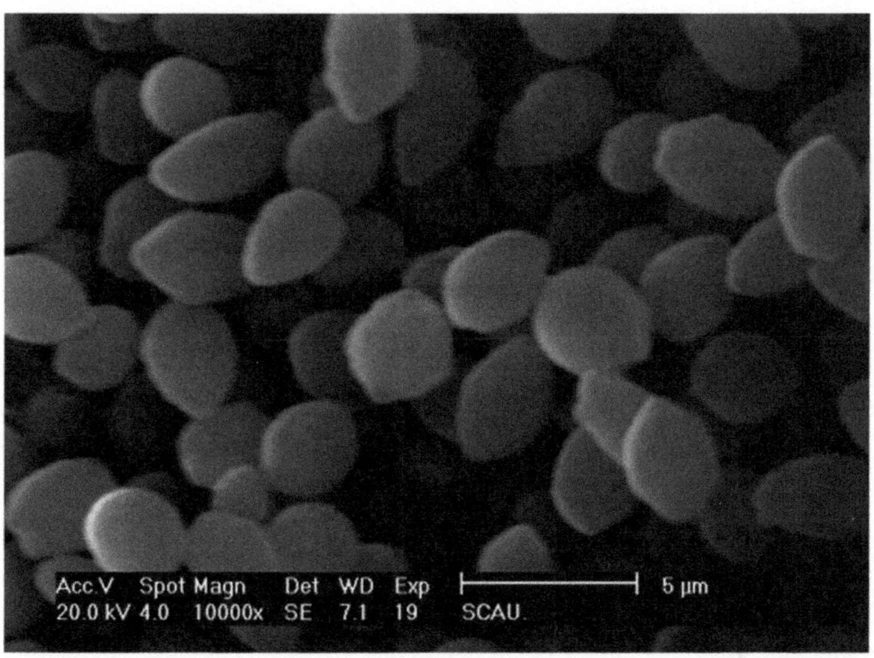

Figure 1. Morphological characteristics of strain ZS-02 under scanning electron microscope (10,000×).

Table 2. Morphological and physio-biochemical characteristics of *Candida pelliculosa* strain ZS-02.

Characteristics	Results	Characteristics	Results
Colonies	small, ivory, round, with entire margin	Cells	oval, 3–4 μm in length and 2–3 μm in width
Oxidase	+	Catalase	+
Glucose	+	Sucrose	+
Trehalose	+	Maltose	+
Glycerol	+	Xylose	+
Raffinose	+	Sorbierite	+
Methyl-α-D- glucopyranoside	+	N-acetyl-D- glucosamine	−
Cellobiose	+	Xylitol	−
Arabinose	−	Adonitol	−
Galactose	−	Inositol	−
Lactose	−	Melezitose	−
2-Keto-D-gluconate	−		

+, tested positive/utilized as substrate; −, tested negative/not utilized as substrate.

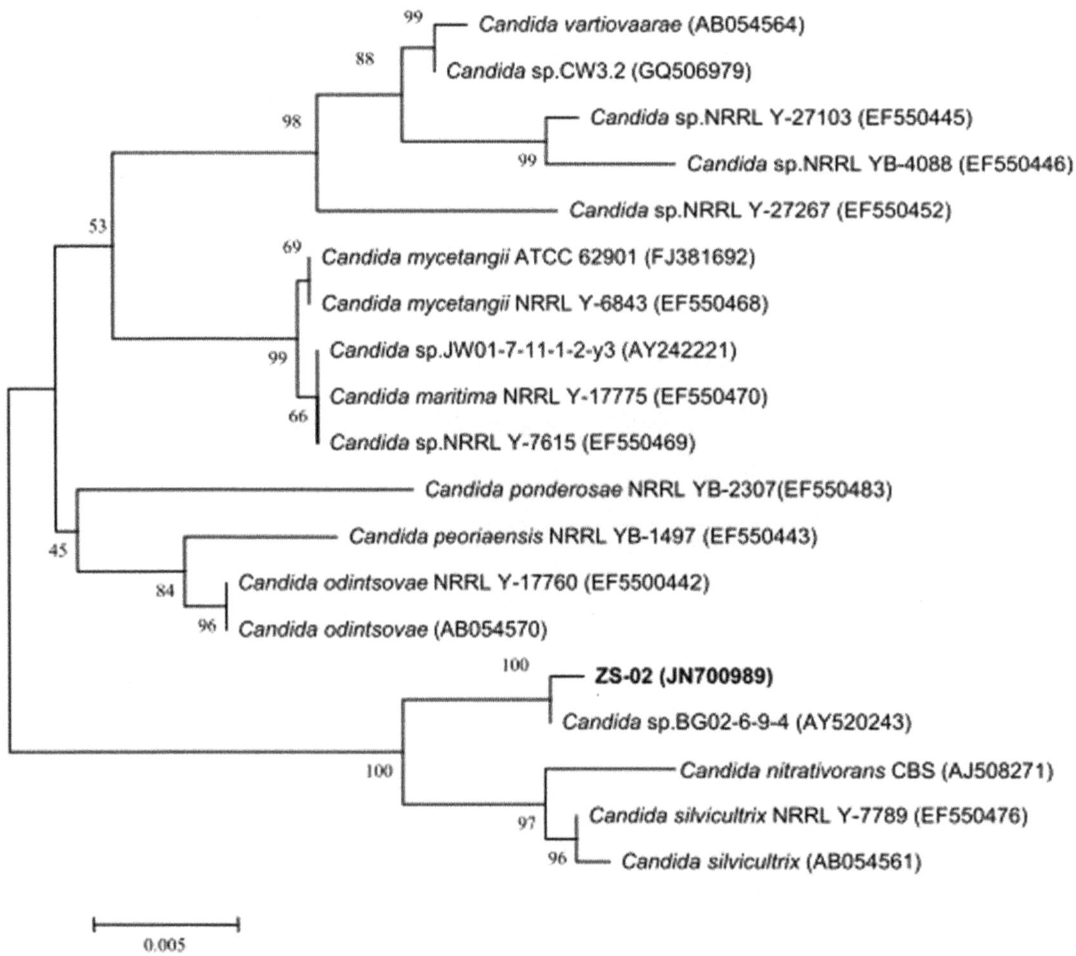

Figure 2. Phylogenetic tree based on the 18S rRNA sequence of strain ZS-02 and related strains. Numbers in parentheses represent the sequences accession number in GenBank. Numbers at the nodes indicate bootstrap values from the neighborhood-joining analysis of 1,000 resampled data sets. Bar represents sequence divergence.

where $Y_{ZS\text{-}02}$ is the predicted bifenthrin degradation (%) by strain ZS-02; X_1, X_2, and X_3 are the coded values for the temperature, pH, and inoculum, respectively.

An R^2 of 0.9770 indicated that approximately 98% of the variability in response could be covered by the model, demonstrating that predicted values of the model were in perfect agreement with the experimental values. The results of the regression parameter estimate revealed that linear and square terms of temperature (X_1) and pH (X_2) values showed significant effects ($p<0.05$) on the bifenthrin degradation by strain ZS-02, but the linear and square terms of inoculum (X_3) and the interaction terms were insignificant ($p>0.05$).

With the value of inoculum (the non-significant variable) fixed at $OD_{600\ nm} = 0.3$, the three-dimensional response surface was plotted to directly display the effects of temperature and pH on the bifenthrin degradation by strain ZS-02 by day 5 (Fig. 3). As shown in Fig. 3, the plot of bifenthrin biodegradation had a theoretical maximum value of 86.56% at the stationary point. At the stationary point, the optimum levels for the two variables of X_1 and X_2 were found to be 0.23 and 0.12 in terms of the coded units, that is, temperature 32.3°C and pH 7.2, respectively. So the optimal conditions for bifenthrin degradation by strain ZS-02 were determined to be 32.3°C, pH 7.2, and inoculum at $OD_{600\ nm} = 0.3$.

Effect of extra carbon source on the growth of strain ZS-02 and degradation of bifenthrin in mineral salt medium (MSM)

Growth of strain ZS-02 monitored by optical density ($OD_{600\ nm}$) and degradation kinetics of bifenthrin in MSM is shown in Fig. 4. After incubation for 7 days, 92.5% of the 50 $mg{\cdot}L^{-1}$ bifenthrin initially added to the medium was degraded by strain ZS-02, while in the same period disappearance rate of bifenthrin in MSM supplemented with 1% glucose was significantly higher ($p<0.05$) and reached 98.9%. Obviously, addition of carbon source enhanced the degradation of bifenthrin by this strain. No significant change in bifenthrin concentration was observed in

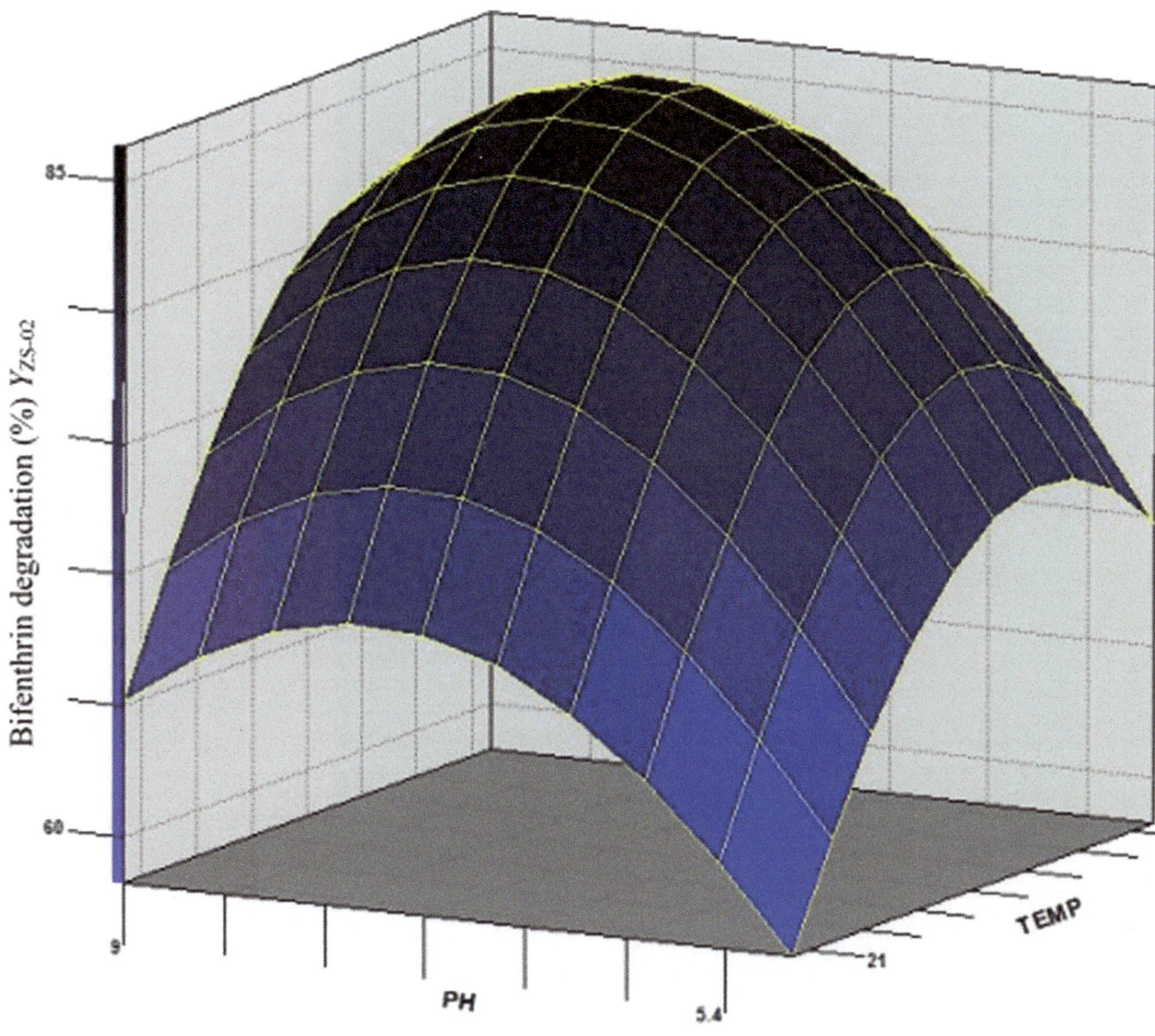

Figure 3. Response surface plot showing the effects of temperature and pH on bifenthrin degradation by strain ZS-02 with inoculum at $OD_{600} = 0.3$.

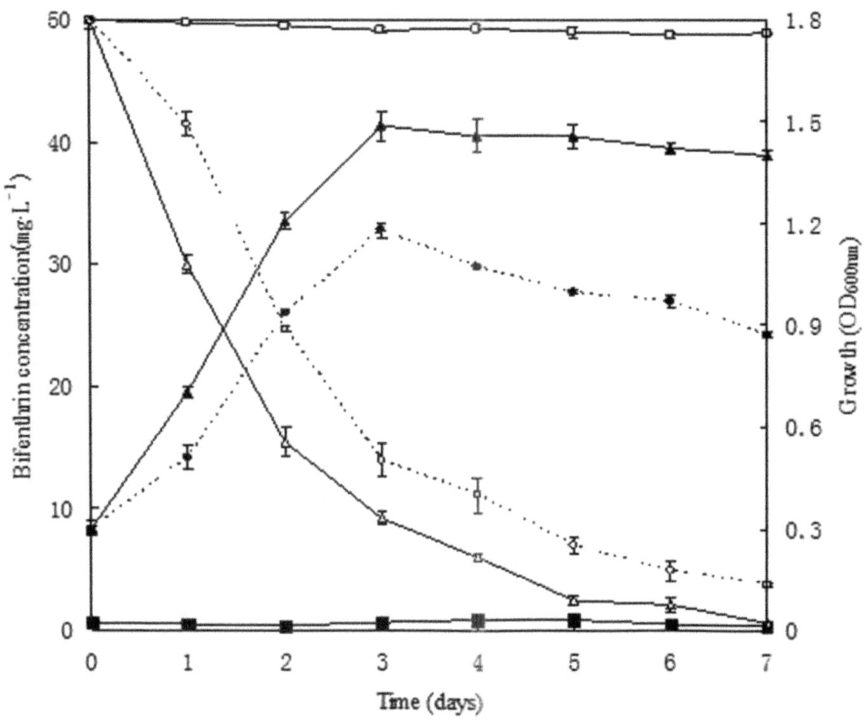

Figure 4. Cell growth of strain ZS-02 and degradation kinetics of bifenthrin during biodegradation studies. ○, degradation kinetics in MSM supplemented with bifenthrin as the sole carbon source; □, degradation kinetics in MSM supplemented with 1% glucose as an additional source of carbon; □, non-inoculated control (degradation); ●, cell growth in MSM supplemented with bifenthrin as the sole carbon source; ▲, cell growth in MSM supplemented with 1% glucose as an additional source of carbon; ■; non-inoculated control (growth). Values are the means of three replicates with standard deviation.

non-inoculated cultures. Bifenthrin removal was associated with a concomitant increase of cell density. Similar to degradation, growth of strain ZS-02 was strongly stimulated in the presence of glucose, and was the most effective within 3 days of incubation whereas $OD_{600\ nm}$ of the culture was significantly increased from 0.3 to 1.5. In contrast, the $OD_{600\ nm}$ of the culture without additional of carbon source was significantly lower ($p<0.05$) and ranged from 0.3 to 1.0 after 3 days of incubation.

Biodegradation of bifenthrin with different initial concentrations

Strain ZS-02 grew on bifenthrin up to the concentration, as high as 600 mg·L^{-1}. As shown in Fig. 5a, the lag phase was extended at higher bifenthrin concentration. Strain ZS-02 completely degraded bifenthrin at concentration of 100 mg·L^{-1} within 5 days. At concentrations of 200, 300, and 400 mg·L^{-1}, the degradation rates reached 97.1%, 95.8%, and 91.3% after 5 days of incubation, respectively. However, only 87.6% and 81.4% degradation were achieved at higher initial concentrations of 500 and 600 mg·L^{-1}, respectively.

The decrease in the specific bifenthrin degradation rate with an increase in the initial bifenthrin concentration implied that bifenthrin acts as an inhibitor to strain ZS-02. Therefore, the substrate inhibition model (Eq.(3)) adapted from Wang et al. [43] was used to explain the degradation kinetics of bifenthrin by strain ZS-02.

$$q=\frac{q_{\max}S}{S+K_s+(S^2/K_i)} \tag{3}$$

where $q_{\max}$ is the maximum specific bifenthrin degradation rate (day^{-1}), K_i is the substrate inhibition constant (mg·L^{-1}), K_s is the half-saturation constant (mg·L^{-1}), S is the substrate concentration (mg·L^{-1}), and S_m is a critical inhibitor concentration of the substrate which decreases degradation.

The kinetic parameters of strain ZS-02 estimated from non-linear regression using matrix laboratory (MATLAB) software (Version 7.8) were $q_{max}=1.7015$ day^{-1} and $K_s=86.2259$ mg·L^{-1}. The inhibitory effect of bifenthrin was considered to occur in a linear manner at $K_i=187.2340$ mg·L^{-1}. The S_m was determined to be 127.06 mg·L^{-1}. The relationship between q and S is shown in Fig. 5b. The value of R^2 was 0.9516 demonstrating that the experimental data were well correlated with the model. As indicated in Fig. 5b, when S were lower than 127.06 mg·L^{-1}, q gradually increased. At higher concentrations, inhibition by bifenthrin became prominent.

Biodegradation of various pyrethroids by strain ZS-02

The abilities of strain ZS-02 to degrade different pyrethroids are shown in Fig. 6. The strain was capable of metabolizing all of the pyrethroids tested. Bifenthrin was the most preferred substrate, with a degradation rate of 95.1% within 5 days of incubation. Cyfluthrin, deltamethrin, fenvalerate, cypermethrin, and fenpropathrin were degraded slower than bifenthrin, with degradation rates of 94.8%, 93.4%, 92.6%, 87.7%, and 51.3%, respectively.

Identification of bifenthrin metabolites

To illuminate the pathway of bifenthrin degradation by strain ZS-02, bifenthrin metabolic products in cell-free filtrates were extracted and confirmed by GC-MS. The degradation products

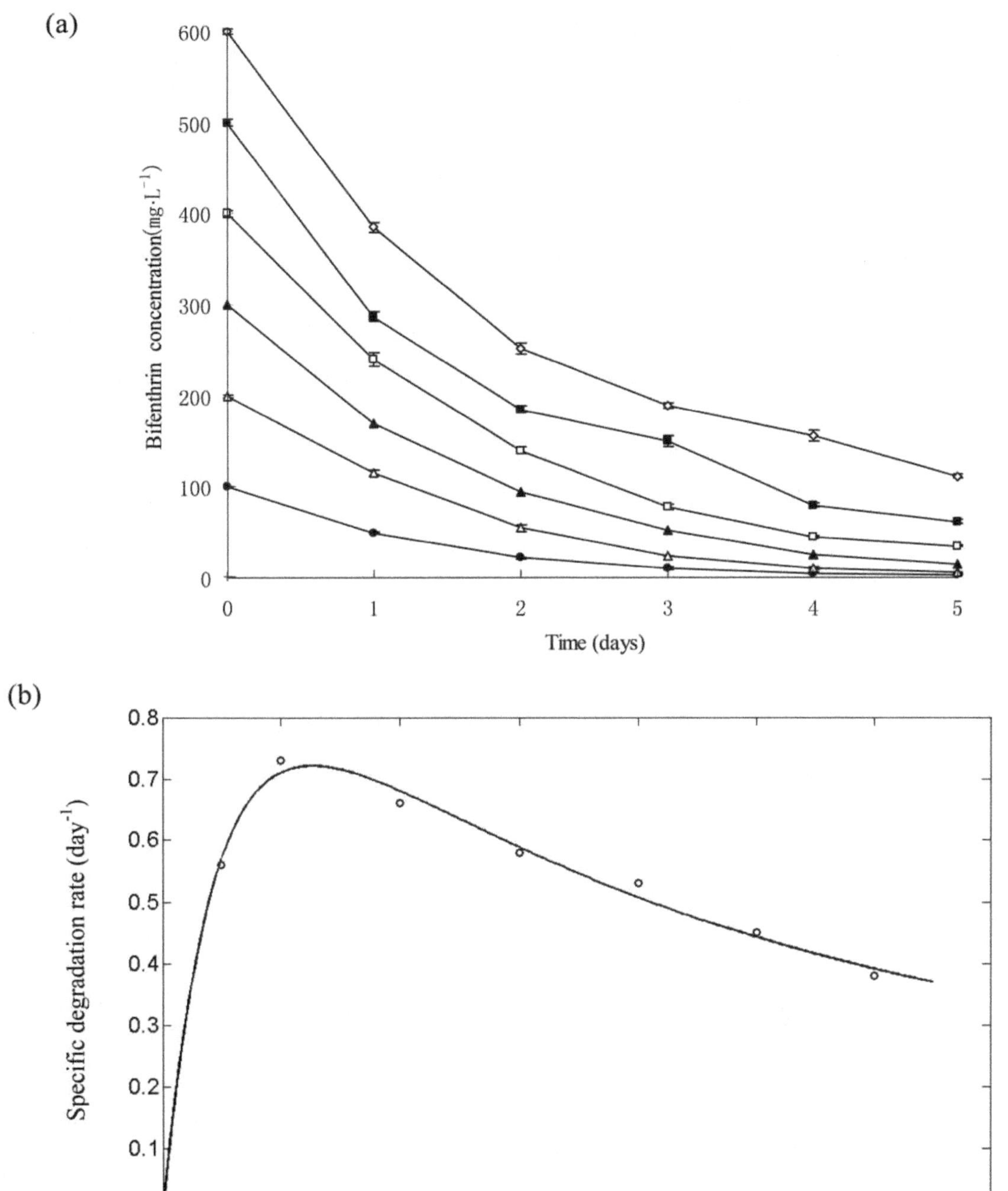

Figure 5. Biodegradation of bifenthrin with different initial concentrations. (a) Degradation kinetics of bifenthrin at different initial concentrations by strain ZS-02. ●, 100 mg·L^{-1}; □, 200 mg·L^{-1}; ▲, 300 mg·L^{-1}; □, 400 mg·L^{-1}; ■, 500 mg·L^{-1}; □, 600 mg·L^{-1}. Values are the means of three replicates with standard deviation. (b) Relationship between specific degradation rate and initial bifenthrin concentration by strain ZS-02.

identified on the basis of mass spectrum were matched with authentic standard compounds from the National Institute of Standards and Technology (NIST, USA) library database.

The GC-MS analysis revealed the presence of five degradation products, as summarized in Table 3. The peak at retention time of 16.391 min corresponded to bifenthrin standard. This peak decreased over time and completely disappeared after 8 days. Bifenthrin [A] disappeared concomitantly with formation of five degradation products (Fig. S1). The five compounds corresponded with those of authentic cyclopropanecarboxylic acid [B], 2-methyl-3-biphenylyl methanol [C], 4-trifluoromethoxy phenol [D], 2-chloro-6-fluoro benzylalcohol [E], and 3,5-dimethoxy phenol [F]. The retention times of these compounds were 9.561, 9.508, 8.785, 11.327, and 12.674 min, respectively. However, the peaks

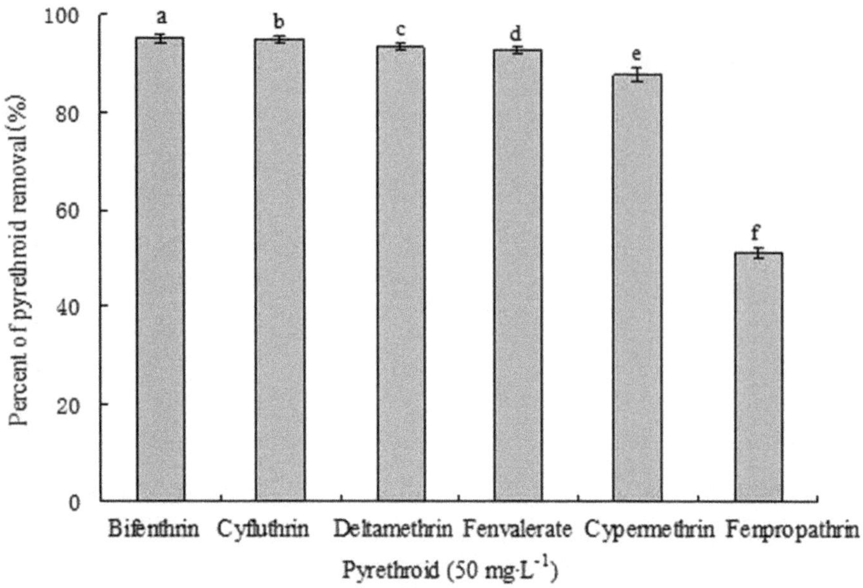

Figure 6. Degradation of various pyrethroids by strain ZS-02 by day 5. Values are the means of three replicates with standard deviation. Different letters indicate significant differences ($P<0.05$, LSD test).

corresponding to these compounds were transient and they disappeared finally. Eventually, no persistent accumulative product was detected.

On the basis of the metabolic products formed, we proposed the degradation pathway for bifenthrin by strain ZS-02 (Fig. 7). In other words, the parent bifenthrin [A] was first metabolized by hydrolysis of the carboxylester linkage to produce cyclopropanecarboxylic acid [B] and 2-methyl-3-biphenylyl methanol [C]. Subsequently, 2-methyl-3-biphenylyl methanol [C] was further transformed by biphenyl cleavage, resulting in formation of 4-trifluoromethoxy phenol [D], 2-chloro-6-fluoro benzylalcohol [E], and 3,5-dimethoxy phenol [F]. Finally, all these metabolites were not detected in the culture medium after 8 days of incubation. These results indicated that the added bifenthrin (50 mg·L^{-1}) was completely degraded by strain ZS-02 without any accumulative metabolites at the end of reaction.

Biodegradation of bifenthrin in soil

The potential of strain ZS-02 to eliminate bifenthrin in contaminated soils was investigated under controlled environmental conditions (temperature and soil humidity). The degradation kinetics are shown in Fig. 8. In the non-sterilized soil introduced with strain ZS-02 (1.0×10^7 CFU·g^{-1} of soil), bifenthrin degradation started rapidly as utilization was observed at the beginning of incubation, apparently there was no lag phase and 75.1% of the 50 mg·kg^{-1} bifenthrin initially added to the soil was removed after 10 days of treatment; whereas in control with indigenous soil microorganisms, this activity decreased only by 8.4%. In case of sterilized soil inoculated with strain ZS-02, 64.7% of added bifenthrin was eliminated within 10 days; while a non-inoculated control showed that the total bifenthrin content decreased only by 3.4%. Obviously, strain ZS-02 could efficiently eliminate bifenthrin in contaminated soils as compared to the non-inoculated control. In addition, the disappearance rate of bifethrin in the non-sterilized soil was higher ($p<0.05$) than that in the sterilized soil.

To confirm the effects on degradation of bifenthrin in different soils by strain ZS-02, the first-order kinetic model (Eq.(4)) adapted from Cycoń et al. [39] was used to describe the rate of total bifenthrin reduction.

$$C_t = C_0 \times e^{-kt} \tag{4}$$

where C_0 is the initial concentration of bifenthrin at time zero, C_t is the concentration of bifenthrin at time t, k and t are the degradation rate constant (day^{-1}) and degradation period in days, respectively.

The half-life ($t_{1/2}$) for bifenthrin in various soils was calculated using the general formula as expressed in Eq.(5).

$$t_{1/2} = \frac{\ln 2}{k} \tag{5}$$

where k is the rate constant (day^{-1}).

The kinetic parameters for all runs calculated from the equations are tabulated in Table 4. Degradation kinetics showed that the process corresponds to first-order kinetics with R^2 ranging from 0.9492 to 0.9860, demonstrating that the experimental data were well correlated with the model. The kinetic constant (k) for the non-sterilized soil with strain ZS-02 was 0.1411 day^{-1}, while

Table 3. Chromatographic properties of metabolites of bifenthrin degraded by strain ZS-02.

Code	Retention time (min)	*m*/*z*	Compounds
A	16.391	458	Bifenthrin
B	9.561	195	Cyclopropanecarboxylic acid
C	9.508	282	2-Methyl-3-biphenylyl methanol
D	8.785	430	4-Trifluoromethoxy phenol
E	11.327	282	2-Chloro-6-fluoro benzylalcohol
F	12.674	281	3,5-Dimethoxy phenol

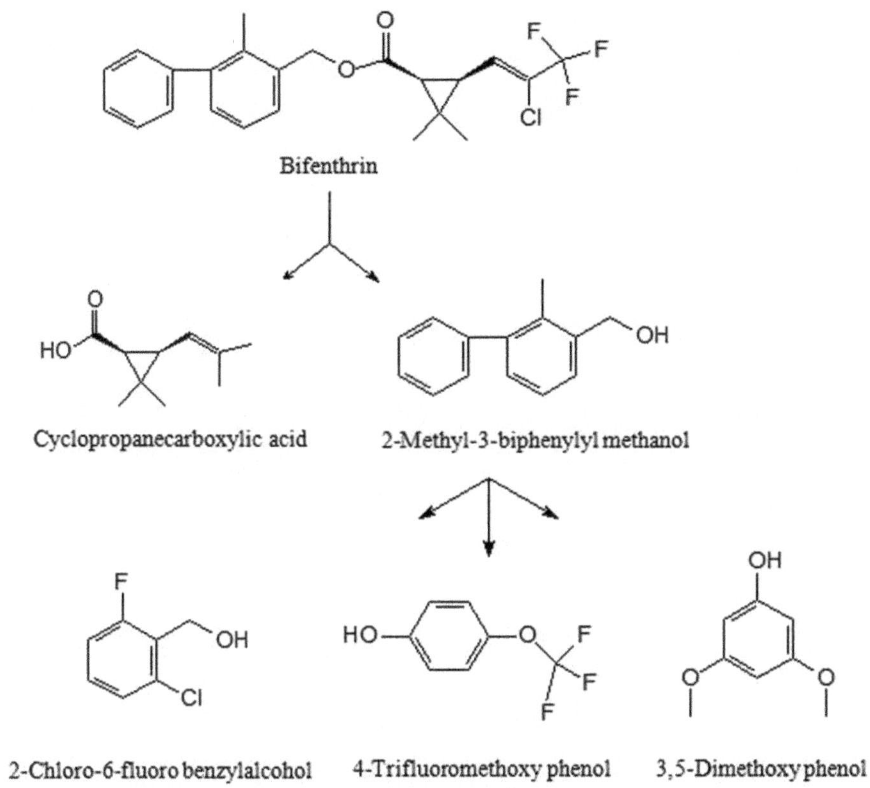

Figure 7. Proposed pathway for the degradation of bifenthrin by strain ZS-02.

the k for the sterilized soil was 0.1096 day^{-1}. In the sterilized soil, longer half-lives ($t_{1/2}$) of 6.3 days were observed, as compared to the non-sterilized soil with a $t_{1/2}$ of 4.9 days. These results revealed that the bifenthrin removal efficiency by strain ZS-02 could be enhanced by the indigenous soil microorganisms. Furthermore, the $t_{1/2}$ for bifenthrin in various soils introduced with strain ZS-02

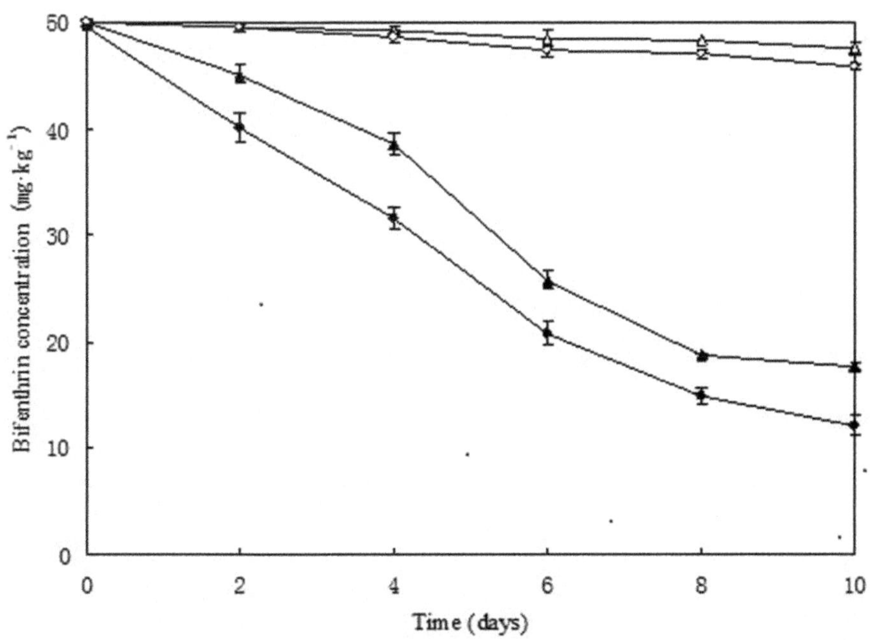

Figure 8. Degradation kinetics of bifenthrin in different contaminated soils by strain ZS-02. □, non-inoculated control (sterilized soil); ○, non-inoculated control with indigenous microbial community (non-sterilized soil); ▲, sterilized soil introduced with strain ZS-02; ●, non-sterilized soil introduced with strain ZS-02. Values are the means of three replicates with standard deviation.

Table 4. Kinetic parameters of degradation of bifenthrin in soils by strain ZS-02.

Soil treatments	Regression equation	k (day^{-1})	$t_{1/2}$ (days)	R^2
nSS+bifenthrin	$C_t = 50.1846e^{-0.0088t}$	0.0088	78.8	0.9799
SS+bifenthrin	$C_t = 50.0841e^{-0.0053t}$	0.0053	130.7	0.9806
nSS+bifenthrin+ZS-02	$C_t = 51.1033e^{-0.1411t}$	0.1411	4.9	0.9860
SS+bifenthrin+ZS-02	$C_t = 52.8446e^{-0.1096t}$	0.1096	6.3	0.9492

Each figure in the table represents the mean of three replicates. All figures were significantly different at $p<0.05$. nSS: non-sterilized soil with indigenous soil microbes; SS: sterilized soil without inoculum; +: introduced with; C_t: residual concentration of bifenthrin (mg·kg^{-1}); t: degradation period (days). R^2: correlation coefficient.

($t_{1/2} = 4.9$–6.3 days) were significantly shortened ($p<0.05$) as compared to that in the non-inoculated control soils ($t_{1/2} = 78.8$–130.7 days).

Discussion

Bifenthrin, a broad-spectrum type I synthetic pyrethroid insecticide, has been heavily used throughout the world for pest control both in agricultural and urban environment during the past two decades, resulting in widespread contamination [2,3,5,8,16]. Bioremediation involves the use of living microorganisms to degrade and detoxify pollutants, and is generally considered to be the most significant process determining the fate and behavior of xenobiotics in the environment [44]. Although several microorganisms capable of degrading pyrethroids have been isolated from diverse geographic locations [26–30], yet no report on the isolation of bifenthrin-degrading strain is available. A possible reason was attributed to that this chemical is persistent and refractory to microbial degradation [45]. Additionally, the existing papers lack the information involved in the biodegradation of pyrethroids by yeast. Yeast possesses the biochemical and ecological capacity to degrade environmental organic chemicals, either by chemical modification or by influencing chemical bioavailability [46]. However, the potential use for yeast in bioremediation of pyrethroids has not received the attention it deserves. This is the first report to our knowledge about bioremediation of pyrethroids by yeast.

In the present study, the screening of bifenthrin-degrading yeast by the method of enrichment procedure from active sludge affected by pyrethroids allowed us to select some potential isolates with a high survivability in the environment and maximal degrading activity towards bifenthrin. It was generally considered that the conditions for environmental microorganisms enrichment and screening are crucial in the selection of isolates not only with the desired degrading enzyme systems but having specific regulation of the degradation pathways as well [47]. One most active yeast, designated ZS-02, able to completely metabolize bifenthrin was identified as *Candida pelliculosa* based on morphological, physio-biochemical characteristics, API 20C AUX test, and 18S rDNA gene analysis. Previous research has shown that the potential pyrethroid-degrading microorganisms were mostly from genera *Bacillus* and *Pseudomonas* [26,27,38], while *C. pelliculosa* strain ZS-02 appeared to be a new species that was found highly effective in degrading bifenthrin and several other pyrethroids. Members of this genus are ubiquitous in the environment and some isolates have antagonistic effect on pathogen due to their production of bioactive secondary metabolites [48]. However, there is limited information concerning the use of these isolates to degrade and detoxify pollutants. The current results expand their use.

Previous studies have shown that temperature and pH are two important factors which strongly influence the degradation ability of microorganisms able to degrade xenobiotic compounds [39,49,50]. Our results indicated that strain ZS-02 engaged in degrading bifenthrin over a wide range of temperatures (20–40°C) and pH (5–9) particularly at low pH. This is an important feature of a microorganism to be employed for bioremediation of variable environments [49]. Furthermore, the optimal conditions for bifenthrin degradation by strain ZS-02 were determined to be 32.3°C and pH 7.2 via response surface methodology (RSM). RSM is an efficient statistical technique that has been successfully applied to optimize degradation conditions for various microorganisms [30,32,33,38,51]. It is a faster and less expensive method for gathering research result than the conventional practice of one factor optimization [51]. Moreover, RSM can distinguish interaction effects from the effects of individual factors through appropriate construction of the process model [30,38]. In our studies, a mathematical model was developed, and this model could be effectively used to predict and optimize the bifenthrin degradation conditions by strain ZS-02 within the limits of chosen factors.

C. pelliculosa strain ZS-02 utilized bifenthrin as the sole carbon for growth as well as co-metabolized it in the presence of extra carbon source, thus suggesting adaptation to oligotrophic and eutrophic conditions. However, this observation did not quite agree with Sørensen et al. [52] who reported isoproturon-degrading strain *Sphingomonas* sp. SRS2 was unable to grow on rich media, thus only adaptation to oligotrophic conditions. Furthermore, the growth of strain ZS-02 was stimulated and degradation of bifenthrin significantly enhanced in the presence of glucose in our studies (Fig. 4). This might be because of the cometabolism with additional carbon source lead to enhanced degradation of bifenthrin by strain ZS-02. Similar accelerated degradation in the presence of extra carbon source was observed by Anwar et al. [50] and Cycoń et al. [39] for different microorganisms degrading organophosphates.

It was noteworthy that this particular strain could tolerate and efficiently degrade bifenthrin up to the concentration, as high as 600 mg·L^{-1}. However, the specific bifenthrin degradation rate decreased with an increase in the initial bifenthrin concentration (Fig. 5b). These findings indicated that increased bifenthrin concentration had a marked effect on degradation performance of strain ZS-02, but did not lead to complete inhibition. These results proved that strain ZS-02 was responsible for bifenthrin degradation. Another important feature which is worth mentioning is that this strain was capable of degrading a variety of of pyrethroids including bifenthrin, cyfluthrin, deltamethrin, fenvalerate, cypermethrin, and fenpropathrin, demonstrating that the pyrethroid hydrolase involved in degradation may have broad-spectrum substrate specificity. Bifenthrin was degraded faster than other pyrethroids tested, suggesting that presence of the cyano group caused relative reduction in the hydrolysis rate, due to either stabilization of the ester bond or toxic effect of the substituent group. Another possible reason for degradation reduction could be attributed to the fact that substitution of the biphenyl for a 3-phenoxybenzyl hinders the pyrethroid hydrolase-substrate interaction in strain ZS-02. However, the result contrasts with previous findings of Wang et al. [31] who reported that replacement of the 3-phenoxybenzyl with a biphenyl decreased the pyrethroid hydrolysis rate in *Ochrobactrum tritici* strain pyd-1. Additionally, cyfluthrin, deltamethrin, fenvalerate, and cypermethrin were degraded at much faster rates than fenpropathrin, indicating that with fluoro, bromo, chloroben, or chloro group on the chrysanthemic acid greatly improved the degradation efficiencies. It might be because of that these substituent groups strongly promoted the hydrolase activity. Therefore, we could conclude that the degradation efficiencies of strain ZS-02 depend on the molecular structure of the pyrethroids. Similar result has been proposed by Wang et al. [31,45] and Guo et al. [29] for various microorganisms degrading pyrethroids.

It was generally suggested that ester hydrolysis by carboxylesterases was the main degradation pathway of pyrethroids in a multitude of species, from mammals, insects to microorganisms [28,30,38,45,53]. In our studies, the degradation pathway of bifenthrin by strain ZS-02 was investigated by metabolite identification, and the proposed pathway is shown in Fig. 7. It was evident from the results that strain ZS-02 first degraded bifenthrin by hydrolysis of ester linkage to produce cyclopropanecarboxylic acid and 2-methyl-3-biphenylyl methanol, leading to loss of its insecticide activity. Subsequently, 2-methyl-3-biphenylyl methanol was further transformed by biphenyl cleavage to form 4-trifluoromethoxy phenol, 2-chloro-6-fluoro benzylalcohol, and 3,5-dimethoxy phenol, resulting in its detoxification. Finally, all these degradation products were not detected by GC-MS after 8 days of incubation. These results indicated that the 50 $mg\cdot L^{-1}$ bifenthrin initially added to the medium was completely degraded by strain ZS-02 without any accumulative metabolites at the end of reaction. To the best of our knowledge, this is the first report of a novel pathway of degradation of bifenthrin by hydrolysis of ester linkage and cleavage of biphenyl in a microorganism. The bifenthrin metabolic pathway by ZS-02, however, appears to be similar to the initial step of cypermethrin degradation by *Micrococcus* sp. strain CPN 1, in which cypermthrin was converted to an acid and an alcohol by hydrolysis [28].

The potential of strain ZS-02 to eliminate bifenthrin in contaminated soils was evaluated to find out whether such strain is suitable for bioremediation purposes. In these experiments soils were inoculated with 1.0×10^7 $CFU\cdot g^{-1}$ of soil and this inoculum density was found to efficiently remove bifenthrin from the environment. As shown in Fig. 8, the introduced strain ZS-02 quickly adapted to the environment and rapidly degraded bifenthrin at the beginning of incubation without any apparent lag phase. After 10 days of incubation, 75.1% and 64.7% of the added bifenthrin (50 $mg\cdot kg^{-1}$) was eliminated in the non-sterilized and sterilized soils, respectively. The $t_{1/2}$ for bifenthrin was remarkably reduced by 63.9 and 124.4 days as compared to the control, implying that strain ZS-02 may have great potential to eliminate bifenthrin in contaminated soils. Furthermore, the removal efficiency of bifenthrin in the non-sterilized soil was significantly higher ($p<0.05$) than that in the sterilized soil, indicating that the indigenous soil microorganisms strongly promoted the ability of strain ZS-02 to eliminate bifenthrin. Enhancement of degradation could be explained by this reason that the introduced strain ZS-02 and soil microorganisms may have a synergistic effect on removal of pollutants. These results are consistent with previous observation [41,42].

In conclusion, the *C. pelliculosa* strain ZS-02 isolated in the present study appeared to be highly efficient in degrading bifenthrin in different contaminated soil and water resources. Degradation of bifenthrin occurred at 20–40°C and pH 5–9. This is an important feature of a microorganism to be employed for bioremediation of variable environments. The yeast utilized bifenthrin as the sole carbon for growth as well as co-metabolized it in the presence of extra carbon source, thus suggesting adaptation to oligotrophic and eutrophic conditions. Another important feature which is worth mentioning is that this strain was proficient in degrading a variety of pyrethroids, indicating that the yeast could be ideal for remediation of soil and water contaminated with pyrethroids. Moreover, the yeast harbors the metabolic pathway for the detoxification of bifenthrin. This is the first report of a novel pathway of degradation of bifenthrin by hydrolysis of ester linkage and cleavage of biphenyl in a microorganism, which we propose plays an important role in the bifenthrin biogeocycle. Finally, strain ZS-02 is the first described yeast capable of degrading bifenthrin and several other pyrethroids.

Supporting Information

Figure S1 GC-MS spectra of metabolites produced from bifenthrin by *Candida pelliculosa* strain strain ZS-02. (a) Bifenthrin; (b) Cyclopropanecarboxylic acid; (c) 2-Methyl-3-biphenylyl methanol; (d) 4-Trifluoromethoxy phenol; (e) 2-Chloro-6-fluoro benzylalcohol; (f) 3,5-Dimethoxy phenol.

Acknowledgments

The authors would like to thank Dr. Muhammad Rizwan-ul-haq for his helpful comments on the manuscript.

Author Contributions

Conceived and designed the experiments: SC MH. Performed the experiments: SC JL MH. Analyzed the data: SC PG YZ. Contributed reagents/materials/analysis tools: JL MH. Wrote the paper: SC MH.

References

1. Laffin B, Chavez M, Pine M (2010) The pyrethroid metabolites 3-phenoxybenzoic acid and 3-phenoxybenzyl alcohol do not exhibit estrogenic activity in the MCF-7 human breast carcinoma cell line or Sprague-Dawley rats. Toxicology 267: 39–44.
2. Hintzen EP, Lydy MJ, Belden JB (2009) Occurrence and potential toxicity of pyrethroids and other insecticides in bed sediments of urban streams in central Texas. Environ Pollut 157: 110–116.
3. Weston DP, Holmes RW, Lydy MJ (2009) Residential runoff as a source of pyrethroid pesticides to urban creeks. Environ Pollut 157: 287–294.
4. Wang LM, Liu WP, Yang CX, Pan ZY, Gan JY, et al. (2007) Enantioselectivity in estrogenic potential and uptake of bifenthrin. Environ Sci Technol 41: 6124–6128.
5. Liu J, Yang Y, Yang Y, Zhang Y, Liu WP (2011) Disrupting effects of bifenthrin on ovulatory gene expression and prostaglandin synthesis in rat ovarian granulosa cells. Toxicology 282: 47–55.
6. Holmes RW, Anderson BS, Phillips BM, Hunt JW, Crane DB, et al. (2008) Statewide investigation of the role of pyrethroid pesticides in sediment toxicity in California's urban waterways. Environ Sci Technol 42: 7003–7009.
7. Riederer AM, Hunter RE, Hayden SW, Ryan PB (2010) Pyrethroid and organophosphorus pesticides in composite diet samples from Atlanta, USA adults. Environ Sci Technol 44: 483–490.
8. Weston DP, Asbell AM, Hecht SA, Scholz NL, Lydy MJ (2011) Pyrethroid insecticides in urban salmon streams of the Pacific Northwest. Environ Pollut 159: 3051–3056.
9. Tewary DK, Kumar V, Ravindranath SD, Shanker A (2005) Dissipation behavior of bifenthrin residues in tea and its brew. Food Control 16: 231–237.
10. Mohapatra S, Ahuja AK, Sharma D (2007) Persistence of bifenthrin residues on mango (Mangifera indica) fruit. Pestic Res J 19: 110–112.
11. Shafer TJ, Meyer DA, Crofton KM (2005) Developmental neurotoxicity of pyrethroid insecticides: critical review and future research needs. Environ Health Persp 113: 123–136.
12. Wolansky MJ, Harrill JA (2008) Neurobehavioral toxicology of pyrethroid insecticides in adult animals: a critical review. Neurotoxicol Teratol 30: 55–78.
13. Dutta R, Das N (2011) Immunomodulation of serum complement (C3) and macrophages by synthetic pyrethroid fenvalerate: *In vitro* study. Toxicology 285: 126–132.

14. Jin YX, Zheng SS, Fu ZW (2011) Embryonic exposure to cypermethrin induces apoptosis and immunotoxicity in zebrafish (*Danio rerio*). Fish Shellfish Immunol 30: 1049–1054.
15. Perry MJ, Venners SA, Barr DB, Xu XP (2007) Environmental pyrethroid and organophosphorus insecticide exposures and sperm concentration. Reprod Toxicol 23: 113–118.
16. Wang C, Chen F, Zhang Q, Fang Z (2009) Chronic toxicity and cytotoxicity of synthetic pyrethroid insecticide *cis*-bifenthrin. J Environ Sci 21: 1710–1715.
17. Ansari AR, Rahman S, Kaur M, Anjum S, Raisuddin S (2011) *In vivo* cytogenetic and oxidative stress-inducing effects of cypermethrin in freshwater fish, *Channa punctata* Bloch. Ecotoxicol Environ Saf 74: 150–156.
18. US EPA (2009) Endocrine Disruptor Screening Program (EDSP). US EPA,Washington, DC.
19. Baskaran S, Kookana RS, Naidu R (1999) Degradation of bifenthrin, chlorpyrifos and imidacloprid in soil and bedding materials at termiticidal application rates. Pestic Sci 55: 1222–1228.
20. Laskowski DA (2002) Physical and chemical properties of pyrethroids. Rev Environ Contam T 174: 49–170.
21. Lee S, Gan JY, Kim JS, Kabashima JN, Crowley DE (2004) Microbial transformation of pyrethroid insecticides in aqueous and sediment phases. Environ Toxicol Chem 23: 1–6.
22. Mohapatra S, Ahuja AK (2009) Effect of moisture and soil type on the degradation of bifenthrin in soil. Pestic Res J 19: 110–112.
23. World Health Organization (WHO) (2009) The WHO Recommended Classification of Pesticides by Hazard and Guidelinesto Classification 2009, WHO, Geneva.
24. Fenlon AK, Jones CK, Semple TK (2011) The effect of soil:water ratios on the induction of isoproturon, cypermethrin and diazinon mineralisation. Chemosphere 82: 163–168.
25. Singh BK, Walker A (2006) Microbial degradation of organophosphorus compounds. FEMS Microbiol Rev 30: 428–471.
26. Maloeny SE, Maule A, Smith ARW (1993) Purification and preliminary characterization of permethrinase from a pyrethroid-transforming strain of *Bacillus cereus*. Appl Environ Microbiol 59: 2007–2013.
27. Grant RJ, Daniell TJ, Betts WB (2002) Isolation and identification of synthetic pyrethroid-degrading bacteria. J Appl Microbiol 92: 534–540.
28. Tallur PN, Megadi VB, Ninnekar HZ (2008) Biodegradation of cypermethrin by *Micrococcus* sp. strain CPN 1. Biodegradation 28: 77–82.
29. Guo P, Wang BZ, Hang BJ, Li L, Ali SW, et al. (2009) Pyrethroid- degrading *Sphingobium* sp. JZ-2 and the purification and characterization of a novel pyrethroid hydrolase. Int Biodeter Biodegr 63: 1107–1112.
30. Zhang C, Jia L, Wang SH, Qu J, Xu LL, et al. (2010) Biodegradation of beta-cypermethrin by two *Serratia* spp. with different cell surface hydrophobicity. Bioresoure Technol 101: 3423–3429.
31. Wang BZ, Ma Y, Zhou WY, Zheng JW, He J, et al. (2011) Biodegradation of synthetic pyrethroids by *Ochrobactrum tritici* strain pyd-1. World J Microbiol Biotechnol 27: 2315–2324.
32. Chen SH, Hu QB, Hu MY, Luo JJ, Weng QF, et al. (2011) Isolation and characterization of a fungus able to degrade pyrethroids and 3-phenoxybenzaldehyde. Bioresource Technol 102: 8110–8116.
33. Chen SH, Lai KP, Li YN, Hu MY, Zhang YB, et al. (2011) Biodegradation of deltamethrin and its hydrolysis product 3-phenoxybenzaldehyde by a newly isolated *Streptomyces aureus* strain HP-S-01. Appl Microbiol Biotechnol 90: 1471–1483.
34. Chen SH, Hu MY, Liu JJ, Zhong GH, Yang L, et al. (2011) Biodegradation of beta-cypermethrin and 3-phenoxybenzoic acid by a novel *Ochrobactrum lupini* DG-S-01. J Hazard Mater 187: 433–440.
35. Barnett JA, Payne RW, Yarrow D (2000) Yeasts: Characteristics and Identification, 3rd ed. Cambridge University Press, Cambridge, UK.
36. Sambrook J, Russell DW (2001) Molecular Cloning: A Laboratory Manual, 3rd ed. Cold Spring Harbor Laboratory, NY.
37. Tamura K, Dudley J, Nei M (2007) MEGA4: molecular evolutionary genetics analysis (MEGA) software version 4.0. Mol Biol Evol 24: 1596–1599.
38. Zhang C, Wang SH, Yan YC (2011) Isomerization and biodegradation of beta-cypermethrin by *Pseudomonas aeruginosa* CH7 with biosurfactant production. Bioresource Technol 102: 7139–7146.
39. Cycoń M, Wojcik M, Piotrowska-Seget Z (2009) Biodegradation of the organophosphorus insecticide diazinon by *Serratia* sp. and *Pseudomonas* sp. and their use in bioremediation of contaminated soil. Chemosphere 76: 494–501.
40. Zhang SB, Yin LB, Liu Y, Zhang DY, Luo XW, et al. (2011) Cometabolic biotransformation of fenpropathrin by *Clostridium* species strain ZP3. Biodegradation 22: 869–875.
41. Chen SH, Yang L, Hu MY, Liu JJ (2011) Biodegradation of fenvalerate and 3-phenoxybenzoic acid by a novel *Stenotrophomonas* sp. strain ZS-S-01 and its use in bioremediation of contaminated soils. Appl Microbiol Biotechnol 90: 755–767.
42. Chen SH, Geng P, Xiao Y, Hu MY (2011) Bioremediation of β-cypermethrin and 3-phenoxybenzaldehyde contaminated soils using Streptomyces aureus HP-S-01. Appl Microbiol Biotechnol, in press.
43. Wang YQ, Zhang JS, Zhou JT, Zhang ZP (2009) Biodegradation of 4-aminobenzenesulfonate by a novel *Pannonibacter* sp. W1 isolated from activated sludge. J Hazard Mater 169: 1163–1167.
44. Singh BK (2009) Organophosphorus-degrading bacteria: ecology and industrial applications. Nature Rev Microbiol 7: 156–163.
45. Wang BZ, Guo P, Hang BJ, Li L, He J, et al. (2009) Cloning of a novel pyrethroid-hadrolyzing carboxylesterase gene from *Sphingobium* sp. strain JZ-1 and characteization of the gene product. Appl Environ Microbiol 75: 5496–5500.
46. Harms H, Schlosser D, Wick LY (2011) Untapped potential: exploiting fungi in bioremediation of hazardous chemicals. Nature Rev Microbiol 9: 177–192.
47. Xu GM, Zheng W, Li YY, Wang SH, Zhang JS, et al. (2008) Biodegradation of chlorpyrifos and 3,5,6-trichloro-2-pyridinol by a newly isolated *Paracoccus* sp. strain TRP. Int Biodeter Biodegr 62: 51–56.
48. Bello GD, Monaco C, Rollan MC, Lampugnani G, Arteta N, et al. (2008) Biocontrol of postharvest grey mould on tomato by yeasts. J Phytopathology 156: 257–263.
49. Singh BK, Walker A, Denis J, Wight DJ (2006) Bioremedial potential of fenamiphos and chlorpyrifos degrading isolates: influence of different environmental conditions. Soil Biol Biochem 38: 2682–2693.
50. Anwar S, Liaquat F, Khan QM, Khalid ZM, Iqbal S (2009) Biodegradation of chlorpyrifos and its hydrolysis product 3,5,6-trichloro-2-pyridinolby *Bacillus pumilus* strain C2A1. J Hazard Mater 168: 400–405.
51. Dritsa V, Rigas F, Doulia D, Avramides EJ, Hatzianestis I (2009) Optimization of culture conditions for the biodegradation of lindane by the polypore fungus *Ganoderma australe*. Water Air Soil Pollut 204: 19–27.
52. Sørensen SR, Ronen Z, Aamand J (2001) Isolation from agricultural soil and characterization of a *Sphingomonas* sp. able to mineralize the phenylurea herbicide isoproturon. Appl Environ Microbiol 67: 5403–5409.
53. Heidari R, Devonshire AL, Campbell BE, Dorrian SJ, Oakeshott JG, et al. (2005) Hydrolysis of pyrethroids by carboxylesterases from *Lucilia cuprina* and *Drosophila melanogaster* with active sites modified by *in vitro* mutagenesis. Insect Biochem Mol Biol 35: 597–609.

7

Ascomycota Members Dominate Fungal Communities during Straw Residue Decomposition in Arable Soil

Anzhou Ma[1], Xuliang Zhuang[1]*, Junmei Wu[1], Mengmeng Cui[1], Di Lv[2], Chunzhao Liu[3], Guoqiang Zhuang[1]*

1 Research Center for Eco-Environmental Sciences, Chinese Academy of Sciences, Beijing, P.R. China, **2** Insitute of Microbiology, Chinese Academy of Sciences, Beijing, P.R. China, **3** National Key Laboratory of Biochemical Engineering, Institute of Process Engineering, Chinese Academy of Sciences, Beijing, P.R. China

Abstract

This study investigated the development of fungal community composition in arable soil during the degradation of straw residue. We explored the short-term responses of the fungal community over 28 days of decomposition in soil using culture-independent polymerase chain reaction in combination with a clone library and denaturing gradient gel electrophoresis (DGGE). Fungal cellobiohydrolase I (*cbhI*) genes in the soil were also characterized, and their diversity suggested the existence of a different cellulose decomposer. The DGGE profiles based on fungal internal transcribed spacer analysis showed different successions of fungal populations during residue decomposition. Members of *Lecythophora* and *Sordariales* were dominant in the early succession, while *Hypocrea* and *Engyodontium* were better adapted in the late succession. The succession of fungal communities might be related to changes of residue quality during decomposition. Collectively, sequences assigned to Ascomycota members were dominant at different stages of the fungal succession during decomposition, revealing that they were key drivers responsible for residue degradation in the arable soil tested.

Editor: Hauke Smidt, Wageningen University, The Netherlands

Funding: This work was funded by National Natural Science Foundation of China (41001151, 41173089).The funders had no role in study design, data collection and analysis, decision to publish, or preparation of the manuscript.

Competing Interests: The authors have declared that no competing interests exist.

* E-mail: gqzhuang@rcees.ac.cn (GZ); xlzhuang@rcees.ac.cn (XZ)

Introduction

Large quantities of lignocellulosic wastes consisting mainly of industrial and agricultural residues are generated worldwide every year [1]. As it is the most abundant renewable resource on the earth, diverse methods for the effective utilization of ligninocellulose have been proposed [1,2]. In sustainable agricultural treatments, crop detritus is incorporated directly into fields, which can improve soil physicochemical properties, increase nutrient ability, and eliminate the harmful influence of agriculture waste to environments [3,4]. The decomposition of crop residues by soil microbial communities is a key step to release inorganic nutrients from plant residues as well as transform this material into soil organic matter. It is essential to elucidate the microbial communities associated with straw residue decomposition in arable soil and understand the decomposition processes.

A broad taxonomic range of microorganisms contributes to the degradation of plant residue in soil, and microbial activities are responsible for most of the turnover of cellulose for the recalcitrant structure of this biopolymer [5]. Fungi are capable of producing a wide range of extracellular enzymes, which make them effective in attacking the intermolecular bonds of the cellulose biopolymer [1]. In fact, soil fungal communities are essential to litter degradation in forest ecosystems, and they are considered the key decomposers [6–8]. Using a cultivation-based isolation method, Osono and colleagues suggested that the succession of decomposer fungal communities is required to complete the litter degradation process [9,10]. Recently, the use of molecular approaches for microbial community analysis provides a more comprehensive investigation of the complex litter decomposition process mediated by various fungal taxa [6]. Compared with forest ecosystems, lignin content in arable soil is relatively low in the chemical constituents of the crop residue input. The differences in the intrinsic quality of the substrate and the surrounding environment could have an influence on the fungal community involved in decomposition process [11–15]. Previous studies of fungal communities during litter decomposition have focused on activities of enzyme, microbial biomass, and phylogenetic groups in the soil based on laboratory and natural ecosystem experiments [16–18]. Some studies have found that members of Ascomycota and Basidiomycota represent the main soil fungal decomposers [19,20], while others have found that members of Basidiomycota are important and are more able to degrade lignocellulose organic matter [21,22]. Results of the general community characteristics have been mixed in early studies, and current knowledge regarding the active and functional fungal communities involved in decomposition is limited.

The application of functional genes can aid in the detection of some functional fungal groups. To this end, Edwards and colleagues recently used the cellubiohydrolase-encoding *cbhI* gene as a functional biomarker to analyze cellulolytic fungal populations, advancing our understanding of microorganisms involved in cellulose degradation [23]. Therefore, the identification of the functional *cbhI* genes of soil together with the analysis of phylogenetic information of fungal communities during straw residue decomposition is useful to gain underlying insights. The main objective of the present study was to investigate the diversity and dynamics of the fungal community throughout the degradation process. To characterize the succession of fungal decomposers, we studied short-term (28 days) responses of the fungal community via constructed microcosms using arable soil and rice straw residue. Using DNA-based denaturing gradient gel electrophoresis (DGGE) methods, we monitored the fungal community changes at different stages during the incubation. We also evaluated the metabolic activity of the fungi community in the soil we sampled through the measurement using Biolog microplates and detection of the cellobiohydrolase gene biomarker.

Materials and Methods

Soil and Preparation of Microcosm

The soil used for the microcosm incubation was taken from a field located in Taoyuan State Key Experimental Station for Ecological Agriculture, Hunan Province. Since 1990, a long-term experiment involving the input of residues into the soil as one organic management practice has been maintained in this field. No specific permits were required for the described field studies. The location is not privately owned, and the field studies did not involve endangered or protected species. Taoyuan State Key Experimental Station for Ecological Agriculture is one of the sites of the Chinese Ecosystem Research Network (CERN), which was implemented by the Chinese Academy of Sciences (CAS) to study environmental issues concerning China. In addition, studies of the C and N biocycle and soil science take place at Taoyuan State Key Experimental Station for Ecological Agriculture. These study fields can be used by the institutes of the CAS freely for research, and all data are shared and can be published. Our institute, Research Center for Eco-Environmental Sciences, is affiliated with CAS; thus, we can perform scientific research in these fields without specific permits. The soil samples were collected in November 2008 at a depth of 0–15 cm. Several small soil cores (approximately 5 cm in diameter) were collected randomly from each plot and mixed together as a representative sample. Each sample was placed in a sterile plastic bag, sealed, and placed on ice while transported to the laboratory. The soil was air-dried in a glass house and then filtered through 2-mm mesh to eliminate plant debris. The characteristics of the soil were as follows: pH, 4.9; total organic C, 23.2 g kg^{-1}; and total N, 2.3 g kg^{-1}.

The soil samples were moistened with sterile deionized water and preincubated at 20°C for two weeks in the dark to reach a biological steady state. The microcosms consisted of 100 g of soil in 500-ml glass universal bottles with screw caps. The cellulose decomposition was investigated by adding small pieces of straw residue (2 g kg^{-1} soil dry weight) to the soil. The microcosms were weighed regularly during the incubation, and water lost through evaporation was replaced through the aseptic addition of sterile deionized water. Soil amended with straw was incubated with duplicate microcosms at 20°C in the dark with weekly aeration. As controls, soil microcosms were incubated without the application of the straw. Subsamples of 10 g soil were sampled from the each treatment group on days 0, 7, 14, and 28, representing different stages of decomposition. The soil was preserved at −20°C in 50-ml Falcon tubes until used. Analysis was carried out on samples from two separate microcosms of the same treatment.

Microbial Activity

The metabolic potential of the soil fungi communities was estimated using the Biolog^{TM} FF plate (Biolog Inc., Hayward, CA, USA) according to the following protocol [24]: 10 g of fresh soil were suspended in an NaCl solution (0.8%) and shaken for 20 min. The sample suspensions were allowed to settle for 10 min, and then 150 μl of the sample supernatant were added to each well of the Biolog plates. Three parallel experiments were performed for the soil. The inoculated microplates were incubated at 25°C in darkness, and the optical density [25] was determined using a Biolog microplate reader for each plate at 12-h intervals

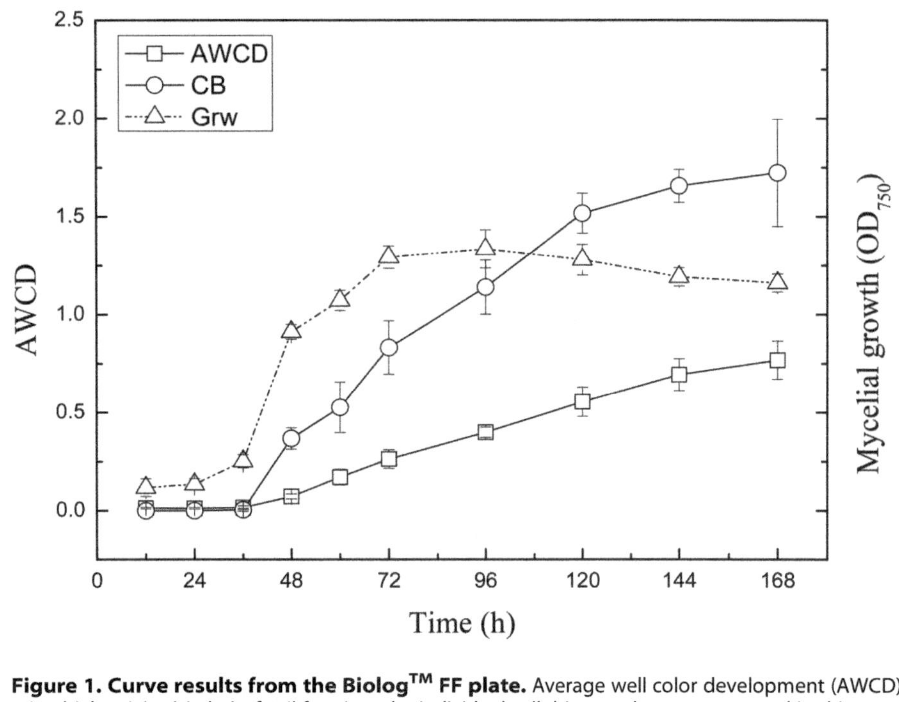

Figure 1. Curve results from the Biolog^{TM} FF plate. Average well color development (AWCD) with time (squares). Growth curve (triangles) and microbial activity (circles) of soil fungi on the individual cellobiose carbon sources used in this experiment. The values represent the means and were calculated from three parallel experiments.

Table 1. Cellobiohydrolase I gene fragments recovered from the arable soil and the putative CBHI protein.

Clone	Size (bp)	Intron 5′ splice site (size [bp])	Protein size (amino acids)	BLASTP (% identity)
C12	512	None	170	*Hypocrea koningii* 1,4-beta-cellobiosidase (99)
C35	585	424(67)	172	*Infundibulicybe gibba* exocellobiohydrolase (81)
C36	585	424(67)	172	*Infundibulicybe gibba* exocellobiohydrolase (81)
C55	561	417(49)	170	*Infundibulicybe gibba* exocellobiohydrolase (78)
C53	582	421(38)	180	*Volvariella volvacea* cellobiohydrolase I-II (75)
C43	561	418(49)	170	*Clitopilus prunulus* exocellobiohydrolase (81)
C62	566	421(50)	171	*Thermoascus aurantiacus* 1,4-beta-cellobiosidase (73)
C73	571	418(58)	170	*Volvariella volvacea* cellobiohydrolase I-II (84)
C14	562	418(49)	170	*Infundibulicybe gibba* exocellobiohydrolase (77)

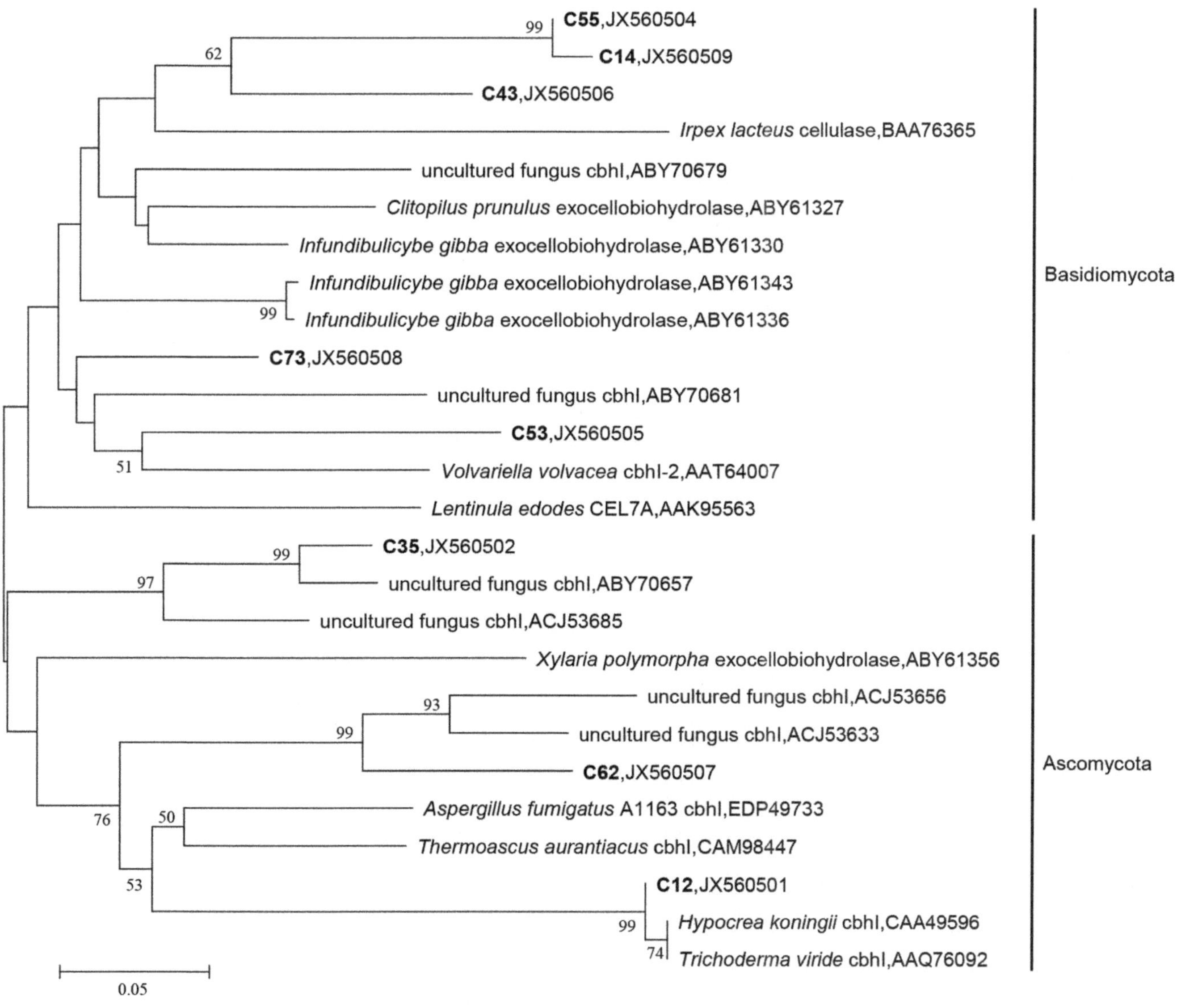

Figure 2. Phylogenetic relationship of the amino acid sequences derived from arable soil cellobiohydrolase I (*cbhI*) sequences. All of the sequences, except for those from this study, were obtained from GenBank. Reference sequences are shown in Roman type, and the sequences generated in this study are in boldface. The numbers on the branches refer to the bootstrap values of 1000, and bootstrap frequencies of >50% are shown.

Table 2. Closest relatives of eukaryotes with the DNA sequences of excised ITS-DGGE bands.

Band ID	Closest species match	Similarity (%)	Phylogenetic affiliation
S7a	Uncultured soil fungus clone 53-40,DQ420800	91	Eukaryote
S7b	Uncultured fungus clone Singleton 134-1201,FJ777879	89	Eukaryote
S14a	*Calluna vulgaris* root associated fungus,FM172810	94	Eukaryote
S14b	Uncultured fungus clone RFLP49,FJ528723	83	Eukaryote
S14c	Uncultured Sordariales clone NHPY79,FJ440938	91	Ascomycota
S14d	*Ustilaginoidea virens*,AB162148	100	Ascomycota
S14e	*Engyodontium album*,AB106650	95	Ascomycota
S14f	*Lecythophora* sp. I179,GU062252	97	Ascomycota
S28a	Uncultured endophytic fungus clone 38-1-11,EF505584	80	Eukaryote
S28b	Uncultured fungus clone LMRF_85,GU078651	90	Eukaryote
S28c	*Hypocrea virens* strain DAOM,EU280090	100	Ascomycota

over a period of 168 h at 590 nm (metabolic activity) and 750 nm (mycelial growth). The data were exported into an Excel spreadsheet for analysis. The average well color developments (AWCD) of the different replicates were calculated, where AWCD equals the sum of the difference between the OD of the blank well (water) and substrate wells divided by 95 (the number of substrate wells in the FF plates). The well color developments and growth curves were also analyzed based on the data collected from the individual cellobiose carbon source wells of the microplate [24].

DNA Extraction and Construction of *cbhI* Gene Clone Library

The soil DNA was extracted using the FastDNA® SPIN Kit for soil (Bio 101 Inc., USA) following the manufacturer's instructions. The DNA was eluted with 50 µl of the solution buffer supplied with the kit and visualized by electrophoresis in a 1% agarose gel. The amount of DNA was quantified using a NanoDrop ND-3300 fluorospectrometer (Thermo Fisher Scientific, Wilmington, DE). The DNA was stored at −20°C until use.

Purified total genomic DNA (~10 ng) was used as a template for the amplification of the cellobiohydrolase I (*cbhI*) gene. Polymerase chain reaction (PCR) was used to amplify the *cbhI* gene using fungcbhIF and fungcbhIR primers according to Edwards and colleagues [23]. PCR amplification was performed in an Eppendorf mastercycler gradient PCR machine using ExTaq DNA polymerase (Takara Biotechnology (Dalian) Co., Ltd., Japan). After purification using EZNA Cycle-Pure Kits (Omega Bio-tek Inc., Doraville, GA, USA), the ampicons were cloned into *Escherichia coli* JM109 using the p-GEMT easy vector (Promega, Madison, WI) in accordance with the manufacturer's instructions. Blue-white screening was used, and white colonies were picked at random. Colony PCR was then performed with pGEM-T primers T7F and SP6R. Clones that had yielded PCR products of the correct sizes, as determined by gel electrophoresis, were selected, and the inserts were sequenced. Sequencing of positive clones was performed with the standard primer T7F on an ABI PRISM 3730 sequencer (Applied Biosystems).

PCR and DGGE Analysis of Fungal ITS

Amplifications of the fungal internal transcribed spacer (ITS) rRNA regions were performed using the primer sets ITS1F/ITS4 and ITS1F-GC/ITS2 using nested PCR [26]. A GC clamp (5′-CGC CCG CCG CGC GCGGCG GGC GGG GCG GGG GCA CGG GGG G-3′) was added to the 5′ end of the ITS1-F primer to prevent complete dissociation of the DNA strands. The reaction was performed in a 50-µl volume that contained approximately 10 ng of DNA, ExTaq buffer, 0.2 µM of dNTPs, 0.2 µM of each primer, and 2 units of ExTaq DNA polymerase. The amplification protocol consisted of an initial denaturation at 95°C for 5 min; followed by 35 cycles of 94°C for 1 min, 55°C for 1 min, and 72°C for 1 min; and a final elongation at 72°C for 5 min. The first round of PCR products was purified using the EZNA Cycle-Pure Kit according to the manufacturer's instructions. A nested PCR was conducted using the purified products as the template and the primers ITS2 and ITS1F-GC using the conditions and parameters described above. All of the reaction mixtures lacking template DNA were performed as negative controls in parallel.

DGGE was carried out using a D-Code universal mutation detection system (Bio-Rad Laboratories) according to the instruction manual and 8% (w/v) polyacrylamide [acrylamide-bisacrylamide (37.5:1)] gels containing denaturing gradients of 20–50% (100% denaturant containing 7 M urea and 40% formamide) for the separation of the PCR products. The gel was run for 17 h at 60 V at 60°C. The gels were stained in 1X TAE buffer containing 1 µg ml^{-1} of ethidium bromide, and the results were visualized under UV light. The dominant bands in the DGGE gels were excised, and the acrylamide slices were crushed and resuspended overnight at 4°C in 50 µl of sterile water to elute the DNA [27]. The recovered DNA fragments were amplified using the primers ITS1F and ITS2. The resulting PCR products were cloned into *E. coli*, and positive clones were sequenced using the same protocols mentioned above.

DNA band positions and intensities were detected using Quantity One software (Bio-Rad Laboratories). Cluster analysis was performed by the unweighted pair group method using arithmetic averages (UPGMA). The Shannon-Weaver index of general diversity (H),$H' = -\sum P_i \ln P_i$, was calculated from the DGGE profiles, where P_i is the relative abundance of species i within a community. The relative band intensity within a profile was calculated by dividing the intensity of an individual band by the total bands intensity (sum of the intensities of all bands in the lane) to minimize the variation in the quantity of loaded PCR product.

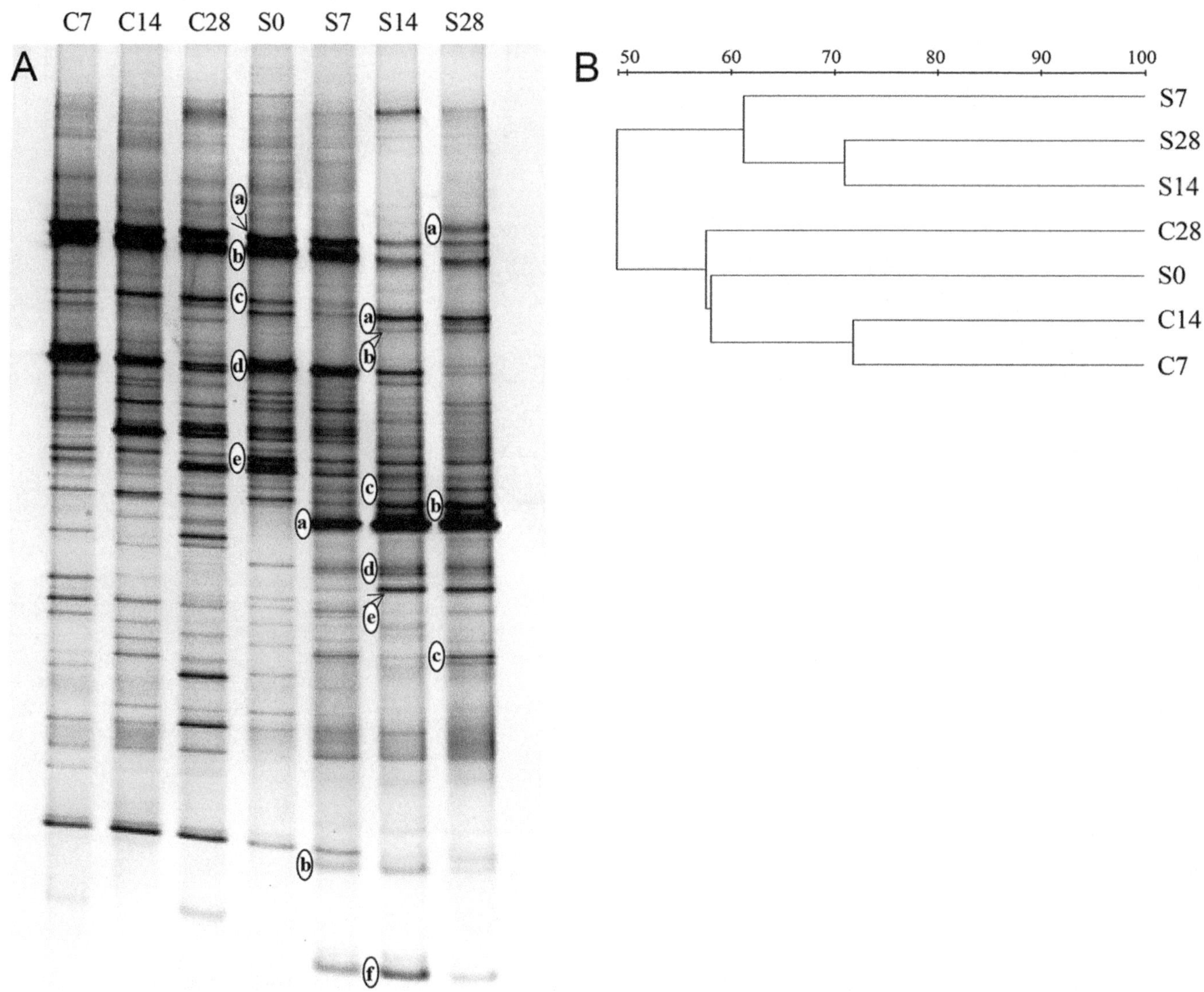

Figure 3. Fingerprints of the soil fungal communities from the different microcosms. (A) Denaturing gradient gel electrophoresis (DGGE) image. (B) UPGMA dendrograms based on the DGGE profiles. The lane numbers correspond to sampling points from the microcosms, and C and S represent the control and straw residue treatment microcosms, respectively. The bands marked with letters to the left of the lane were excised and sequenced.

Sequence Analysis

Raw sequence data was assembled and checked with DNAStar (Madison, WI, USA). NetGene2 was used to detect possible intron splice sites [28]. The sequences were compared with the GenBank database sequences using BLASTN and BLASTP for ITS and amino acid sequences deduced from the *cbhI* DNA sequences, respectively, and the highest matched sequences were obtained from the GenBank database (http://www.ncbi.nlm.nih.gov/BLAST/). The ITS sequences were screened for putative chimeras, as proposed by Nilsson and colleagues [29]. The final sequences were then aligned with the sequences downloaded from GenBank using ClustalW, and manual adjustments were made to the alignment where necessary. The neighbor-joining trees were constructed using MEGA version 4.0 [Molecular Evolutionary Genetics Analysis (http://megasoftware.net)] with 1000 bootstrap replicates [30].

Nucleotide Sequence Accession Numbers

The *cbhI* gene sequences of the clone library reported here were deposited in GenBank under the accession numbers JX560501 to JX560509. The sequences of the DGGE bands were submitted to GenBank under the accession numbers JX560485 to JX560500.

Results and Discussion

Metabolic Activity of Fungal Community

Fungi are known to secrete hydrolytic enzymes involved in biopolymer degradation, such as cellobiohydrolase and glucosidase. Using these enzymes, fungi can hydrolyze cellulolytic biomass [1]. To assess the heterotrophic soil fungal communities, we applied Biolog FF carbon-source-utilization-based optical density data analysis during an incubation of 168 h (Fig. 1). The average well color development values increased slowly during the incubation. Compared with AWCD, the Biolog values, based on the sole cellobiose carbon source, exhibited a significant increase,

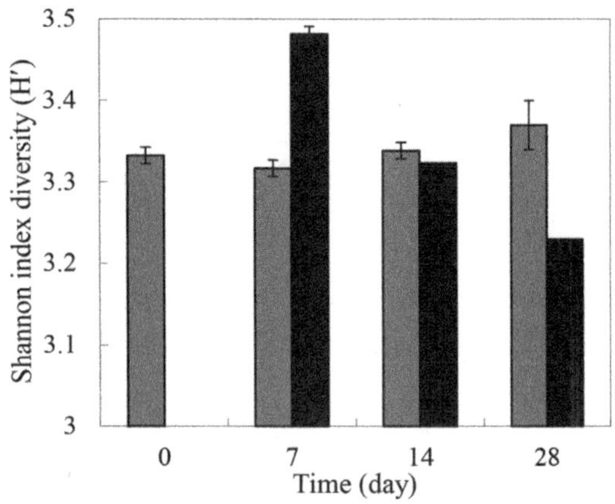

Figure 4. The Shannon index diversity (H) from the DGGE profiles of the ITS fragments. The ITS fragments were amplified from the DNA of the microcosm soil with (black column) or without (gray column) the straw residue at different time points. Error bars refer to the standard error between the replicates.

indicating that the soil fungi have a higher metabolic activity on cellobiose. Moreover, the growth of the soil fungi using cellobiose carbon is also observed in the growth curve (Fig. 1). The potential cellobiose carbon source utilization provides an independent estimate of the existence of β-glucosidase.

CBHI cellobiohydrolases are key extracellular cellulase enzymes mediating cellulose decomposition to release cellobiose [31]. To characterize the *cbhI* functional gene diversity of the fungal community, we cloned and sequenced PCR products from the soil. Sequences ranging from 512–585 bp were obtained from the soil library (Table 1). Possible intron splice site analysis revealed that the majority of the sequences contained a single intron, except for C12. The translation of putative amino acid sequences ranging in length from 170–180 residues and all of the putative amino acid sequences resulted in BLASTP matches (73–99% identity) to known fungi CBHI sequences (Table 1). Pairwise similarity of the putative protein sequence ranged from 54.2–99.4%. Only C12 matched best to the known fungi *Hypocrea koningii* CBHI sequence (99%). Given that few known *cbhI* genes are deposited in public databases, it was not surprising that some of these environmental *cbhI* genotypes were not annotated to known species. The taxonomic affinity of the environmental clones were displayed by phylogenetic analysis (Fig. 2). The deduced CBHI protein fragments recovered from the soil libraries were clustered in the Ascomycota and Basidiomytoca group members (Fig. 2). In combination, the metabolic activity of β-glucosidase and the diversity of the *cbhI* functional gene reveal the cellulose degradation potential of the fungi community in arable soil.

Fungal Community Dynamics during Decomposition

To analyze the dynamics of the fungal community during cellulose residue decomposition, DGGE was performed on fungal ITS fragments derived from the microcosm at 7, 14, and 28 days for both treatments and on the initial soil (day 0) (Fig. 3). Although the duplication is insufficient for proper statistical assessment, reproducible profiles were obtained from both replicates for all soil treatments (Text S1). After 7 days, the DGGE profile generated from the DNA of the straw residue microcosm differed from that obtained from the DNA of the control microcosm with the appearance of new bands (S7a, S7b, and S14f), and a slight increase in the relative intensity of the existing band S0e was observed. Novel bands (S14a, S14c, S14d, S14e, S14f, and S7b) of the cellulose microcosm appeared, as compared with the DGGE profile of the control microcosm sampled at day 14. At day 28, new bands (S28a, S28c, S14a, S14e, S14f, and S7b) of the straw treatment microcosm appeared compared with the DGGE pattern of the control microcosm. In contrast, band S0c disappeared, and a decrease in the relative intensity was observed in bands S0a, S0b, and S0d. The specific band S7a of the cellulose microcosm always appeared throughout the incubation period; however, it was nearly undetectable in the DGGE profile of the control microcosm. Almost all of the bands that newly appeared in the treatment microcosms at different stages belonged to the Ascomycota based on the phylogenetic analysis (Table 2). Detailed analysis of the fungal phylogeny revealed that they were assigned to *Ustilaginoidea*, *Lecythophora*, *Engyodontium*, and *Hypocrea*. Their potential role during decomposition is discussed below. Changes in the DGGE patterns of the fungal ITS were also observed in the control treatment, which might be due to the utilization of the soil basal organic matter.

The DGGE patterns derived from the cellulose microcosm were different from the control microcosm at each sampling time point (Fig. 3A). The straw residue incorporation affected the structure of the fungal communities in the soil, which is consistent with the cluster analysis of the fungal ITS fingerprints. Cluster analysis revealed two major clusters: one formed by the residue microcosm and the other consisting of the control microcosm (Fig. 3B). The different clusters indicated that relatively similar fungal communities were involved in the cellulose microcosm and control microcosm. The fungal ITS fingerprints obtained from the cellulose residue microcosm exhibited specific profiles at each incubation time (from 0–28 days) (Fig. 3). This highlighted the existence of a succession of the microbial community during the degradation process, as described in recent studies on plant litter decomposition based on phospholipid fatty acid analyses [32]. The Shannon-Wiener indices for the samples with and without straw residue revealed that there was not a significantly greater biological diversity between the treated samples and the control. For the control microcosm, the average value was 3.35; for the cellulose residue treatment, the average value was 3.34. The diversity recovered in the control and enriched microcosms was higher than that reported in the soil environments [4]. The diversity remained similar, while the straw residue treatment significantly affected the microbial community structure and diversity (Fig. 4). The H value (3.48) was highest 7 days following the addition of the cellulose residue, suggesting an increase in the diversity of the fungal population. Interestingly, a lower diversity was observed over the time course in the enriched microcosm (Fig. 4). Similarly, a decrease in the active bacterial diversity has been observed after the incorporation of the wheat residue into the soil [33]. This was surprising, because an addition of a large amount of straw residue might be expected to increase the community diversity and dominance of specific fungal decomposers within the decomposition process. One portion of the carbon compounds of the straw residue, easily utilizable by soil microorganisms, might stimulate the growth of the soil fungi in early stages of the decomposition process [13]. These results suggested that the addition of the straw residue stimulated a great diversity of fungal populations affiliated with the *Ustilaginoidea*, *Lecythophora*, *Engyodontium*, and *Hypocrea* at early stages of decomposition process (Fig. 3 and Fig. 5). However, with the disappearance of an increasing proportion of labile components and the concomitant accumulation of remaining recalcitrant organic matter, most fungal populations with a limited ability to degrade recalcitrant compounds could not grow well. Within the

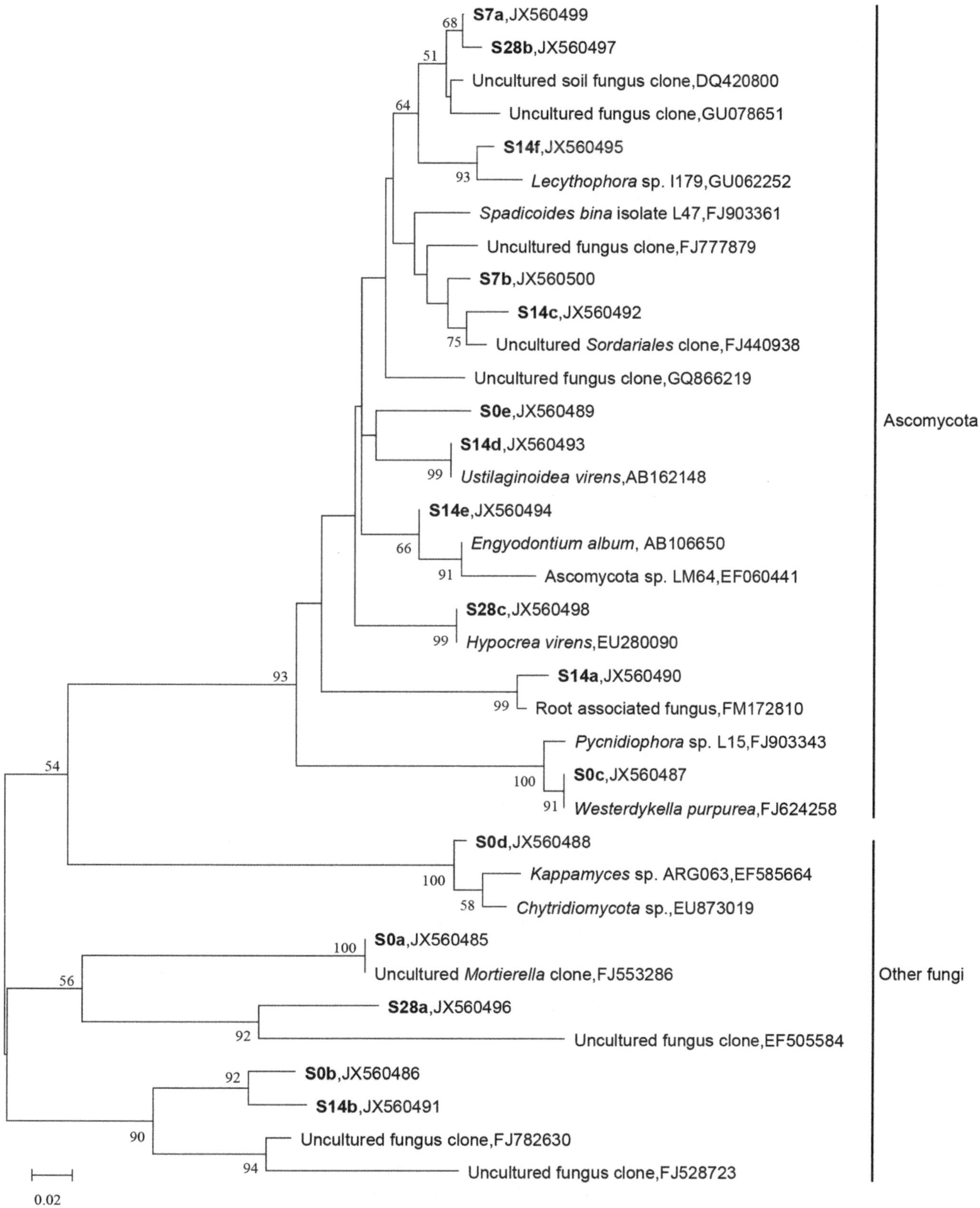

Figure 5. Phylogenetic relationships between the sequences of the sequenced ITS DGGE bands. The band numbers correspond to those presented in Fig. 3. The reference sequences available in GenBank are shown in Roman type, and the sequences generated in this study are in boldface. The scale bar represents the number of base substitutions per site. The bootstrap values are shown for the major branches, which had >50% support in a bootstrap analysis of 1000 replicates.

decomposition process, the disappearing proportion of the labile component and the accumulation of the remaining recalcitrant components would regroup the related decomposers, and the novel niche with a more uniform environment might decrease the diversity of the microbial community [6].

Fungal Genetic Structure during Decomposition

The DGGE profiles for the microcosm samples showed a clear succession of fungal populations during the straw residue decomposition. The representative bands were excised for further sequence analysis. The closest relatives of the characteristic DGGE bands (Fig. 3) mentioned above are listed in Table 2. The majority of the sequences had high (>90%) similarity to the fungal ITS sequences in the databases, except for two sequences (S14b and S28a) with low sequence similarities (<85%). The Ascomycota populations were the major active soil fungal decomposers (Table 2). The DGGE band sequences consistently clustered to the Ascomycota phylum in the phylogenetic analysis (Fig. 5). Interestingly, Osono and colleagues proposed that fungi from the Ascomycota phylum were dominant at early stages of the litter degradation process. Fungi from the Basidiomycete phylum are important in forest soil during the degradation of plant residue, especially in litter with high lignin content [25]. In crop residue in which lignin content is low, this might make fungi from the Ascomycota phylum the key decomposers in the agricultural soil tested.

The early successional group increased in the beginning stage followed by a decrease in the later stages. Bands S7b and S14f in the straw treatment were indicative of the early succession during the decomposition (Fig. 3). They were all affiliated with Ascomycota group, as shown in the phylogenic analysis (Fig. 5). Band S14f corresponded to *Lecythophora*, and some members of this genus are known to degrade lignocellulose and produce cellulase and xylanase, indicating that the fungi may contribute to the degradation of cellulose [34]. Band S7b was related to the uncultured *Sordariales* clone, an ectomycorhizosphere member [35], which remained a possibility for the production of cellulolytic enzymes that can break glycosidic bonds present in the plant cell wall. Moreover, *Sordariaceae* is one of the most well-known and best-studied families, as it consists of some notable taxa such as *Neurospora crassa*, which contains highly similar putative proteins to those shown to be associated with plant decomposition [36]. The characteristics of the related microorganisms suggest that these fungi might be involved in the primary assimilation of the cellulose during incubation.

The late successional community members gradually increased and remained at a relatively high level at the end of the incubation (Fig. 3). The late successional consumers (S7a, S14c, S14e, and S28c) were also members of Ascomycota group in the phylogenetic analysis (Fig. 5). S14e has a high similarity to *Engyodontium album*, which has ligninolytic activity and can degrade complex and recalcitrant aromatic organic matter [37,38]. S28c shares 99% similarity with *Hypocrea virens* (syn. *Gliocladium virens*), which secretes cellulase and xylanase to *decompose* ligninocellulose, showing the ability to utilize straw residue [39]. S0a and S0b existed without significant changes over the time course of the incubation; therefore, they could be considered the non-succession group (Fig. 5). They affiliate with cellulolytic *Mortierellales* members and probably have the ability to degrade lignocellulose [40]. The ability to decompose cellulose aerobically is also widely recognized among fungi, especially among members of the Ascomycota and Basidiomycota groups. Although the Basidiomycota *cbhI* genes were detected in the sampled soil, it was interesting that members of Basidiomycota were rarely found in the sequence analysis of ITS bands during straw residue degradation in our study. Ascomycota and Basidiomycota represent the main classical fungal decomposers in different soils [19]; however, members of Ascomycota have a limited ability to degrade the recalcitrant lignin-containing litter material [41]. When soil was amended with straw, Ascomycota could use the easily degradable fraction of residues for fast-growing fungal populations. With decomposition progressing, the dissappearance of labile components and accumulation of more recalcitrant compounds would stimulate the related Basidiomycota decomposers [41,42].

This study demonstrates the dynamics of fungal communities involved at different stages in residue decomposition. The fungal community succession can be divided into early succession, late succession, and non-succession. Importantly, Ascomycota fungi were dominant in the early and late succession communities, suggesting they were key drivers of arable soil decomposition process in our study. Furthermore, the distribution of soil fungi *cbhI* functional gene in Ascomycota and Basidiomycota suggested that the decomposition of the crop residue is a more complex and long-term process for the complete transformation of plant biopolymers. However, further research, for example, the use of stable isotope probing approaches *in situ*, is needed to identify the actual decomposers and their contribution during degradation.

Acknowledgments

We thank Wuren Huang for help with field sampling. We also acknowledge Juanli Yun and Hongguang Guo for assistance in the laboratory. We are very grateful to anonymous reviewers for their helpful comments on the manuscript.

Author Contributions

Conceived and designed the experiments: AM XZ GZ. Performed the experiments: AM DL JW MC. Analyzed the data: AM XZ GZ. Contributed reagents/materials/analysis tools: AM DL JW MC CL. Wrote the paper: AM XZ CL GZ.

References

1. Lynd LR, Weimer PJ, van Zyl WH, Pretorius IS (2002) Microbial cellulose utilization: Fundamentals and biotechnology. Microbiol Mol Biol Rev 66: 739–739.
2. Kwon YJ, Ma AZ, Li Q, Wang F, Zhuang GQ, et al. (2011) Effect of lignocellulosic inhibitory compounds on growth and ethanol fermentation of newly-isolated thermotolerant *Issatchenkia orientalis*. Bioresour Technol 102: 8099–8104.
3. Hadas A, Kautsky L, Goek M, Kara EE (2004) Rates of decomposition of plant residues and available nitrogen in soil, related to residue composition through simulation of carbon and nitrogen turnover. Soil Biol Biochem 36: 255–266.
4. Wakelin SA, Colloff MJ, Harvey PR, Marschner P, Gregg AL, et al. (2007) The effects of stubble retention and nitrogen application on soil microbial community structure and functional gene abundance under irrigated maize. FEMS Microbiol Ecol 59: 661–670.
5. Wilson DB (2011) Microbial diversity of cellulose hydrolysis. Curr Opin Microbiol 14: 259–263.
6. Voriskova J, Baldrian P (2012) Fungal community on decomposing leaf litter undergoes rapid successional changes. ISME J 7: 477–486.
7. Lindahl BD, Ihrmark K, Boberg J, Trumbore SE, Hogberg P, et al. (2007) Spatial separation of litter decomposition and mycorrhizal nitrogen uptake in a boreal forest. New Phytol 173: 611–620.

8. Kellner H, Zak DR, Vandenbol M (2010) Fungi unearthed: transcripts encoding lignocellulolytic and chitinolytic enzymes in forest soil. Plos One 5: e10971.
9. Osono T (2006) Role of phyllosphere fungi of forest trees in the development of decomposer fungal communities and decomposition processes of leaf litter. Can J Microbiol 52: 701–716.
10. Osono T, Ishii Y, Takeda H, Seramethakun T, Khamyong S, et al. (2009) Fungal succession and lignin decomposition on *Shorea obutsa* leaves in a tropical seasonal forest in northern Thailand. Fungal Divers 36: 101–119.
11. Dilly O, Munch JC (2004) Litter decomposition and microbial characteristics in agricultural soils in northern, central, and southern Germany. Soil Sci Plant Nutr 50: 843–853.
12. Johnson JMF, Barbour NW, Weyers SL (2007) Chemical composition of crop biomass impacts its decomposition. Soil Sci Soc Am J 71: 155–162.
13. Bertrand I, Prevot M, Chabbert B (2009) Soil decomposition of wheat internodes of different maturity stages: Relative impact of the soluble and structural fractions. Bioresour Technol 100: 155–163.
14. Wu JP, Liu ZF, Wang XL, Sun YX, Zhou LX, et al. (2011) Effects of understory removal and tree girdling on soil microbial community composition and litter decomposition in two Eucalyptus plantations in South China. Funct Ecol 25: 921–931.
15. Zhao J, Wang XL, Shao YH, Xu GL, Fu SL (2011) Effects of vegetation removal on soil properties and decomposer organisms. Soil Biol Biochem 43: 954–960.
16. Dilly O, Bartsch S, Rosenbrock P, Buscot F, Munch JC (2001) Shifts in physiological capabilities of the microbiota during the decomposition of leaf litter in a black alder (*Alnus glutinosa* (Gaertn.) L.) forest. Soil Biol Biochem 33: 921–930.
17. Nicolardot B, Bouziri L, Bastian F, Ranjard L (2007) A microcosm experiment to evaluate the influence of location and quality of plant residues on residue decomposition and genetic structure of soil microbial communities. Soil Biol Biochem 39: 1631–1644.
18. Han X, Cheng ZH, Meng HW (2012) Soil properties, nutrient dynamics, and soil enzyme activities associated with garlic stalk decomposition under various conditions. Plos One 7.
19. Vandenkoornhuyse P, Baldauf SL, Leyval C, Straczek J, Young JPW (2002) Evolution - Extensive fungal diversity in plant roots. Science 295: 2051–2051.
20. Bastian F, Bouziri L, Nicolardot B, Ranjard L (2009) Impact of wheat straw decomposition on successional patterns of soil microbial community structure. Soil Biol Biochem 41: 262–275.
21. Osono T, Takeda H (2006) Fungal decomposition of Abies needle and Betula leaf litter. Mycologia 98: 172–179.
22. Yelle DJ, Ralph J, Lu FC, Hammel KE (2008) Evidence for cleavage of lignin by a brown rot basidiomycete. Environ Microbiol 10: 1844–1849.
23. Edwards IP, Upchurch RA, Zak DR (2008) Isolation of fungal cellobiohydrolase I genes from sporocarps and forest soils by PCR. Appl Environ Microbiol 74: 3481–3489.
24. Friedl MA, Schmoll M, Kubicek CP, Druzhinina IS (2008) Photostimulation of *Hypocrea atroviridis* growth occurs due to a cross-talk of carbon metabolism, blue light receptors and response to oxidative stress. Microbiology 154: 1229–1241.
25. Blackwood CB, Waldrop MP, Zak DR, Sinsabaugh RL (2007) Molecular analysis of fungal communities and laccase genes in decomposing litter reveals differences among forest types but no impact of nitrogen deposition. Environ Microbiol 9: 1306–1316.
26. Anderson IC, Campbell CD, Prosser JI (2003) Diversity of fungi in organic soils under a moorland–Scots pine (*Pinus sylvestris* L.) gradient. Environ Microbiol 5: 1121–1132.
27. Mondani L, Benzerara K, Carriere M, Christen R, Mamindy-Pajany Y, et al. (2011) Influence of uranium on bacterial communities: A comparison of natural uranium-rich soils with controls. Plos One 6: e25771.
28. Jacobsen J, Lydolph M, Lange L (2005) Culture independent PCR: an alternative enzyme discovery strategy. J Microbiol Methods 60: 63–71.
29. Nilsson RH, Abarenkov K, Veldre V, Nylinder S, De Wit P, et al. (2010) An open source chimera checker for the fungal ITS region. Mol Ecol Resour 10: 1076–1081.
30. Tamura K, Dudley J, Nei M, Kumar S (2007) MEGA4: Molecular evolutionary genetics analysis (MEGA) software version 4.0. Mol Biol Evol 24: 1596–1599.
31. Ma AZ, Hu Q, Qu YB, Bai ZH, Liu WF, et al. (2008) The enzymatic hydrolysis rate of cellulose decreases with irreversible adsorption of cellobiohydrolase I. Enzyme Microb Technol 42: 543–547.
32. Snajdr J, Cajthaml T, Valaskova V, Merhautova V, Petrankova M, et al. (2011) Transformation of *Quercus petraea* litter: successive changes in litter chemistry are reflected in differential enzyme activity and changes in the microbial community composition. FEMS Microbiol Ecol 75: 291–303.
33. Bernard L, Mougel C, Maron PA, Nowak V, Leveque J, et al. (2007) Dynamics and identification of soil microbial populations actively assimilating carbon from 13C-labelled wheat residue as estimated by DNA- and RNA-SIP techniques. Environ Microbiol 9: 752–764.
34. Lopez MJ, Vargas-Garcia MD, Suarez-Estrella F, Nichols NN, Dien BS, et al. (2007) Lignocellulose-degrading enzymes produced by the ascomycete *Coniochaeta ligniaria* and related species: Application for a lignocellulosic substrate treatment. Enzyme Microb Technol 40: 794–800.
35. Phosri C, Polme S, Taylor AFS, Koljalg U, Suwannasai N, et al. (2012) Diversity and community composition of ectomycorrhizal fungi in a dry deciduous dipterocarp forest in Thailand. Biodivers Conserv 21: 2287–2298.
36. Tian CG, Beeson WT, Iavarone AT, Sun JP, Marletta MA, et al. (2009) Systems analysis of plant cell wall degradation by the model filamentous fungus *Neurospora crassa*. Proc Natl Acad Sci U S A 106: 22157–22162.
37. Ryckeboer J, Mergaert J, Vaes K, Klammer S, De Clercq D, et al. (2003) A survey of bacteria and fungi occurring during composting and self-heating processes. Ann Microbiol 53: 349–410.
38. Artham T, Doble M (2010) Biodegradation of physicochemically treated polycarbonate by fungi. Biomacromolecules 11: 20–28.
39. Gomes J, Gomes I, Esterbauer H, Kreiner W, Steiner W (1989) Production of cellulases by a wild strain of *Gliocladium virens* - optimization of the fermentation medium and partial characterization of the enzymes. Appl Microbiol Biotechnol 31: 601–608.
40. Ruan ZH, Zanotti M, Wang XQ, Ducey C, Liu Y (2012) Evaluation of lipid accumulation from lignocellulosic sugars by *Mortierella isabellina* for biodiesel production. Bioresour Technol 110: 198–205.
41. Lundell TK, Makela MR, Hilden K (2010) Lignin-modifying enzymes in filamentous basidiomycetes - ecological, functional and phylogenetic review. J Basic Microbiol 50: 5–20.
42. Duong LM, McKenzie EHC, Lumyong S, Hyde KD (2008) Fungal succession on senescent leaves of *Castanopsis diversifolia* in Doi Suthep-Pui National Park, Thailand. Fungal Divers 30: 23–36.

8

Overexpression of *phyA* and *appA* Genes Improves Soil Organic Phosphorus Utilisation and Seed Phytase Activity in *Brassica napus*

Yi Wang[1,2], Xiangsheng Ye[1], Guangda Ding[1], Fangsen Xu[1]*

1 National Key Laboratory of Crop Genetic Improvement, and Microelement Research Center, Huazhong Agricultural University, Wuhan, China, **2** College of Resources and Environmental Sciences, Henan Agricultural University, Zhengzhou, China

Abstract

Phytate is the major storage form of organic phosphorus in soils and plant seeds, and phosphorus (P) in this form is unavailable to plants or monogastric animals. In the present study, the phytase genes *phyA* and *appA* were introduced into *Brassica napus* cv Westar with a signal peptide sequence and CaMV 35S promoter, respectively. Three independent transgenic lines, P3 and P11 from *phyA* and a18 from *appA*, were selected. The three transgenic lines exhibited significantly higher exuded phytase activity when compared to wild-type (WT) controls. A quartz sand culture experiment demonstrated that transgenic *Brassica napus* had significantly improved P uptake and plant biomass. A soil culture experiment revealed that seed yields of transgenic lines P11 and a18 increased by 20.9% and 59.9%, respectively, when compared to WT. When phytate was used as the sole P source, P accumulation in seeds increased by 20.6% and 46.9% with respect to WT in P11 and a18, respectively. The P3 line accumulated markedly more P in seeds than WT, while no significant difference was observed in seed yields when phytate was used as the sole P source. Phytase activities in transgenic canola seeds ranged from 1,138 to 1,605 U kg^{-1} seeds, while no phytase activity was detected in WT seeds. Moreover, phytic acid content in P11 and a18 seeds was significantly lower than in WT. These results introduce an opportunity for improvement of soil and seed phytate-P bioavailability through genetic manipulation of oilseed rape, thereby increasing plant production and P nutrition for monogastric animals.

Editor: Tianzhen Zhang, Nanjing Agricultural Universitym, China

Funding: This work was supported by the grants from National Basic Research and Development Program (2011CB100301 (http://www.973.gov.cn/Default_3.aspx) and National Natural Science Foundation of China (31172019), China (http://www.nsfc.gov.cn/Portal0/default152.htm). The funders had no role in study design, data collection and analysis, decision to publish, or preparation of the manuscript.

Competing Interests: The authors have declared that no competing interests exist.

* E-mail: fangsenxu@mail.hzau.edu.cn

Introduction

Phosphorus (P) is an essential macroelement required for plant growth. Poor availability of P in soils and consequent P-deficiency are major constraints to crop production globally [1]. Phosphorus is taken up by plants as orthophosphate (P_i). However, in most cultivated soils for agriculture, organic P comprises 30–80% of the total P, and approximately 60–80% of organic P exists in the form of phytate and is not directly available to plants [2,3]. Thus, improving phytate-P bioavailability is important for plant P nutrition, and for sustainable agricultural development due to the exhausting P ore resources worldwide [4].

Phytase is a type of phosphatase enzyme that catalyses the hydrolysis of phytic acid into inorganic phosphate (P_i) and myo-inositol [5]. Phytases have been most commonly detected and characterised in fungi but have also been found to occur in animals, plants and bacteria [6]. Because phytate in plant seeds is largely indigestible by monogastric animals, much attention has been given to the improvement of phytate-P bioavailability in animal feed by overexpressing phytase genes in plant seeds [4]. The transgenic production of high phytase activity in plant seeds would reduce the costs of producing the enzyme which is currently achieved by fermentation technology [7]. Plant overexpressing either phytase gene *phyA* or *appA* significantly increased phytase activity in wheat grains [8], soybean seeds [9,10], canola seeds [11,12] and maize grains [13].

Plants have developed a variety of mechanisms to increase the availability of soil P, including changes in root morphology and architecture, symbiosis with mycorrhizal fungi, improvement of internal phosphatase activity, secretion of organic acids and phosphatases, and up-regulation of high affinity phosphate transporters [3,1]. Phosphatases are required for the mineralisation of organic P to release phosphate into the soil [14]. Phytases have been identified in roots and root exudates in plants [15,16,17]. However, it has been suggested that the enzymatics activity in root exudates is not enough for effective utilisation of organic P [4,17]. When exogenous phytase was added into the media, phytate became available for plant growth [16,18,19]. Therefore, in recent years, more attention has been focused on the potential for overexpressing phytase in plant roots by genetic manipulation for the improvement of organic P use from soil. Overexpression of *phyA* in transgenic *Arabidopsis* with the *Pht1;2* promoter increased extracellular phytase activity and phytate-P use efficiency, and the transgenic plants were able to grow in a medium containing phytate as the sole P source [20,21]. Transgenic *Trifolium subterraneum* expressing *phyA* showed a 77-fold average increase in exuded phytase activity and took up 1.3- to 3.6-fold more P compared to wild-type plants [22]. Transgenic expression of the phytase gene *MtPHY1* isolated from the model legume *Medicago truncatula* improved phytate-P utilization and plant growth in *Arabidopsis* [23] and white clover [24]. Li *et al.* (2009) expressed the phytase gene *AfPhyA* in soybean roots driven by the *Pky10* promoter and found that the extracellular phytase activity was 6-fold higher than in control roots [25]. Moreover, transgenic cotton overexpressing *phyA* had an improved ability to utilise organic P [26].

These results suggest that overexpression of phytase genes containing a secretion signal improve phytate-P utilisation. *Brassica napus* is an important oilseed crop worldwide, but it is sensitive to P deficiency [27]. Moreover, the widespread use of oilseed meal in livestock feed and the inability to utilise phytate-P can lead to serious environmental consequences [1]. In the present study, phytase genes *phyA* and *appA* from *Aspergillus niger* and *Escherichia coli*, respectively, were used to generate transgenic *Brassica napus* plants that could secrete phytase as an extracellular enzyme. The results showed that transgenic plants not only significantly increased their capacity of P acquisition from phytate, but they also produced stable recombinant phytase in seeds. This study could have significant implications for the improvement of phytate-P bioavailability in soil and seeds, for increased plant production and for P-nutrition for monogastric animals from a single transgenic plant.

Results

Generation of transgenic lines overexpressing phytase genes

The plasmids pBI121-phyA and pBI121-appA (Fig. 1A, B) were used for the transformation of *B. napus* cv Westar. A total of 28 independent transformation events were obtained, 22 with pBI121-appA and 6 with pBI121-phyA. First, the presence of *A. niger phyA* or *E. coli appA* in the transformed lines was confirmed by genomic PCR analysis (data not shown). The positive T_0 plants were self-crossed to ultimately obtain T_3 transgenic lines for further analysis. Three canola transgenic lines, P3 and P11 with *phyA* and a18 with *appA*, were selected based on their high phytase activity in root extracts and lack of segregation with kanamycin-resistant screening. Southern blotting of the three T_3 transgenic lines demonstrated that the exogenous *phyA* and *appA* were correctly integrated into the genome of transgenic canola plants (Fig. 1C, D). P3 and P11 had one and three copies of *phyA* gene, respectively (Fig. 1C), and the a18 line had two copies of the *appA* gene (Fig. 1D). The mRNA expression of *A. niger phyA* and *E. coli appA* in the roots of the three transgenic lines was analysed by northern blotting (Fig. 1E, F). Each transgenic line showed high accumulation of either *phyA* or *appA* transcripts. By contrast, no hybridisation bands corresponding to *A. niger phyA* or *E. coli appA* were detected in wild-type (WT) plants.

Phytase activity in transgenic *Brassica napus*

There was no significant difference in phytase activity in leaf extracts between transgenic lines and WT with the exception of P11 in Phy-P treatment (Fig. 2A). Phytase activities were higher in *ex::phyA* and *ex::appA* root extracts than in WT plants regardless of P availability (Fig. 2B). When phytate was added as a P source, phytase activity significantly increased compared to other P treatments, and the activity in the P11 line was nearly two-fold higher than in WT.

Exuded phytase activity was markedly higher in transgenic lines than in WT when grown in P-deficient conditions (Fig. 3A). With Phy-P treatment, exuded phytase activities were increased approximately 3-fold in *ex::phyA* and *ex::appA* lines when compared to WT plants. The exuded phytase activity was very low at high P (HP) supply, and this was not significantly different between transgenic lines and WT (Fig. 3A). In low phosphorus (LP) treatment, exuded acid phosphatase (APase) activity was significantly higher in transgenic lines than in WT (Fig. 3B). However, when grown in Phy-P, no significant difference in exuded APase activity was observed between transgenic lines and WT (Fig. 3B).

Growth and P nutrition of transgenic lines in sand culture experiment

A quartz sand culture experiment was performed to evaluate the response of transgenic plants to the supply of phytate-P in the growth medium. Vegetative growth phenotypes indicated that the three transgenic lines had larger shoots than WT in the three treatments (Fig. 4A). When grown in Phy-P-treated sand, transgenic lines showed much larger leaf sizes and more leaves compared to WT plants (Fig. 4B). Moreover, WT plants exhibited serious P-deficiency symptoms in their old leaves, while no evident P-deficiency was observed in P3 leaves with Phy-P treatment (Fig. 4B).

When phytate was used as the sole P source, the three transgenic lines P3, P11 and a18 had more P accumulation in the shots than WT by 70.8%, 36.8% and 19.6%, respectively (Table 1). Moreover, shoot biomass increased from 14.7% to 36.8% in the transgenic lines compared to WT. In HP treatment, P accumulation in transgenic shoots increased from 41.2% to 70.9% and shoot biomass increased from 28.2% to 53.4% compared to WT. And in LP treatment, the P3 line exhibited a 44.4% increase in shoot biomass and a 167% increase in total P content compared to WT (Table 1).

Seed yields and P accumulation in transgenic lines

To detect the transgenic plant growth in P-deficient soil during the entire growth period, we conducted a pot culture experiment. Under Phy-P treatment, transgenic lines P11 and a18 had significant increases in seed yields of 20.9% and 59.9%, respectively, compared to WT (Table 2). In LP treatment, only the a18 line improved seed yields, with a 46.1% increase with respect to WT. Seed yields from P11 and a18 lines with HP treatment were 62.5% and 55.5% higher than in WT, respectively (Table 2). However, no significant difference was observed between the seed yields of the P3 line and WT in the three P treatment soils (Table 2). In HP treatment, P accumulation in seeds of transgenic lines P11 and a18 increased by 41.8% and 59.9%, respectively, compared with WT. Transgenic line P3, P11 and a18 respectively accumulated 11.5%, 20.6% and 46.9% more P in seeds than WT when plants were grown in P-deficient soil to which only phytate was applied (Table 2). However, in LP treatment, only the a18 line accumulated more P in seeds than WT (Table 2). Overexpression of *phyA* and *appA* in *B. napus* can improve the seed yields of transgenic plants and accumulated more P in seeds when phytate was the sole P source.

Phytase activity and phytic acid content in seeds

T_3 seeds harvested from the field by self-cross were assayed for phytase activities and phytic acid contents. Phytase activities for P3, P11 and a18 were 1,225, 1,605 and 1,138 U kg^{-1} seeds, respectively (Fig. 5A). However, no phytase activity was detected in WT seeds. The phytic acid content in WT seeds was approximately 31.6 mg g^{-1} seeds, and it was reduced to 22.5 and 25.5 mg g^{-1} seeds in P11 and a18 transgenic seeds,

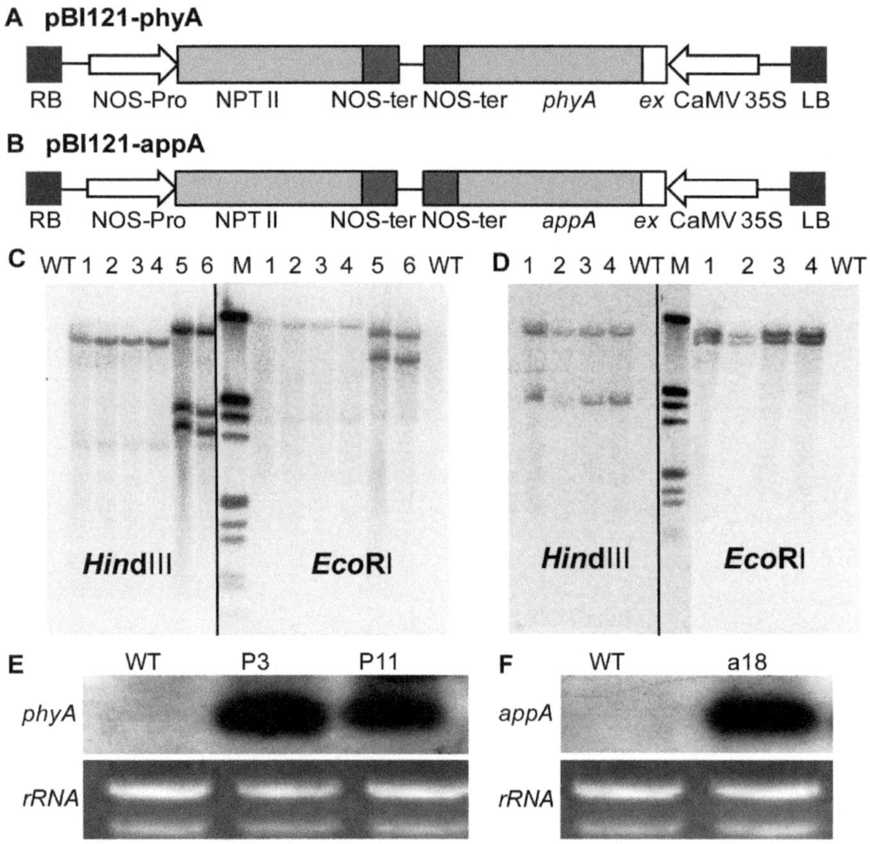

Figure 1. Construction of transgenic plants overexpressing *A. niger phyA* or *E. coli appA*. A and **B**: Schematic representation of binary vector pBI121-phyA (**A**) and pBI121-appA (**B**) for *B. napus* plant transformation. Phytase genes (*A. niger phyA* and *E. coli appA*) under the control of the CaMV 35S promoter and nos-terminator were modified for extracellular secretion by inclusion of an extracellular targeting sequence from the carrot extensin (*ex*) gene. The construct has a selectable marker for the *NPTãÀ* gene. **C** and **D**: Southern blotting analysis of *ex::phyA* (**C**) and *ex::appA* (**D**) transgenic lines, using *phyA* and *appA* probes, respectively. Genomic DNA was digested by *Hind*III and *Eco*RI. **C**, M, Marker; WT, wild-type canola; 1–4,T_3 transgenic plants of P3 line; 5–6, T_3 transgenic plants of P11 line; **D**, 1–4, T_3 transgenic plants of a18 line. **E** and **F**: Northern blotting analysis of *phyA* (**E**) and *appA* (**F**) expression in roots of transgenic lines using *phyA* and *appA* probe, respectively.

respectively (Fig. 5B). However, no significant difference in phytic acid content was observed between the P3 line and WT seeds.

The results of the soil pot culture experiment indicated that phytase activity in the three transgenic lines ranged from 344.7 to 1251.6 U kg^{-1} seeds when grown in the presence of HP, LP and Phy-P treatments (Table 2). However, no phytase activity was detected in WT seeds. The phytic acid content in WT seeds was approxiamtely 30 mg g^{-1} seed, while it was reduced to 19.2 and 19.9 mg g^{-1}, respectively, in P11 and a18 transgenic seeds with HP treatment (Table 2). The phytic acid content in P11 and a18 seeds was reduced by 46.2% and 30.7%, respectively, compared to WT seeds when phytate was the sole P source (Table 2). However, the phytic acid content of P3 seeds was not reduced compared with WT in either of the two treatments, which was consistent with the results in Figure 8B. When grown in LP-treated soil, phytic acid content was significantly reduced in the three transgenic lines compared to WT.

Discussion

In total, 30% to 80% of the total P in agricultural soils exists as organic phosphate, and up to 80% of organic phosphate is in the form of phytate-P [2,28]. Because phytate-P is not directly available to plants, low P availability becomes one of the limiting factors for plant growth. Phytases were identified in plant roots and root exudates [15,16]; however, the activity of these enzymes in roots is inadequate for efficient utilisation of organic P from soil [16,17]. In the present study, the exudate phytase activities were very low in WT plants when grown in HP treatment. However, the secreted phytase activity increased in WT plants in P-stressed conditions (LP and Phy-P) (Fig. 3A). This suggested that P starvation induced canola to exude phytase into the rhizosphere depending on P availability. Hayes et al. (2000) and Richardson et al. (2000) showed that a low intrinsic level of phytase activity in plant roots exudates is a critical factor that limits the ability of plants to obtain P from phytate [16,17]. Idriss et al. (2002) reported that the presence of purified FZB45 phytase promoted the growth of maize seedlings when phytate was the sole P source [18]. In the present study significant increase in extracellular phytase activity was observed in transgenic lines expressing *ex::phyA* or *ex::appA* compared to WT in P deficient conditions, which is consistent with Richardson et al. (2001) and George et al. (2004) [20,22]. No significant difference in extracellular phytase activity was observed between the transgenic lines and WT plants in HP treatment (Fig. 3A), even though the exogenous *ex::phyA* and *ex::appA* gene were highly expressed (Fig. 1E, F). However, the difference was significant under P-stressed conditions (Fig. 3A). Phytase activity was much higher in root extracts from the

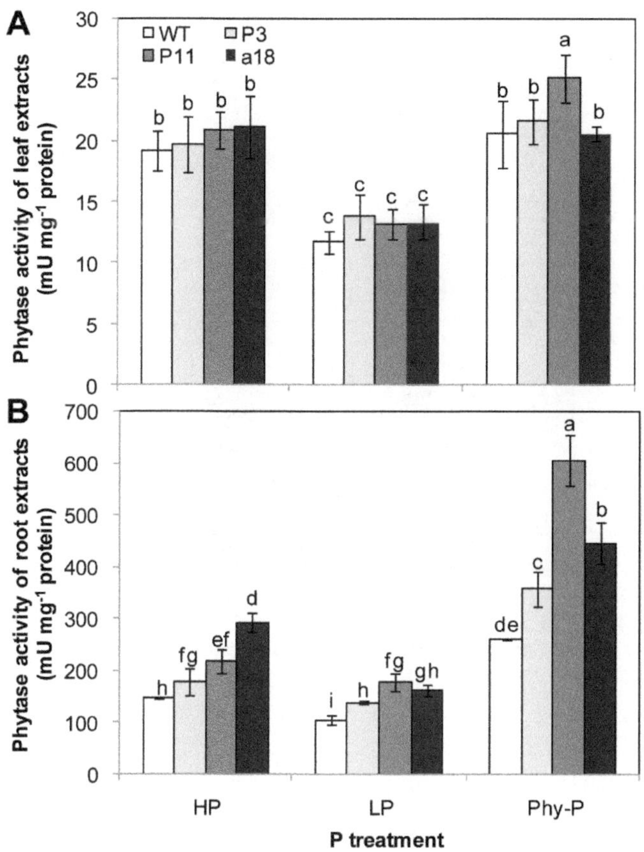

Figure 2. Phytase activity in leaf (A) and root (B) extracts of transgenic lines when grown in different P conditions. HP: plants were grown in high P (250 μM KH_2PO_4) nutrient solution for 18 days; LP: plants were grown in low P (5 μM KH_2PO_4) nutrient solution for 18 days; Phy-P: plants were grown in 250 μM P in the form of phytate for 18 days. Each column is the mean of four replicates with SD. Different letters represent significant differences at the $p<0.05$ level.

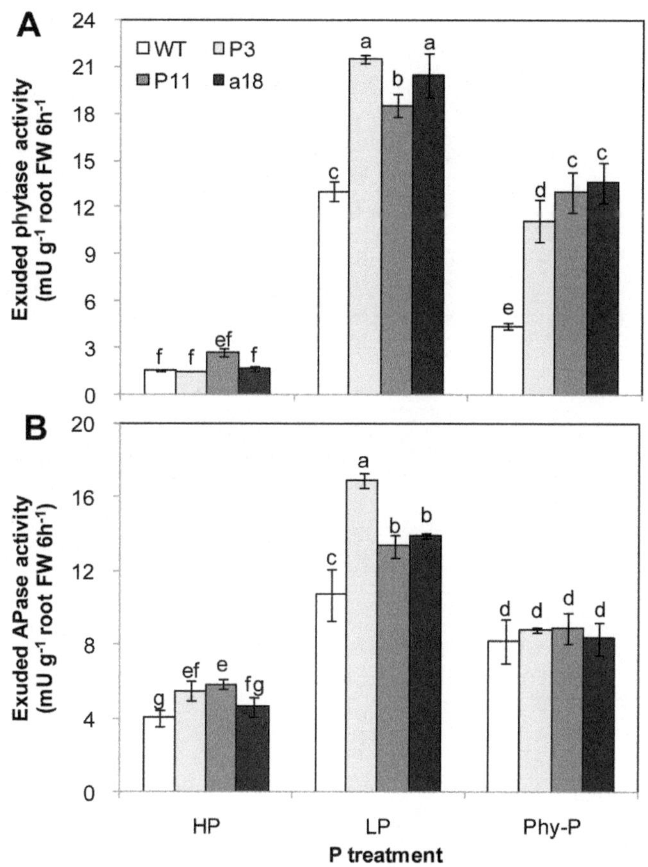

Figure 3. Phytase activity (A) and acid phosphatase activity (B) exuded from roots of transgenic lines expressing *ex::phyA/appA* and WT. Plants were grown in different P treatment solutions. Each column is the mean of four replicates with SD. Different letters represent significant differences at the $p<0.05$ level.

transgenic lines than from WT (Fig. 2B), suggesting that the exudation of phytase into the rhizosphere is induced by P limitation. The phytase secretion mechanism induced by P deficiency together with the extracellular targeted signal peptide sequence may have contributed to phytase exudation in the transgenic lines.

Enhanced exudation of phytase by overexpressing *ex::phyA* or *ex::appA* in transgenic *B. napus* conferred the ability to acquire P from phytate. All *ex::phyA/appA* lines exhibited increased P uptake and shoot biomass production in comparison with WT when grown in quartz sand supplied with phytate as the only source of P (Table 1). The increase in shoot biomass was comparable to the 30% reported by Zimmermann et al. (2003) in transgenic potato [29], the 23% increase observed by Mudge et al. (2003) in *Arabidopsis* [21] and the 27% increase observed by George et al. (2004) in *Trifolium subterraneum* L. [22]. The observed increase in total P content of shoots in *ex::phyA/appA B. napus* corresponds well to the 40% increase in P uptake observed in transgenic potato [29], the 3-fold increase observed in transgenic *Arabidopsis* [21] and the 1.3- to 3.6-fold increase observed in transgenic *T. subterraneum* [22]. Moreover, the dry weight of the transgenic plants increased from 28.2% to 53.4% and total P content increased from 50.0% to 167% compared to WT when grown in HP and LP treatments in sand culture.

In HP and Phy-P treatment soils, seed yield and phytic acid content in P11 were significantly higher than in WT plants. However, no difference was observed between P3 and WT (Table 2). This might de due to the difference in copy number of the *phyA* gene in the two transgenic lines. Hong et al. (2004) found that transgenic lines 8FN-8 and 8AN-14, which contain three and four copies of their transgenes, respectively, showed the highest phytase activity compared to other transgenic lines. They suggested that multiple copies of transgenes could, but may not necessarily, contribute to higher expression levels of phytase in germinated rice seeds [30]. The bacterial phytases have broader pH optima and higher specific activities compared to the fungal phytase [31], which may partly explain the difference in seed yield, phytase activity and phytic acid content between *phyA* lines and *appA* line. However, this speculation needs to be further verified by using abundant independent transgenic lines with different copies of the *phyA* and *appA* genes.

Phytate is the main storage form of P in plant seeds, which are a major component of animal feed [32]. Phytic acid is largely indigestible by monogastric animals because they have no or limited phytase activity in their digestive tract [4]. Thus, in addition to improving P uptake from soil, transgenic expression of phytase in plant seeds offers prospects for improving phytate-P bioavailability in monogastric animals. Previous studies indicated that the overexpression of phytase genes in soybean [9,10], maize [13], canola [11,12], rice [30] and wheat [7,8] seeds improved

Figure 4. Growth response of transgenic plants and WT plants in sand culture. A: Phenotype of transgenic lines (P3, P11 and a18) and WT plants grown for 60 d in quartz sand culture with different P sources. **B:** Leaf number and size of transgenic lines and WT plants when grown in quartz sand with phytate as the sole P source. The leaves arranged from left to right ranged from the oldest to youngest according to the growth order in the plant.

phosphorus availability and reduced the impact of animal production on the environment. In the present study, in addition to improving phosphorus uptake from phytate, overexpression of *ex::phyA* or *ex::appA* in canola produced stable recombinant phytase and significantly reduced phytic acid content in transgenic seeds. Phytase activities in transgenic seeds were 1,138 to 1,605 U kg^{-1} seeds (Fig. 5A), which were close to the 2,000 U kg^{-1} seeds observed by Chen *et al.* (2008) in transgenic maize [13]. Moreover, nearly no phytase activity was detected in WT canola seeds. The role of the signal sequence in carrot extensin (*ex*) is to transport extensins through the cell membrane to reach the cell wall [33,34]. In transgenic plants expressing a phytase genes with the *ex* signal peptide sequence, researchers have historically only focused on the extracellular phytase activity and soil phytate-P uptake but have not noted the changes in seed phytase activity [20,21,22,35]. Our results demonstrate that the overexpression of *ex::phyA* or *ex::appA* in *B. napus* not only increased extracellular phytase activity but also accumulated phytase in seeds. This result indicated that expression of the phytase gene with the signal peptide sequence from *ex* gene may improve both the P uptake in roots and the bioavailability of phytate-P in seeds in one transgenic plant.

In conclusions, transgenic *B. napus* plants expressing *ex::phyA* or *ex::appA* significantly improved P uptake, plant biomass and seed yields when phytate was supplied as the sole P source. Moreover,

Table 1. Shoot biomass, P concentration and P accumulation of transgenic lines and WT grown in quartz sand.

P treatment	Line	Dry weight	P concentration	P accumulation
		g $plant^{-1}$	%	mg $plant^{-1}$
HP	WT	1.03±0.07	0.42±0.04	4.30±0.21
	P3	1.32±0.16*	0.47±0.04	6.07±0.81**
	P11	1.58±0.21**	0.49±0.05	7.22±0.88**
	a18	1.39±0.23*	0.57±0.03*	7.35±1.23**
LP	WT	0.09±0.01	0.14±0.01	0.12±0.01
	P3	0.13±0.01*	0.24±0.01*	0.32±0.03**
	P11	0.13±0.02*	0.17±0.02	0.18±0.01*
	a18	0.10±0.02	0.16±0.00	0.18±0.01*
Phy-P	WT	0.68±0.04	0.31±0.04	2.09±0.43
	P3	0.78±0.08*	0.43±0.04*	3.57±0.34**
	P11	0.93±0.05**	0.31±0.03	2.86±0.34*
	a18	0.93±0.08**	0.27±0.01	2.50±0.21*

Values represent means ± SD of four replicates. A significant difference (*, ** at $P<0.05$ and $P<0.01$, respectively) between WT and transgenic lines in the same P treatment was calculated by LSD test.

Table 2. Effects of different P sources on seed yield, P concentration, P accumulation, phytase activity and phytic acid content of WT and transgenic lines grown in P deficient soil.

P treatment	Line	Seed yield	P concentration	P accumulation	phytase activity	phytic acid content
		g plant^{-1}	%	mg plant^{-1}	U kg^{-1} seeds	mg g^{-1} seeds
HP	WT	4.99±0.60	0.81±0.01	40.55±4.31	n.d.	29.57±1.08
	P3	4.66±0.08	0.83±0.06	35.81±2.44	938.4±32.8	30.20±0.86
	P11	8.11±0.71**	0.71±0.04*	57.51±7.91**	871.9±34.1	19.20±2.81**
	a18	7.76±0.60**	0.82±0.04	64.85±6.74**	1251.6±35.5	19.89±3.12**
LP	WT	1.80±0.12	0.47±0.07	8.84±1.07	n.d.	23.13±0.82
	P3	1.08±0.19	0.45±0.07	4.63±0.06*	620.1±50.7	20.44±1.33*
	P11	1.37±0.26	0.42±0.04	5.25±0.68*	672.5±38.4	11.33±1.51**
	a18	2.63±0.25*	0.35±0.03*	9.31±1.66*	344.7±13.5	16.92±1.25**
Phytate-P	WT	4.32±0.28	0.65±0.03	27.01±1.59	n.d.	24.24±3.54
	P3	3.79±0.16	0.80±0.03*	30.09±0.69*	981.8±79.5	24.44±2.39
	P11	5.21±0.17*	0.62±0.04	32.57±2.97*	1116.1±88.9	13.04±1.38**
	a18	6.91±0.48**	0.62±0.07	39.69±3.97**	707.2±42	16.80±1.92**

Values represent means ± SD of four replicates. A significant difference (*, ** at $P<0.05$ and $P<0.01$, respectively) between the WT and transgenic lines in the same P treatment was calculated by LSD test. HP treatment: 200 mg P_2O_5 kg^{-1} soil with KH_2PO_4; LP treatment: 20 mg P_2O_5 kg^{-1} soil with KH_2PO_4; Phy-P treatment: 200 mg P_2O_5 kg^{-1} soil with phytate. n.d., means not detectable.

transgenic seeds showed high phytase activity and significantly reduced phytic acid content. For the first time in one transgenic plant, these results indicate that the transgenic approach demonstrates great potential for improving plant P acquisition and animal feed supplement.

Materials and Methods

Plant transformation

Expression cassettes of *Aspergillus niger* phytase (*phyA*, Accession no. M94550.1) and *Escherichia coli* phytase (*appA*, Accession no. M58708.1) including a signal peptide sequence from the carrot extensin gene were excised from the 35S-phyA and 35S-appA plasmids (provided by Profs. Ping Wu and Huixia Shou, Zhejiang University, China) with *Hind*III and *EcoR*I, respectively. It was then subcloned into the pBI121 binary vector from which the *GUS* expression cassette had been previously removed with *Hind*III and *EcoR*I resulting in the transformation vectors pBI121-phyA and pBI121-appA. The two microbial phytase genes, under the control of the CaMV 35S promoter and nos terminator, were modified for extracellular secretion by inclusion of a signal peptide sequence (Fig. 1A, B). The pBI121-phyA and appA plasmids were transformed into *B. napus* cv Westar plants by an *Agrobacterium tumefaciens* (LBA4404) mediated method described by Wang *et al.* (2012) [36]. All tissue culture materials were maintained in a controlled growth chamber (24°C/16 h of light, 23°C/8 h of dark, and a light intensity of 80 μmol m^{-2} s^{-1}).

Molecular identification of transformants

For Southern blotting analyses, genomic DNA was isolated from leaves using the cetyltrimethyl ammonium bromide extraction method [37]. Thirty micrograms of genomic DNA digested

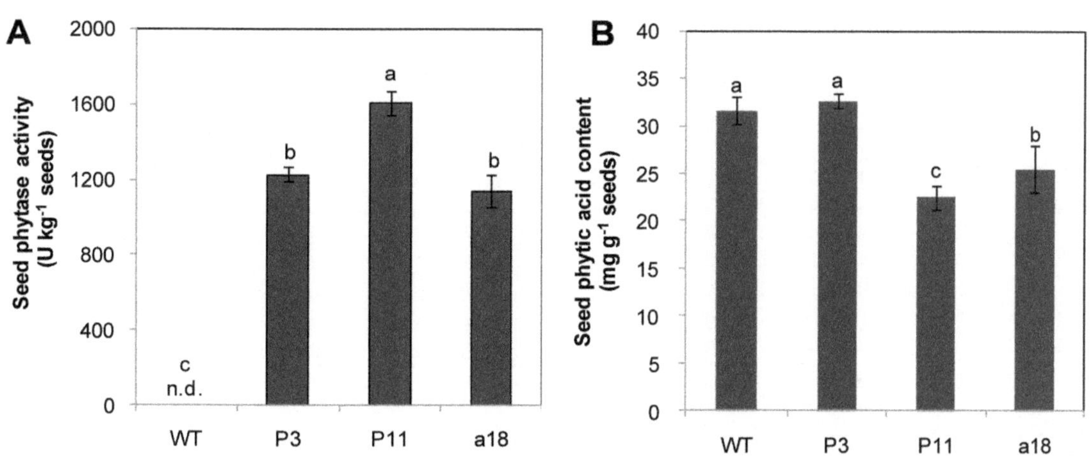

Figure 5. Phytase activity (A) and phytic acid content (B) in seeds of transgenic T_3 lines and WT. Each column is the mean of four replicates with SD. Different letters represent significant differences at the $p<0.05$ level. The seeds were harvested from the field by self-cross. n.d., means not detectable.

with *Eco*RI or *Hin*dIII was separated on a 0.8% agarose gel and transferred to nylon membranes, which were hybridised with probes from *A. niger phyA* and *E. coli appA* α-P^{32} dCTP-labelled PCR products. The 441 bp PCR product was amplified with *A. niger phyA* specific primers 5′-ACAGACACGAAGTGACC-TACC-3′ and 5′-CCAAGCAGACGAGAATCC-3′. Specific primers 5′-GAACGTGACTGACGCGATCC-3′ and 5′-ACGC-CAGCGTTCAAACACC-3′ were used to amplify 565 bp of *E. coli appA*.

For northern blotting analysis, total RNA was extracted from roots with TRIzol reagent (Invitrogen, CA, USA) according to the manufacturer's instruction, and separated by electrophoresis on agarose formaldehyde denaturing gels. The separated RNA was then transferred to nylon membranes, which were hybridised with the probes from the α-P^{32} dCTP-labelled PCR products of *A. niger phyA* and *Ecoli appA*.

Solution culture experiment

Seeds from WT and transgenic canola lines P3, P11 and a18 were surface-sterilised for 1 min using 10% (*w/v*) sodium hypochlorite and washed four times in deionised water. The surface-sterilised seeds were then germinated on moistened gauze that was fixed to a black plastic tray filled with deionised water at 25°C. After 5 to 6 days, uniformly sized seedlings were transferred to a 10 L black plastic tray (53.0 cm × 34.0 cm×7.5 cm) filled with Hoagland's nutrient solution modified to one-fourth strength [38]. After 5 days, the nutrient solution was replaced with half strength nutrient solution, which was replaced again with full strength nutrient solution after another 5 days. The composition of the full strength nutrient solution was 5.04 mM KNO_3, 0.25 mM KH_2PO_4, 1.99 mM $MgSO_4{\cdot}7H_2O$, 5.0 mM $Ca(NO_3)_2{\cdot}4H_2O$, 0.32 μM $CuSO_4{\cdot}5H_2O$, 0.77 μM $ZnSO_4{\cdot}7H_2O$, 9.14 μM $MnCl_2{\cdot}4H_2O$, 0.37 μM $NaMoO_4{\cdot}2H_2O$, 46 μM H_3BO_3 and 0.07 mM EDTA-Fe. The pH of the nutrient solution was controlled to 5.5±0.2 with 1 M NaOH or HCl every 2 days. The seedlings were grown in an illuminated culture room at a cycle of 16 h/24°C day and 8 h/22°C night and a light intensity of 300–320 μmol m^{-2} s^{-1} with a relative humidity of 65–80%.

In the low P stress experiment, three P levels were designed with four replicates per level. Both WT and transgenic canola seedlings were grown for 18 days in high P (HP) and low P (LP) nutrient solutions containing 250 μM and 5 μM P, respectively. For Phy-P treatment, the seedlings were grown in 250 μM P ($C_6H_6O_{24}P_6Na_{12}$, phytate, P3168, Sigma) for 18 days. In the LP and Phy-P solutions, KH_2PO_4 was replaced by K_2SO_4.

Root exudate collection was performed according to the method described by George et al. (2004) and Richardson et al. (2001) with some modifications [22,20]. The seedlings (2 seedlings per biological replicate) from different P treatments were transferred to 15 mL of 15 mM MES (2-(*N*-morpholino)-ethane-sulfonic acid monohydrate) buffer (pH 5.5) containing 0.5 mM $CaCl_2$ (MES/Ca buffer) for 6 h. The fresh weight of shoot and root material was measured and frozen in liquid nitrogen. The exudate was filled to 15 mL with the MES/Ca buffer and stored at 4 °C prior to the enzyme assays. All treatments contained four biological replicates and two plants in each biological replicate.

For the enzyme extraction in leaves and roots, approximately 0.4 g of leaves or roots was ground in liquid nitrogen and homogenised with 2 mL MES/Ca buffer containing 1 mM EDTA. Extracts were then centrifuged (10 min, 12,000 g) and subsequently used to determine phytase activities. All treatments contained four biological replicates and two plants in each biological replicate.

Determination of enzyme activities

Phytase and APase activities were determined using $C_6H_6O24_4P_6Na_{12}$ (phytate, P3168, Sigma) and *p*-nitrophenyl phosphate (*p*NPP), respectively. The activities were expressed in either mU g^{-1} root FW or mU mg^{-1} protein, where 1 U is defined as the release of 1 mmol of P_i (phytase) or *p*-nitrophenol (*p*NP) (APase) min^{-1} under the assay conditions described.

Phytase activity in extracts was measured as described by Richardson et al. (2000) with some modifications [17]. The phytase assay was conducted by incubating 500 μL extract in a total volume of 1 mL MES/Ca buffer (pH 5.5) at 37°C for 60 min in the presence of 1 mM $C_6H_6O24_4P_6Na_{12}$ (pH 5.5). The reaction was terminated by the addition of an equal volume of 10% TCA (trichloroacetic acid). Phytase activity was calculated from the release of P_i in centrifuged samples (12,000 g, 10 min) and was determined using the molybdate-blue assay [39]. The protein content of root extracts was measured using bovine serum albumin standards, according to Bradford (1976) [40].

Extracellular phytase and APase activity analysis were performed as described by Richardson et al. (2000) with some modifications [17]. Root exudates (950 μL) were combined with 50 μL substrates (final concentration 1 mM for phytate and 0.5 mM for *p*NPP) and incubated at 37°C for 60 min (phytase) or 30°C for 30 min (APase). Reactions were terminated with an equal volume of either 10% TCA (phytase) or 0.25 M NaOH (APase) at the end of the incubation. APase activity was calculated from the release of *p*NP, as determined spectrophotometrically at 405 nm relative to standard solutions. Phytase activity was calculated as described above.

Quartz sand culture experiment

For the quartz sand culture experiment, seeds of WT and transgenic lines were surface-sterilised and planted into a plastic box (17 cm×8 cm×17 cm) filled with sterilised quartz sand. For inhibiting moss growth, a black plastic membrane was used to cover the surface of the quartz sand. The plants were fertilised with the modified Hoagland's nutrient solution described above with different P concentrations. In high P (HP) and low P (LP) treatments, nutrient solutions contained 250 μM and 5 μM P (KH_2PO_4), respectively. In Phy-P treatments, phytate ($C_6H_6O_{24}P_6Na_{12}$, P3168, Sigma) was added as the sole P source and the P concentration was 250 μM. Each treatment contained four biological replicates with two plants in each replicate. Sixty days after germination, the plants were harvested for shoot biomass and P content determination. The total P was determined according to the vanadomolybdate method described by Westerman (1990) after digesting the samples in concentrated H_2SO_4-$HClO_4$ using a digestion block system [41].

Soil culture experiment

Seeds of WT and transgenic lines were surface sterilised and then sowed into a pot filled with 7.0 kg of soil from the Huazhong Agricultural University campus (Hubei province, China, 30.47°S 114.35°E). Soil characteristics were as follows: pH 6.19, total nitrogen 0.85 g kg^{-1}, total phosphate 0.21 g kg^{-1}, Olsen-P 2.68 mg kg^{-1} and organic matter 5.14 g kg^{-1}. Three treatments were employed including 20 mg P_2O_5 kg^{-1} soil (LP), 200 mg P_2O_5 kg^{-1} soil (HP) (added as KH_2PO_4) and 200 mg P_2O_5 kg^{-1} soil added as $C_6H_6O24_4P_6Na_{12}$ (phytate, P3168, Sigma). Nitrogen (N) and potassium (K) fertilisers were mixed with the soil before planting at a concentration of 200 mg N and K_2O kg^{-1} soil in the form of $(NH_4)_2SO_4$ and KCl. Other nutrients were supplied as a nutrient solution containing 25 mg kg^{-1} $MgSO_4{\cdot}7H_2O$, 2.86 mg kg^{-1} H_3BO_3, 1.81 mg kg^{-1} $MnCl_2{\cdot}4H_2O$, 0.22 mg kg^{-1}

$ZnSO_4 \cdot 7H_2O$, 0.08 mg kg^{-1} $CuSO_4 \cdot 5H_2O$ and 0.02 mg kg^{-1} $(NH_4)_6MO_7O_2 \cdot 4H_2O$. Eight months after germination, plants were harvested for yield evaluation, seed phytase activity and phytic acid content analysis. Total P in seeds was determined as described above. Each treatment contained four biological replicates and two plants in each replicate.

Determination of phytase activity and phytic acid content in seeds

Seeds were powdered in liquid nitrogen, and 100 mg of sample was extracted in 2 mL of enzyme buffer (0.1 M sodium acetate, pH 5.5, 1 mM $CaCl_2$, and 0.1 g L^{-1} Tween 20) by thorough vortex mixing, followed by incubation on ice for 10 min. Samples were centrifuged for 5 min at 15,000 g, 4°C [42]. The supernatants were transferred into fresh tubes and used for phytase activity measurements with the method described by Wyss et al. (1999) [5]. Briefly, 200 µL supernatant was mixed with 700 µL of 0.1 M sodium acetate buffer and 100 µL of 1 mM phytate and incubated at 37°C for 30 min. The reaction was stopped by adding 1 mL of 10% (v/v) TCA. As a control, TCA was added to the supernatant, followed by the phytate substrate and incubated in the same conditions. Phytase activity was calculated as described above.

Phytic acid content was assayed according to the methods from Latta and Eskin (1980) and Kwanyuen and Burton (2005) with some modifications [43,44]. Samples of 100 mg were extracted with 2 mL of 0.4 M HCl–15% TCA at 23°C for 3 h. After centrifugation at 3,000 g for 15 min, 20 µL of the supernatant was mixed with 280 µL of 36.3 mM NaOH and 200 µL of 0.03% (w/v) $FeCl_3 \cdot 6H_2O$–0.3% sulfosalicylic acid. The mixture was centrifuged at 2,000 g for 10 min. The absorbance of the supernatant at 500 nm was measured using an Ultrospec 1100 *pro* spectrophotometer (Biochrom Ltd, England).

Statistical analysis

The data were analysed statistically using the SAS program (Statistical Analysis Systems Institute, version 8.1).

Acknowledgments

We thank Professors Ping Wu and Huixia Shou (Zhejiang University, China) for kindly providing for the vectors 35S-phyA and 35S-appA containing the *A. niger phyA* and *E.coli appA* expression cassettes, respectively.

Author Contributions

Conceived and designed the experiments: YW FX. Performed the experiments: YW GD XY. Analyzed the data: YW XY GD FX. Contributed reagents/materials/analysis tools: YW XY GD. Wrote the paper: YW XY GD FX.

References

1. Vance CP, Uhde-Stone C, Allan DL (2003) Phosphorus acquisition and use: critical adaptations by plants for securing a nonrenewable resource. New Phytol 157: 423–447.
2. Dalal RC (1977) Soil organic phosphorus. Adv Agron 29: 83–117.
3. Schachtman DP, Reid RJ, Ayling SM (1998) Phosphorus uptake by plants: from soil to cell. Plant Physiol 116: 447–453.
4. Brinch-Pedersen H, Sørensen LD, Holm PB (2002) Engineering crop plants: getting a handle on phosphate. Trends Plant Sci 7: 118–125.
5. Wyss M, Pasamontes L, Friedlein A, Rémy R, Tessier M, et al. (1999) Biophysical characterization of fungal Phytases (*myo-inositol* hexakisphosphate phosphohydrolases): molecular size, glycosylation pattern, and engineering of proteolytic resistance. Appl Environ Microb 65: 359–366.
6. Lei X-G, Porres JM, Mullaney EJ, Brinch-Pedersen H (2007) Phytase: source, structure and application. In: Polaina J, MacCabe AP, editors. Industrial Enzymes. Springer Netherlands. pp. 505–529.
7. Brinch-Pedersen H, Hatzack F, Stöger E, Arcalis E, Pontopidan K, et al. (2006) Heat-stable phytases in transgenic wheat (*Triticum aestivum* L.): deposition pattern, thermostability, and phytate hydrolysis. J Agric Food Chem 54: 4624–4632.
8. Brinch-Pedersen H, Olesen A, Rasmussen SK, Holm PB (2000) Generation of transgenic wheat (*Triticum aestivum* L.) for constitutive accumulation of an *Aspergillus* phytase. Mol Breeding 6: 195–206.
9. Chiera JM, Finer JJ, Grabau EA (2004) Ectopic expression of a soybean phytase in developing seeds of *Glycine max* to improve phosphorus availability. Plant Mol Biol 56: 895–904.
10. Bilyeu KD, Zeng P, Coello P, Zhang ZJ, Krishnan HB, et al. (2008) Quantitative conversion of phytate to inorganic phosphorus in soybean seeds expressing a bacterial phytase. Plant Physiol 146: 468–477.
11. Ponstein AS, Bade JB, Verwoerd TC, Molendijk L, Storms J, et al. (2002) Stable expression of phytase (*phyA*) in canola (*Brassica napus*) seeds: towards a commercial product. Mol breeding 10: 31–44.
12. Peng RH, Yao QH, Xiong AS, Cheng ZM, Li Y (2005) Codon-modifications and an endoplasmic reticulum-targeting sequence additively enhance expression of an *Aspergillus* phytase gene in transgenic canola. Plant Cell Rep 25: 124–132.
13. Chen RM, Xue GX, Chen P, Yao B, Yang WZ, et al. (2008) Transgenic maize plants expressing a fungal phytase gene. Transgenic Res 17: 633–643.
14. Raghothama KG (1999) Phosphate Acquisition. Annu Rev Plant Physiol Plant Mol Biol 50: 665–693.
15. Li M, Osaki M, Madhusudana Rao I, Tadano T (1997) Secretion of phytase from the roots of several plant species under phosphorus-deficient conditions. Plant Soil 195: 161–169.
16. Hayes J, Simpson R, Richardson A (2000) The growth and phosphorus utilisation of plants in sterile media when supplied with inositol hexaphosphate, glucose 1-phosphate or inorganic phosphate. Plant Soil 220: 165–174.
17. Richardson AE, Hadobas PA, Hayes JE (2000) Acid phosphomonoesterase and phytase activities of wheat (*Triticum aestivum* L.) roots and utilization of organic phosphorus substrates by seedlings grown in sterile culture. Plant Cell Environ 23: 397–405.
18. Idriss EE, Makarewicz O, Farouk A, Rosner K, Greiner R, et al. (2002) Extracellular phytase activity of *Bacillus amyloliquefaciens* FZB45 contributes to its plant-growth-promoting effect. Microbiol 148: 2097–2109.
19. Unno Y, Okubo K, Wasaki J, Shinano T, Osaki M (2005) Plant growth promotion abilities and microscale bacterial dynamics in the rhizosphere of *Lupin* analysed by phytate utilization ability. Environ Microbiol 7: 396–404.
20. Richardson AE, Hadobas PA, Hayes JE (2001) Extracellular secretion of *Aspergillus* phytase from Arabidopsis roots enables plants to obtain phosphorus from phytate. Plant J 25: 641–649.
21. Mudge SR, Smith FW, Richardson AE (2003) Root-specific and phosphate-regulated expression of phytase under the control of a phosphate transporter promoter enables *Arabidopsis* to grow on phytate as a sole P source. Plant Science 165: 871–878.
22. George TS, Richardson AE, Hadobas PA, Simpson RJ (2004) Characterization of transgenic *Trifolium subterraneum* L. which expresses *phyA* and releases extracellular phytase: growth and P nutrition in laboratory media and soil. Plant Cell Environ 27: 1351–1361.
23. Xiao K, Harrison MJ, Wang ZY (2005) Transgenic expression of a novel *M. truncatula* phytase gene results in improved acquisition of organic phosphorus by *Arabidopsis*. Planta 222: 27–36.
24. Ma XF, Wright E, Ge Y, Bell J, Xi Y, et al. (2009) Improving phosphorus acquisition of white clover (*Trifolium repens* L.) by transgenic expression of plant-derived phytase and acid phosphatase genes. Plant Science 176: 479–488.
25. Li G, Yang S, Li M, Qiao Y, Wang J (2009) Functional analysis of an *Aspergillus* ficuum phytase gene in *Saccharomyces cerevisiae* and its root-specific, secretory expression in transgenic soybean plants. Biotechnol Lett 31: 1295–1301.
26. Liu JF, Wang XF, Li QL, Li X, Zhang GY, et al. (2011) Biolistic transformation of cotton (*Gossypium hirsutum* L.) with the *phyA* gene from *Aspergillus ficuum*. Plant Cell Tissue Organ Cult 106: 207–214.
27. Yang M, Ding GD, Shi L, Feng J, Xu FS, et al. (2010) Quantitative trait loci for root morphology in response to low phosphorus stress in *Brassica napus*. Theor Appl Genet 121: 181–193.
28. George TS, Richardson AE (2008) Potential and limitations to improving crops for enhanced phosphorus utilization. In: PJ White, JP Hammond (Eds.), Ecophysiology of Plant-Phosphorus Interactions, Springer, Netherlands (2008), pp. 247–270.
29. Zimmermann P, Zardi G, Lehmann M, Zeder C, Amrhein N, et al. (2003) Engineering the root-soil interface via targeted expression of a synthetic phytase gene in trichoblasts. Plant Biotechnol J 1: 353–360.
30. Hong CY, Cheng KJ, Tseng TH, Wang CS, Liu LF, et al. (2004) Production of two highly active bacterial phytases with broad pH optima in germinated transgenic rice seeds. Transgenic Res 13: 29–39.
31. Golovan S, Wang G, Zhang J, Forsberg CW (1999) Characterization and overproduction of the *Escherichia coli appA* encoded bifunctional enzyme that exhibits both phytase and acid phosphatase activities. Canadian J Microbiol 46: 59–71.

32. Lottl JNA, Ockenden I, Raboy V, Batten GD (2000) Phytic acid and phosphorus in crop seeds and fruits: a global estimate. Seed Sci Res 10: 11–33.
33. Chen J, Varner JE (1985) An extracellular matrix protein in plants: characterization of a genomic clone for carrot extensin. EMBO J 4: 2145–2151.
34. Kieliszewski MJ, Lamport DTA, Kieliszewski MJ, Lamport DTA (1994) Extensin: repetitive motifs, functional sites, post-translational codes, and phylogeny. Plant J 5: 157–172.
35. George TS, Simpson RJ, Hadobas PA, Richardson AE (2005) Expression of a fungal phytase gene in *Nicotiana tabacum* improves phosphorus nutrition of plants grown in amended soils. Plant Biotechnol J 3: 129–140.
36. Wang Y, Xu H, Kou JJ, Shi L, Zhang CY, et al. (2013) Dual effects of transgenic *Brassica napus* overexpressing CS gene on tolerances to aluminum toxicity and phosphorus deficiency. Plant Soil 362: 231–246.
37. Murray MG, Thompson WF (1980) Rapid isolation of high molecular weight plant DNA. Nucl Acids Res 8: 4321–4326.
38. Hoagland DR, Arnon DI (1950) The water-culture method for growing plants without soil. California Department of Agriculture experimental station circular 347.
39. Murphy J, Riley JP (1962) A modified single solution method for the determination of phosphate in natural waters. Anal Chim Acta 27: 31–36.
40. Bradford MM (1976) A rapid and sensitive method for the quantitation of microgram quantities of protein utilizing the principle of protein-dye binding. Anal Biochem 72: 248–254.
41. Westerman RL (1990) Soil testing and plant analysis. Soil Science Society of America book series no. 3.
42. Bilyeu KD, Zeng P, Coello P, Zhang ZJ, Krishnan HB, et al. (2008) Quantitative conversion of phytate to inorganic phosphorus in soybean seeds expressing a bacterial phytase. Plant physiol 146: 468–477.
43. Latta M, Eskin M (1980) A simple and rapid colorimetric method for phytate determination. J Agric Food Chem 28: 1313–1315.
44. Kwanyuen P, Burton JW (2005) A simple and rapid procedure for phytate determination in soybeans and soy products. J Am Oil Chem Soc 82: 81–85.

9

Variations in Soil Microbial Communities and Residues Along an Altitude Gradient on the Northern Slope of Changbai Mountain, China

Bin Zhang[1,2], Chao Liang[3], Hongbo He[1]*, Xudong Zhang[1,4]*

1 State Key Laboratory of Forest and Soil Ecology, Institute of Applied Ecology, Chinese Academy of Sciences, Shenyang, P. R. China, **2** Graduate University of Chinese Academy of Sciences, Beijing, P. R. China, **3** Department of Soil Science, University of Wisconsin-Madison, Madison, Wisconsin, United States of America, **4** National Field Observation and Research Station of Shenyang Agroecosystems, Shenyang, P. R. China

Abstract

Altitudinally-defined climate conditions provide specific vegetation types and soil environments that could influence soil microbial communities, which in turn may affect microbial residues. However, the knowledge is limited in terms of the degree to which microbial communities and residues present and differ along altitude. In this study, we examined the soil microbial communities and residues along the northern slope of Changbai Mountain, China using phospholipid fatty acid (PLFA) and amino sugar analysis, respectively. Soil samples were taken from five different vegetation belts defined by climates. Principal component analysis (PCA) revealed substantial differences in soil microbial community composition among study sites, appeared to be driven primarily by soil pH and C/N ratio on the first principal component (PC1) which accounted for 50.7% of the total sample variance. The alpine tundra was separated from forest sites on the second principal component (PC2) by a signifiscantly higher amount of fungal PLFA (18:2ω6,9). Soil pH and C/N ratio were also correlated with the ratios of Gram-positive to Gram-negative bacteria (Gm^+/Gm^-), glucosamine to galactosamine (GluN/GalN), and glucosamine to muramic acid (GluN/MurA). Both total PLFAs and amino sugars were positively correlated with soil organic carbon, inorganic nitrogen, available phosphorus and potassium. We concluded that soil pH and C/N ratio were the most important drivers for microbial community structure and amino sugar pattern, while substrate availability was of great importance in determining the concentrations of microbial communities and residues. These findings could be used to facilitate interpretation of soil microbial community and amino sugar data derived from measurements in latitude or managed forests.

Editor: Hauke Smidt, Wageningen University, The Netherlands

Funding: This work was supported by the National Key Basic Research Foundation of China (2011CB403202), the Natural Science Foundation of China (41130524) and the International Partnership Program of Chinese Academy of Sciences (KZCX2-YW-T06). The funders had no role in study design, data collection and analysis, decision to publish, or preparation of the manuscript.

Competing Interests: The authors have declared that no competing interests exist.

* E-mail: hehongbo@iae.ac.cn (HH); xdzhang@iae.ac.cn (XZ)

Introduction

Soil microorganisms are of great importance to carbon (C) and nitrogen (N) cycling and storage [1], [2], ecosystem functioning [3], and global climate change [4]. Soil microbial communities have consequently received great interests for decades. It is well documented that a number of biotic and abiotic factors, such as vegetation type [5–8], temperature [9], water content [10], pH [11], [12], soil type [13], and soil depth [14], influence soil microbial communities, which in turn may affect the turnover and accumulation of soil microbial residues. Our understanding of soil microorganisms and their interactions with environmental factors is improving; however, the influences of climatic regimes on soil microbial communities and residues are still insufficiently investigated, with some existing studies on latitude [15–17] but little attention to altitude. The altitudinally-defined vegetation belts on mountain slopes are counterparts to the latitudinally-controlled climatic zones. Temperature gradients in mountains may also represent an analogue to those related to latitude since the mean annual temperature (MAT) decreases with increasing in both altitude and latitude. This makes mountain regions well-suited for the study of climate impacts because of the pronounced climatic gradients on a comparatively small scale [18].

The altitude-induced environmental conditions have been reported to influence soil microbial communities in mountains. For example, a decrease in fungal biomass [19] and diversity [20] with increasing altitude were observed in the Austrian Central Alps. A negative correlation between bacterial population and altitude was reported by Ma et al. [21] in the cold temperate Kalasi Lake and by Giri et al. [22] in a tropical dry deciduous forest. Margesin et al. [19] pointed out several shifts in microbial community composition with altitude in the Austrian Central Alps, such as a significant increase in the relative abundance of fungi and Gram-negative (Gm^-) bacteria. In contrast, Männistö et al. [23] reported that altitude-varied changes in microbial community composition were controlled by pH rather than temperature fluctuations in Arctic fjelds of Finnish Lapland. Shen et al. [24] also found that soil pH drives the spatial distribution of bacterial communities along elevation on Changbai Mountain with a bar-coded pyrosequencing technique. Moreover, Djukic et

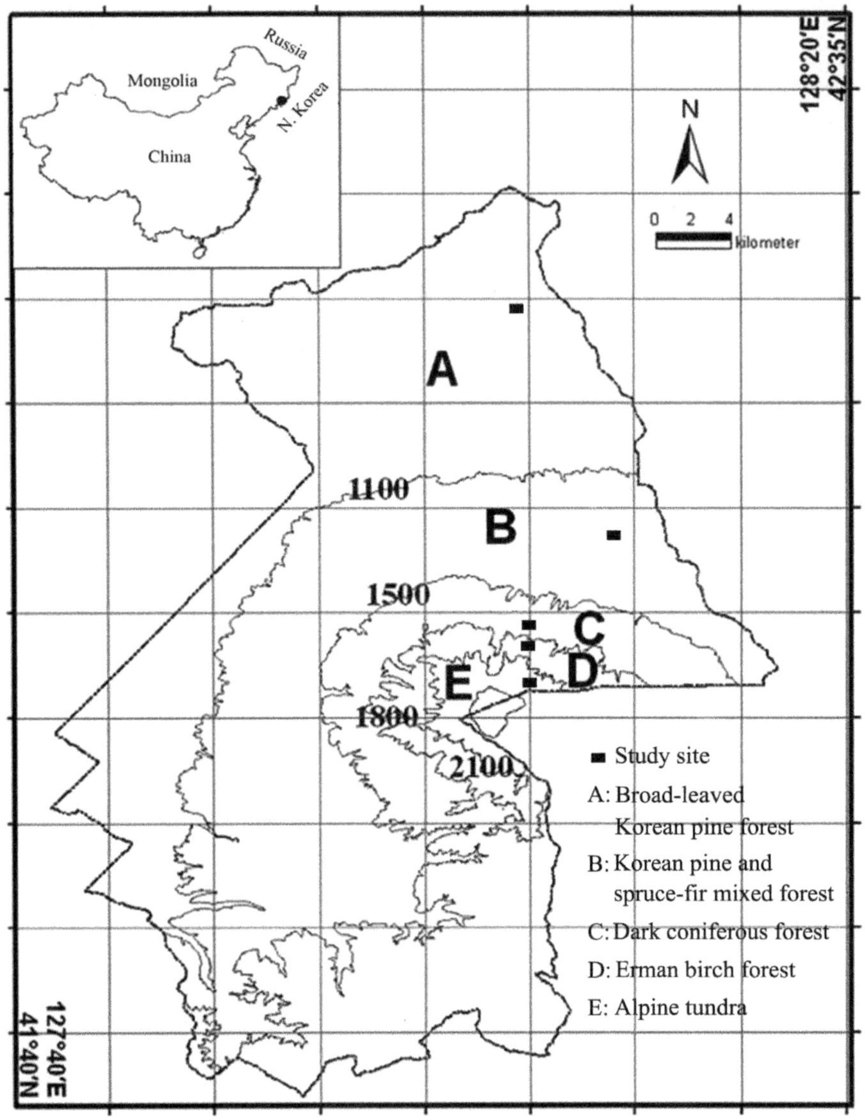

Figure 1. Map of study area and the five vertical vegetation belts along the northern slope of Changbai Mountain, China. The locations of the study sites were marked with rectangles.

al. [25] indicated that the microbial community structure was connected with decomposition conditions and changes in vegetation composition along an elevation gradient in the Austrian Limestone Alps. These studies suggest that soil microbial communities can resist simple prediction and are linked to a wide range of factors than the mere altitude. On the other hand, anabolic activity of microbial communities contributed to soil organic matter (SOM) pool by introducing a mass of the successive refractory residues. The accumulation of microbial residues is highly dependent on the quantity and quality of substrate inputs during laboratory incubation [26–28], and several field studies have reported that microbial residues could be preferentially decomposed in soils with poor substrate availability [29], [30]. Different plant species can also specifically influence the contents and patterns of microbial residues in both laboratory and field experiments [29], [31]. Amelung et al. [17] demonstrated a parabolic relationship between the microbial residues and MAT in native grassland soils along a climosequence in North America. Because these factors act in concert along altitude, disentangling their interactions is important for understanding and predicting how soil microbial communities and residues respond to future climate change in mountain soils.

Various methods are available to characterize soil microorganisms. Phospholipid fatty acid (PLFA) analysis is widely accepted as a sensitive tool to indicate viable microbial biomass and fingerprint microbial community composition [32]. Amino sugar analysis is also routinely used to indicate the storage of microbial residues [33]. Furthermore, different origins of individual amino sugars could provide important information on the relative contribution of fungi and bacteria to SOM turnover and accumulation [34–36]. The chitin of fungal cell walls is the major source of glucosamine (GluN) in soil, although bacterial cell walls and the exoskeletons of soil invertebrates also make some contribution [37]. Muramic acid (MurA) originates uniquely from bacteria [33], [37]. Despite galactosamine (GalN) accounts for 30–50% of the amino sugar pool, little is known about its origin [28], [36]. Combining viable microbial biomass indicator (PLFAs) and microbial cell wall residues (amino sugars) might provide important information about microbial significance in SOM cycling and storage [38]. However, few studies have considered both methods simultaneously.

The objectives of this study were: (1) to examine soil microbial communities by PLFAs and microbial residues by amino sugars along an altitude gradient of Changbai Mountain; (2) to determine if and how differences in soil microbial communities and residues are correlated with environmental factors; and (3) to explore the control and feedback between PLFAs and amino sugars.

Materials and Methods

Study area and soil sampling

This study was conducted on the northern slope of Changbai Mountain in Jilin Province, northeastern China (Figure 1). The permission is issued by Changbai Mountain Administrative Committee. According to the Köppen Climate Classification, the climate of this region is humid continental. From the lowest part of the mountain at 740 m to the summit at 2691 m, the MAT decreases from 2.8 to −7.3°C, the mean annual precipitation (MAP) increases from 750 to 1340 mm [39]. Climatic and topographic variations along the altitude gradient result in five vertical vegetation belts on the northern slope (Figure 1). Therefore, five study sites (one in each vegetation belt) were selected to represent a climosequence from the montane ecosystems to the subalpine and alpine zones (Figure 1, Table 1). The influence of human activities was minimized by selecting sites that have not been disrupted.

Four replicate soil monoliths (25×25 cm) were collected from each of the five sites on July 20, 2009. Samples were taken from the A horizon after the litter and humus layer was removed carefully. The horizon thickness varied from 5 to 11 cm. Field-moist samples were placed in a plastic bag and kept cool until processed in the laboratory. After removal of visible fresh roots and plant material, the soils were homogenized and passed through a 2-mm sieve. One sub-sample was air-dried for soil chemical analyses. Another sub-sample was freeze-dried and used for PLFA extraction.

Soil chemical analyses

Soil total C and N were determined by dry combustion on ground samples (100-mesh) using a C/N analyzer (LECO Corporation, MI, USA). Because these soils are free of carbonates, the total C content is equivalent to soil organic C (SOC) content. Soil pH was measured in a 1:2.5 soil/water suspension. Available phosphorus (P) and potassium (K) were determined, respectively, by Bray-1 method and ammonium acetate extraction method [40]. NH_4 and NO_3 were extracted with 2 M KCl and analyzed on a TRAACS 2000 autoanalyzer by using the Berthelot reaction method and cadmium reduction method, respectively.

Amino sugar analysis was conducted according to Zhang and Amelung [41]. Briefly, samples were hydrolyzed with 6 M HCl at 105 °C for 8 h, and then the solution was filtered and purified by neutralization. After drying of the supernatant, amino sugars were washed out from the residues using methanol, transformed into aldononitrile derivatives, and then extracted from the aqueous solution with dichloromethane. The amino sugar derivatives were separated on an Agilent 6890A gas chromatography (GC, Agilent Technologies, USA) equipped with an HP-5 capillary column (30 m×0.32 mm×0.25 μm) and a flame ionization detector (FID). Amino sugars were quantified based on the internal standard myo-inositol which was added prior to purification. Methyl-glucamine was used as a recovery standard before derivatization to monitor recovery efficiency. We calculated the total amino sugar content as the sum of the four amino sugars determined. We used GluN as the biomarker for fungal cell-wall residues and MurA for bacterial cell-wall residues. Mannosamine (ManN) is not considered alone due to its ambiguous origin and trace amounts.

Phospholipid fatty acid analysis

PLFA extraction was conducted for each sample following the procedure of Bligh and Dyer [42] after modifications by Bossio et al. [13]. Briefly, lipids were extracted in a single-phase chloroform-methanol-citrate buffer (1:2:0.8) system. Phospholipids were separated from neutral lipids and glycolipids on silica solid phase extraction columns (Supelco, Inc., Bellefonte, USA). After methylation of the polar lipids, PLFA methyl esters were analyzed by an Agilent 6890A GC equipped with an HP-5 capillary column (30 m×0.32 mm×0.25 μm) and a FID. Nonadecanoic acid methyl ester (19:0, Sigma-Aldrich) was added as an internal standard when the samples were dissolved in 150 μL of hexane before GC analysis. Super purified nitrogen was used as the carrier gas with a flow rate of 0.8 mL min^{-1}. The Supelco 37 Component FAME Mix and Bacterial Acid Methyl Esters (Sigma-Aldrich) were used for peak identification and quantification. A total of thirty-four different PLFAs including saturated, monounsaturated, polyunsaturated, cyclopropyl, and methyl fatty acids were identified. Fifteen PLFAs (14:0, i15:0, a15:0, 15:0, i16:0, 16:1ω7c, 16:0, i17:0, cy17:0, 17:0, 18:2ω6,9, 18:1ω9c, 18:1ω9t, 18:0, 20:0) consistently presented in the samples were used for data analysis. The fatty acid signatures 14:0, i15:0, a15:0, 15:0, i16:0, 16:1ω7c, i17:0, cy17:0, and 17:0, which are considered to be of bacterial origin [32], were used as biomarkers for bacterial biomass. The fatty acid 18:2ω6,9, which is known to correlate well with ergosterol, was used as an indicator for fungal biomass [32]. We used fatty acids i15:0, a15:0, i16:0, and i17:0 to represent Gram-positive (Gm^+) bacteria, whereas cy17:0, 16:1ω7c, 14:0, 15:0, and 17:0 to represent Gm^- bacteria [25]. The sum of all PLFAs was used to represent total microbial lipid biomass.

Statistical analyses

To explore variation in soil microbial community composition among study sites, the mole percentages (mol%) of individual PLFAs were subjected to principal component analysis (PCA) after standardizing to unit variance. To explore the relationships between individual PCs and environmental variables, redundancy analysis (RDA) was carried out with the rda function in the 'vegan' library in R. One-way analysis of variance (ANOVA) procedures, with Tukey's honestly significant difference (HSD) as post hoc, were used to test significant differences in soil properties, sums and ratios of various microbial lipid groups, and concentrations and patterns of amino sugars among study sites. Correlations between variables were calculated with the Pearson correlation coefficients. Statistical analysis was performed using the software package SPSS 13.0 for Windows (SPSS Inc. Chicago, USA). Figures were generated by Sigmaplot 10.0 (Systat Software Inc.) and the R package.

To explore the amino sugar data explained by a linear model of individual PLFAs and soil properties, RDA was carried out again. The ordination of the response variables (amino sugar data) was constrained by a multiple regression on the explanatory variables (individual PLFAs and soil properties).The explanatory PLFAs were selected by three rules: (1) significantly correlated with individual amino sugars; (2) indicative of specific microbial groups; and (3) higher in mol% (>5%). NO_3, K, and P were excluded from the explanatory variables as they were significantly correlated with SOC. The significance of the RDA results was tested by permutation test (999 permutations). Since $P<0.001$, we presented

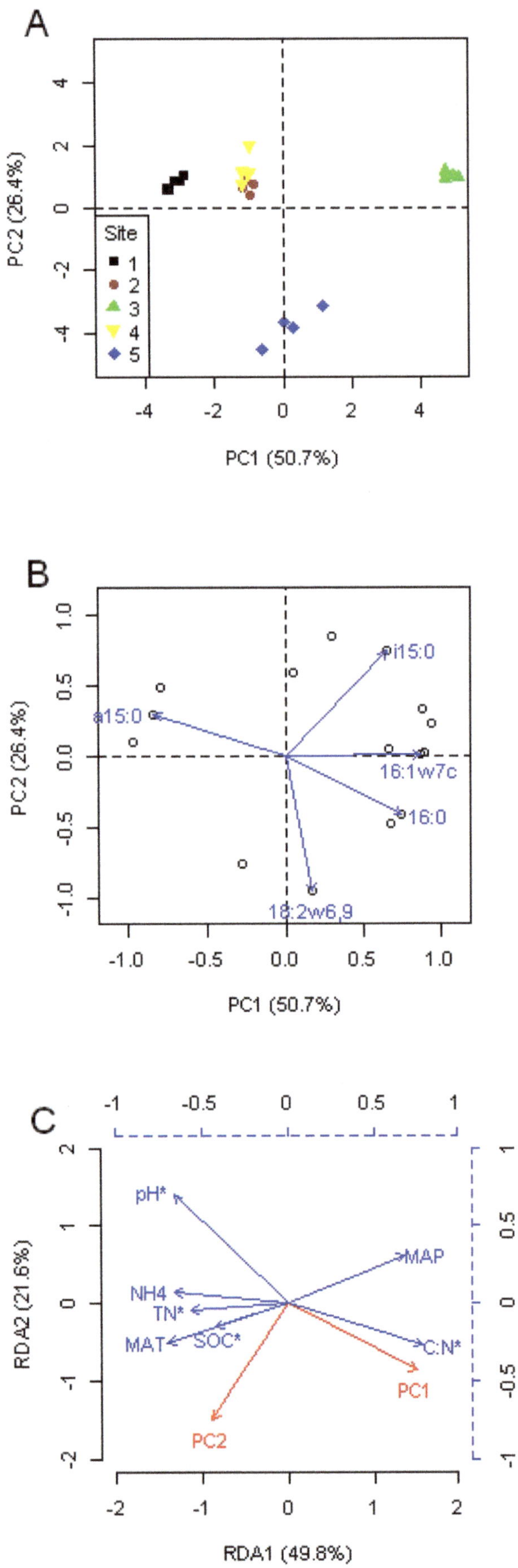

Figure 2. Multivariate statistical analysis of the phospholipids fatty acids (PLFAs) data. A) Principal component analysis (PCA) of PLFAs (mol%) in different study sites along an altitude gradient on the northern slope of Changbai Mountain, China. PC1 explains 50.7% of the variance in the PLFA data, PC2 explains 26.4%. Sites 1–5 represent a climosequence from the montane to the subalpine and alpine vegetation zones. B) Loading plot for individual PLFAs. The PLFAs most responsible for the variations in soil microbial community composition among study sites were presented by vectors. C) Redundancy analysis (RDA) ordination biplot showing relationships between the first two PCs and environmental variables. The environmental variables followed by an asterisk indicate significant influences on the PCs. Environmental variables are scaled by the blue dashed axes. SOC, soil organic carbon; TN, total nitrogen; MAT, mean annual air temperature; MAP, mean annual precipitation.

the ordination biplot which shows sites as points, amino sugars, individual PLFAs, and soil properties as vectors. The angles in the biplot between response and explanatory variables, and between response variables themselves or explanatory variables themselves, reflect their correlations. The proportion of explained variation was calculated by using adjusted R-squared values as described by Peres-Neto et al. [43]. The biplot was generated by the R package.

Results

Soil characteristics

Selected soil characteristics were significantly different among study sites ($P<0.05$, Table 1). Although MAT decreases and MAP increases with increasing elevation, there was no altitudinal changing trend for these soil properties. However, if we excluded site 2 from the study sites, a significant decrease was observed in SOC with increasing altitude ($P<0.05$, Table 1). The C/N ratios varied from 14.9 to 21.2, with the lowest value found at the site that is dominated by broadleaf trees (site 1) and the highest at the site with the highest percentage of conifers (site 3). All study sites were acidic (Table 1). The lowest soil pH (3.95) was found at the site with the highest C/N ratio (site 3) and the highest (5.10) at the site with the lowest C/N ratio (site 1). Soil available P, K, and inorganic N ($NH_4 + NO_3$) ranged from 8.10–27.2, 163–300, and 53.0–123 mg kg^{-1}, respectively (Table 1).

Phospholipid fatty acids

Total microbial lipid biomass was significantly higher in site 1 (302 nmol g^{-1} soil) and significantly lower in sites 2, 4, and 5 (125–129 nmol g^{-1} soil) in comparison with site 3 (195 nmol g^{-1} soil, $P<0.05$). Total microbial lipid biomass was positively correlated with SOC ($r=0.99$, $P<0.01$), inorganic N ($r=0.80$, $P<0.01$), available P ($r=0.81$, $P<0.01$) and K ($r=0.92$, $P<0.01$).

Principal component analysis of the PLFA data suggested substantial differences in soil microbial community composition among study sites (Figure 2A). The first principal component (PC1) explained 50.7% and the second (PC2) 26.4% of the total variance in the PLFAs. PC1 was negatively correlated with soil pH ($r=-0.92$, $P<0.001$) and positively with C/N ratio ($r=0.82$, $P<0.001$) (Figure 2C). The site with the highest pH and lowest C/N ratio (site 1) was found on the left-hand side of Figure 2A, the sites with intermediate pH values and C/N ratios (sites 2, 4, and 5) in the middle portion of Figure 2A, and the site with the lowest pH and highest C/N ratio (site 3) on the right-hand side of Figure 2A. Along the PC2 axis, the alpine tundra (site 5) showed negative scores, while the forest sites (sites 1–4) showed positive scores. The PCA plot also showed that data points for sites 2 and 4 were intermixed (Figure 2A). For PC1, lipid signatures i15:0, 16:0, and 16:1ω7c had higher positive loading scores while a15:0 had lower negative loading scores (Figure 2B). For PC2, the fungal biomarker 18:2ω6,9 had large negative loading scores and appeared to become less abundant in sites 1–4 than site 5 (Figure 2B).

The absolute abundance of specific microbial groups showed different patterns along the altitude gradient. Site 1 contained 31.9% higher ($P<0.05$) bacterial PLFAs than site 3, while it had similar concentration of fungal PLFA as site 3 (Figure 3). Both bacterial and fungal PLFAs in sites 2 and 4 were similar ($P>0.05$), but significantly lower than those in sites 1 and 3 ($P<0.05$, Figure 3). Site 5 had the highest fungal PLFA and lowest bacterial PLFAs, which resulted directly in significantly higher fungi to bacteria (F/B) ratio than other sites ($P<0.05$, Figure 3). Significantly higher F/B ratios were also observed in sites 2 and 3 than sites 1 and 4 ($P<0.05$, Figure 3). The ratios of Gm^+/Gm^- were significantly lower in sites 3 and 5 than sites 1, 2, and 4 ($P<0.05$, Figure 3).

Amino sugars

For the study sites, the concentrations of individual amino sugars varied from 1793 to 7205 μg GluN g^{-1} soil, from 586 to 2431 μg GalN g^{-1} soil, and from 105 to 413 μg MurA g^{-1} soil (Figure 4). All three amino sugars had significantly greater concentrations at site 1 ($P<0.05$). GluN was significantly higher in sites 3 and 4 and significantly lower in site 5 compared with site 2 ($P<0.05$), while GalN decreased in the order site 4> site 3> site 5> site 2 (Figure 4). MurA was not significantly different among sites 2–5 ($P>0.05$, Figure 4). Both GluN/GalN and GluN/MurA ratios were significantly higher in sites 2 and 3 than sites 1 and 5 ($P<0.05$, Figure 4). Site 4 had similar GluN/GalN ratio as sites 1 and 5 and similar GluN/MurA ratio as sites 2 and 3 ($P>0.05$, Figure 4). Total amino sugars varied in the same way as GluN did, with values ranging from 2656 to 10357 μg g^{-1} soil, and were positively correlated with SOC ($r=0.97$, $P<0.01$), inorganic N ($r=0.91$, $P<0.01$), available P ($r=0.84$, $P<0.01$), and available K ($r=0.97$, $P<0.01$). Total amino sugar amounts in SOC (AS/SOC) were significantly higher in sites 1 and 4 (60.5 and 59.5 mg g^{-1} respectively) and significantly lower in sites 3 and 5 (40.1 and 44.7 mg g^{-1} respectively) in comparison with site 2 (54.0 mg g^{-1}, $P<0.05$).

Relationships between amino sugars and PLFAs

RDA showed that the first and second canonical axes explained 28.6% and 7.3%, respectively, of the total variance in the amino sugar data (Figure 5). Because ecological data are generally quite noisy, we can be confident that the major trends have been modeled in this analysis. Furthermore, the first unconstrained eigenvalue is comparatively small, which means that it does not display any important residual structure of the amino sugar data. The amino sugar data appeared to fall into two groups in the ordination biplot, which can be described as: (1) fungal-derived GluN; and (2) GalN and bacterial-derived MurA. Both groups seemed positively correlated with a15:0, i17:0, SOC, and TN. Permutation test by terms revealed a15:0, i15:0, i16:0, i17:0, 16:1ω7c and SOC had significant influence on amino sugar data. In addition, the study sites in the ordination biplot were clearly separated.

Discussion

In this study, we examined soil microbial communities and residues along an elevation gradient of Changbai Mountain. We found neither soil microbial communities nor residues showed a consistent altitudinal change along the studied elevation gradient. Similar altitudinal trends in microbial communities were found in Austrian Limestone Alps [25] and Bornean tropical forest ecosystems [44]. In contrast, Margesin et al. [19] observed a decrease of microbial (bacterial and fungal) biomass with increasing altitude in the Austrian Central Alps. One possible explanation for the apparent disparity between studies lies in site selection. For example, Djukic et al. [25] selected six study sites in

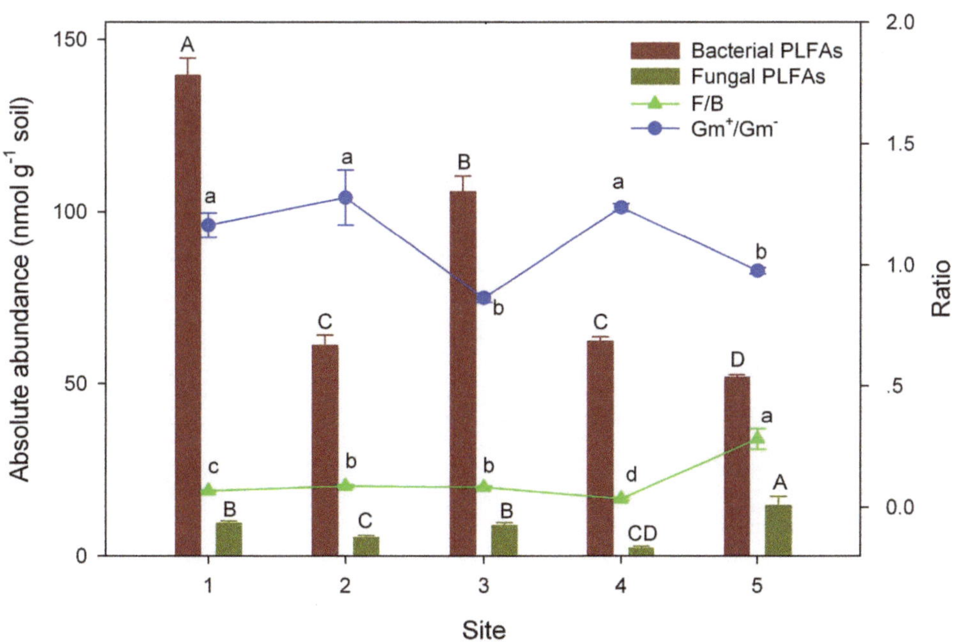

Figure 3. Sums and ratios of phospholipid fatty acids (PLFAs) of various microbial groups in different study sites along an altitude gradient on the northern slope of Changbai Mountain, China. Different letters within each group indicate significant differences among study sites ($P<0.05$, Tukey's HSD). Error bars show standard errors (n = 4). F/B, the ratio of fungal to bacterial PLFAs; Gm^+/Gm^-, the ratio of Gram-positive to Gram-negative bacteria.

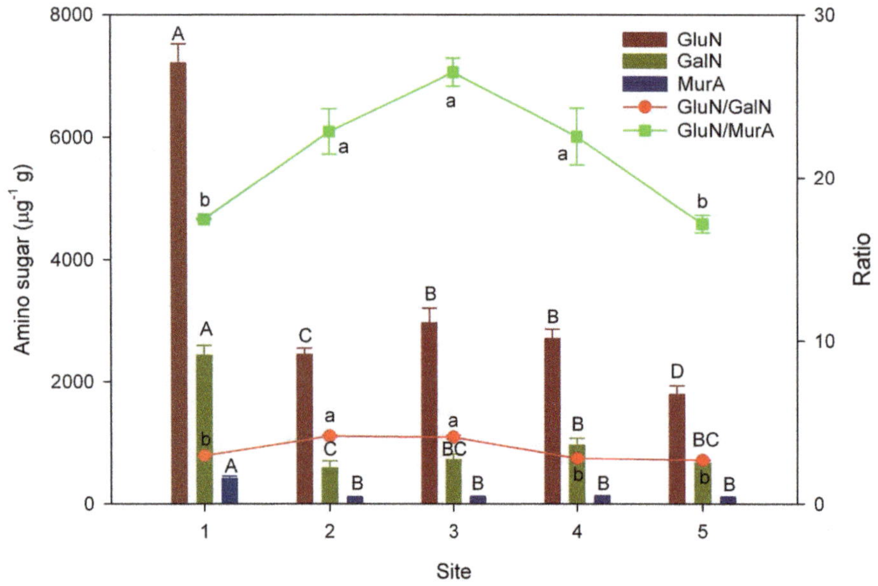

Figure 4. Concentrations and patterns of amino sugars in different study sites along an altitude gradient on the northern slope of Changbai Mountain, China. Different letters within each group indicate significant differences among study sites ($P<0.05$, Tukey's HSD). Error bars show standard errors (n = 4). GluN, glucosamine; GalN, galactosamine; MurA, muramic acid.

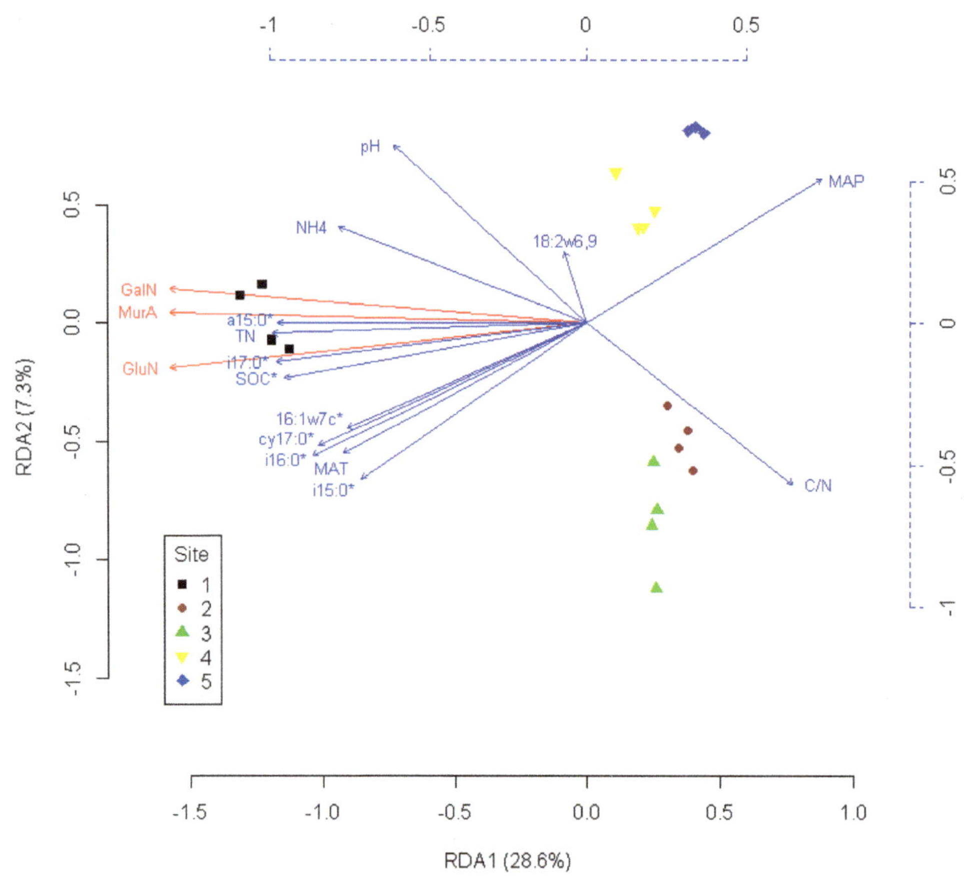

Figure 5. Redundancy analysis (RDA) of the amino sugar data for 20 soil samples using 7 microbial lipids and 7 environmental properties as explanatory variables. The amino sugars and sites are scaled by the black solid axes (bottom and left), and the explanatory variables are scaled by the blue dashed axes (top and right). The explanatory variables followed by an asterisk indicate significant influences on the amino sugar data.The proportion of explained variation was calculated by using adjusted R-squared values as described by Peres-Neto et al. [43]. GluN, glucosamine; GalN, galactosamine; MurA, muramic acid; SOC, soil organic carbon; TN, total nitrogen; MAT, mean annual air temperature; MAP, mean annual precipitation.

200-m elevation increments from 900 to 1900 m, whereas Margesin et al. [19] compared microbial communities only between alpine (2300–2530 m) and subalpine (1500–1900 m) soils. Because there was no available study regarding microbial residues along altitude, we compared our amino sugar data with the results from latitudinal studies. While Sowden [45] did not find any correlation between amino sugars and climatic elements in Canadian soils from different climate and vegetation zones, Amelung et al. [17] showed a parabolic relationship between amino sugars and MAT along a climosequence in North America. It seems that the effect of altitude on soil microbial communities and residues was indirect and mainly through its control on vegetation inputs and microbial habitat in our study. Alternatively, it could be that the control mechanism of altitude was more complex than simple linear relationship.

This study indicated that soil pH and C/N ratio were the most important drivers for soil microbial community composition and residue pattern along the northern slope of Changbai Mountain. Similar forces were found to shape microbial community composition in boreal forest soils [46] and alpine environments [25]. This is also in agreement with the results observed by ribosomal RNA gene based analysis on the same mountain [24]. The site with the lowest pH and highest C/N ratio (site 3) was structurally distinct from the other sites as a consequence of higher relative abundance of i15:0, 16:1ω7c, and 16:0 and lower relative abundance of a15:0. Significantly higher i15:0/a15:0 ratio (3.75) in this site than other sites (1.57–1.71) indicated niche stress for microbial growth. This is in line with report of increased i15:0 and decreased a15:0 in low pH oak forest soils [47]. However, Nilsson et al. [47] also found the PLFA 16:1ω7c was associated with high pH in forest soils, which seems to contrast with our results. Disparities might be explained if other methods determining microbial community composition at higher resolution were included, such as ribosomal RNA gene based analysis [24]. The lower Gm^+/Gm^- ratio in this site probably contributed to its higher GluN/MurA ratio because cell wall of Gm^- bacteria contains thinner murein layers than that of Gm^+ bacteria, leading to lower MurA concentrations in living Gm^- than Gm^+ bacteria [48], [49].

It is generally held that soils with high pH and low C/N ratio are more favourable to bacteria than fungi, and vice versa [50], [51]. This supports our result of fungal dominance at sites 2 and 3 and bacterial dominance at sites 1 and 4 (Figure 3, Table 1). Our observations are consistent with reports of increased F/B ratios in low pH coniferous forest soils than in high pH beech forest soils [32]. The influence of soil pH and C/N ratio on the relative abundance of fungi and bacteria in the studied forest sites was well reflected in GluN/GalN ratio (Figure 3, 4). An increasing GluN/GalN ratio has been used to indicate an increasing fungal contribution to SOM humification in forests [37]. Liang et al. [38] reported that of three forest soils, the one with the highest F/B ratio showed the highest fungal contribution to the soil amino sugar pool, which is in line with our study. However, such reflection was not observed in heavy metal polluted soils [52] or agricultural soils under contrast tillage practices [53]. Our results substantiate the "memory effect" of microbial residues which integrate microbial community structure over time [28], [54].

A negative correlation was found between C/N ratios and the ratios of total amino sugars to total PLFAs ($r = -0.86$, $P<0.01$). This corresponds well with the study of Liang et al. [38] in which a decrease in the ratio of amino sugars to PLFAs was associated with an increase in C/N ratio. This result could be explained by selective preservation which is a common humification pathway in forest soils [55]. For example, site 3 contained a relatively higher

Table 1. Site information and general soil characteristics.

Site	Altitude (m)	Coordinates	Vegetation	MAT[a] (°C)	MAP[a] (mm)	SOC[a] (mg g^{-1})	C/N	pH (H_2O)	NH_4+NO_3 (mg kg^{-1})	P (mg kg^{-1})	K (mg kg^{-1})
1	807	42°23′ N 128°05′ E	Broadleaved and Korean pine forest	2.58	691	171a[b]	14.9c	5.10a	123a	27.2a	300a
2	1204	42°09′ N 128°09′ E	Korean pine and spruce-fir mixed forest	0.27	811	59.6d	19.4b	4.57c	53.0c	13.2c	173c
3	1707	42°04′ N 128°04′ E	Dark coniferous forest	−2.29	967	96.5b	21.2a	3.95d	55.8c	11.3d	173c
4	1974	42°03′ N 128°04′ E	Erman's Birch forest	−3.31	1038	65.4c	15.4c	4.59c	79.3b	8.10e	163d
5	2295	42°02′ N 128°04′ E	Alpine tundra	−4.84	1154	59.5d	18.6b	4.78b	57.9c	16.2b	180b

[a]MAT, mean annual air temperature; MAP, mean annual precipitation; SOC, soil organic carbon.
[b]Different letters within each column indicate significant differences among study sites ($P<0.05$, Tukey's HSD)

living biomass (Figure 3). However, due to the poor quality of substrate (C/N ratio, 21.2), amino sugars might be decomposed and served as a preferred substrate for microorganisms to meet their growth needs. This is also the case for sites 2 and 5 with C/N ratios of 19.4 and 18.6, respectively, resulting in significantly lower microbial contribution to SOM pools (AS/SOC) in those sites than sites 1 and 4.

The significant correlation between nutrient availability and PLFAs as well as amino sugars indicated a strong substrate control over microbial communities and their residues. Substrate availability was also found to influence soil microbial communities in Bornean tropical forests [44] and soil microbial residues in a Michigan old-growth forest [29]. Soil microorganisms are usually considered to be C limited [56], even in coniferous forest soils with high soil C/N ratios [57] such as site 3 of our study. In this study, individual amino sugars contributed to the total in the order GluN > GalN > MurA, which is consistent with results from other forest soils [29], [58], native grassland soils [17], and agricultural soils [30], [59].

Despite a few studies found significant tree species-specific effects on soil microbial communities [5], [6] and residues [29], we put less focus on vegetation since Ushio et al. [60] pointed out tree species influence soil microbial community mainly through their effects on soil pH, total C and N. Nevertheless, the fungal dominance of the coniferous forests (sites 2 and 3) in our study could be partially explained by vegetation types. Fungi are thought to be more efficient in the decomposition of recalcitrant phenolic compounds [6] which have been identified with high quantities in coniferous forest [61]. In addition, ectomycorrhizal fungi could partially contribute to the fungal biomass since they exist in a symbiotic relationship with various species of coniferous trees [62].

The findings of this research also provide some insight into the relationships between soil microbial communities and their residues, as indicated by the relationships between the PLFA and amino sugar signatures in the ordination biplot. The positive correlations of amino sugars with bacterial-derived PLFAs suggest that variations in bacterial communities could lead to changes in amino sugars. It seems that bacteria, especially Gm^+ bacteria, played a more important role in the turnover and accumulation of amino sugars. The significant correlation between Gm^+/Gm^- and AS/SOC ($r=0.83$, $P<0.01$) supports this point. Fungal PLFA 18:2ω6,9, on the other hand, did not show a significant influence on amino sugar data despite the fungal-derived GluN accounted 67.5–76.5% of the total amino sugar pool. The significant influence of SOC on amino sugar data is consistent with the substrate control discussed above. Nevertheless, it remains a significant challenge to determine how the changes in microbial communities that occur under different environmental conditions could be manifested in changes in microbial residues.

Conclusions

In this study, we examined soil microbial communities and residues along an elevation gradient of Changbai Mountain by analyzing PLFA and amino sugar signatures. We found both soil microbial communities and residues differ between sites, thus reflecting variations between sites in several biotic and abiotic factors. However, the differences were not related to altitude, suggesting either the control of altitude is indirect or the mechanism is complex than simple linear relationship with soil microbial communities and residues. We found soil pH and C/N ratio were the most important drivers for microbial community structure and amino sugar pattern, while substrate availability was of great importance in determining the concentrations of microbial communities and residues. Redundancy analysis indicated bacteria were more important in controlling the feedback with amino sugar pools than fungi.

Acknowledgments

We would like to thank Xiaobo Li, Qiuxiang Tian, and Yang Wang for their help with the soil sampling.

Author Contributions

Conceived and designed the experiments: BZ XDZ. Performed the experiments: BZ. Analyzed the data: BZ CL. Contributed reagents/materials/analysis tools: HBH XDZ. Wrote the paper: BZ.

References

1. Lucas RW, Casper BB, Jackson JK, Balser TC (2007) Soil microbial communities and extracellular enzyme activity in the New Jersey Pinelands. Soil Biol Biochem 39: 2508–2519.
2. Liang C, Cheng G, Wixon DL, Balser TC (2011) An Absorbing Markov Chain approach to understanding the microbial role in soil carbon stabilization. Biogeochemistry 106: 303–309.
3. Stroud JL, Paton GI, Semple KT (2007) Microbe-aliphatic hydrocarbon interactions in soil: Implications for biodegradation and bioremediation. J Appl Microbiol 102: 1239–1253.
4. Singh BK, Bardgett RD, Smith P, Reay DS (2010) Microorganisms and climate change: Terrestrial feedbacks and mitigation options. Nat Rev Microbiol 8: 779–790.
5. Bach LH, Grytnes JA, Halvorsen R, Ohlson M (2010) Tree influence on soil microbial community structure. Soil Biol Biochem 42: 1934–1943.
6. Hackl E, Pfeffer M, Donat C, Bachmann G, Zechmeister-Boltenstern S (2005) Composition of the microbial communities in the mineral soil under different types of natural forest. Soil Biol Biochem 37: 661–671.
7. Weand MP, Arthur MA, Lovett GM, McCulley RL, Weathers KC (2010) Effects of tree species and N additions on forest floor microbial communities and extracellular enzyme activities. Soil Biol Biochem 42: 2161–2173.
8. Liang C, Jesus EC, Duncan DS, Jackson RD, Tiedje JM, et al. (2012) Soil microbial communities under model biofuel cropping systems in southern Wisconsin, USA: Impact of crop species and soil properties. Appl Soil Ecol 54: 24–31.
9. Pettersson M, Bååth E (2003) Temperature-dependent changes in the soil bacterial community in limed and unlimed soil. FEMS Microb Ecol 45: 13–21.
10. Brockett BFT, Prescott CE, Grayston SJ (2012) Soil moisture is the major factor influencing microbial community structure and enzyme activities across seven biogeoclimatic zones in western Canada. Soil Biol Biochem 44: 9–20.
11. Bååth E, Anderson TH (2003) Comparison of soil fungal/bacterial ratios in a pH gradient using physiological and PLFA-based techniques. Soil Biol Biochem 35: 955–963.
12. Rousk J, Brookes PC, Bååth E (2010) The microbial PLFA composition as affected by pH in an arable soil. Soil Biol Biochem 42: 516–520.
13. Bossio DA, Scow KM, Gunapala N, Graham KJ (1998) Determinants of soil microbial communities: Effects of agricultural management, season, and soil type on phospholipid fatty acid profiles. Microb Ecol 36: 1–12.
14. Fierer N, Schimel JP, Holden PA (2003) Variations in microbial community composition through two soil depth profiles. Soil Biol Biochem 35: 167–176.
15. Wu Y, Ma B, Zhou L, Wang H, Xu J, et al. (2009) Changes in the soil microbial community structure with latitude in eastern China, based on phospholipid fatty acid analysis. Appl Soil Ecol 43: 234–240.
16. Yergeau E, Bokhorst S, Huiskes AHL, Boschker HTS, Aerts R, et al. (2007) Size and structure of bacterial, fungal and nematode communities along an Antarctic environmental gradient. FEMS Microb Ecol 59: 436–451.
17. Amelung W, Zhang X, Flach KW, Zech W (1999) Amino sugars in native grassland soils along a climosequence in North America. Soil Sci Soc Am J 63: 86–92.
18. Diaz HF, Grosjean M, Graumlich L (2003) Climate variability and change in high elevation regions: Past, present and future. Climatic Change 59: 1–4.
19. Margesin R, Jud M, Tscherko D, Schinner F (2009) Microbial communities and activities in alpine and subalpine soils. FEMS Microb Ecol 67: 208–218.
20. Schinner F, Gstraunthaler G (1981) Adaptation of microbial activities to the environmental conditions in alpine soils. Oecologia 50: 113–116.
21. Ma X, Chen T, Zhang G, Wang R (2004) Microbial community structure along an altitude gradient in three different localities. Folia Microbiologica 49: 105–111.

22. Giri DD, Shukla PN, Kashyap S, Singh P, Kashyap AK, et al. (2007) Variation in methanotrophic bacterial population along an altitude gradient at two slopes in tropical dry deciduous forest. Soil Biol Biochem 39: 2424–2426.
23. Männistö MK, Tiirola M, Häggblom MM (2007) Bacterial communities in Arctic fjelds of Finnish Lapland are stable but highly pH-dependent. FEMS Microb Ecol 59: 452–465.
24. Shen C, Xiong J, Zhang H, Feng Y, Lin X, et al. (2013) Soil pH drives the spatial distribution of bacterial communities along elevation on Changbai Mountain. Soil Biol Biochem 57: 204–211.
25. Djukic I, Zehetner F, Mentler A, Gerzabek MH (2010) Microbial community composition and activity in different Alpine vegetation zones. Soil Biol Biochem 42: 155–161.
26. Engelking B, Flessa H, Joergensen RG (2007) Shifts in amino sugar and ergosterol contents after addition of sucrose and cellulose to soil. Soil Biol Biochem 39: 2111–2118.
27. Liang C, Zhang X, Balser TC (2007b) Net microbial amino sugar accumulation process in soil as influenced by different plant material inputs. Biol Fertil Soils 44: 1–7.
28. He H, Zhang W, Zhang X, Xie H, Zhuang J (2011) Temporal responses of soil microorganisms to substrate addition as indicated by amino sugar differentiation. Soil Biol Biochem 43: 1155–1161.
29. Liang C, Fujinuma R, Wei L, Balser TC (2007a) Tree species-specific effects on soil microbial residues in an upper Michigan old-growth forest system. Forestry 80: 65–72.
30. Ding XL, Zhang B, Zhang XD, Yang XM, Zhang XP (2011) Effects of tillage and crop rotation on soil microbial residues in a rainfed agroecosystem of northeast China. Soil Till Res 114: 43–49.
31. Liang C, Zhang X, Rubert IV KF, Balser TC (2007c) Effect of plant materials on microbial transformation of amino sugars in three soil microcosms. Biol Fertil Soils 43: 631–639.
32. Frostegård A, Bååth E (1996) The use of phospholipid fatty acid analysis to estimate bacterial and fungal biomass in soil. Biol Fertil Soils 22: 59–65.
33. Amelung W (2001) Methods using amino sugars as markers for microbial residues in soil. In: Lal R, Kimble JM, Follett RF, Stewart BA, editors. Assessment Methods for Soil Carbon. Boca Raton: CRC/Lewis Publishers. pp. 223–270.
34. Zhang X, Amelung W, Yuan Y, Zech W (1998) Amino sugar signatures of particle size fractions in soils of the native prairie as affected by climate. Soil Sci 63: 220–229.
35. Zhang X, Wang J, Xie HT, Wang JK, Zech W (2003) Comparison of organic compounds in the particle-size fractions of earthworm casts and surrounding soil in humid Laos. Appl Soil Ecol 23: 147–153
36. Joergensen RG, Mäder P, Flieβbach A (2010) Long-term effects of organic farming on fungal and bacterial residues in relation to microbial energy metabolism. Biol Fertil Soils 46: 303–307.
37. Parsons JW (1981) Chemistry and distribution of amino sugars in soils and soil organisms. In: Paul EA, Ladd JN, editors. Soil Biochemistry. New York: Marcel Dekker. pp. 197–227.
38. Liang C, Fujinuma R, Balser TC (2008) Comparing PLFA and amino sugars for microbial analysis in an Upper Michigan old growth forest. Soil Biol Biochem 40: 2063–2065.
39. He HS, Hao Z, Mladenoff DJ, Shao G, Hu Y, et al. (2005) Simulating forest ecosystem response to climate warming incorporating spatial effects in north-eastern China. J Biogeogr 32: 2043–2056.
40. Lu RK (2000) Methods of Soil and Agro-chemistry. Beijing: Chinese Agricultural Science and Technology Press.
41. Zhang X, Amelung W (1996) Gas chromatographic determination of muramic acid, glucosamine, mannosamine, and galactosamine in soils. Soil Biol Biochem 28: 1201–1206.
42. Bligh EG, Dyer WJ (1959) A rapid method of total lipid extraction and purification. Can J Biochem Phys 37: 911–917.
43. Peres-Neto PR, Legendre P, Dray S, Borcard D (2006) Variation partitioning of species data matrices: estimation and comparison of fractions. Ecology 87: 2614–2625.
44. Wagai R, Kitayama K, Satomura T, Fujinuma R, Balser T (2011) Interactive influences of climate and parent material on soil microbial community structure in Bornean tropical forest ecosystems. Ecol Res 26: 627–636.
45. Sowden FJ (1977) Distribution of nitrogen in representative Canadian soils. Can J Soil Sci 57: 445–456.
46. Högberg MN, Högberg P, Myrold DD (2007) Is microbial community composition in boreal forest soils determined by pH, C-to-N ratio, the trees, or all three? Oecologia 150: 590–601.
47. Nilsson LO, Bååth E, Falkengren-Grerup U, Wallander H (2007) Growth of ectomycorrhizal mycelia and composition of soil microbial communities in oak forest soils along a nitrogen deposition gradient. Oecologia 153: 375–384.
48. Millar WN, Casida LE (1970) Evidence for muramic acid in soil. Can J Microb 16: 299–304.
49. Eudy LW, Walla MD, Morgan SL, Fox A (1985) Gas chromatographic-mass spectrometric determination of muramic acid content and pyrolysis profiles for a group of gram-positive and gram-negative bacteria. The Analyst 110: 381–385.
50. Alexander M (1977) Introduction to Soil Microbiology, second edition. New York: Wiley.
51. Sterner RW, Elsner JJ (2002) Ecological Stoichiometry: The Biology of Elements from Molecules to the Biosphere. Princeton: Princeton University Press.
52. Kandeler E, Tscherko D, Bruce KD, Stemmer M, Hobbs PJ, et al. (2000) Structure and function of the soil microbial community in microhabitats of a heavy metal polluted soil. Biol Fertil Soils 32: 390–400.
53. van Groenigen KJ, Bloem J, Bååth E, Boeckx P, Rousk J, et al. (2010) Abundance, production and stabilization of microbial biomass under conventional and reduced tillage. Soil Biol Biochem 42: 48–55.
54. Glaser B, Turrión MB, Alef K (2004) Amino sugars and muramic acid - Biomarkers for soil microbial community structure analysis. Soil Biol Biochem 36: 399–407.
55. Kögel-Knaber I (1993) Biodegradation and humification processes in forest soils. In: Bollag JM, Stotsky G, editors. Soil Biochemistry. New York: Marcel Dekker Inc. pp. 101–135.
56. Wardle DA (1992) A comparative assessment of factors which influence microbial biomass carbon and nitrogen levels in soil. Biol Rev 67: 321–358.
57. Allen AS, Schlesinger WH (2004) Nutrient limitations to soil microbial biomass and activity in loblolly pine forests. Soil Biol Biochem 10: 215–221.
58. Moritz LK, Liang C, Wagai R, Kitayama K, Balser TC (2009) Vertical distribution and pools of microbial residues in tropical forest soils formed from distinct parent materials. Biogeochemistry 92: 83–94.
59. Guggenberger G, Frey SD, Six J, Paustian K, Elliott ET (1999) Bacterial and fungal cell-wall residues in conventional and no-tillage agroecosystems. Soil Sci Soc Am J 63: 1188–1198.
60. Ushio M, Wagai R, Balser TC, Kitayama K (2008) Variations in the soil microbial community composition of a tropical montane forest ecosystem: Does tree species matter? Soil Biol Biochem 40: 2699–2702.
61. Gallet C, Pellissier F (1997) Phenolic compounds in natural solutions of a coniferous forest. J Chem Ecol 23: 2401–2412.
62. Smith SE, Read DJ (1997) Mycorhizal symbiosis. San Diego: Academic Press.

Differentiating Intracellular from Extracellular Alkaline Phosphatase Activity in Soil by Sonication

Shuping Qin[1], Chunsheng Hu[1]*, Oene Oenema[2]

1 Key Laboratory of Agricultural Water Resources, Center for Agricultural Resources Research, Institute of Genetic and Developmental Biology, The Chinese Academy of Sciences, Shijiazhuang, Hebei, China, 2 Wageningen University and Research, Alterra, Wageningen, The Netherlands

Abstract

Differentiating intracellular from extracellular enzyme activity is important in soil enzymology, but not easy. Here, we report on an adjusted sonication method for the separation of intracellular from extracellular phosphatase activity in soil. Under optimal sonication conditions [soil:water ratio = 1/8 (w/v) and power density = 15 watt ml^{-1}], the activity of alkaline phosphomonoesterase (phosphatase) in a Haplic Cambisol soil increased with sonication time in two distinct steps. A first plateau of enzyme activity was reached between 60 and 100 s, and a second higher plateau after 300 s. We also found that sonication for 100 s under optimal conditions activated most (about 80%) of the alkaline phosphatase that was added to an autoclaved soil, while total bacteria number was not affected. Sonication for 300 s reduced the total bacteria number by three orders of magnitude but had no further effects on enzyme activity. Our results indicate that the first plateau of alkaline phosphatase activity was derived from extracellular enzymes attached to soil particles, and the second plateau to the combination of extracellular and intracellular enzymes after cell lysis. We conclude that our adjusted sonication method may be an alternative to the currently used physiological and chloroform-fumigation methods for differentiating intracellular from extracellular phosphatase activity in soil. Further testing is needed to find out whether this holds for other soil types.

Editor: Hauke Smidt, Wageningen University, The Netherlands

Funding: This work was supported by the National Natural Science Foundation of China (31270554) and (30970534). The funders had no role in study design, data collection and analysis, decision to publish, or preparation of the manuscript.

Competing Interests: The authors have declared that no competing interests exist.

* E-mail: cshu@sjziam.ac.cn

Introduction

Phosphomonoesterase (phosphatase) activities are widely investigated in soils since they catalyze the hydrolysis of ester–phosphate bonds, releasing the inorganic phosphate, which can be assimilated by plants and micro-organisms [1–3]. The overall phosphatase activities in soils are composed of extracellular activity from accumulated phosphatase adhered to soil particles and intracellular phosphatase in living microorganisms [4]. As yet, it is still a challenge to discriminate intracellular enzyme activity from extracellular enzyme activity in soil [5–10]. Among the numerous attempts to address this issue, the physiological method [11] and the chloroform-fumigation method [9] were most widely used. A critical assumption in the physiological method is that the results from the enzyme assays are a combination of intracellular and extracellular activity [11]. However, a direct correlation between the enzyme activity and the size of the microbial biomass is sometimes absent [12]. The chloroform-fumigation method assumes that the enzymatic activity measured before chloroform-fumigation originates from the extracellular enzyme activity, while that after chloroform-fumigation from the combination of extracellular and intracellular activity [9]. However, the released intracellular enzyme might be partially degraded by protease following chloroform-fumigation [13]. A rapid fumigation method (5 min) was proposed to minimize proteolysis [10].

We explored the potentials of sonication to discriminate between intracellular versus extracellular phosphatase activity in soil. Sonication is a wave generated by vibration [14]. It generally causes both heating and cavitation in liquid surroundings [15]. Sonication is widely used for the determination of particle-size distribution [16] and fractionation of organic matter [17]. Short-time (30 s) sonication with low-power intensity was reported to not significantly decrease the total coliform number of wastewater particles until the power exceeded 30 watt (W) [18]. In contrast, long-time sonication (150 s) was reported to be able to lyse soil microbial cells and release the intracellular compounds [19]. The above results suggest that sonication might be a promising tool for differentiating between extracellular and intracellular enzyme activity in soil. Earlier, De Cesare et al. [20] explored the potential of sonication for evaluating the activity of immobilized acid phosphatase. Acid phosphatase activity was increased in a clay loam soil after 120 s sonication, due to both of breakdown of soil aggregates and detachment of extracellular enzyme from soil particle surface, rather than the release of intracellular enzyme from proliferating cells. In this study, we further investigated the potential of sonication for discriminating extracellular versus intracellular alkaline phosphatase activity in a Haplic Cambisol.

We hypothesized that short-time sonication would activate the extracellular enzymes bonded with soil particles, while long-time sonication would lyse the microbial cell and release the intracellular enzymes.

Materials and Methods

Soil collection and processing

Soil samples were collected from the Luancheng Agro-ecosystem Station (37°90'N, 114°67'E, elevation 50 m) from the Chinese Academy of Sciences. The soil is classified as a silt loam Haplic Cambisol (FAO classification system; 13.8% sand, 66.3% silt, and 19.9% clay, pH 8.2, bulk density 1.44 g cm^{-3}, organic carbon 12.05 mg g^{-1} and total nitrogen 1.12 mg g^{-1}). The cropping system is winter wheat (mid-October to early-June) with summer corn (early-June to late September). After removing the crop residues, five soil cores (4.8 cm × 10 cm depth) were randomly sampled and thoroughly mixed, ground through a 2-mm sieve and stored at 4°C. Before each experiment, soils were pre-incubated at 25°C for five days to activate microbial activity following the cold storage.

Sonicator testing

Aliquots of 100 ml deionized water were sonicated for 100 s at power densities of 0, 5, 10, 15, 20 W ml^{-1}, following the procedure of De Cesare et al. [20]. The increase of water temperature with the sonication time was monitored by a temperature sensor in the sonicator (JYD-650, 20 kHz, ZhiSun Instrument Co., Shanghai, China). The effects of the sonicator on temperature were analyzed statistically via regression analysis.

Ultrasonic effects on soil dispersion

This experiment was carried out to find the optimal soil-water ratio (w/v). Different ratios (1:4, 1:6, 1:8, 1:10, w/v) were subjected to sonication at different power densities (0, 5, 10, 15, 20, 25, 30 W ml^{-1}). Moist soil samples equivalent to 2.50, 1.66, 1.11 and 1.00 g dry weight and portions of 10 ml 50 mM borax-borate buffer (pH 8.2) were mixed in 20 ml cylindrical glass jars each. The jars were put in an ice bath to prevent heating during sonication. After 10 s gentle stirring and 180 s equilibration, the mixtures were sonicated (probe at 15 mm depth, 2 s burst and 2 s rest) for 300 s at power density of 0, 5, 15, 20, 25, 30 W ml^{-1}, respectively. If the temperature of soil suspension exceeded 15°C, the temperature sensor will stop the sonicator automatically until the temperature decreased to below 13°C. The sonicated samples were equilibrated at 37°C for 10 min under gentle agitation. Then the suspensions were centrifuged at 12,000 *g* for 2 min. The absorbance of the supernatants were measured by a spectrophotometer at 410 nm.

Ultrasonic effects on alkaline phosphatase and total bacterial number

Moist soil samples (1.11 g, dry weight) were mixed with 10 ml 50 mM borax-borate buffer (pH 8.2) in 20 ml cylindrical glass vessels and sonicated (probed at 15 mm depth) at power density of 10, 15, and 20 W ml^{-1} for 0, 20, 60, 80, 100, 150, 200, 250, 300, 350, 400 and 500 s, respectively.

Alkaline phosphatase activity was spectrophotometrically determined at 410 nm using *p*-Nitrophenyl-phosphate (40 mM) as substrates, according to the procedure of De Cesare et al. [20], with slight modifications. Instead of 0.5 M acetate buffer (pH 5) we used 50 mM borax-borate buffer (pH 8.2). Two controls were included: one was to remove the absorbance of chromophores that were released during the sonication, the other was to remove the *p*-Nitrophenyl-phosphate-hydrolyzing activity of non-enzymatic components in soil.

Total bacteria numbers (Log_{10} CFU g^{-1} dry soil) of the sonicated and the non-sonicated soil were determined using the most probable number (MPN) methods with beef extract–peptone substrate [21].

Ultrasonic effects on commercial alkaline phosphatase activity

In order to simulate the ultrasonic effects on extracellular alkaline phosphatase activity, 30 μl 50 mU alkaline phosphatase from *Escherichia coli* (Sigma No. P-4252, St. Louis, USA) was added into 10 ml 50 mM borax-borate buffer (pH 8.2). Autoclaved (121 °C, 1 h) soil samples (1.11 g, dry weight), which had no alkaline phosphatase activity, were added into the buffer. Then the slurry was shaken for 30 s, equilibrated for 15 min at 25°C, and sonicated at a power density of 15 W ml^{-1} for 0, 20, 60, 80, 100, 150, 200, 250, 300, 350, 400 and 500 s, respectively. Control samples (without autoclaved soil) were identically and synchronously treated. The sonication procedure and determination of alkaline phosphatase activity (mU ml^{-1}) were the same as described above.

Statistics

Differences between means were analyzed statically using the software SPSS 13.0 (SPSS Inc, 2004), and were compared by the LSD test at $P<0.05$

Results

Temperature change by sonication

The water temperature increased nearly linearly with the power density of the sonicator ($P < 0.05$) (Fig. 1); for each power unit (W) increase, temperature increased 0.21 °C after sonication for 100 s. Increasing sonication time increased temperature also linearly (not shown).

Soil dispersion by sonication

The absorbance of soil supernatants (after centrifugation at 12,000 *g* for 2 min) increased significantly when the ultrasonic power density increased from 5 to 15 W ml^{-1} (Fig. 2). A further increase of the ultrasonic power density from 15 to 30 W ml^{-1} did not contribute much to a further increase. The highest absorbance was found for a soil-water ratio of 1:8 (w:v), but differences between soil-water ratios were relatively small, though statistically significant (Fig. 2).

Changes in phosphatase activity by sonication

At relatively low power density (5 W ml^{-1}), alkaline phosphatase activity steadily increased with sonication time until a plateau was reached at 1.8 to 1.9 μmol *p*-nitrophenol h^{-1} g^{-1} dry soil (Figure 3). Increasing the power density from 5 to 10, 15 and 20 W ml^{-1} decreased the time needed to reach the plateau at 1.8 to 1.9 μmol *p*-nitrophenol h^{-1} g^{-1} dry soil. Continued sonication further increased phosphatase activity at high power density, until a second plateau was reached at about 2.5 μmol *p*-nitrophenol h^{-1} g^{-1} dry soil (Figure 3). This step-wise increase of the phosphatase activity with sonication time is in line with our hypothesis and suggest that the activity related to the first plateau is from extracellular phosphatase and that the activity related to the second plateau is from extracellular plus intracellular phosphatase.

Changes in extracellular phosphatase activity by sonication

The activity of added extracellular phosphatase to an autoclaved borax-borate buffer (control treatment) was quite stable against sonication (Fig. 4). In contrast, the activity of added

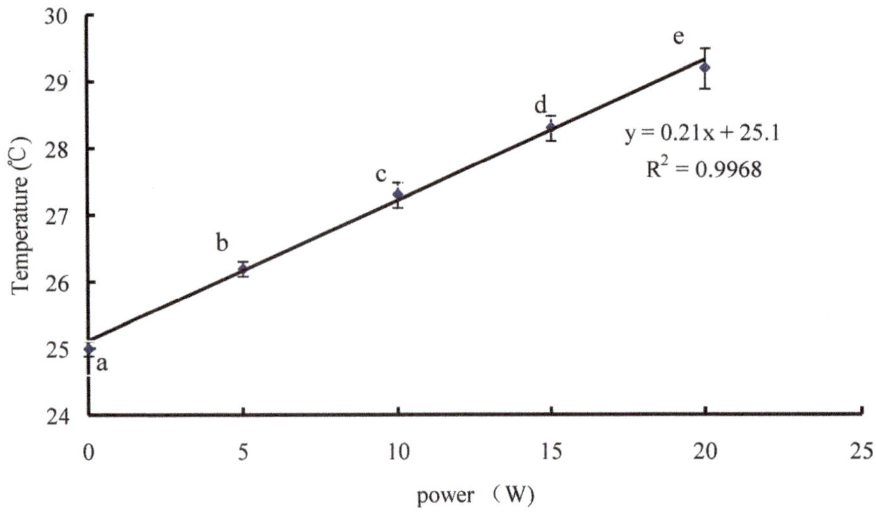

Figure 1. Temperature of 100 ml deionized water as a function of sonication power.

extracellular phosphatase to autoclaved soil increased significantly with sonication time, until a plateau was reached (Fig. 4). Hence, sonication activated the extracellular external phosphatase from soil particles.

The total counts of bacteria (Log CFU g^{-1} dry soil) remained rather constant with an increase in sonication time until about 150 s. Thereafter, bacteria counts decreased, especially between 200 and 300 s. Total counts of bacteria decreased by three orders of magnitude when sonication time increased to 500 s (Fig. 5).

Discussion

Ultrasonic effects on soil dispersion

The increase in heat absorbed by the water was linearly related to the energy output of the sonicator under all power densities (Fig. 1), indicating the reliability of the sonicator applied. Also, sonication time and energy could be easily adjusted, and our temperature-regulated sonication prevented excessive heating of the samples.

Some spectrophotometrically active chromophores may be released from soil colloids during sonication treatment [20], and the increase in the absorbance of chromophores indirectly reflects the degree of soil dispersion [20]. Our results showed that the soil:water ratio was not very critical in the range of 1:4 to 1:10 (w:v) (Figure 2). Because of the highest absorbance at a soil:water ratio of 1:8 (Figure 2), we adopted this ratio in the subsequent experiments.

Ultrasonic effects on phosphatase activity

The soil native alkaline phosphatase activity increased in two steps with an increase in sonication time, when sonication energy was in the range of 10–20 W ml^{-1} (Figure 3). This result contrasts with the results obtained with De Cesare et al. [20], who observed only one plateau of acid phosphatase activity. At low energy

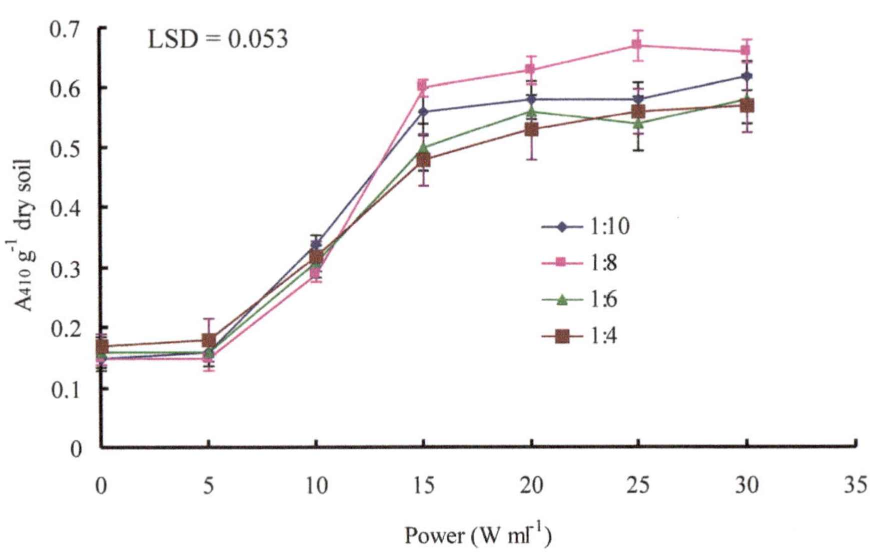

Figure 2. Absorbance (410 nm) of supernatants of soil slurries after 300 s sonication at different extraction ratios as a function of power density.

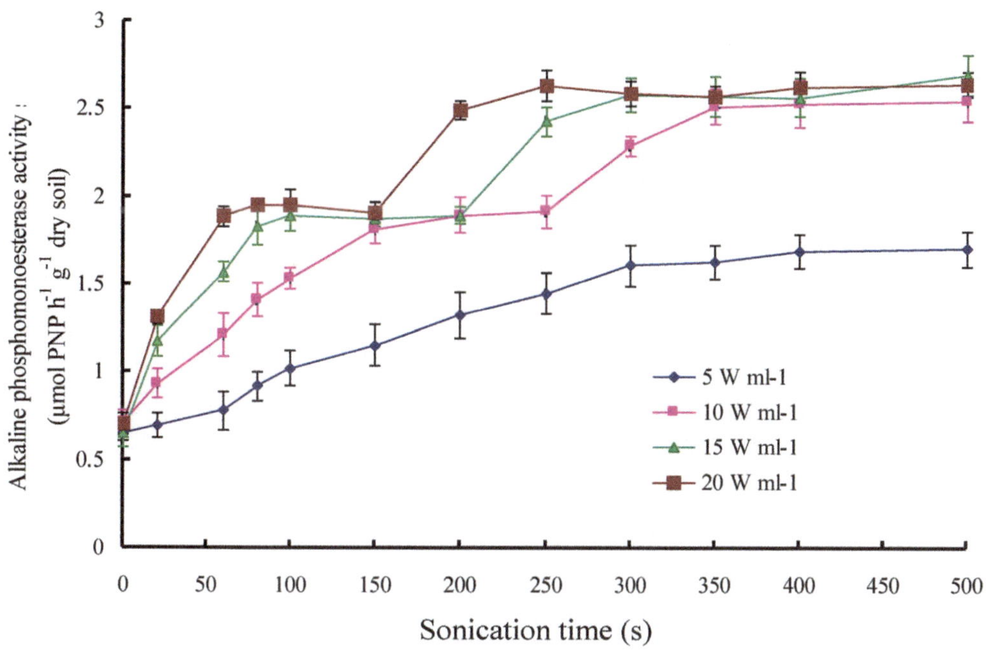

Figure 3. Activity of alkaline phosphatase of the field-moist soil as a function of sonication time. Data are means ± SD (n = 3).

sonication, i.e. 5 W ml^{-1}, we observed only one plateau (the lower one), and at high energy sonication (20 W ml^{-1}) we observed that the first and second plateaus were reached soon (Figure 3).

Two independent mechanisms have been proposed to explain the ultrasonic effects on alkaline phosphatase activity. The first relates to the activation of the immobilized extracellular phosphatase in soil particles following its exposure [22], and the second to the release of intracellular phosphatase after cell lysis [20]. Short-time sonication (less than 200 s) appears effective for extracting living bacteria from soil [18,23,24]. In this study, 200 s sonication at a power density of 15 W ml^{-1} did not significantly decrease bacterial counts (Fig. 5), indicating that 200 s sonication at the power density of 15 W ml^{-1} did not cause cell lysis and very likely did not release intracellular enzymes. The first plateau of phosphatase activity in this study was thus probably related to the activation of extracellular enzyme in soil aggregates. Independent proof for this hypothesis was obtained from the experiment in which extracellular phosphatase was added to autoclaved soil. About 80 % of the added phosphatase was activated after 100 to 200 s sonication at a power density of 15 W ml^{-1} (Fig. 4).

Alkaline phosphatase can be excreted by both plant roots and soil microbes [25-26]. Microbes, especially bacteria, are able to produce and release large amounts of extracellular phosphatase due to their large biomass, high metabolic rate and short lifecycles [27]. The total number of bacteria in this study decreased by three orders of magnitude at high-energy sonication (4,500 J ml^{-1}) for 150 to 500 s (Fig. 5). Stevenson [23] also reported that 300 s sonication significantly decreased the number of soil bacteria at a soil:water ratio of 1:10 (both soil type and energy not described).

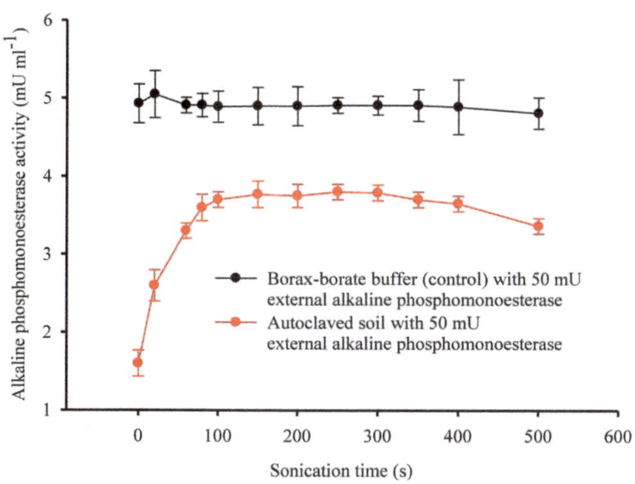

Figure 4. Activity of alkaline phosphatase of the autoclaved soil and the borax-borate buffer (control) with 50 mU external alkaline phosphatase as functions of sonication time.

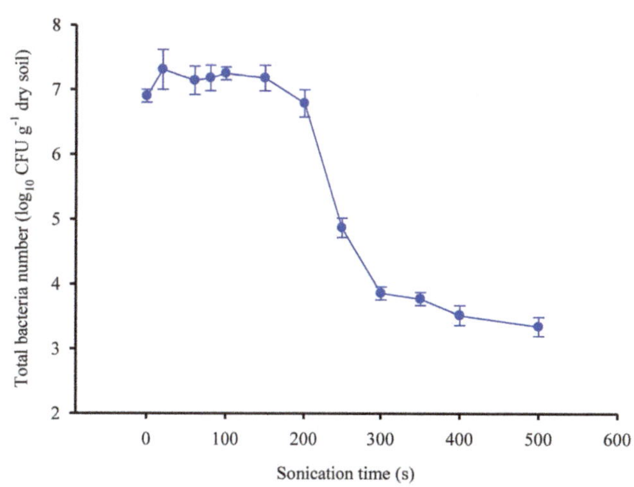

Figure 5. Total bacteria number of field-moist soil as a function of sonication time.

Vargas et al. [28] reported that 240 s sonication with an amplitude of 20 μm caused cell disruption and release of invertase from a fungi (*Aspergillus niger*) in a liquid culture. Those results clearly suggest that sonication at high energy-level may destroy soil microbial cells and thereby may release intracellular alkaline phosphatase. Therefore, the second plateau of phosphatase activity observed in this study after sonication at relatively high-energy sonication of long duration was very likely related to a combination of extracellular phosphatase plus intracellular phosphatase released after microbial cell lysis.

Potential of sonication to differentiate extracellular from intracellular phosphatase

Our results showed that low-energy sonication was able to activate the majority of adscititious (simulated extracellular) alkaline phosphatase but did not decrease the bacteria number. In contrast, high-energy sonication significantly decreased the number of bacteria (Fig. 5). These results indicate that extracellular alkaline phosphatase may be discriminated from intracellular alkaline phosphatase by differentiating the energy-level and duration of sonication. In this study, the extracellular activity could be discriminated from the intracellular alkaline phosphatase activity by step-wise sonication for 500 s at energy levels of 10 to 20 W ml^{-1} (Figure 3). It is worthy to note that the sonication effects on soil bacterial cell lysis was not comprehensively considered since the MPN method only determine 1-10% of soil microorganisms.

Sonication is a promising alternative for differentiating extracellular from intracellular phosphatase activity in soils. It is worthy to note that the extracellular alkaline phosphatase activity was two times higher after sonication for 300 s than before sonication (Fig. 3 and 4). This was probably due to the activation of entrapped phosphatase in soil aggregates [22]. The activation of entrapped phosphatase was considered a disadvantage of sonication for assaying extracellular enzyme activities [8]. However, soil enzyme activity reflects the potential rather than the actual *in situ* activity [29–31].

Conclusions

A sonication pretreatment offers the potential to quantify two levels of phosphatase activity, i.e., (i) potential extracellular activity, measured after high-energy sonication for 100-150 s., and (ii) potential extracellular plus intracellular activity, measured after high-energy sonication for >300 s. Applied level of energy and the duration of the sonication as well as the time-resolution of the enzyme activity measurements are critical. Further experimentations are needed to test whether sonication is also suitable for discriminating intracellular from extracellular enzyme activities in other soils.

Acknowledgments

We thank Dr. Xinhua He for improving the preliminary version of the paper.

Author Contributions

Conceived and designed the experiments: SQ CH. Performed the experiments: SQ. Analyzed the data: SQ . Wrote the paper: SQ CH OO.

References

1. Nannipieri P, Giagnoni L, Landi L, Renella G (2011) Role of phosphatase enzymes in soil. In Bunemann E K, Oberson A, Frossard E (eds) Phosphorus in Action. Soil Biology 26. Springer Verlag, Berlin Heidelberg, pp 215–241.
2. Wang J, Chen Z, Chen L, Zhu A, Wu Z (2011) Surface soil phosphorus and phosphatase activities affected by tillage and crop residue input amounts. Plant, Soil and Environment 57: 251–257.
3. Guénon R, Gros R (2013) Frequent-wildfires with shortened time-since-fire affect soil microbial functional stability to drying and rewetting events. Soil Biology and Biochemistry 57: 663–674.
4. Saha S, Prakash V, Kundu S, Kumar N, Mina BL (2008) Soil enzymatic activity as affected by long term application of farm yard manure and mineral fertilizer under a rainfed soybean–wheat system in NW Himalaya. European Journal of Soil Biology 44: 309–315.
5. Nannipieri P, Giagnoni L, Renella G, Puglisi E, Ceccanti B, et al. (2012) Soil enzymology: classical and molecular approaches. Biology and Fertility of Soils 48: 743–762.
6. Dilly O, Nannipieri P (1998) Intracellular and extracellular enzyme activity in soil with reference to elemental cycling. Zeitschrift Fur Pflanzenernahrung Und Bodenkunde 161: 243–248.
7. Grierson PF, Adams MA (2000) Plant species affect acid phosphatase, ergosterol and microbial P in a Jarrah (Eucalyptus marginata Donn ex Sm.) forest in south-western Australia. Soil Biology & Biochemistry 32: 1817–1827.
8. Nannipieri P, Kandeler E, Ruggiero P (2002) Enzyme activities and microbiological and biochemical processes in soil. Enzymes in the environment Marcel Dekker, New York: 1–33.
9. Klose S, Tabatabai MA (2002) Response of phosphatasephosphatases in soils to chloroform fumigation. Journal of Plant Nutrition and Soil Science-Zeitschrift Fur Pflanzenernahrung Und Bodenkunde 165: 429–434.
10. Margon A, Fornasier F (2008) Determining soil enzyme location and related kinetics using rapid fumigation and high-yield extraction. Soil Biology & Biochemistry 40: 2178–2181.
11. Nannipieri P, Sastre I, Landi L, Lobo MC, Pietramellara G (1996) Determination of extracellular neutral phosphatase activity in soil. Soil Biology & Biochemistry 28: 107–112.
12. Speir T, Ross D (1978) Soil phosphatase and sulphatase. Soil enzymes 203.
13. Renella G, Landi L, Nannipieri P (2002) Hydrolase activities during and after the chloroform fumigation of soil as affected by protease activity. Soil Biology & Biochemistry 34: 51–60.
14. Mamba S, Kratochvil B (1995) Application of ultrasound to dissolution of environmental samples for elemental analysis. International journal of environmental analytical chemistry 60: 295–302.
15. Yachmenev VG, Blanchard EJ, Lambert AH (2004) Use of ultrasonic energy for intensification of the bio-preparation of greige cotton. Ultrasonics 42: 87–91.
16. Stemmer M, Gerzabek MH, Kandeler E (1998) Organic matter and enzyme activity in particle-size fractions of soils obtained after low-energy sonication. Soil Biology & Biochemistry 30: 9–17.
17. Murage EW, Voroney PR, Kay BD, Deen B, Beyaert RP (2007) Dynamics and turnover of soil organic matter as affected by tillage. Soil Science Society of America Journal 71: 1363–1370.
18. Ormeci B, Linden KG (2005) Comparison of physical and chemical methods for extraction of coliform from wastewater particles and flocs. Environmental Engineering Science 22: 459–471.
19. Yeates C, Gillings M, Davison A, Altavilla N, Veal D (1998) Methods for microbial DNA extraction from soil for PCR amplification. Biological procedures online 1: 40–47.
20. De Cesare F, Garzillo AMV, Buonocore V, Badalucco L (2000) Use of sonication for measuring acid phosphatase activity in soil. Soil Biology & Biochemistry 32: 825–832.
21. Kong C, Wang P, Zhao H, Xu X, Zhu Y (2008) Impact of allelochemical exuded from allelopathic rice on soil microbial community. Soil Biology and Biochemistry 40: 1862–1869.
22. Boyd SA, Mortland M (1990) Enzyme interactions with clays and clay-organic matter complexes. Soil biochemistry 6: 1–28.
23. Stevenson I (1958) The effect of sonic vibration on the bacterial plate count of soil. Plant and Soil 10: 1–8.
24. Ramsay AJ (1984) Extraction of bacteria from soil: efficiency of shaking or ultrasonication as indicated by direct counts and autoradiography. Soil Biology & Biochemistry 16: 475–481.
25. Nakas J, Gould W, Klein D (1987) Origin and expression of phosphatase activity in a semi-arid grassland soil. Soil Biology & Biochemistry 19: 13–18.
26. Tarafdar JC, Claassen N (1988) Organic Phosphorus-Compounds as a Phosphorus Source for Higher-Plants through the Activity of Phosphatases Produced by Plant-Roots and Microorganisms. Biology and Fertility of Soils 5: 308–312.
27. Kramer S, Green DM (2000) Acid and alkaline phosphatase dynamics and their relationship to soil microclimate in a semiarid woodland. Soil Biology & Biochemistry 32: 179–188.
28. Vargas LHM, Piao ACS, Domingos RN, Carmona EC (2004) Ultrasound effects on invertase from Aspergillus niger. World Journal of Microbiology & Biotechnology 20: 137–142.

29. Klose S, Moore J, Tabatabai M (1999) Arylsulfatase activity of microbial biomass in soils as affected by cropping systems. Biology and Fertility of Soils 29: 46–54.
30. Gianfreda L, Ruggiero P (2006) Enzyme activities in soil. In Nannipier P, Smalla K (eds) Nucleic acids and proteins in soil. Soil Biology 8. Springer Verlag, Berlin Heidelberg, pp 257–311.
31. Udawatta RP, Kremer RJ, Garrett HE, Anderson SH (2009) Soil enzyme activities and physical properties in a watershed managed under agroforestry and row-crop systems. Agriculture Ecosystems & Environment 131: 98–104.

11

Factors Affecting Soil Fauna Feeding Activity in a Fragmented Lowland Temperate Deciduous Woodland

Jake E. Simpson[1,2]*, Eleanor Slade[3], Terhi Riutta[4], Michele E. Taylor[1]

1 Centre for Ecology and Hydrology, Natural Environment Research Council, Wallingford, Oxfordshire, United Kingdom, **2** Department of Biology, Imperial College London, London, United Kingdom, **3** Wildlife Conservation Research Unit, Department of Zoology, University of Oxford, Oxford, Oxfordshire, United Kingdom, **4** Environmental Change Institute, School of Geography and the Environment, University of Oxford, Oxford, Oxfordshire, United Kingdom

Abstract

British temperate broadleaf woodlands have been widely fragmented since the advent of modern agriculture and development. As a result, a higher proportion of woodland area is now subject to edge effects which can alter the efficiency of ecosystem functions. These areas are particularly sensitive to drought. Decomposition of detritus and nutrient cycling are driven by soil microbe and fauna coactivity. The bait lamina assay was used to assess soil fauna trophic activity in the upper soil horizons at five sites in Wytham Woods, Oxfordshire: two edge, two intermediate and one core site. Faunal trophic activity was highest in the core of the woodland, and lowest at the edge, which was correlated with a decreasing soil moisture gradient. The efficiency of the assay was tested using four different bait flavours: standardised, ash (*Fraxinus excelsior* L.), oak (*Quercus robur* L.), and sycamore (*Acer pseudoplatanus* L.). The standardised bait proved the most efficient flavour in terms of feeding activity. This study suggests that decomposition and nutrient cycling may be compromised in many of the UK's small, fragmented woodlands in the event of drought or climate change.

Editor: Frederick R. Adler, University of Utah, United States of America

Funding: The Centre for Ecology and Hydrology funded this project. Drs E. Slade and T. Riutta were funded by the HSBC Climate Partnership Programme through the Earthwatch Institute. This work was also supported by the Environmental Change Network, funded through the Centre for Ecology and Hydrology (Natural Environment Research Council) (http://www.ceh.ac.uk/). The funders had no role in study design, data collection and analysis, decision to publish, or preparation of the manuscript.

Competing Interests: The authors have declared that no competing interests exist.

* E-mail: jakesimpson3000@gmail.com

Introduction

Leaf litter decomposition is a key step in the cycling of carbon and nutrients, and is achieved by the activity of soil microbes and fauna working in tandem. Decomposition by soil microbes is facilitated by the mechanical and metabolic degradation of litter by soil invertebrates [1–4]. Trophic activity of soil fauna is constrained by physical and biological factors: soil temperature and moisture [5], soil profile structure [6], the presence of above ground leaf litter [7], soil chemistry (following fertilisation, liming or pollution) [8,9], substrate quality [10,11] and soil fauna community composition [12]. Soil moisture and temperature are particularly variable with proximity to woodland edges [13], and therefore may influence the feeding activity of soil invertebrates.

Since the advent of modern agriculture and industrial and urban development, British forests have been extensively fragmented. They now cover approximately 12% of the UK land area compared to 75% of continuous forest 6000 years ago, with 75% of wooded areas being less than 2 ha and consisting of largely edge habitat [14,15]. The forest microclimate at woodland edges is drier and warmer than forest interior, this is brought about by both the direct and indirect effects (i.e. greater rate of transpiration) of increased solar radiation and air turbulence [13,16,17]. This microclimate is known to affect some invertebrate communities in woodland areas up to 1 km from the edge [18] therefore it should not be assumed that ecosystem functions (e.g. soil turnover and nutrient cycling) are uniform across the whole woodland.

Soil fauna feeding activity can be used as an indicator of the biological status of soil [7]. Given the complexity of soil fauna ecology whereby the same group can function at different trophic levels, general patterns in community composition which influence decomposition in a predictable way are yet to be discovered [1]. For simplicity, a general index of biological activity can be measured using soil fauna feeding activity alone (rather than attempting to measure decomposition by all the component organisms in the system). To measure soil fauna activity as a function of decomposition, the bait lamina assay was developed by Von Törne [19]. This technique has been used in microcosm [12], mesocosm [3] and *in situ* investigations [7], with great success in ecotoxicological studies [8,20]. The first bait lamina studies acknowledged that it was difficult to disentangle the effects of fauna and micro-organisms on feeding activity. However, recent studies both in microcosms and in the field have shown that macrofauna, such as earthworms [20], mesofauna, such as enchytraeids, and microarthropods, such as collembola [12] and acari [7] are the main feeders on bait lamina. Specifically, these studies have concluded that while some microbial activity does occur, it is small compared with the faunal activity [12]. Moreover, Gongalsky et al. [6] suggest that micro-organisms cannot contribute to perforation during the short durations typically used in bait lamina studies.

Gestel et al. [3] found that while alternative methods, such as litter bags or cotton sticks, are mainly indicative of the activity of soil microbes the bait lamina method is the most direct measure of the activity of the soil fauna. Moreover, with bait lamina sticks faunal feeding activity can be measured across different depth profiles.

Soil fauna have demonstrated preferential feeding for a nettle-based substrate in laboratory experiments [12], whereas no significant preference for different grass-based bait was found *in situ* in arid grassland in Canada [10]. Leaf litter from the dominant tree species can be incorporated into bait to examine the *in situ* relationship between local dominant tree species and faunal feeding preference [21]. Lumbricid worms have been shown to prefer leaf litter from *Fraxinus* species to *Tilia* species, and their abundance can be correlated with the palatability of the leaf litter [22]. However, it is unknown whether baits made from local tree species are more efficient indicators of trophic activity than the standardised-bait in temperate broadleaf woodland ecosystems.

To the best of our knowledge, this study is the first to experimentally investigate the effect of changes in abiotic conditions at the forest edge on the feeding activity of soil invertebrates using the bait lamina method. It is hypothesised that feeding activity will differ significantly at the woodland edges compared to the core, where soil moisture and temperature differ correspondingly. This study is also the first to compare standardised and bait made from local tree leaf litter in the bait lamina assay. Bait is hypothesised to be preferentially fed upon if it is made of the local dominant tree leaf litter compared to the standardised-bait. Furthermore, trophic activity is expected to be greatest towards the soil surface, where food is most abundant for detritivores.

Materials and Methods

Site description

The experiment was conducted at Wytham Woods (51°46′N, 001°20′W, UK National Grid SP 461 081), Oxfordshire. The woodland is approximately 400 ha and is characterised by ancient and secondary woodland. The woodland is a 'mixed deciduous' woodland community W8 according to the National Vegetation Classification (*Fraxinus excelsior* L.- *Acer campestre* L. - *Mercurialis perennis* L. woodland) [23,24]. Meteorological records for the site from 1992 to 2009 show the mean annual temperature to be 10.1°C, and an average precipitation of 730 mm (Environmental Change Network data, obtained from Centre for Ecology and Hydrology, Wallingford, UK, 2010), which is typical of the Central England climate.

Experimental design

Bait lamina sticks (Terra Protecta GmbH, Berlin, Germany) are thin PVC strips (1 mm×6 mm×120 mm) with sixteen apertures (1.5 mm diameter) bored at 5 mm intervals [21]. The apertures are filled with either standardised bait or bespoke bait, whereby bran flakes are substituted with local dried and ground leaf material. The strips are placed vertically into the soil, and feeding activity is assessed by comparing the number of apertures that have been perforated versus the number left untouched during the exposure period [21]. The bait lamina sticks are sold with a standardised-bait containing cellulose powder, bran flakes and activated carbon (70:27:3). Ash (*Fraxinus excelsior*), oak (*Quercus robur*), and sycamore (*Acer pseudoplatanus*) leaves were collected in the autumn of 2008 using raised nets in the woods, and were ground using a ball grinder into a fine-grade powder. Three bait flavours were prepared using the leaf powder, cellulose powder and activated carbon (to the same ratio as the standardised bait with ground leaves replacing the bran flakes). Distilled water was added to the bait powder to form a thick paste. The apertures on the sticks were filled manually (by compressing the paste between the finger and thumb, then running the stick through it) and then allowed to dry at room temperature. A second application was required because the bait contracts on drying. Each aperture was verified as being filled completely by holding the sticks against a light source after the bait had dried following the second application. In total 360 sticks were used (90 of each flavour).

Five sites were selected: the southern edge (SE), 100 m from the southern edge towards the core (S100m), within the woodland core (Core), the northern edge (NE), and 100 m from the northern edge (N100m). Sites were considered to be spatially independent; the distance between the core and the northern and southern sites was 0.9 km and 1.4 km respectively, and between the northern and southern edges 2.1 km. Edge sites were located within 5 m of the perimeter fence of the woods. As soil type has a significant effect on soil fauna activity [5], each site is on the same Denchworth series clay soil, a surface water gley soil in the England and Wales soil Survey Classification [25]. Results from a 18 ha forest monitoring plot (within which the core plot is sited) indicated there is little variability in physical properties or chemical analysis across the Denchworth soil series [26]. At each of the edge and 100 m edge sites, three replicate plots (40 cm×40 cm) were marked approximately15 m apart along a transect parallel to the woodland edge. In the core site [27] the three plots were marked randomly within a 1 ha area, approximately 25 m apart. The leaf litter was left undisturbed, leaving the natural litter depth at each plot. Each plot was divided into a grid of four columns and three rows, forming 12 cells, with two replicate sticks in each. Each column contained one of four bait flavour treatments (standard, sycamore, ash or oak), and the rows signified the exposure time of the bait lamina sticks (9, 19 or 34 days), although time data were pooled together for analysis. To insert the sticks into the soil without dislodging the bait in the apertures, or breaking the sticks, a modified metal knife created a slit 10 cm deep, with the same width and thickness as the bait lamina sticks. The shallowest aperture was at 1.2 cm, and the deepest at 8.7 cm. The sticks were installed on 14th October 2009. To test whether soil invertebrates are the agents responsible for making the perforations in the sticks a control test was conducted in the woods. A container was filled with defaunated soil (freeze treated to kill invertebrates), and 36 sticks (9 per flavour) were placed in the soil and exposed for the same periods as the experimental sticks. There were no perforations of any of the apertures of these sticks after any of the time intervals verifying the method.

Environmental data were collected throughout the exposure periods. On installation of the bait lamina sticks, soil moisture and temperature were measured with a hand held 12 cm TDR probe (HydroSense, Campbell Scientific Ltd., Shepshed, U.K.) and temperature probe at five locations around the plot to give a mean reading for each plot (15 readings per site). The soil temperature and moisture were not measured within the plot to avoid disturbing the sampling area. Soil moisture and temperature readings were taken on seven subsequent occasions during the exposure period. Continuous data for air temperature, rainfall and soil temperature (10 cm) for Wytham were obtained from an automatic weather station (AWS, Didcot Instruments Co., Abingdon, U.K.) at a nearby grassland site. Soil pH was measured after the exposure period by taking ten 10 cm deep soil samples from each site, mixing them together and taking three separate pH readings on the field moist samples, on 1: 2.5 extracts in water [28] to achieve a mean pH for the site.

Statistical Analysis

The environmental conditions were analysed across the five sites for soil temperature, moisture, and pH separately using general linear models. A Spearman's rank test was used (because the data were non-parametric) to test if soil moisture and temperature were correlated.

Feeding activity was analysed as the proportion of holes perforated and the data was arcsine square-root transformed. To account for temporal pseudoreplication (the same sites were revisited at different intervals) the average number of holes perforated per flavour per plot over the three separate time periods was taken. Using generalized linear additive models, location and bait flavour were modelled as fixed effects with plot within location as a random effect and soil moisture as a covariate. Soil temperature was excluded as a covariate as because no significant difference was observed between sites. Depth was included in the model with a cubic regression spline smoother due to the nonlinear relationship between feeding activity and depth. To account for the heterogeneity of variances among locations and bait flavours a separate variance structure for each location and flavour was applied. Alternative models were compared using likelihood ratio tests and AIC values (a model with simple variance structure vs. different variances per strata; a model with one general smoother for depth vs. smoothers by individual locations vs. smoothers by groups of locations). The optimal model with different variances for location and flavour and three smoothers for depth (Core; N100m and S100m; NE and SE) was selected. Analyses were carried out using R(GUI) [29] nlme and mcgv packages.

Results

Environmental Conditions

Over the 34 day sampling period mean soil temperature did not vary significantly among the sites ($F_{5,\ 121} = 1.55$, $P = 0.18$). The average soil moisture was significantly higher at the core than the 100 m and edge plots ($F_{5,\ 121} = 21.86$, $P<0.001$, figure 1). Soil moisture and temperature were negatively correlated ($\rho = -0.58$, $P<0.001$). Mean soil pH (6.85 ± 0.51) was anti-logged and did not differ significantly across sites ($F_{4,\ 5} = 2.54$, $P = 0.061$).

The air temperature at Wytham for the preceding month and during the sampling period was similar to the long term average. Rainfall though was particularly low in September 2009 (9.0 mm compared to 65.4 mm, 1992–2008), but average for October and November combined.

Factors affecting soil fauna feeding activity

Feeding activity varied between 0–75% (perforations per stick), with an overall mean feeding activity of 8.5% for all flavours and all sites. Moisture, location and bait flavour and depth were the principal factors influencing feeding activity (table 1). Trophic activity was lowest at the edges and highest in the core.

Feeding preferences were different among the sites. Overall, standard bait was preferred, especially in the core and 100 m sites, while edges showed more variability (figure 2, table 1). The standard bait represented 9% of holes perforated at the northern edge, compared to its highest proportion of 62% at the N100m site.

Feeding activity tended to be highest near the surface, with an average of 18% and 14% holes perforated at 1.2 cm and 1.7 cm respectively (figure 3). The locations could be divided into three groups based on the feeding activity pattern: Feeding activity had the steepest depth gradient in the core, intermediate at the 100 m sites and the least pronounced at the edge sites.

Discussion

Edge effects on feeding activity

The results demonstrate that feeding activity decreases towards the edge of the woodland. Soil moisture was a significant term in

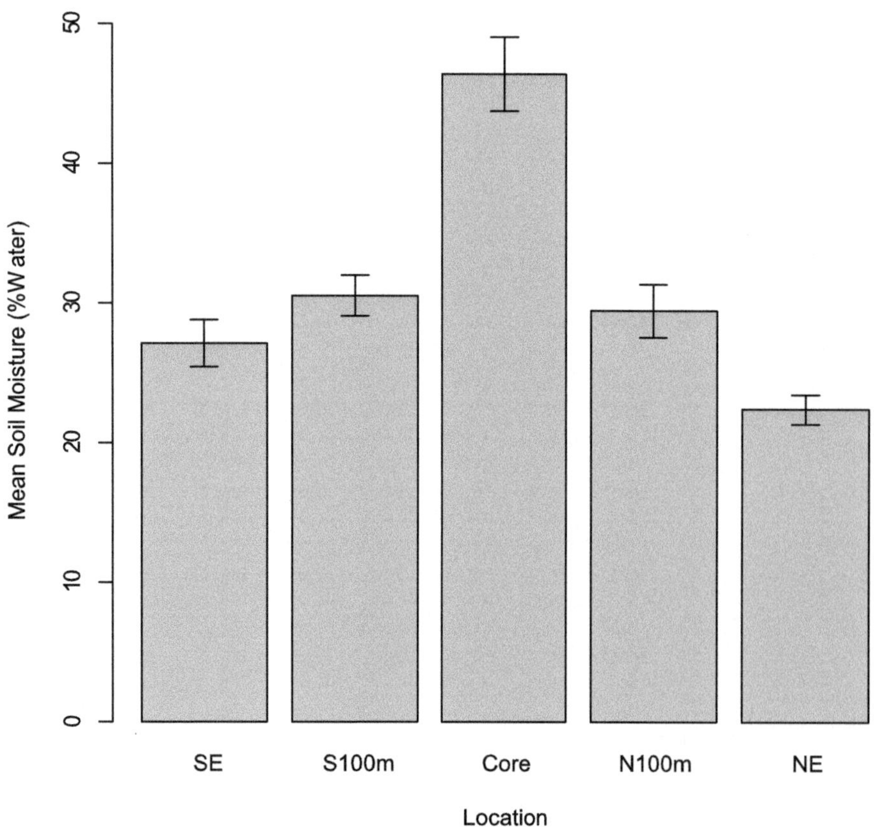

Figure 1. Mean soil moisture across sites (±SE). The bars represent the soil moisture data taken on 8 occasions during 34 day sampling period.

Table 1. Optimal generalized additive mixed model: feeding as a function of moisture, location, bait flavour and depth.

Explanatory variable	Degrees of freedom	F value	P value
Moisture	1	8.132	<0.001
Location	4	2.229	0.064
Flavour	3	10.598	<0.001
Location*Flavour	12	14.562	<0.001
Smoother Core	3.720	9.339	<0.001
Smoother N100m and S100m	1.004	40.468	<0.001
Smoother NE and SE	3.968	12.752	<0.001

Depth was modelled with cubic regression spline smoothers. Note that the degrees of freedom and P values for the smoothers are estimates. The higher the degrees of freedom the more non-linear the smoother (df = 1 indicates a linear smoother).

the minimal adequate model. Feeding activity was strongly correlated with the decreasing moisture gradient which occurs with proximity to the edge [16]. Gongalsky et al. [5] found soil temperature to have a greater influence on feeding than soil moisture, but when soil temperature does not significantly vary between sites, soil moisture has been shown to be the principle abiotic driver of trophic activity.

As feeding activity at 100 m was similar to the core, it is suggested that the feeding is not significantly affected by woodland edge effects at this distance. Chen et al. [30] found soil moisture variability was highest within 15 m from the edge. Most of the feeding activity occurred within 2 cm of the soil surface, as found by similar studies using bait lamina sticks [7,8,10], and feeding at the edges was found to be especially concentrated at the surface, compared with at the core where higher levels of feeding occurred at all depths. This may be due to lower soil moisture at the edges which causes the clay soil to harden, in turn limiting the vertical movement of soil invertebrates. Eggleton et al. [31] demonstrated that a reduction in soil moisture has a significant negative effect on the abundance of epigeic (surface feeding) and endogeic (sub-terrestrial feeding) species.

The edge effect is not limited to the direct abiotic factors considered in this study, but also includes a variety of indirect biotic factors which are beyond the scope of this study [32]. Using our model, the spatial difference in soil fauna feeding activity can be attributed to the moisture component of the edge effect. The results have demonstrated that feeding at the edge is significantly lower at the edges than sites 100 m towards the woodland interior and beyond. As soil fauna play a key role in nutrient cycling and decomposition processes [4,11] this has important applications to

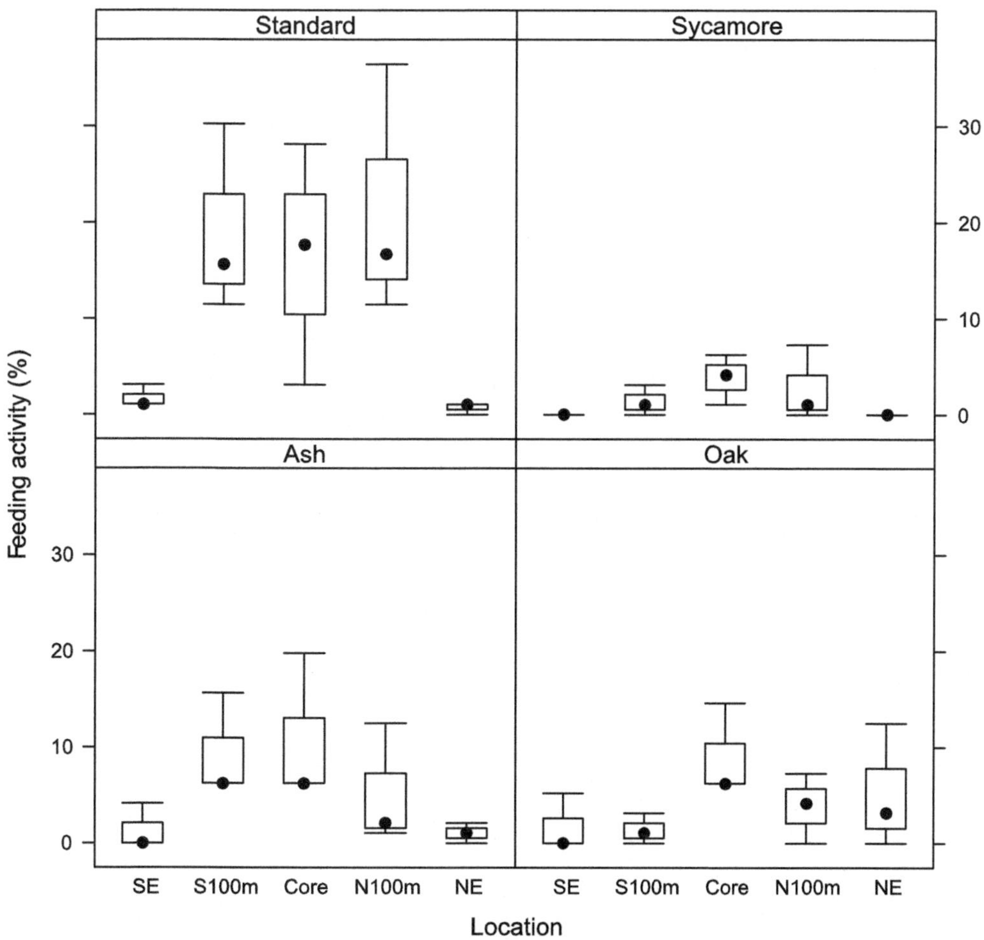

Figure 2. Feeding activity for each bait flavour across the five sampling sites in Wytham Woods. The black dots represent the median, with the boxes representing the 25th and 75th percentiles. The range is given by the whiskers.

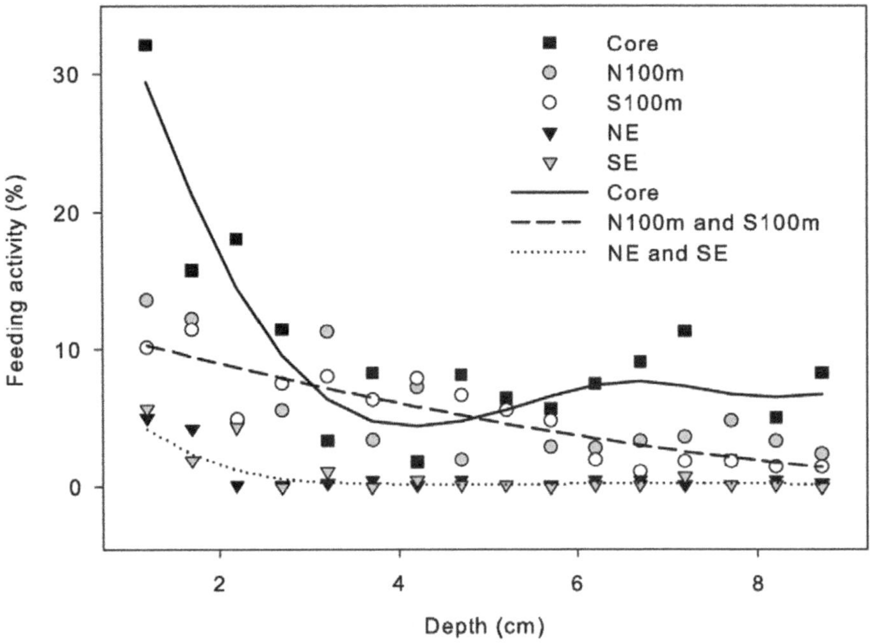

Figure 3. Feeding activity at different depths below the soil surface. The proportion is total number of holes perforated at each depth, averaged across bait flavours and plots within each site. The smoothers are derived from the generalised additive mixed model (see Statistical analysis). The model validation showed that the N100m and S100m sites could be described with one smoother as could the SE and NE sites.

small wooded areas which have a high edge to core ratio. Therefore, these processes may be significantly reduced as soil fauna feeding activity decreases towards the drier woodland edges.

Bait flavours and feeding activity

The standard bait proved to be the best indicator of trophic activity compared to the bespoke baits (mean feeding activity of 16% and 6% respectively). Field studies using the standard bait alone have tended to report overall higher trophic activity than those using bespoke baits. An average of 3% activity over 65 days was observed in Canadian soils using only bespoke baits [10], however it is difficult to draw comparisons between previous studies due to abiotic and biotic differences in sampling sites. For the purposes of assessing spatial heterogeneity in feeding activity efficiently and comparably, it is suggested that only the standardised bait is used.

Effect of the environmental conditions

Overall feeding activity was lower than expected according to the bait lamina manufacturers [33], but Von Törne [19] observed comparable feeding activity of 4–16% with standard bait in German soils. The month leading up to the exposure period was comparatively dry, causing the clay soil to harden and crack. These conditions are not conducive to invertebrate activity, especially for earthworms that favour moist soil and thus, and could have caused the overall low feeding activity [31]. The relatively dry period preceding the study may have extenuated the moisture limiting condition at the woodland edge, but these short drought periods are not out of the ordinary and climate models predict drier summers and/or increased frequency of summer droughts in the UK [34,35].

Feeding activity is known to occur down to temperatures of 5°C [5] which is lower than the minimum observed soil temperature in this experiment (8.5°C). While soil temperature was demonstrated to have an important effect on feeding activity in laboratory studies [5], the mean soil temperature between sites did not differ significantly, and therefore its effects could not be attributed to the variance observed in this study.

The soil at the surface, if not insulated by leaf litter, is very prone to desiccation which in turn may have pronounced effects in reducing nutrient turnover as a result of decreased trophic activity [4,7,11]. While small visual differences in the depth of leaf litter were noted between the sites, leaf litter manipulation studies have demonstrated that additional layers (i.e. increasing the depth of leaf litter) have little effect on soil temperature and moisture compared to natural litter depths [36]. Therefore, any differences in leaf litter depth between the sites are unlikely to have affected feeding activity. Soil acidity is known to affect the distribution of both earthworms and soil arthropods [9,37,38], but analysis of soil pH across the five sampling sites and the control show no significant difference, and therefore it is assumed that soil pH did not contribute to differences in feeding activity.

This study emphasises the possible effects of climate change on soil fauna feeding activity with reduced soil moisture, particularly in small, fragmented woodlands. Consistent with our results, studies conducted using the litter bag method have shown that low precipitation and moisture deficit reduces litter decomposition rates [39–42].Litter decomposition is a key process in ecosystem carbon cycle and is estimated to contribute up to 70% to the annual carbon efflux [43,44]. Given our results, future estimates of ecosystem functions (i.e. litter decomposition by soil fauna) and services (i.e. climate regulation) of temperate forests in a fragmented landscape should account for heterogeneity in soil fauna feeding activity as a result of the edge effect.

Acknowledgments

We thank Oxford University for permission to use Wytham Woods in this investigation.

Author Contributions

Conceived and designed the experiments: JES MET. Performed the experiments: JES. Analyzed the data: JES MET ES TR. Wrote the paper: JES MET TR ES.

References

1. Hattenschwiler S, Tiunov AV, Scheu S (2005) Biodiversity and litter decomposition interrestrial ecosystems. Annu Rev Ecol Evol Syst 36: 191–218.
2. Lavelle P (1997) Faunal activities and soil processes: Adaptive strategies that determine ecosystem function. Advances in Ecological Research Vol 27 27: 93–132.
3. Gestel CAM, Kruidenier M, Berg MP (2003) Suitability of wheat straw decomposition, cotton strip degradation and bait-lamina feeding tests to determine soil invertebrate activity. Biol Fertility Soils 37(2): 115–123.
4. Vetter S, Fox O, Ekschmitt K, Wolters V (2004) Limitations of faunal effects on soil carbon flow: Density dependence, biotic regulation and mutual inhibition. Soil Biol Biochem 36(3): 387–397.
5. Gongalsky KB, Persson T, Pokarzheuskii AD (2008) Effects of soil temperature and moisture on the feeding activity of soil animals as determined by the bait-lamina test. Applied Soil Ecology 39(1): 84–90.
6. Gongalsky KB, Pokarzhevskii AD, Filimonova ZV, Savin FA (2004) Stratification and dynamics of bait-lamina perforation in three forest soils along a north-south gradient in russia. Applied Soil Ecology 25(2): 111–122.
7. Rombke J, Hofer H, Garcia MVB, Martius C (2006) Feeding activities of soil organisms at four different forest sites in central amazonia using the bait lamina method. J Trop Ecol 22: 313–320.
8. Filzek PDB, Spurgeon DJ, Broll G, Svendsen C, Hankard PK, et al. (2004) Metal effects on soil invertebrate feeding: Measurements using the bait lamina method. Ecotoxicology 13(8): 807–816.
9. Geissen V, Brummer GW (1999) Decomposition rates and feeding activities of soil fauna in deciduous forest soils in relation to soil chemical parameters following liming and fertilization. Biol Fertility Soils 29(4): 335–342.
10. Hamel C, Schellenberg MP, Hanson K, Wang H (2007) Evaluation of the "bait-lamina test" to assess soil microfauna feeding activity in mixed grassland. Applied Soil Ecology 36(2–3): 199–204.
11. Osler GHR, Sommerkorn M (2007) Toward a complete soil C and N cycle: Incorporating the soil fauna. Ecology 88: 1611–1621.
12. Helling B, Pfeiff G, Larink O (1998) A comparison of feeding activity of collembolan and enchytraeid in laboratory studies using the bait-lamina test. Appl Soil Ecol 7(3): 207–212.
13. Davies-Colley RJ, Payne GW, van Elswijk M (2000) Microclimate gradients across a forest edge. N Z J Ecol 24(2): 111–121.
14. Forestry Commission (2009) Forestry facts and figures 2009: A summary of statistics about woodland and forestry.
15. Watts K (2006) British forest landscapes: The legacy of woodland fragmentation. Quarterly Journal of Forestry 100(4): 273–279.
16. Herbst M, Rosier PTW, Morecroft MD, Gowing DJ (2008) Comparative measurements of transpiration and canopy conductance in two mixed deciduous woodlands differing in structure and species composition. Tree Physiol 28(6): 959–970.
17. Young A, Mitchell N (1994) Microclimate and vegetation edge effects in a fragmented podocarp-broadleaf forest in new zealand. Biol Conserv 67(1): 63–72.
18. Ewers RM, Didham RK (2008) Pervasive impact of large-scale edge effects on a beetle community. Proc Natl Acad Sci U S A 105(14): 5426–5429.
19. Von Torne E (1990) Assessing feeding activities of soil living animals. bait lamina tests. Pedobiologia 34: 89–101.
20. Forster B, Van Gestel CAM, Koolhaas JE, Nentwig G, Rodrigues JML, et al. (2004) Ring-testing and field-validation of a terrestrial model ecosystem (TME) - an instrument for testing potentially harmful substances: Effects of carbendazim on organic matter breakdown and soil fauna feeding activity. Ecotoxicology 13(1–2): 129–141.
21. Kratz W (1998) The bait-lamina test - general aspects, applications and perspectives. Environmental Science and Pollution Research 5(2): 94–96.
22. Hendriksen NB (1990) Leaf litter selection by detritivore and geophagous earthworms. Biology and Fertility of Soils 10(1): 17–21.
23. Environmental Change Network (2010) ECN sites: Wytham. 31/01/2010.
24. Rodwell JS (1991) British plant communities. volume I. woodlands and scrub. Cambridge, UK: Cambridge University Press. 395 p.
25. Claydon B, Hollis JM (1984) Criteria for differentiating soil series.
26. Butt N, Campbell G, Malhi Y, Morecroft M, Fenn K, et al. (2009) Initial results from establishment of a long-term broadleaf monitoring plot at wytham woods.
27. Fenn K, Malhi Y, Morecroft M, Lloyd C, Thomas M (2010) Comprehensive description of the carbon cycle of an ancient temperate broadleaved woodland. Biogeosciences Discussions 7: 3735–3763.
28. Sykes JM, Lane AMJ (1996) The united kingdom environmental change network: Protocols for standard measurements at terrestrial sites. London, UK: HMSO. 219 p.
29. R Development core team (2010) R: A language and environment for statistical computing. 2.11.1.
30. Chen JQ, Franklin JF, Spies TA (1995) Growing-season microclimatic gradients from clear-cut edges into old-growth douglas-fir forests. Ecol Appl 5(1): 74–86.
31. Eggleton P, Inward K, Smith J, Jones D, Sherlock E (2009) A six year study of earthworm (lumbricidae) populations in pasture woodland in southern england shows their responses to soil temperature and soil moisture. Soil Biol Biochem 41(9): 1857–1865.
32. Berg B, Laskowski R (2009) Origin and structure of secondary organic matter and sequestration of C and N. In: Anonymous Advances in Ecological Research 38: Litter Decomposition: a Guide to Carbon and Nutrient Turnover Academic Press. pp 185–226.
33. terra protecta GmbH (1999) The bait-lamina test. http://www.terra-protecta.de/frame-de.htm. Accessed 2011 May 1.
34. Jenkins G, Murphy J, Sexton D, Lowe J, Jones P, et al. (2009) UK climate projections; briefing report.
35. Solomon S, Qin D, Manning M, Chen Z, Marquis M, et al. (2007) Climate change 2007: The physical science basis. contribution of working group I to the fourth assessment report of the intergovernmental panel on climate change. summary for policymakers.. Geneva, Switzerland: Intergovernmental Panel on Climate Change (IPCC). 18 p.
36. Sayer EJ (2006) Using experimental manipulation to assess the roles of leaf litter in the functioning of forest ecosystems. Biol Rev 81(1): 1–31.
37. Joschko M, Fox CA, Lentzsch P, Kiesel J, Hierold W, et al. (2006) Spatial analysis of earthworm biodiversity at the regional scale. Agric Ecosyst Environ 112(4): 367–380.
38. Mulder C, Van Wijnen HJ, Van Wezel AP (2005) Numerical abundance and biodiversity of below-ground taxocenes along a pH gradient across the netherlands. J Biogeogr 32(10): 1775–1790.
39. O'Neill E, Johnson D, Ledford J, Todd D (2003) Acute seasonal drought does not permanently alter mass loss and nitrogen dynamics during decomposition of red maple (acer rubrum L.) litter. Global Change Biol 9(1): 117–123.
40. Taylor A, Wolters V (2005) Responses of oribatid mite communities to summer drought: The influence of litter type and quality RID B-4635-2010. Soil Biol Biochem 37(11): 2117–2130.
41. Staley JT, Hodgson CJ, Mortimer SR, Morecroft MD, Masters GJ, et al. (2007) Effects of summer rainfall manipulations on the abundance and vertical distribution of herbivorous soil macro-invertebrates RID E-8814-2010. Eur J Soil Biol 43(3): 189–198.
42. Vasconcelos SS, Zarin DJ, Silva da Rosa MB, Oliveira FdA (2007) Leaf decomposition in a dry season irrigation experiment in eastern amazonian forest regrowth. Biotropica 39(5): 593–600.
43. Aerts R (2006) The freezer defrosting: Global warming and litter decomposition rates in cold biomes. J Ecol 94(4): 713–724.
44. Lal R (2004) Soil carbon sequestration impacts on global climate change and food security. Science 304(5677): 1623–1627.

12

Spatial Covariance between Aesthetic Value & Other Ecosystem Services

Stefano Casalegno*, Richard Inger, Caitlin DeSilvey, Kevin J. Gaston

Environment and Sustainability Institute, University of Exeter, Penryn, Cornwall, United Kingdom

Abstract

Mapping the spatial distribution of ecosystem goods and services represents a burgeoning field of research, although how different services covary with one another remains poorly understood. This is particularly true for the covariation of supporting, provisioning and regulating services with cultural services (the non-material benefits people gain from nature). This is largely because of challenges associated with the spatially specific quantification of cultural ecosystem services. We propose an innovative approach for evaluating a cultural service, the perceived aesthetic value of ecosystems, by quantifying geo-tagged digital photographs uploaded to social media resources. Our analysis proceeds from the premise that images will be captured by greater numbers of people in areas that are more highly valued for their aesthetic attributes. This approach was applied in Cornwall, UK, to carry out a spatial analysis of the covariation between ecosystem services: soil carbon stocks, agricultural production, and aesthetic value. Our findings suggest that online geo-tagged images provide an effective metric for mapping a key component of cultural ecosystem services. They also highlight the non-stationarity in the spatial relationships between patterns of ecosystem services.

Editor: Tobias Preis, University of Warwick, United Kingdom

Funding: This work was supported by the University of Exeter. The funders had no role in study design, data collection and analysis, decision to publish, or preparation of the manuscript

Competing Interests: The authors have declared that no competing interests exist.

* E-mail: stefano@casalegno.net

Introduction

Key to the successful maintenance and management of environmental resources is an understanding of their spatial distribution. Recognition of the significance of, and threats to, ecosystem services (the benefits that humans gain from ecosystems), has thus been associated with growth in attempts to map their patterns of variation and covariation [1–4]. In many instances, however, this remains challenging. Of the four main categories of services (supporting, provisioning, regulating and cultural [5]), this is particularly true of cultural services: the non-material benefits people gain from ecosystems (including aesthetic, recreational and spiritual benefits). Therefore, most studies have tended to focus on provisioning and regulating services, rather than cultural ones [6,7]. Indeed, whilst there has been substantial improvement in understanding of the fundamental importance of cultural services, their effective integration into the application of ecosystem service frameworks has been limited by the challenges of quantifying, valuing and mapping them [8].

Efforts to quantify and map cultural services have concentrated foremost on estimates of the relative numbers of recreational visitors to particular areas [1,2,9,10]. Other measures include the number of tourist attractions, tax value of summer cottages, number of reported sightings of rare species [11], tourist expenditure [12], accessibility to natural areas [13], days spent fishing [14] and indices combining multiple such variables [15–17]. When compared, these measures have, in the main, been found to be weakly correlated with spatial variation in other ecosystem services, which has significant implications for identifying and managing priority areas for their maintenance [1,2,11]. However, although these data are useful, a much broader portfolio of metrics and proxies for cultural services, and an understanding of how these covary with measures of other ecosystem services, is urgently required.

Quantification of the aesthetic value that people place on different parts of the landscape represents an innovative development in the mapping of cultural services. One potential measure of aesthetic value can be found in the spatial distribution of photographs of the natural environment that people post online, working from the premise that areas more highly valued for their aesthetic attributes will generate 'hotspots' of activity. Particularly useful are Internet platforms which specifically facilitate posting of geo-tagged digital images, and which are populated by an increasing number of users worldwide. Here we exploit one such resource, the "Panoramio" web platform (www.panoramio.com), to provide a valuable additional metric of cultural ecosystem services (aesthetic value), and document its relationship with two other ecosystem goods and services: a provisioning service (agricultural production) and a supporting/regulating service (carbon stocks in soil).

Materials and Methods

The study was carried out in Cornwall, UK, as a regional approach is considered most appropriate for the analysis of cultural activity [18]: tourism and agriculture are the largest components of the economy of this particular region [19]; and both residents and visitors have extensive (albeit not universal) access to digital cameras and the internet [20–21] that are required for the use of posted geo-tagged images.

Aesthetic value (cultural service)

Panoramio hosts photos of "places of the world", with a particular focus on images of landscapes, natural features (such as woodlands) and animals in their natural environment [22]. Images that have as their central subject people, machines, vehicles or the interiors of structures, or that depict public events such as fairs or concerts, are excluded from the platform [22]. The semantic content of Panoramio's images makes it better adapted to measure the perceived aesthetic value of ecosystems than other geo-tagged web platforms, that do not focus on landscape and environment. The number of individuals per unit area (1 km^2) uploading photographs to Google Earth via the Panoramio web platform was used as our measure of aesthetic value. This measure is more appropriate than the total number of photographs uploaded in each area, which reflects the level of activity of individual photographers rather than the overall value placed on a site by visitors.

Soil carbon (supporting and regulating service)

Data on carbon storage in soil (1 km^2 resolution map) were obtained from the European Commission Joint Research Centre [23,24]. These data are especially accurate for England, as detailed ground survey verification has been carried out.

Agricultural production (provisioning service)

Following Anderson et al. [1] and Eigenbrod et al. [25], we calculated an overall measure of agricultural production by summing the gross margins for all major crops/livestock. As inputs we used agricultural census data [26] at ward level, the CORINE land cover map [27], and gross margin estimates [28]. Agricultural production was expressed in units of £ per ha, and processed at 100 m resolution and then resampled at 1 km resolution. We improved on the original methodology by computing an averaged agricultural value from 2000 to 2005 (instead of using one year of data); differentiating gross margins according to lowlands, disadvantaged and severely disadvantaged areas; and using more precise input land cover data to achieve a higher resolution.

Analysis

Ecosystem service data were normalised to a 0–100 scale for comparison. We tested potential bias of aesthetic value by population density and by coastal/non-coastal locations. Cornish population densities were obtained from the 2011 census [29]. We selected 55 centres including the main towns (population greater than 3,000) and the populations coincident with hotspots of aesthetic value (>40 photographers per grid cell: the upper 99th quantile).

Spatially autocorrelated data violate the assumption of sample independence for a traditional test of significance. Therefore, we tested the presence of spatial autocorrelation using Moran's I coefficient [30] and quantified correlations between ecosystem services using the CRH-method [31,32], which correct correlation statistics for spatial autocorrelation [33–35]. Because of the distribution of our data we rank transformed inputs as outlined in Zar (2007) [36] correctly to apply the CRH method [33,34]. To determine the influence of spatial extent on results, we also divided Cornwall into six separate zones: coastal, Lizard peninsula, west, north-east, south-east and central Cornwall (Figure 1). We performed comparison tests for the overall study area and within separate zones, and corrected results for multiple test significance by applying Benjamin-Hochberg corrections on p-values [37].

The aesthetic and agricultural production surfaces were generated using Bash-Awk and GRASS [38]. Statistical analyses were performed in R [39] using "SpatialPack" and "raster" libraries for spatial related statistics.

Results

A total of 113,686 photographs were uploaded by 15,413 users in Cornwall from 2005 to 2011; 9,632 photographers in coastal, 1,414 in west, 1,411 in north-east, 1,385 in central, 1,227 in south-east and 344 in the Lizard peninsula (Figure 1). Hotspots of aesthetic value (35 of 3,843 grid cells; Figure 2) were all located in coastal areas: seven were in coastal towns (population>3000), 17 close to sparsely populated settlements (<3000 inhabitants), and 11 in unpopulated areas (beaches or touristic coastal sites).

There was a negative correlation between population density and aesthetic value (CRH correlation $= -0.56$, $n = 55$, *p-value*$<0{,}001$).

Soil carbon storage was highest in four main areas, located in the west and in three inland parts of Cornwall (Figure 1). The zonal statistics showed low carbon storage in coastal areas and highest in west and north-east Cornwall (Figure 1). Agricultural production was highest in west Cornwall and the Lizard peninsula, and lowest in coastal zones and in the north-east region.

We found positive spatial autocorrelation in all ecosystem service layers (Figure 3): soil carbon had the highest while aesthetic value had more dispersed spatial patterns. Compared to the overall study area, regional zones such as Coastal, North-East and West Cornwall had weaker patterns of spatial autocorrelation. Across the whole region, agricultural production was negatively correlated with both aesthetic value and soil carbon storage, and the latter two were themselves weakly negatively correlated (Table 1). However, within zones the relationships were quite variable in strength (Table 1).

Discussion

Most mapping of geographic variation in ecosystem services, and analysis of the patterns of covariation between services, has been conducted by ecologists and conservation biologists [6]. This has tended to result in a heavy reliance on conventional data sources with an established geographically explicit component. This is particularly the case for cultural ecosystem services, for which measures have in consequence been quite restricted, and for which novel approaches will need to be explored. Here we present one such approach, using data derived from social media to capture variation in the value that people place on different parts of the landscape.

The emerging field of computational social science exploits the capacity to collect and analyse data about human interactions on an unprecedented scale [40]. The potential for online digital data sets to provide valuable information for scientific studies has been well recognised [41] and applied to diverse disciplines [42–47], but to our knowledge they have not previously been employed in the context of ecosystem services. The research shares its approach with diverse computational social sciences applications [42–44,46–47] which aim to identify relationships between people's behaviour online and real world quantities.

We find substantial variation in our measure of aesthetic value across the study region. The peaks are distinct from the centres of human population, contrasting with the findings of previous studies that have measured recreational usage [1,25], but reflecting a priori expectation.

Our findings lend further support to the general conclusion that spatial variation in cultural services tends to be poorly or

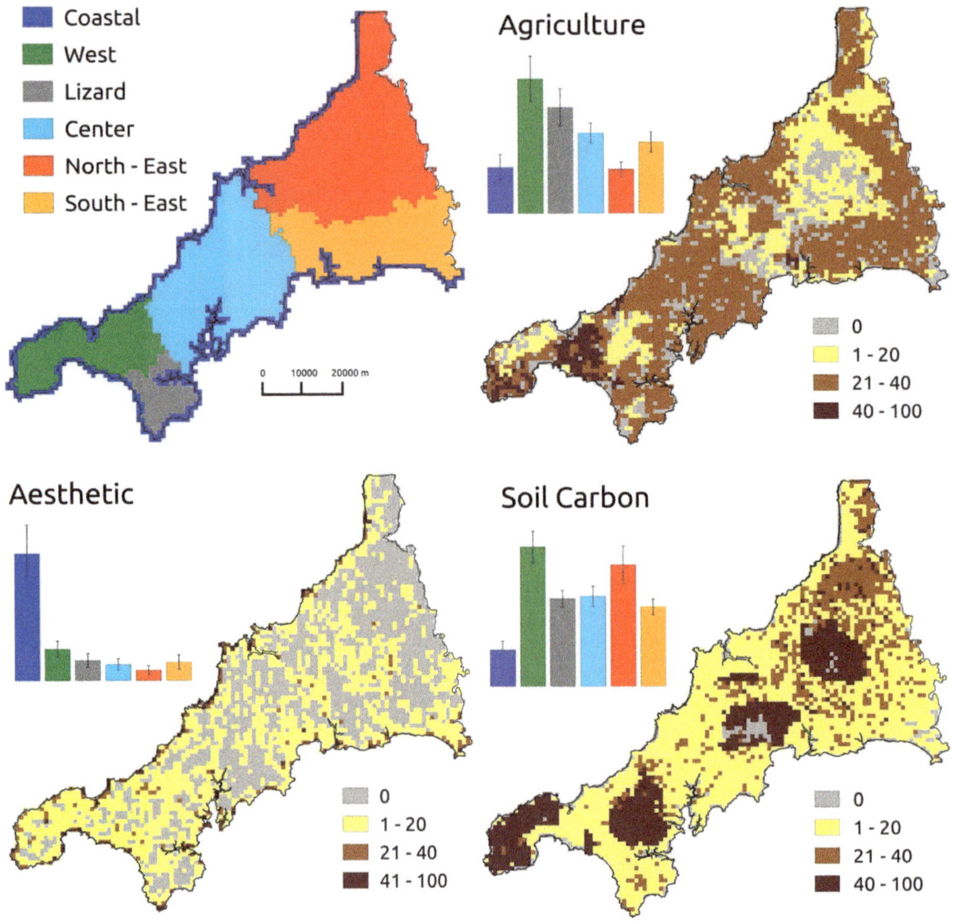

Figure 1. Study area and ecosystem services distribution. Geographical zonation of Cornwall (upper left), the distribution of agriculture, aesthetics and soil carbon (other maps; variation scaled from 0–100), and the mean value of each ecosystem service within each geographical zone (histograms).

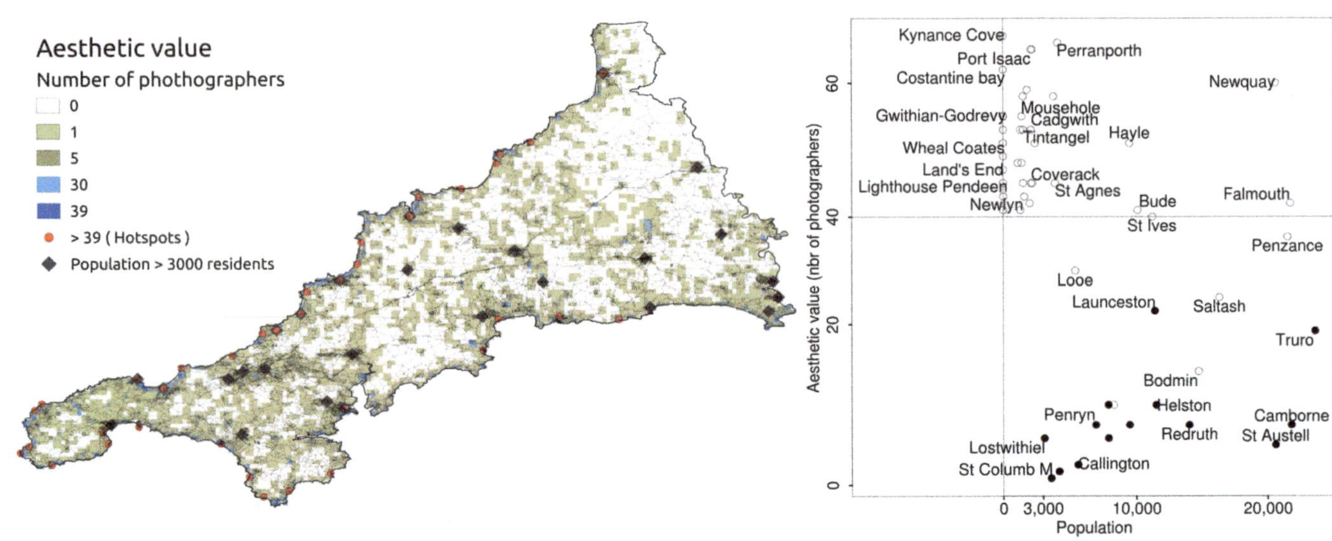

Figure 2. Aesthetic value and population of Cornwall. Aesthetic value map of Cornwall (left) and spatial covariance with the most populated places in Cornwall (right). Population data: Office for National Statistics, mid year estimates 2010. White dots: coastal locations; black dots: inland locations (not all location labels are shown).

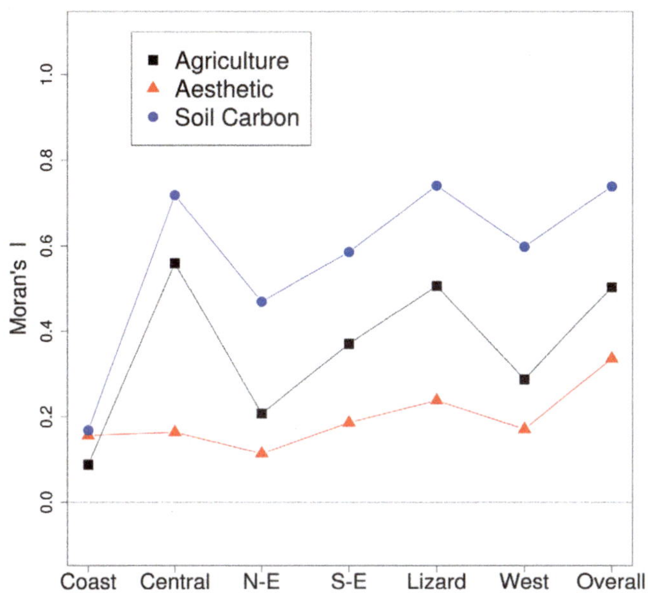

Figure 3. Test of spatial autocorrelation for agriculture, aesthetics and soil carbon data in Cornwall and within each geographical zone.

negatively correlated with that in many other ecosystem services [1,2,11], albeit we consider only single supporting/regulating and provisioning services. This would seem to highlight the need both to explore and test a variety of measures of cultural services, and to identify some such measures that are particularly well-founded and robust. Any failure to do so will likely serve to hinder attempts to ensure that cultural services get fuller consideration when planning for the maintenance of ecosystem service provision both at present and in the future.

We also find that the extent over which patterns of covariation in ecosystem services are determined can have a marked influence on the outcome, with these patterns typically becoming much more variable among smaller areas. Such an outcome has previously been documented [1,3] but over yet larger extents than those examined here. This will make generalizing about the relationships between different services challenging, and raises the spectre that observed relationships between variation in different ecosystems services is highly context-specific. Indeed, one can see an analogous situation arising to that which has developed in the context of attempts to understand spatial relationships between patterns of species richness and environmental variables, in which two schools of thought exist, a prevailing one that continues largely to ignore such complexities and another which focuses on the challenges of statistically handling non-stationary processes [48–50].

Table 1. CRH correlation test between different ecosystem services across Cornwall, and within different regions thereof.

	Coastal	Central	North-E	South-E	Lizard	West	Overall
Agriculture vs Aesthetics							
n	602	419	160	872	1,167	623	3,843
CRH	−0.02	−0.26*	−0.07	−0.07*	−0.12	−0.23*	−0.16*
Agriculture vs Soil Carbon							
n	308	417	160	836	1,154	603	3,478
CRH	0.10	−0.52**	−0.03	−0.07*	−0.45***	−0.16***	−0.22***
Soil Carbon vs Aesthetics							
n	308	417	160	836	1,154	603	3,478
CRH	0.04	0.06	−0.13	0.03	0.02	0.08	−0.11***

Multiple significance tests adjusted using Benjamin-Hochberg's (Benjamin et al. 1995) corrections: *p-value<0.05; ** p-value<0.01; *** p-value<0.001.

Author Contributions

Conceived and designed the experiments: SC KJG. Performed the experiments: SC. Analyzed the data: SC RI KJG. Contributed reagents/materials/analysis tools: SC. Wrote the paper: SC RI CD KJG.

References

1. Anderson BJ, Armsworth PR, Eigenbrod F, Thomas CD, Gillings S, et al. (2009) Spatial covariance between biodiversity and other ecosystem service priorities. J. Appl. Ecol. 46: 888–896. doi:10.1111/j.1365-2664.2009.01666.x.
2. Eigenbrod F, Armsworth PR, Anderson BJ, Heinemeyer A, Gillings S, et al. (2010) Error propagation associated with benefits transfer-based mapping of ecosystem services. Biol. Conserv.143: 2487–2493. doi:10.1016/j.biocon.2010.06.015.
3. Holland RA, Eigenbrod F, Armsworth PR, Anderson BJ, Thomas CD, et al. (2011) The influence of temporal variation on relationships between ecosystem services. Biodivers. Conserv. 20: 3285–3294. doi:10.1007/s10531-011-0113-1.
4. Egoh B, Reyers B, Rouget M, Bode M, Richardson D (2009) Spatial congruence between biodiversity and ecosystem services in South Africa. Biological Conservation 142: 553–562. doi:10.1016/j.biocon.2008.11.009.
5. Millennium Ecosystem Assessment (2005) Ecosystems and human well-being: synthesis. Reid WV, editor. Washington DC:Island Press. 137 p.
6. Seppelt R, Dormann CF, Eppink FV, Lautenbach S, Schmidt S (2011) A quantitative review of ecosystem service studies, approaches, shortcomings and the road ahead. J. Appl. Ecol.48: 630–636. doi:10.1111/j.1365-2664.2010.01952.x.
7. Martın-Lopez B, Iniesta-Arandia I, Garcıa-Llorente M, Palomo I, Casado-Arzuaga I, et al. (2012) Uncovering ecosystem service bundles through social preferences. PLoS ONE 7(6): e38970. doi:10.1371/journal.pone.0038970.
8. Daniel TC, Muhar A, Arnberger A, Aznar O, Boyd JW, et al. (2012) Contributions of cultural services to the ecosystem services agenda. P. Natl. Acad. Sci. USA 109: 8812–8819. doi:10.1073/pnas.1114773109.
9. Gimona A, Horst D (2007). Mapping hotspots of multiple landscape functions, a case study on farmland afforestation in Scotland. Landscape Ecol.22: 1255–1264. doi:10.1007/s10980-007-9105-7.
10. Hill GW, Courtney PR (2006) Demand analysis projections for recreational visits to countryside woodlands in Great Britain. Forestry 79: 185–200. doi:10.1093/forestry/cpl005.
11. Raudsepp-Hearne C, Peterson GD, Bennett EM (2010) Ecosystem service bundles for analyzing tradeoffs in diverse landscapes. P. Natl. Acad. Sci.USA 107: 5242–5247.
12. O'Farrell PJ, De Lange WJ, Le Maitre DC, Reyers B, Blignaut JN, et al. (2011) The possibilities and pitfalls presented by a pragmatic approach to ecosystem service valuation in an arid biodiversity hotspot. J. Arid Environ.75: 612–623. doi:10.1016/j.jaridenv.2011.01.005.
13. Chan KM, Shaw MR, Cameron DR, Underwood EC, Daily GC (2006) Conservation planning for ecosystem services. PLoS Biology 4: 2138–2152.
14. Chan KM, Hoshizaki L, Klinkenberg B (2011) Ecosystem services in conservation planning, targeted benefits vs. co-benefits or costs? PloS ONE 6: e24378. doi:10.1371/journal.pone.0024378.
15. Joyce K, Sutton S (2009) A method for automatic generation of the recreation opportunity spectrum in New Zealand. Appl Geogr 29: 409–418. doi:10.1016/j.apgeog.2008.11.006.
16. Maes J, Paracchini ML, Zulian G (2011) A European assessment of the provision of ecosystem services, towards an atlas of ecosystem services. Tech. rep. European Commission Joint Research Centre. doi:10.2788/63557.
17. Norton L, Inwood H, Crowe A, Baker A (2012) Trialling a method to quantify the cultural services of the English landscape using countryside survey data. Land Use Policy 29: 449–455. doi:10.1016/j.landusepol.2011.09.002.
18. Hein L, Vankoppen K, Degroot R, Vanierland E (2006) Spatial scales, stakeholders and the valuation of ecosystem services. Ecol Econ 57: 209–228. doi:10.1016/j.ecolecon.2005.04.005.
19. Cornwall Council (2009) Economic evidence review. Appendix 1. Green paper (economic priorities and strategic intent). Community Intelligence Chief

Executives Department, Technical report. Available: http://www.cornwall.gov.uk/idoc.ashx?docid = d82fc54b-1d94-4c3e-ae92-b8a445bdef55&version = -1 Accessed 2012 July 3.
20. International Telecommunications Unions website (2012) Percentage of individuals using the internet 2000–2010. Available: http://www.itu.int/ITU-D/ict/statistics/material/excel/2010/IndividualsUsingInternet_00-10.xls.Accessed 2012 July 3.
21. International Telecommunications Unions website (2012) Key global telecom indicators for the world telecommunication service sector. Available: http://www.itu.int/ITU-D/ict/statistics/at_glance/keytelecom.html. Accessed 2012 July 3.
22. Panoramio website (2013) Panoramio acceptance policy for Google Earth and Google Maps. Available: http://www.panoramio.com/help/acceptance_policy. Accessed 2013 May 3.
23. European Commission Joint Research Centre website (2012) The map of organic carbon content in topsoil in Europe: version 1.2. Available: http://eusoils.jrc.ec.europa.eu/ESDB_Archive/octop/octop_data.html. Accessed 2012 July 3.
24. Jones RJA, Hiederer R, Rusco E, Loveland PJ, Montanarella L (2005) Estimating organic carbon in the soils of Europe for policy support. European J. Soil Sci.56: 655–671.
25. Eigenbrod F, Anderson BJ, Armsworth PR, Heinemeyer A, Jackson SF, et al. (2009) Ecosystem service benefits of contrasting conservation strategies in a human-dominated region. Proc. R. Soc. B.276: 2903–11. doi:10.1098/rspb.2009.0528.
26. Departement for Environment, Food and Rural Affairs website (2012) June survey of agriculture 2000. Available: http://www.defra.gov.uk/statistics/foodfarm/landuselivestock/junesurvey/junesurveyresults/. Accessed 2012 July 3.
27. European Environment Agency website (2012) Corine land cover 2000 raster data. Available: http://www.eea.europa.eu/data-and-maps/data/corine-land-cover-2000-raster-1 .Accessed 2012 July 3.
28. Farm Management Handbook(2005) Agricultural Economics Unit. Exeter, IK:University of Exeter. 117 p.
29. Cornwall Council website (2012) Office for national statistics mid year estimates 2010: lower layer super output areas. Available: http://www.cornwall.gov.uk/default.aspx?page = 26436. Accessed July 2012 July 3.
30. Moran PAP (1950) Notes on continuous stochastic phenomena. *Biometrika* 37:17–23.
31. Clifford P, Richardson S, Hemon D (1989) Assessing the significance of the correlation between two spatial processes. *Biometrics* 45: 123–134.
32. Dutilleul P (1993) Modifying the t test for assessing the correlation between two spatial processes. *Biometrics* 49: 305–314.
33. Thomson JD, Weiblen BA, Thomson BA, Alfaro S, Legendre P (1996) Untangling multiple factors in spatial distributions: Lilies, Gophers, and Rocks. *Ecology* 77: 1698–1715.
34. Carroll C, Zielinski W, Noss R (1999) Using presence-absence data to build and test spatial habitat models for the fisher in the Klamath region, U.S.A. *Cons. Biol.* 13: 1344–1359.
35. Dormann CF, McPherson JM, Araujo MB, Bivand R, Bolliger J, et al. (2007) Methods to account for spatial autocorrelation in the analysis of species distributional data: a review. *Ecography* 30: 609–628. doi: 10.1111/j.2007.0906-7590.05171.x.
36. Zar JH (2007) Biostatistical analysis. Upper Saddle River, NJ:Prentice Hall .960 p.
37. Benjamini Y, Hochberg Y (1995) Controlling the false discovery rate: a practical and powerful approach to multiple testing. *J. Roy. Statist. Soc. Ser. B* 57: 289–300.
38. GRASS development team (2012) Geographic Resources Analysis Support System (GRASS) Software. v.6.4.2. Open Source Geospatial Foundation Project. Available: http://grass.osgeo.org. Accessed 2013 Jun 7.
39. R Core Team (2013) R: A Language and Environment for Statistical Computing. R foundation for statistical computing. Available: http://www.R-project.org. Accessed 2013 Jun 7.
40. Lazer D, Pentland A, Adamic L, Aral S, Barabási AL, et al. (2009) Computational social science. *Science* 323: 721–723 doi: 10.1126/science.1167742.
41. Bainbridge WS (2007) The scientific research potential of virtual worlds. *Science* 317 472–476 doi:10.1126/science.1146930.
42. Brownstein CA, Brownstein JS, Williams DS, Wicks P, Heywood JA (2009) The power of social networking in medicine. Nat. biotechnol. 27: 888–90. doi:10.1038/nbt1009–888.
43. Malizia A, Bellucci A, Diaz P, Aedo I, Levialdi S (2011) eStorys, A visual storyboard system supporting back-channel communication for emergencies. J. Visual Lang.Comput. 22: 150–169. doi:10.1016/j.jvlc.2010.12.003.
44. Signorini A, Segre AM, Polgreen PM (2011) The use of Twitter to track levels of disease activity and public concern in the U.S. during the influenza A H1N1 pandemic. *PLoS ONE* 6: e19467. doi:10.1371/journal.pone.0019467.
45. Gonzalez MC, Hidalgo CA, Barbasi AL (2008) Understanding individual human mobility patterns. *Nature* 453: 779–782. doi:10.1038/nature06958.
46. Preis T, Moat HS, Stanley HE (2013) Quantifying trading behavior in financial markets using google trends. *Sci. Rep.*3: 1684. doi:10.1038/srep01684.
47. Preis T, Moat HS, Stanley HE, Bishop SR (2012) Quantifying the advantage of looking forward. *Sci. Rep.* 2: 350. doi:10.1038/srep00350.
48. Casalegno S, Amatulli G, Bastrup-Birk A, Durrant TH, Pekkarinen A (2011) Modelling and mapping the suitability of European forest formations at 1-km resolution. *Eur. J. For. Res.* 130: 971–981. doi:10.1007/s10342-011-0480-x.
49. Bradford AH (2012) Eight (and a half) deadly sins of spatial analysis. *J. Biogeogr.* 39: 1–9. doi:10.1111/j.1365–2699.2011.02637.x.
50. Miller JA (2012) Species distribution models: Spatial autocorrelation and non-stationarity. *Progress in Phys. Geogr.* 36: 681–692.

13

Effect of Sugarcane Burning or Green Harvest Methods on the Brazilian Cerrado Soil Bacterial Community Structure

Caio T. C. C. Rachid[1,2], Adriana L. Santos[2], Marisa C. Piccolo[1], Fabiano C. Balieiro[3], Heitor L. C. Coutinho[3], Raquel S. Peixoto[2], James M. Tiedje[4], Alexandre S. Rosado[2]*

1 Center for Nuclear Energy in Agriculture, Universidade de São Paulo, Piracicaba, São Paulo, Brazil, **2** Institute of Microbiology Paulo de Góes, Federal University of Rio de Janeiro, Rio de Janeiro, Rio de Janeiro, Brazil, **3** Embrapa Solos, Rua Jardim Botânico, Rio de Janeiro, Rio de Janeiro, Brazil, **4** Center for Microbial Ecology, Michigan State University, East Lansing, Michigan, United States of America

Abstract

Background: The Brazilian Cerrado is one of the most important biodiversity reservoirs in the world. The sugarcane cultivation is expanding in this biome and necessitates the study of how it may impact the soil properties of the Cerrado. There is a lack of information especially about the impacts of different sugarcane management on the native bacterial communities of Cerrado soil. Therefore, our objective was to evaluate and compare the soil bacterial community structure of the Cerrado vegetation with two sugarcane systems.

Methods: We evaluated samples under native vegetation and the impact of the two most commonly used management strategies for sugarcane cultivation (burnt cane and green cane) on this diversity using pyrosequencing and quantitative PCR of the rrs gene (16S rRNA).

Results and Conclusions: Nineteen different phyla were identified, with Acidobacteria (≈35%), Proteobacteria (≈24%) and Actinobacteria (≈21%) being the most abundant. Many of the sequences were represented by few operational taxonomic units (OTUs, 3% of dissimilarity), which were found in all treatments. In contrast, there were very strong patterns of local selection, with many OTUs occurring only in one sample. Our results reveal a complex bacterial diversity, with a large fraction of microorganisms not yet described, reinforcing the importance of this biome. As possible sign of threat, the qPCR detected a reduction of the bacterial population in agricultural soils compared with native Cerrado soil communities. We conclude that sugarcane cultivation promoted significant structural changes in the soil bacterial community, with Firmicutes phylum and Acidobacteria classes being the groups most affected.

Editor: Gabriele Berg, Graz University of Technology (TU Graz), Austria

Funding: This work received funding from the Conselho Nacional de Desenvolvimento Científico e Tecnológico (CNPq), Fundação de Amparo à Pesquisa do Estado do Rio de Janeiro (FAPERJ), Coordenação de Aperfeiçoamento de Pessoal de Nível Superior (CAPES) and Fundação de Amparo à Pesquisa do Estado de São Paulo (Fapesp). The funders had no role in study design, data collection and analysis, decision to publish, or preparation of the manuscript.

Competing Interests: The authors have declared that no competing interests exist.

* E-mail: asrosado@micro.ufrj.br

Introduction

Brazilian Cerrado (neotropical savanna vegetation) covers more than 20% of the country's surface and is one of the most diverse biomes (major biotic community) in the world. In addition, this biome is classified as a high-priority area for biodiversity conservation [1–3].

However, around 40% of the Cerrado land has been converted for agricultural and livestock production, a process that has been very active during the last two decades [3,4]. Agricultural expansion has been stimulated by incentives from the Brazilian Federal government, e.g. the Cerrado Development Program, the Cattle-Raising Development Council and the Alcohol Program and multilateral agencies such as the Nippo-Brazilian Cooperation Program for the Development of the Cerrado [5]. This scenario has resulted in the Cerrado region producing 49% of the grain, 41% of the milk and 40% of the beef in Brazil [6].

Recently, the expansion of sugarcane production in the Cerrado has received renewed attention because sugarcane is an important material for agroenergy and bioethanol production [7]. The cultivated areas in the states of Goiás and Mato Grosso do Sul have grown more than 300% in the last 5 years [8]. This expansion is the outcome of the Brazilian biofuel program and occurs mainly through the replacement of established agricultural areas (pastures and soy and corn fields) by new sugarcane areas [9,10].

The impact of sugarcane on different components of the agroecosystem (water, soil structure, greenhouse emissions, enzyme activities, soil carbon stocks and fertility) is well documented, although mainly for the Brazilian Atlantic Forest biome [11–16]. Briefly, these works show that residue management (the use of burnt cane techniques, in which, pre-harvest burning accompanies manual harvest, vs. green sugarcane, in which fire is not used and which is associated with mechanical harvest) has little effect on C stocks in the short term (2–4 years). However, these studies have shown that labile organic matter fractions, such as microbial biomass, particulate organic matter and soil enzyme activity, and some physical measures, such as the mean weight diameter of water-stable soil aggregates, are more sensitive to these soil management methods.

Land use change can significantly influence the bacterial community structure and abundance, which, in turn, is correlated with soil biogeochemical processes and ecosystem productivity [17–20]. Therefore, understanding the microbial community structure and its relationship with the changes in land use and management are fundamental to understanding ecosystems.

There is a lack of information about the bacterial diversity of the Cerrado soil, and the impacts of different management strategies on native bacterial communities, or even more comprehensives studies, such as clone library screening [21,22], to document bacterial community structure. Next-generation sequencing technology such as 454 pyrosequencing provides much greater capacity to improve the knowledge about the soil microbial diversity of this biome. Only one recent work used this technology to study the Cerrado bacterial diversity, however, without evaluating the impact of agriculture on soil microbiology [23].

Our goal was to evaluate the bacterial community structure of Cerrado soil samples under native vegetation and the impact of the two most commonly used management practices of sugarcane cultivation, burnt cane and green cane, on this community using pyrosequencing and quantitative real-time PCR.

Materials and Methods

Field Site and Sampling

The study area (17° 55′ 35″ S, 50° 08′ 36″ W) was located at the municipality of Porteirão, state of Goiás, Brazil. The regions climate is classified as Aw (Köppen), with annual average rainfalls exceeding 1500 mm year^{-1} and annual average air temperatures of 23.1°C. The soil is a eutrophic Latossolo vermelho (Ferralsols), which is characterised by high levels of base saturation (>50%). Although the area is flat, petroplinthites (lateritic nodules or concretions) are found in the subsurface, which may restrict drainage and exhibit concretionary characteristics (Oliveira et al. 1992).

The field had been previously used for cotton, soybean and sunflower production, and was converted to sugarcane cultivation in 2002. The samples were collected in September 2008, during the sugarcane growth stage, approximately 7 to 8 months after bud germination (after six yearly harvest cycles). The field was divided into three regions (split-plot) in which three different regimes were applied:

(i) Burnt sugarcane – Before harvest, the sugarcane was ignited to remove the leaves. The stem was then manually harvested. After harvest, the soil remained uncovered (17°55′32.05″S 50° 8′50.64″W).

(ii) Green sugarcane – Harvest was performed using a machine that separates the sugarcane leaves from the stems. The leaves are then returned to the soil. After harvest, the soil remained covered by the vegetal residues (17°55′35.83″S 50° 8′41.10″W).

(iii) Cerrado – covered by a dense formation of trees up to 4 metres tall, as a typical Cerradão formation [24]. It represents the soil in a more natural condition, and received no addition of fertilizers (17°55′32.52″S 50° 8′37.92″W).

The sugarcane treatments had 6 years of application before the sampling. The sizes of the burnt sugarcane, green sugarcane and Cerrado areas were 23.5, 9.9 and 2.9 ha, respectively. The three treatments were very close to one another, less than 300 m apart. To allow replication, per treatment, three 5×5 m subplots were defined randomly (approximately 10 m of distance from each other). The soil was collected, five points per subplot (which were pooled) approximately to 10 cm depth, using a core borer. All soil was collected in the same day (approximately 2.5 kg), mixed and transported to the operational base in normal temperature within 2 hours. Then, smaller amounts of the soil (10 g) were separated, stored in centrifuge tubes and frozen. The soil was kept frozen until DNA extraction. This present study is the continuation of a previous work, which used the same soil samples for physico-chemical and biological analysis. Information about the physical and chemical properties of the sites has been previously described [25], and the general soil characterization can be found in the supplementary material (Table S1).

Real-time PCR Analysis of the rrs Gene

Quantitative PCR was performed on the ABI PRISM® SDS 7000 (PE Applied Biosystems). Amplification reactions were performed with the SYBR Green PCR master mix (Applied Biosystems) using the primers 357F (5′ CTA CGG GRS GCA G 3′) and 529R (5′ CGC GGC TGC TGG CAG 3′) [26] specific for bacteria, at a concentration of 300 nM each, and a DNA template volume of 1 µl (≈ 20 ng) was added to 24 µl of PCR master mix in MicroAmp Optical 96-well reaction plates. The real-time PCR thermocycling steps for all primer sets were as follows: 50°C for 2 min; 95°C for 10 min; and 40 cycles of 95°C for 1 min, 50°C for 1 min and 60°C for 1 min. In all experiments, appropriate negative controls containing no template DNA were subjected to the same procedure to exclude or detect any possible contamination or carryover. Melting curves were also routinely checked to confirm the purity of the amplified products.

Standard curves were obtained by plotting Ct (threshold cycle) as a function of the log of the copy number of the target DNA. Tenfold serial dilutions of the plasmids containing the standard sequences, ranging from 10^1 to 10^8 serial dilutions, were used as the target DNA. All measurements were performed in triplicate per sample. To test the difference among the treatments, the data we performed one anova with log transformed data.

Bacterial Pyrosequencing and Sequence Processing

Soil (0.5 g) DNA was extracted from triplicate samples using the FastDNA® Spin Kit for Soil and the FastPrep® equipment (Bio 101, CA, USA) according to the manufacturer's instructions. The extracted DNA was submitted to PCR amplification, targeting hyper-variable region 4 of the rrs gene (16S rRNA) using the primers 563F and 802R [27]. The reactions were performed as described by [28]. Equimolar amounts of the PCR products were submitted to pyrosequencing on a Genome Sequencer FLX system (454 Life Sciences, USA) at Michigan State University. The sequences are available at the NCBI Sequence Read Archive under the following accession numbers: SRR077401.2, SRR077402.2, SRR077403.2, SRR077404.2, SRR077405.2, SRR077406.2, SRR077407.2, SRR077408.2, and SRR077409.2.

The raw sequences were processed by using Mothur v. 1.26.0. [29]. The flowgrams were submitted to Pyronoise [30] to reduce error in the retained data set. Additionally all sequences missing the forward primer and/or had a length smaller than 190 bases were removed. The high quality sequences were then aligned using Silva reference database and the chimeras were detected and eliminated. The overall quality processing removed around 3,600 sequences, and resultant alignments file with only high quality sequences served as inputs to construct the distance matrix and to cluster the sequences into OTUs. To avoid the bias of analysing data with different sequences numbers, we performed all the analyses (with exception of the taxonomic assignment for relative abundance) with a normalised number of sequences for all treatments. The clusters were constructed at a 3% dissimilarity and were used to generate predictive rarefaction models and make calculations using the richness indices Ace and Chao1 [31] and Shannon's diversity index [32]. Mothur was also used to perform the NMDS ordination of the samples and to test the significance of the differences among the treatments based on the Weight Fast UniFrac and Amova tests.

To obtain the taxonomic assignment and relative abundance of the different bacterial groups the sequences were submitted to RDP-II Classifier using an 80% confidence threshold [33,34].

Results and Discussion

A total of 22,424 high-quality sequences were obtained from all samples. The minimum and maximum number of sequences for each sample ranged from 1322 (GC2) to 5164 (GC1), respectively. When analysed together, the treatments showed more than 3000 different operational taxonomic units (3% dissimilarity). The number of OTUs was similar in all analysed samples, with no significant differences among the treatments (Table 1). The most abundant OTUs were composed mainly by uncultured bacteria from different phyla. The summary of them can be found in the supplementary material (Table S2). There were also no differences ($p>0.05$) for the Chao1, Ace and Shannon diversity estimators among the treatments (Table 1), nor in the rarefaction curve (Figure S1). Although the pyrosequencing did not detect a significant influence in the bacterial diversity index cited above, the qPCR did reveal a significant decrease in the number of rrs gene (16S rRNA) copies in the cultivated sites, GC $\approx$ 3.3 $10^8 \pm 6.10^7$ (average and standard deviation, per gram of dry soil) and BC $\approx$ 2.2 $10^8 \pm 8.10^7$, compared with the natural area CE $\approx$ 6.3 10^8, $\pm 6.10^7$ suggesting an overall suppression of the bacterial community in the cultivated areas.

With respect to the phylogenetic composition of the samples, many of the sequences from all treatments (from 32% to 43%) could only be assigned to the Bacteria domain and had no classification at the phylum level. This percentage of unclassified bacteria is higher than that already described in other environments such as grassland, pasture, and agricultural systems in Texas (5% to 10% unclassified) [35]; German forest and grassland soils (10 to 25%) [36]; oak forest soil (18 to 22%) [37] and Amazonian dark earth (26 to 36%) [38]. The high fraction of unclassified sequences may have two possible explanations: first, the sequence size may be too short for accurate classification using the bootstrap cut-off value of 80%, and second, the database may not be complete enough to include all different components within each phylum and, as a result, be missing some comparative elements with which to classify all microbial diversity. The second reason is more plausible to explain the high level of unclassified sequences because the sequences length were larger than 200 bp, which is enough to classify known bacteria at the phylum level with 99.5% accuracy using the Classifier tool [34]. Furthermore, the high level of unclassified bacteria emphasises the high degree of undiscovered biodiversity found in the "hotspot" Cerrado biome.

Among the classified sequences, the most abundant phyla found in all treatments were (in rank order): Acidobacteria, Proteobacteria, Actinobacteria, Verrucomicrobia, Gemmatimonadetes, Firmicutes, Planctomycetes and Bacteroidetes (Figure 1). Eleven other phyla were found in at least one sample (Chloroflexi, TM7, Armatimonadetes, Chlamydiae, WS3, OD1, Nitrospira, Cyanobacteria/Chloroplast, Spirochaetes, BRC1 and Fusobacteria). These data are in accordance with other studies that used pyrosequencing or clone libraries which also found these phyla as the most abundant in soils [21,23,37,39–41].

However, the predominance of Acidobacteria compared with the other phyla has not been observed in most other studies. In general, Proteobacteria have the highest frequency in previous studies [21,37,39–41], with exception of another Cerrado soil survey that found Acidobacteria as the most dominant group [23], highlighting a pattern for this biome. Moreover, the relative frequency of Acidobacteria has been shown to range from 2.4% to 78.5% in different soils, varying mainly due to differences in pH, with Acidobacteria more abundant in acid soils, but also in correlation with other soil properties [42].

Our data suggest that the role of the phylum Bacteroidetes is of minor relevance in our area because it accounts for only a very

Table 1. Estimated OTU richness and diversity index for rrs gene of Cerrado soil samples under native vegetation and sugarcane cultivation.

Treatments	Sequence number*	OTUs	Estimated OTU richness		Shannon
			CHAO1	ACE	
CE1	1306	595	1422(1216;1698)	2433(2197;2704)	5.85(5.78;5.92)
CE2	1306	639	1764(1486;2135)	3082(2800;3401)	5.99(5.92;6.05)
CE3	1306	596	1296(1122;1527)	2062(1863;2292)	5.88(5.81;5.95)
GC1	1306	613	1415(1217;1678)	2286(2071;2534)	5.95(5.89;6.02)
GC2	1306	578	1300(1112;1554)	1868(1696;2066)	5.91(5.85;5.97)
GC3	1306	624	1530(1305;1830)	2247(2033;2494)	5.99(5.93;6.06)
BC1	1306	705	2033(1721;2440)	3840(3496;4226)	6.11(6.04;6.17)
BC2	1306	565	1228(1054;1463)	1684(1527;1867)	5.87(5.80;5.93)
BC3	1306	598	1556(1312;1882)	2644(2394;2928)	5.88(5.81;5.95)
Total	11754				

*Normalized number of sequences.
Inside brackets, the lower and upper limits of values between the 95% confidence intervals).

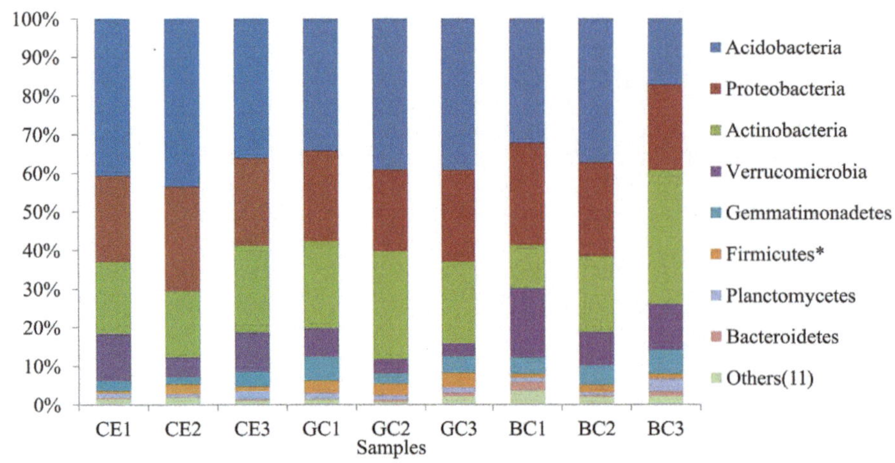

Figure 1. The relative frequencies of the different phyla found in Cerrado (CE), Green Cane (GC) and Burnt Cane (BC) samples.

small fraction of the entire bacterial diversity (0.1 to 3% of the sequences), in contrast with other studies in which this phylum is very abundant and accounts for up to 20% of the classified sequences [39,40]. The phylum Cyanobacteria, represented by five sequences, was only found in the BC treatment in the present study.

Among the most abundant phyla, only Firmicute sequences had significantly different distributions among the treatments ($p<0.01$), with a higher frequency in the GC treatment than in the others. Despite not being significant, the relative frequency of Acidobacteria decreased in the cultivated areas, mainly in the burnt cane plots, as can be noted by comparing the average values of the frequency of this phylum in CE (41%) to those in GC (38%) and BC (29%). However, the average relative frequency of Verrucomicrobia tended to be higher in the BC area (13%) than in CE (9.3%) and GC (5%).

The Acidobacteria phylum was represented by eight different classes (Figure 2A), which are, in the order of decreasing abundance, Gp6> Gp4> Gp1> Gp3> Gp7> Gp2> Gp22 and Gp25. The highest difference among the treatments was found in these classes. The burnt cane responded with a decrease in the relative frequency of Gp6 ($p<0.001$) and with an increase in the relative abundances of Gp1 ($p<0.001$) and Gp3 ($p<0.001$). These alterations can be explained by small variations in the soil pH. Jones and colleagues [42] showed that Gp6 correlated positively with pH, whereas Gp1 and Gp3 correlated negatively. The burnt cane exhibited a slight soil acidification, pH 5.8 ± 0.2 (average and standard deviation) compared with the green cane and Cerrado treatments, ph 6.6 ± 0.08 and 6.4 ± 0.06 respectively, and this change in the pH was correlated with the variation in the relative frequencies of these groups. Additionally, some other differences were observed, most specifically with the class Gp4 being significantly different between CE and GC and the class Gp25 being different between CG and BC.

The Proteobacteria phylum (Figure 2B) was fully dominated by the Alphaproteobacteria class, accounting for 70% (CE2) to 84% (GC1) of the relative frequencies, whereas Betaproteobacteria, Gammaproteobacteria and Deltaproteobacteria were also found but in low frequencies. Only Deltaproteobacteria differed significantly, being higher in the CE ($p=0.01$) than in the other treatment areas, revealing a higher sensitivity to cropping. These data differ from those found by Roesch and colleagues [39], who studied different agricultural soils in North America (including a sugarcane field) and Brazil using pyrosequencing. These authors found that Betaproteobacteria was the most abundant class of Proteobacteria in all places except in Brazil, where Gammaproteobacteria was more frequent. However, other studies found a predominance of Alphaproteobacteria followed by Betaproteobacteria [36,37].

The third most abundant phylum, Actinobacteria, was composed basically by the sub-classes Rubrobacteridae and Actinobacteridae, but Acidimicrobidae was also found at a low frequency (Figure 2C). This finding is in accordance with that of Janssen [41], who described the presence of these three sub-classes in some clone libraries, with a higher average frequency of the former two classes than of Acidimicrobidae.

Only Bacilli and Clostridia comprise the Firmicute phylum, with 70% to 100% of the sequences belonging to the Bacilli class (Figure S2A). Using pyrosequencing, Teixeira et al. [28] showed that Firmicute was an important phylum in rhizospheric Antarctic soil, but the major component (approximately 80%) of this phylum in this environment was Clostridia. However, Janssen (2006) in a review of clone libraries showed that Bacilli are more common than Clostridia in the majority of soils.

Finally, Verrucomicrobia was composed mainly of the Spartobacteria class in all treatments, with a Subdivision 3 class and a few sequences of Opitutae and Verrucomicrobiae occurring (Figure S2B).

To understand the distribution of OTUs among the different treatments, the sequences of all three samples from all treatments were clustered together. The results showed that only 39 out of the 3000 OTUs were shared by all samples (Figure 3), indicating that few species had a broad distribution. Despite the low number, these shared OTUs accounted for 27% of all sequences, revealing that these species were very dominant. However, there were many OTUs that occurred only in one sample (Figure 3), and the number of OTUs occurring exclusively in one sample ranged between 175 (GC2) to 337 (BC1). The number of exclusive OTUs was much higher than those shared between treatments. This result reveals a high level of local microniche selection. However, these OTUs were represented by a very small proportion of sequences, indicating that they were basically composed of rare organisms.

The taxonomic assignment of the OTUs shared by all samples reveals a contrasting frequency of phyla compared with those that

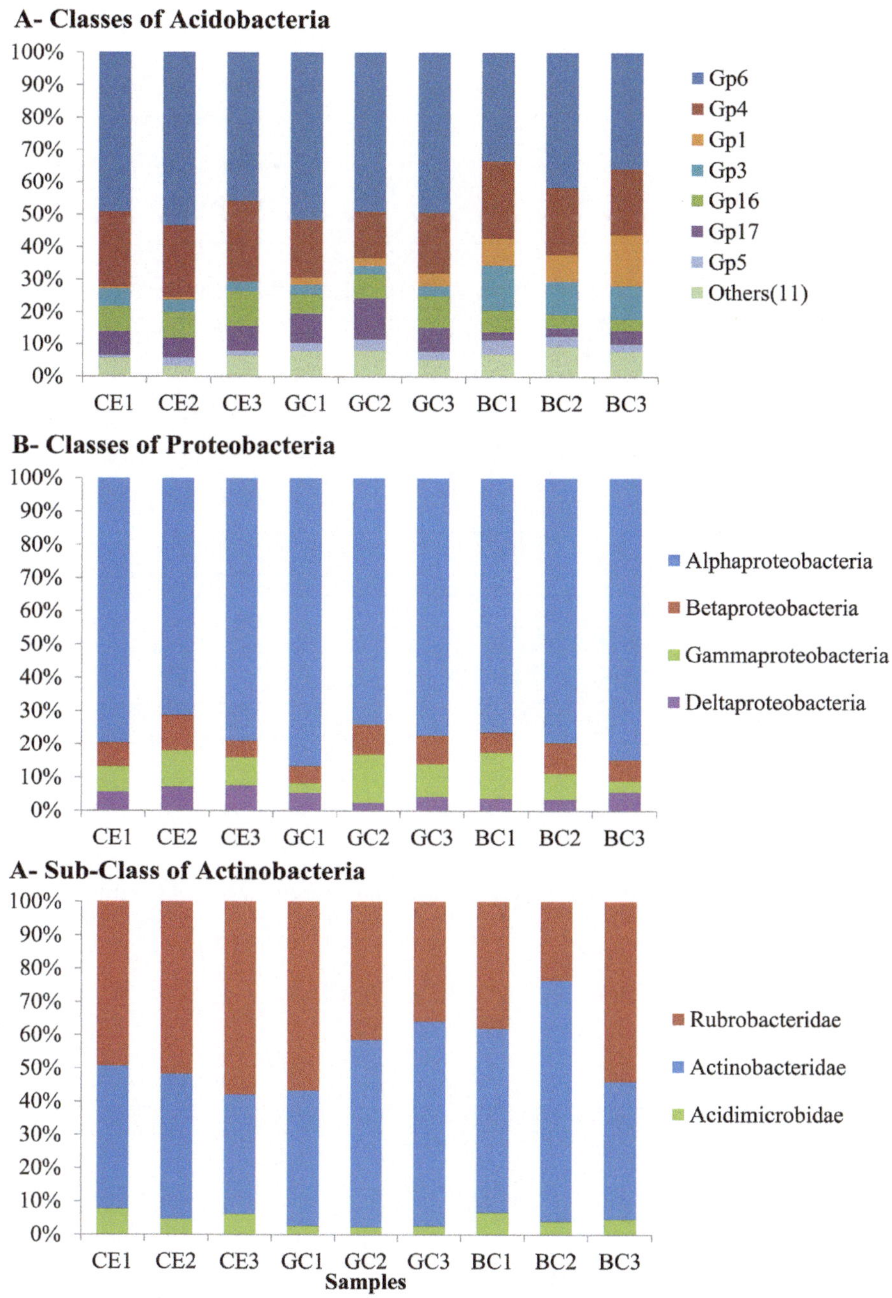

Figure 2. The relative frequencies of the different classes found in Cerrado (CE), Green Cane (GC) and Burnt Cane (BC) samples for the following phyla: A – Acidobacteria, B – Proteobacteria, C – Actinobacteria.

occurred in only one sample. The OTUs with a broad distribution belonged to only five phyla and had a higher predominance of Acidobacteria, whereas the taxonomic assignment of the OTUs that occurred only in one sample revealed the presence of at least ten different phyla for each sample. The relative frequencies of the taxonomic assignment of those samples were also different. There was a predominance of Proteobacteria for all samples except BC3 and, interestingly, a high frequency of Planctomyces in many samples, especially in BC3 and CE3. Curiously, no sequences belonging to Planctomyces were found among the OTUs with a broad distribution.

The land management had a significant influence on the microbial community, as revealed by the non-metric multidimensional scaling (NMDS) analysis of the pyrosequencing data (Figure 4), despite the low variation in the relative abundance of the different phyla among the treatments. A clear clustering of each treatment in a distinct region of the plot could be detected. To test the significance of this clustering, the Weight Unifrac and Amova tests were used. Both tests showed significant differences when all treatments were analysed together ($p = 0.001$), or in pairwise comparisons. Therefore, the NMDS analysis suggests that the cultivation of sugarcane plays an important role in the transformation of the microbial community. Additionally, the microbial community selected for by green cane management seems to be more similar to the microbial community observed in areas under native vegetation

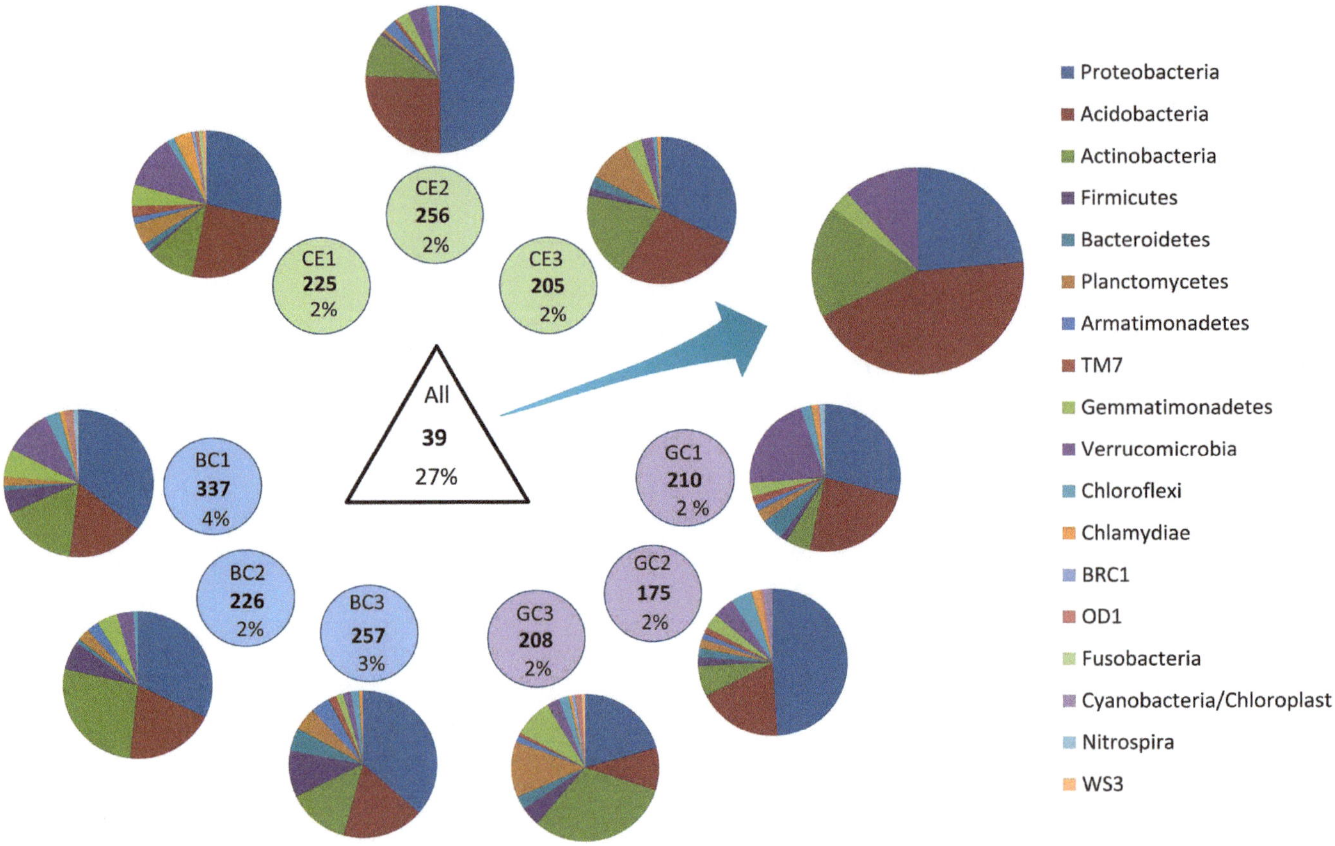

Figure 3. The general distribution of different OTUs among and within the samples and the relative frequencies of the phyla found in each situation (only the classified sequences). The written information represents the code of the sample, the number of OTUs found and the proportion of the sequences that these OTUs represent. Each written circle shows the data for the OTUs found only in that sample. The triangle shows the data for the OTUs found in all samples.

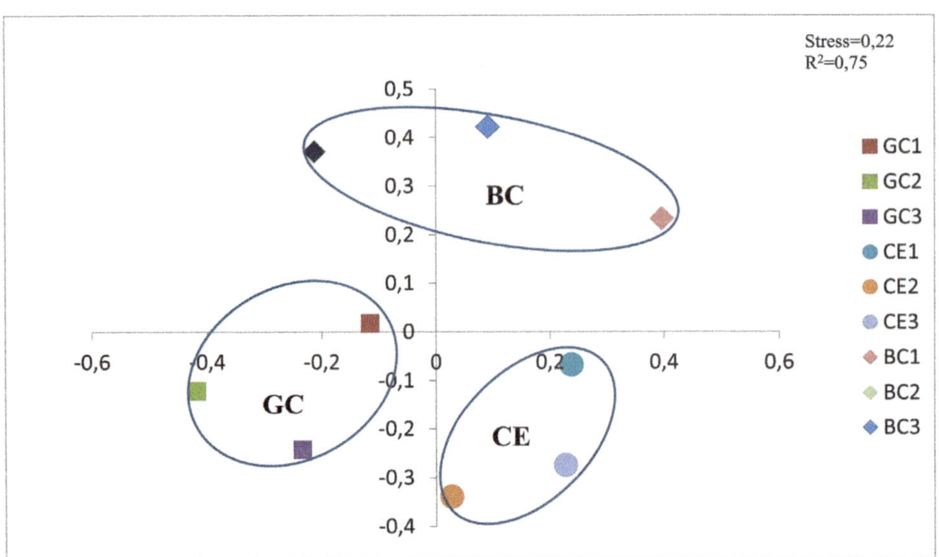

Figure 4. The NMDS ordination of the pyrosequencing data of Cerrado (CE), Green Cane (GC) and Burnt Cane (BC) samples.

compared than that selected by burnt cane, indicating a lower impact of green cane management on microbial bioindicators.

Conclusion

Cerrado soil is a very complex ecosystem with a great bacterial diversity, of which a substantial fraction remains undiscovered. The pyrosequencing of Cerrado soils has revealed a high level of unclassified bacteria. Sugarcane cultivation reduced the bacterial population on ferralsols-sampled soil, with Acidobacteria classes being the most influenced by this land use. In all areas, up to nineteen phyla were identified, with Acidobacteria, Proteobacteria and Actinobacteria being the most abundant. Firmicute sequences exhibited significantly different distributions among the treatments ($p<0.01$), with a higher frequency in the GC treatment. In contrast, Deltaproteobacteria occurred at higher levels on CE ($p = 0.01$), revealing a higher sensitivity to cropping. Finally, significant structural changes of the community were observed, with the burnt cane management having a greater impact than green cane management on the native Cerrado soil communities. In this work, we demonstrated the impact in the microbial community resulting from the use of the Cerrado to cultivation of the sugarcane. However, due to the great variability of the Cerrado ecosystem, further research is required to confirm these findings, with soil samples from different sites and seasons, in order to address the impact due to changes in management over the years.

Supporting Information

Figure S1 A rarefaction curve of the Cerrado (CE), Green Cane (GC) and Burnt Cane (BC) samples, constructed with Mothur and using 3% of dissimilarity.

Figure S2 The relative frequencies of the different classes found in Cerrado (CE), Green Cane (GC) and Burnt Cane (BC) samples for the following phyla: A – Firmicutes, B – Verrucomicrobia.

Table S1 Average values of soil properties.

Table S2 Closest relatives, and classification at phylum level of the most abundant OTUs (more than 1% of the sequences) from each sample. Representative sequences of each OTU were selected using Mothur, and the Ribossomal Database Project Seqmatch tool was used to establish the closest match for each OTU.

Author Contributions

Conceived and designed the experiments: CTCCR MCP FCB HLCC RSP ASR JMT. Performed the experiments: CTCCR ALS. Analyzed the data: CTCCR ALS MCP FCB HLCC JMT RSP ASR. Contributed reagents/materials/analysis tools: JMT RSP ASR. Wrote the paper: CTCCR ALS MCP FCB HLCC JMT RSP ASR.

References

1. Olson DM, Dinerstein E (1998) The Global 200: A Representation Approach to Conserving the Earth's Most Biologically Valuable Ecoregions. Conservation Biology 12: 502–515.
2. Myers N, Mittermeier RA, Mittermeier CG, Fonseca GAB, Kent J (2000) Biodiversity hotspots for conservation priorities. Nature 403: 853–858.
3. Conservation International (2007) Biodiversity hotspots: Cerrado. Available: http://www.biodiversityhotspots.org/xp/hotspots/cerrado/Pages/default.aspx. Accessed 2012 Jan 10.
4. WWF (2011) Cerrado, the Brazilian Savanna. Available: http://wwf.panda.org/what_we_do/where_we_work/cerrado/. Accessed 2012 Jan 10.
5. Sano EE, Rosa R, Brito JLS, Ferreira LG (2008) Mapeamento semidetalhado do uso da terra do Bioma Cerrado. Pesquisa Agropecuária Brasileira 43: 153–156.
6. Pereira PAA, Jr GBM, Santana CAM, Alves E (2012) The development of Brazilian agriculture: future technological challenges and opportunities. Agriculture & Food Security 1: 1–12.
7. Macedo IC, Seabra JEA, Silva EAR (2008) Green house gases emissions in the production and use of ethanol from sugarcane in Brazil: The 2005/2006 averages and a prediction for 2020. Biomass and Bioenergy 32: 582–595.
8. CANASAT (2011) Sugarcane crop mapping in Brazil by Earth observing satellite images. Available: http://www.dsr.inpe.br/laf/canasat/en/index.html. Accessed 2012 Jan 10.
9. Loarie SR, Lobell DB, Asner GP, Mu Q, Field CB (2011) Direct impacts on local climate of sugar-cane expansion in Brazil. Nature Climate Change 1: 105–109.
10. Gauder M, Graeff-Hönninger S, Claupein W (2011) The impact of a growing bioethanol industry on food production in Brazil. Applied Energy 88: 672–679.
11. Campos DVB, Machado PLOA, Braz SP, Santos GA, Lima E, et al. (2004) Decomposition of soil carbon derived from forest in an ultisol under sugar cane or Brachiaria sp. in the atlantic forest region of Brazil. XII International Meeting of International Humic Substances Society - IHSS 12: 647–649.
12. Resende AS, Xavier RP, Oliveira OC, Urquiaga S, Alves BJR, et al. (2006) Long-term effects of pre-harvest burning and nitrogen and vinasse applications on yield of sugar cane and soil carbon and nitrogen stocks on a plantation in Pernambuco, N.E. Brazil. Plant and Soil 281: 339–351.
13. Silva AJN, Ribeiro MR, Carvalho FG, Silva VN, Silva LESF (2007) Impact of sugarcane cultivation on soil carbon fractions, consistence limits and aggregate stability of a Yellow Latosol in Northeast Brazil. 94: 420–424.
14. Sant'anna SAC, Fernandes MF, Ivo WMPM, Costa JLS (2009) Evaluation of soil quality indicators in sugarcane management in sandy loam soil. Pedosphere 19: 312–322.
15. Galdos MV, Cerri CC, Cerri CEP (2009) Soil carbon stocks under burned and unburned sugarcane in Brazil. Geoderma 153: 347–352.
16. Pinheiro ÉFM, Lima E, Ceddia MB, Urquiaga S, Alves BJR, et al. (2010) Impact of pre-harvest burning versus trash conservation on soil carbon and nitrogen stocks on a sugarcane plantation in the Brazilian Atlantic forest region. Plant and Soil 333: 71–80.
17. Alef K, Nannipieri P (1995) Methods in applied soil microbiology and biogeochemistry. 1st ed. Alef K, Nannipieri P, London. Academic.
18. Valpassos MAR, Cavalcante ESG, Cassiolato AMR, Alves MC (2001) Effects of soil management systems on soil microbial activity. Pesq agropec bras 36: 1539–1545.
19. Aboim MCR, Coutinho HLC, Peixoto RS, Barbosa JC, Rosado AS (2008) Soil bacterial community structure and soil quality in a slash-and-burn cultivation system in Southeastern Brazil. Plant Biotechnology 38: 100–108.
20. Da C Jesus E, Marsh TL, Tiedje JM, De S Moreira FM (2009) Changes in land use alter the structure of bacterial communities in Western Amazon soils. The ISME journal 3: 1004–1011.
21. Quirino BF, Pappas GJ, Tagliaferro AC, Collevatti RG, Leonardecz E, et al. (2009) Molecular phylogenetic diversity of bacteria associated with soil of the savanna-like Cerrado vegetation. Microbiol Res 164: 59–70.
22. De Castro AP, Quirino BF, Allen H, Williamson LL, Handelsman J, et al. (2011) Construction and validation of two metagenomic DNA libraries from Cerrado soil with high clay content. Biotechnology letters. 33: 2169–75.
23. Araujo JF, De Castro AP, Costa MMC, Togawa RC, Júnior GJP, et al. (2012) Characterization of soil bacterial assemblies in brazilian savanna-like vegetation reveals acidobacteria dominance. Microbial ecology 64: 760–770.
24. Eiten G (1972) The Cerrado Vegetation of Brazil. Botanical Review 38: 201–341.
25. Rachid CT, Piccolo MC, Leite DC a, Balieiro FC, Coutinho HLC, et al. (2012) Physical-chemical and microbiological changes in Cerrado Soil under differing sugarcane harvest management systems. BMC microbiology 12: 170.
26. Muyzer G, Dewaal EC, Uitterlinden AG (1993) Profiling of complex microbial-populations by denaturing gradient gel-electrophoresis analysis of polymerase chain reaction-amplified genes-coding for 16s ribosomal-RNA. Applied and Environmental Microbiology 59: 695–700.
27. Sul WJ, Cole JR, Jesus EDC, Wang Q, Farris RJ, et al. (2011) Bacterial community comparisons by taxonomy-supervised analysis independent of sequence alignment and clustering. Proceedings of the National Academy of Sciences of the United States of America 108: 14637–14642.
28. Teixeira LCRS, Peixoto RS, Cury JC, Sul WJ, Pellizari VH, et al. (2010) Bacterial diversity in rhizosphere soil from Antarctic vascular plants of Admiralty Bay, maritime Antarctica. The ISME journal 4: 989–1001.
29. Schloss PD, Westcott SL, Ryabin T, Hall JR, Hartmann M, et al. (2009) Introducing mothur: open-source, platform-independent, community-supported software for describing and comparing microbial communities. Applied and environmental microbiology 75: 7537–7541.
30. Quince C, Lanzen A, Davenport RJ, Turnbaugh PJ (2011) Removing noise from pyrosequenced amplicons. BMC bioinformatics 12: 38.

31. Chao A, Bunge J (2002) Estimating the number of species in a stochastic abundance model. Biometrics 58: 531–539.
32. Shannon CE (1948) The mathematical theory of communication. 1963. The Bell System Techinical Journal 27: 379–423.
33. Cole JR, Wang Q, Cardenas E, Fish J, Chai B, et al. (2009) The Ribosomal Database Project: improved alignments and new tools for rRNA analysis. Nucleic acids research 37: D141–5.
34. Wang Q, Garrity GM, Tiedje JM, Cole JR (2007) Naive Bayesian classifier for rapid assignment of rRNA sequences into the new bacterial taxonomy. Applied and environmental microbiology 73: 5261–5267.
35. Acostamartinez V, Dowd S, Sun Y, Allen V (2008) Tag-encoded pyrosequencing analysis of bacterial diversity in a single soil type as affected by management and land use. Soil Biology and Biochemistry 40: 2762–2770.
36. Nacke H, Thürmer A, Wollherr A, Will C, Hodac L, et al. (2011) Pyrosequencing-based assessment of bacterial community structure along different management types in German forest and grassland soils. PloS one 6: e17000.
37. Uroz S, Buée M, Murat C, Frey-klett P, Martin F (2010) Pyrosequencing reveals a contrasted bacterial diversity between oak rhizosphere and. Environmental Microbiology 2: 281–288.
38. Navarrete A a., Cannavan FS, Taketani RG, Tsai SM (2010) A molecular survey of the diversity of microbial communities in different amazonian agricultural model systems. Diversity 2: 787–809.
39. Roesch LFW, Fulthorpe RR, Riva A, Casella G, Km A, et al. (2007) Pyrosequencing enumerates and contrasts soil microbial diversity. Soil and Water 1: 283–290.
40. Acosta-Martínez V, Dowd SE, Sun Y, Wester D, Allen V (2010) Pyrosequencing analysis for characterization of soil bacterial populations as affected by an integrated livestock-cotton production system. Applied Soil Ecology 45: 13–25.
41. Janssen P (2006) Identifying the dominant soil bacterial taxa in libraries of 16S rRNA and 16S rRNA genes. Applied and Environmental Microbiology 72: 1719–1728.
42. Jones RT, Robeson MS, Lauber CL, Hamady M, Knight R, et al. (2009) A comprehensive survey of soil acidobacterial diversity using pyrosequencing and clone library analyses. The ISME journal 3: 442–453.

14

The Accumulation of Organic Carbon in Mineral Soils by Afforestation of Abandoned Farmland

Xiaorong Wei[1]*, Liping Qiu[1], Mingan Shao[1], Xingchang Zhang[1], William J. Gale[2]

1 State Key Laboratory of Soil Erosion and Dryland Farming in the Loess Plateau, Northwest A & F University, Yangling, China, **2** College of Resources and Environment, Northwest A & F University, Yangling, China

Abstract

The afforestation of abandoned farmland significantly influences soil organic carbon (OC). However, the dynamics between OC inputs after afforestation and the original OC are not well understood. To learn more about soil OC dynamics after afforestation of farmland, we measured the soil OC content in paired forest and farmland plots in Shaanxi Province, China. The forest plots had been established on farmland 18, 24, 48, 100, and 200 yr previously. The natural ^{13}C abundance of soil organic matter was also analyzed to distinguish between crop- and forest-derived C in the afforested soils. We observed a nonlinear accumulation of total OC in the 0–80 cm depth of the mineral soil across time. Total soil OC accumulated more rapidly under forest stands aged 18 to 48 yr than under forest stands aged 100 or 200 yrs. The rate of OC accumulation was also greater in the 0–10 cm depth than in the 10–80 cm depth. Forest-derived OC in afforested soils also accumulated nonlinearly across time, with the greatest increase in the 0–20 cm depth. Forest-derived OC in afforest soils accounted for 52–86% of the total OC in the 0–10 cm depth, 36–61% of the total OC in the 10–20 cm depth, and 11–50% of the total OC in the 20–80 cm depth. Crop-derived OC concentrations in the 0–20 cm depth decreased slightly after afforestation, but there was no change in crop-derived OC concentrations in the 20–80 cm depth. The results of our study support the claim that afforestation of farmland can sequester atmospheric CO_2 by increasing soil OC stocks. Changes in the OC stocks of mineral soils after afforestation appear to be influenced mainly by the input of forest-derived C rather than by the loss of original OC.

Editor: Ben Bond-Lamberty, DOE Pacific Northwest National Laboratory, United States of America

Funding: This study was supported by the National Natural Science Foundation of China (40801111, 40901145), Special Foundation for State Major Basic Research Program of China (SQ2012FY4910023), Science and Technology Development Research Program in Shaanxi Province (2011KJXX25), the Program for Youthful Talents in Northwest A & F University, and the Program from Northwest A & F University (QN2011147). The funders had no role in study design, data collection and analysis, decision to publish, or preparation of the manuscript.

Competing Interests: The authors have declared that no competing interests exist.

* E-mail: xrwei78@163.com

Introduction

Changes in land use have important effects on ecological processes and global climate change [1,2,3]. During the past two decades, many studies have focused on the effects of land use change on the carbon content of ecosystems, particularly in soils. These effects are not well understood due to large differences in climatic conditions, soil properties, and the type of land use change [2,4–12].

Forest ecosystems perform important ecological functions and are essential for maintaining life on local and global scales [13]. The afforestation of abandoned farmland has occurred in many parts of the world due to an increasing demand for timber and wood products, the high economic returns from forestry, the desire to protect soil and water resources, and the potential of forests to sequester carbon to counter climatic change [13,14–16]. For example, the afforestation of large areas throughout New Zealand have created a major C sink, offsetting about 59% of the country's greenhouse gas emissions (CO_2-equivalent) from energy and industrial uses and 29% of country's total gas emissions in 2006 [16]. Similarly, more than 20.7 million ha of abandoned farmland have been afforested in China through the "Grain-for-Green Project", which aims to reduce soil loss and land degradation in the western part of the country [17].

The large potential for CO_2 sequestration in soil after the afforestation of farmland has been widely acknowledged [5,11,18]. Generally, soil organic carbon (OC) declines in the first years after afforestation and then gradually increases [11,19,20]. However, the stand age in most studies has been less than 60 yrs. This is insufficient for a thorough understanding of OC dynamics after the afforestation of farmland.

Assessing the response of soil OC to land use change requires an understanding of the dynamics of OC derived from both the previous and the current land use forms. Many researchers have reported on C dynamics after land use change [21–25]; however, knowledge about the dynamics of crop-derived OC after the afforestation of farmland is lacking. Such information is essential for accurately evaluating OC changes in soils and precisely predicting C sequestration by afforestation. Additionally, the dynamics of soil OC in mineral soils after the afforestation of farmland is unclear and needs further attention. In this experiment, we combined a chronosequence approach with differences in the ^{13}C natural abundance of C_3 (forest) and C_4 (crop) plants to study the dynamics of original crop-derived OC and newly input forest-derived OC in afforested farmland.

Materials and Methods

Ethics Statement

The permit for soil sampling was obtained from the Huanglongshan Forest Bureau, which is responsible for the protection and management of the Huanglongshan Forest Area.

Study site

We performed this study in the Huanglongshan Forest Area, which is located in central Shaanxi Province, China (109°38′49″–110°12′47″E, 35°28′49″–36°02′01″N, 1100 to 1300 m above sea level). The area has a continental, monsoon climate with an average annual temperature of 8.6°C. Mean monthly temperatures range from −22.5°C in January to 36.7°C in July. The average annual precipitation is 612 mm, and the average frost-free period is 175 days. The loess-derived soil in the study area is classified by Chinese scientists as a cinnamon soil, which belongs to the Cambisol soil group according to the FAO system. The soil texture is silt loam. The particle size distribution is shown in Fig. 1. The soil profile is free of stones to a depth of 50 m.

Agricultural activity in the Huanglongshan Forest Area has a history of more than 2000 yrs. The main crops are maize (*Zea mays L*) and millet (*Panicum miliaceum L*) (both C_4 plants). Some of the farmland was converted to forest in the past, either through natural succession or through ecological restoration programs. The main tree species in these forests is Chinese pine (*Pinus tabuliformis*).

Field investigation and soil sampling

We used the space-for-time substitution method to estimate the effects of afforestation on soil OC dynamics. The method is an important and often necessary tool for studying the temporal dynamics of plant communities and soil development across multiple timescales [26]. The study consisted of fifteen paired forest and farmland sites. The forests had been converted from farmland 18, 24, 48, 100, and 200 yrs previously. There were three replications of each age sequence, making a total of 30 sites (5 ages×2 land use types×3 replicates). The replications were 2 to 5 km apart. The forests with stand ages of 100 and 200 yrs were established by natural succession and the forest age was determined by Yang and Hou (2005) [27]. The forests with stand ages of 18, 24, and 48 yrs were man-made and identified with the help of the Huanglongshan Forest Bureau. The paired forest and farmland sites adjoined each other; therefore, we attributed differences in soil OC concentration and OC stock between paired farmland and forest to afforestation.

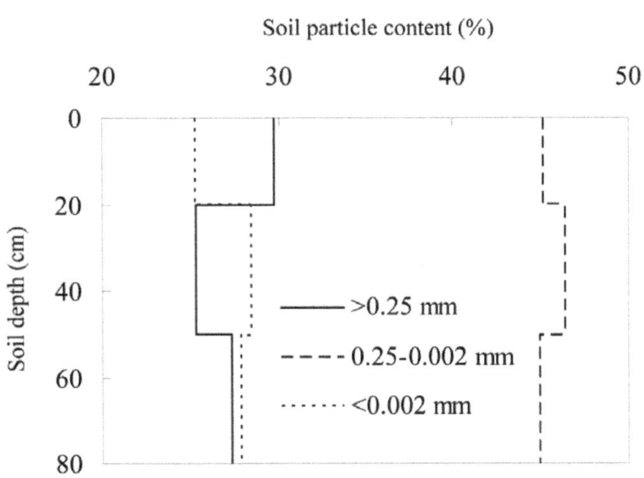

Figure 1. Soil particle distribution in the study site.

In this study, we assumed that there has been no significant change in soil OC concentrations of farmland across time (i.e., the soil OC of current farmland is equal to that when the farmland was abandoned 48, 100 or 200 yrs ago). We made this assumption even though there have been changes in fertilization and tillage practices over time (e.g., chemical fertilizer replaced manure as the primary source of plant nutrients about 50 yr ago). A preliminary study indicated that there had been no significant change in the soil OC of farmland during the past 26 yrs (data not shown); however, there is no direct way to quantify soil OC concentrations 100 or 200 yrs ago.

Three sampling plots were established within each treatment in August 2009. The sampling plots were 20×20 m in afforested areas and 5×5 m farmland areas. The sampling area of the afforested plots was larger than that of the farmland plots because soil variability is greater in afforested areas than in cropped areas. Each sampling plot was at least 40 m from the boundary to reduce the possibility of tree litter being added to farmland plots. The tree density and vegetation biomass was determined for each forest plot. The dominant plants on the forest floor were bunge needlegrass (*Stipa bungeana Trin.*) and Dahurian bush clover (*Lespedeza daurica (Laxm.) Schindl.*). Five trees were randomly selected to determine height, diameter at breast height (DBH, measured at 1.35 m), and crown density. The main characteristics of the vegetation in the afforested plots are presented in Table 1.

In each sampling plot, a pit 1.0 m long×0.7 m wide×1.0 m deep was dug for measuring soil bulk density. The soil was sampled using a stainless-steel cutting ring (5.0 cm diam.×5.0 cm high) at depths of 0–10, 10–20, 20–40, 40–60, and 60–80 cm. Three representative soil profiles were randomly selected in each plot, and the organic layers were removed. Samples of the mineral soil were collected with a tube auger (5.0 cm diam.) at the same depth increments as the bulk density samples. The soil samples were taken in 10 cm increments to avoid compaction. Visible pieces of undecomposed organic matter were removed. The moist soil samples were brought to the laboratory, air-dried, and ground for laboratory analysis. Forest litter was collected in each plot, brought to the laboratory, oven dried at 65°C, and ground for $\delta^{13}C$ analysis.

Laboratory and data analyses

Soil OC concentrations were analyzed with a VARIO EL III CHON analyzer (Elementar, Germany) at the Testing and Analysis Center, Northwest University, China. The soils were treated with HCl to remove the carbonates before analysis for soil OC. The OC stock was calculated as follows:

$$SOC_i = D_i \times BD_i \times OC_i/10 \qquad (1)$$

where SOC_i, D_i, BD_i, and OC_i represent the OC stock (Mg ha^{-1}), thickness (cm), bulk density (g cm^{-3}), and OC concentration (g kg^{-1}), respectively, of the *i*th layer of soil.

The natural abundance of ^{13}C in the soil organic matter and forest litter was analyzed with a MAT253 Stable Isotope Ratio Mass Spectrometer (Thermo Fisher Scientific, American) at the State Key Laboratory of Hydrology-Water Resources and Hydraulic Engineering, Hohai University, China. The $\delta^{13}C$ values of the soil OC and litter were calculated as follows:

$$\delta^{13}C = \left(R_{sample}/R_{PDB} - 1\right) \times 1000 \qquad (2)$$

Table 1. Characteristics of the forest stands at Huanglongshan.

Stand age (yrs)	Tree density (trees/ha)	DBH[a] (cm)	Height (m)	Crown density[b]	Vegetation biomass (tons ha^{-1})
18	4500	8	6	>90%	0.61
24	4000	10	8	>90%	0.69
48	3000	21	15	>95%	0.74
100	2400	32	18	>98%	0.41
200	2400	41	20	>98%	0.46

[a]DBH, diameter at breast height.
[b]Crown density was estimated visually based on the amount of light passing through the canopy, with larger values indicating a more closed canopy.

where R is the ratio of ^{13}C to ^{12}C, and PDB is the belemnite carbonate standard of the Peedee Formation, South Carolina.

For afforested soils, the proportion (f) of OC derived from forest litter (C_3-C) at each soil depth was calculated according to the following equation:

$$f=(\delta^{13}C_F-\delta^{13}C_C)/(\delta^{13}C_L-\delta^{13}C_C) \quad (3)$$

where $\delta^{13}C_F$ is the $\delta^{13}C$ value of soil organic matter in afforested soil, $\delta^{13}C_C$ is the $\delta^{13}C$ value of soil organic matter in farmland soil, and $\delta^{13}C_L$ is the $\delta^{13}C$ value in forest litter (we used a mean value of −27.2‰ for the calculation). Finally, the forest-derived C (C_3-C) and original crop-derived C (C_4-C) in afforested soils were calculated as follows:

$$OC_F=OC\times f \quad (4)$$

$$OC_C=OC\times(1-f) \quad (5)$$

$$SOC_F=SOC\times f \quad (6)$$

$$SOC_C=SOC\times(1-f) \quad (7)$$

where OC_F and OC_C are the concentrations of forest-derived C and crop-derived C, respectively; SOC_F and SOC_C are the stocks of forest-derived C and crop-derived C, respectively; and OC and SOC are the total OC concentrations and OC stocks in afforested soil, respectively.

To assess the dynamics of original farmland C after afforestation, we calculated the mean residence time (MRT) and half-life ($T_{1/2}$) of the original farmland C in afforested soils. The mean residence time was calculated as [28]:

$$MRT=-(t-t_0)/\ln(C_t/C_0) \quad (8)$$

where t is the time of sampling (yr), t_0 is the time of afforestation (yrs), ln is the natural logarithm, C_t is crop-derived OC in the afforested soil at the time of sampling (Mg ha^{-1}), and C_0 is the crop-derived OC at the time of afforestation (Mg ha^{-1}). We used the soil OC of the paired adjacent farmland as C_0 and regressed $t-t_0$ versus $\ln(C_t/C_0)$. The slope of the resulting line was equal to the MRT.

The $T_{1/2}$ was calculated as:

$$T_{1/2}=\ln 2\times MRT \quad (9)$$

The analyses of variance, correlation, and regression were conducted using SAS software (Version 8.0) [29]. Three-way analysis of variance was used to test the effects of afforestation, stand age, and soil depth on OC concentration and stock. Two-way analysis of variance was used to test the effects of stand age and soil depth on crop- and forest-derived OC concentrations and OC stocks. Correlation analysis was used to assess the relationships between forest-derived OC and the net increase in the OC of the afforested soil, and between the original OC of the afforested soil (C_4-C) and the OC of current farmland. Regression analysis was used to establish the relationships between stand age and different variables, including OC accumulation rate and the proportion and accumulation rates of forest-derived OC in afforested soils at each soil depth.

Results

Increase in total OC

The afforestation of abandoned farmland significantly increased total OC concentrations and stocks in mineral soils. The amount of increase varied with forest age and soil depth (Table 2). The increase in total OC after afforestation mainly occurred in the 0–10 cm depth. For the 0–10 cm depth, the increase in OC

Table 2. The ANOVA results of soil OC concentration and OC stock as affected by afforestation, stand age, and soil depth.

		OC concentration		OC stock	
	DF[a]	F	P	F	P
Afforestation	1	742	<0.0001	541	<0.0001
Soil depth	4	709	<0.0001	101	<0.0001
Afforestation×Soil depth	4	237	<0.0001	75	<0.0001
Stand age	4	69	<0.0001	18	<0.0001
Afforestation×Stand age	4	58	<0.0001	9	<0.0001
Soil depth×Stand age	16	23	<0.0001	8	<0.0001
Afforestation×Soil depth×Stand age	16	27	<0.0001	6	<0.0001

[a]DF, degrees of freedom.

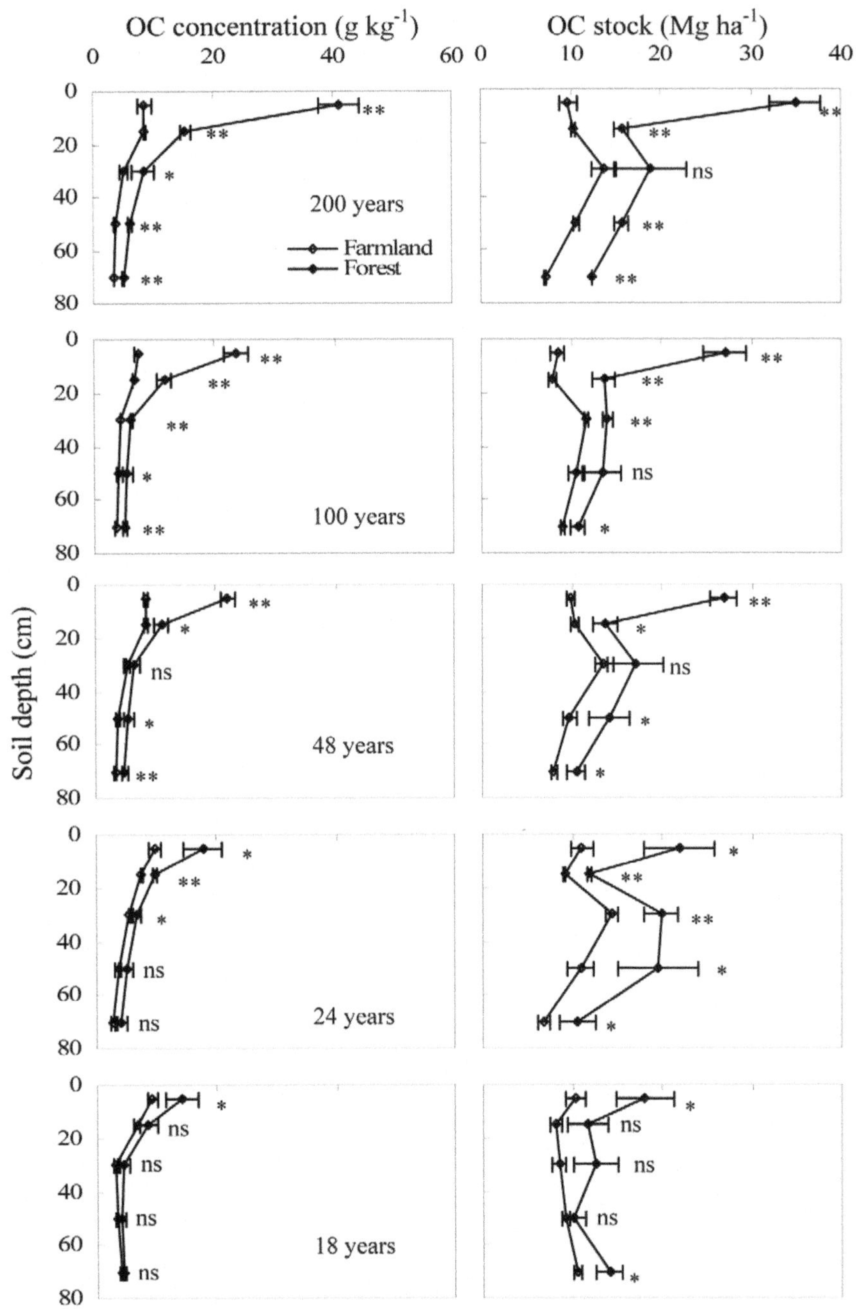

Figure 2. Effects of afforestation of abandoned farmland on soil OC concentration and stock at different stand ages (*: the difference between forest and farmland was significant at P<0.05; **: the difference was significant at P<0.01; ns: P>0.05; the error bars are standard deviation of the means).

concentrations ranged from 5.0 to 32 g kg^{-1} and the increase in OC stocks ranged from 8.0 to 25 Mg ha^{-1}. For the 10–80 cm depth, the increase in OC concentrations ranged from 0.3 to 6.8 g kg^{-1} and the increase in OC stocks ranged from 1.2 to 5.8 Mg ha^{-1} (Fig. 2).

Although total soil OC in the 0–80 cm depth increased with stand age, accumulation rates were generally highest in recently afforested soils (Fig. 3). Specifically, soil OC accumulation rates in stands less than 48 yr old ranged from 0.35 to 0.45 Mg ha^{-1} yr^{-1} for the 0–10 cm depth and from 0.06 to 0.23 Mg ha^{-1} yr^{-1} for the 10–80 cm depth. In comparison, accumulation rates in the 100 and 200 yr stands ranged from 0.13 to 0.18 Mg ha^{-1} yr^{-1} for the 0–10 cm depth and from 0.02 to 0.06 Mg ha^{-1} yr^{-1} for the 10–80 cm depth. The change in the accumulation rate with stand age was described by a power function (Fig. 3), indicating a nonlinear increase in soil OC after afforestation of farmland. The observation that accumulation rates in the 0–10 cm depth were significantly greater than those in the 10–80 cm depth across all stand ages suggests that OC inputs in the afforested soil were primarily derived from forest litter rather than tree roots.

Increase in forest-derived OC

Forest-derived OC in afforested soils increased with stand age, with most of the increase occurring in the 0–20 cm depth (Fig. 4).

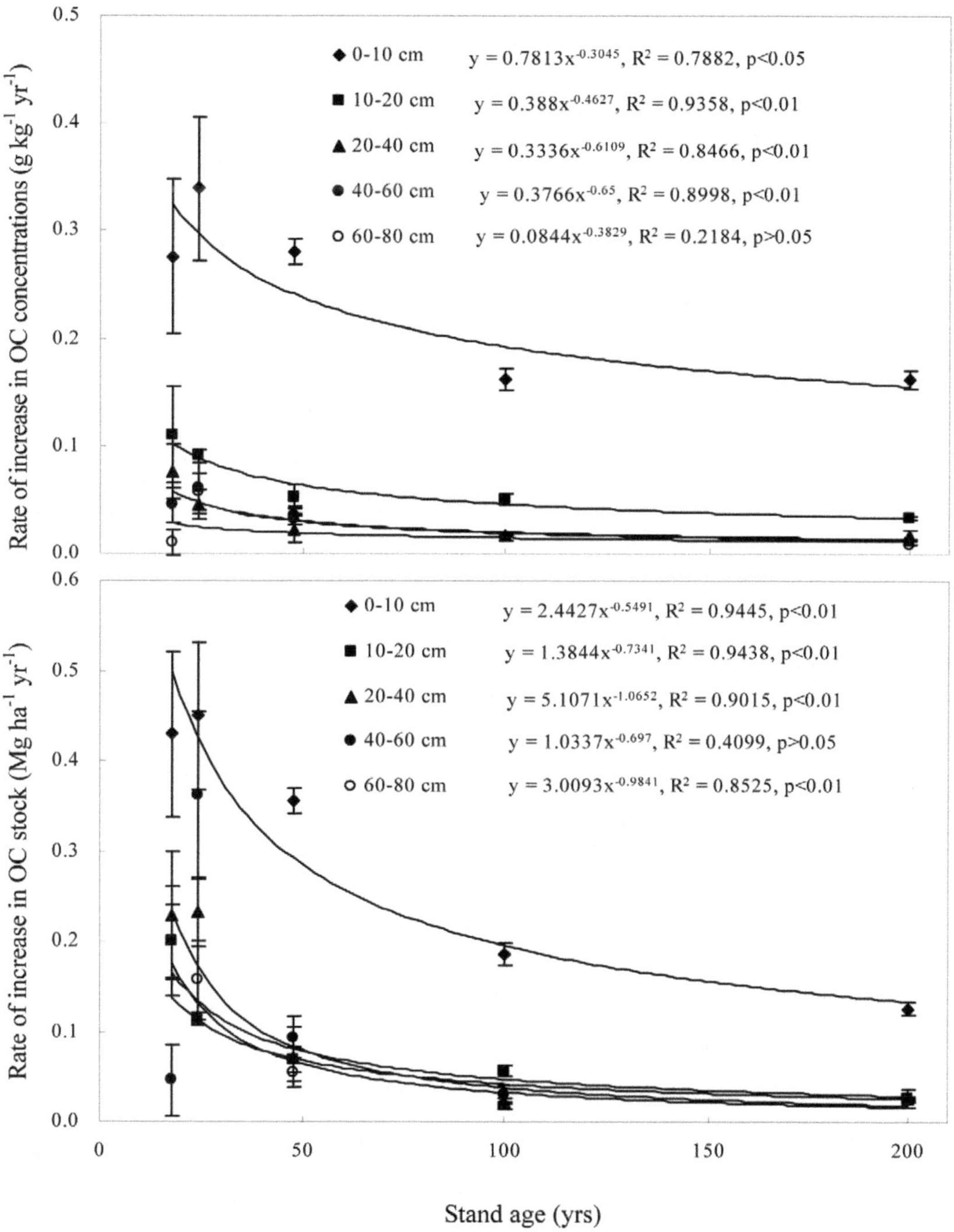

Figure 3. Relationships between stand age and the rate of increase in total OC concentrations and stocks in mineral soils (the error bars are standard deviation of the means).

Forest-derived OC comprised 52–86% of the OC in 0–10 cm depth, 36–61% of the OC in the 10–20 cm depth, and 11–50% of the OC in the 20–80 cm depth. In the 0–20 cm depth, the proportion of forest-derived C to total OC increased nonlinearly as stand age increased (Fig. 5).

We assumed that the OC concentrations and OC stocks in current farmland are equal to those when the forests were first established. If this is true, then the difference in OC between forest and farmland (i.e. net increase in OC due to afforestation) should be equal to the amount of forest-derived OC determined by ^{13}C analysis. We observed a highly significant linear relationship between forest-derived OC and the net increase of OC in afforested soil (Fig. 6). The slopes of the relationship were 1.1 for soil OC concentrations and 1.2 for soil OC stocks. This supports our supposition that the OC of the current farmland is equal to the OC of the farmland at the time of afforestation. The forest-derived OC concentration was on average slightly higher than the net increase in OC concentration in the 0–20 cm depth but about the same as the net increase in the OC concentration in the 20–80 cm depth. In comparison, the crop-derived OC concentrations of afforested soil were slightly less than that of current farmland at the 0–20 cm depth but about the same as that of current farmland at

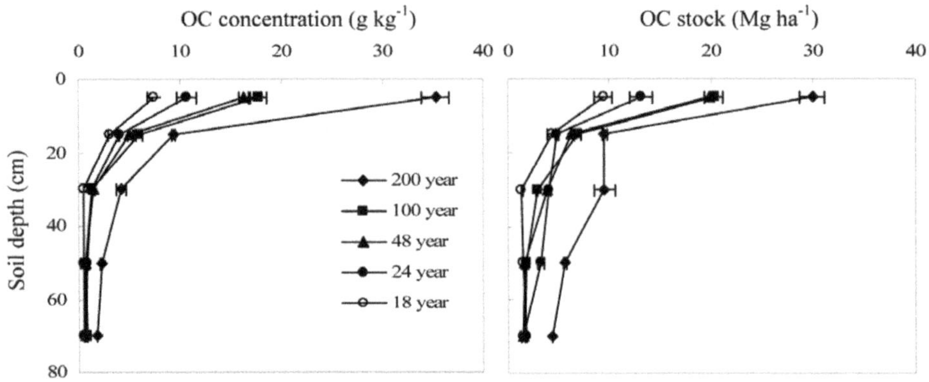

Figure 4. Vertical distribution of forest-derived OC (C_3-C) concentrations and OC stocks in afforested farmland (the error bars are standard deviation of the means).

the 20–80 depth. These results suggest a slight decline in the crop-derived OC concentration of the 0–20 cm depth but no change in the crop-derived OC concentration of the 20–80 cm depth after the afforestation of farmland.

The rate of increase in forest-derived OC concentrations decreased nonlinearly with stand age (Fig. 7). For example, the rate of increase in forest-derived OC concentrations in the 0–10 cm depth slowed from 0.41 g kg^{-1} yr^{-1} in 18 yr stands to 0.18 g kg^{-1} yr^{-1} in 100 yr stands and 0.17 g kg^{-1} yr^{-1} in 200 yr stands. For the 10–20 cm depth, the rate of increase slowed from 0.18 g kg^{-1} yr^{-1} in 18 yr stands to 0.06 g kg^{-1} yr^{-1} in 100 yr stands and 0.05 g kg^{-1} yr^{-1} in 200 yr stands. For the 40–60 cm depth, the rate of increase slowed from 0.031 g kg^{-1} in 18 yr stands to 0.012 g kg^{-1} in 100 yr stands and 0.011 g kg^{-1} in 200 yr stands. For the 60–80 cm depth, the rate of increase slowed from 0.029 g kg^{-1} yr^{-1} in 18 yr stands to 0.012 g kg^{-1} yr^{-1} in 100 yr stands and 0.008 g kg^{-1} yr^{-1} in 200 yr stands. The accumulation rate of forest-derived OC stocks followed a similar pattern with that of forest-derived OC concentrations. The changes described above suggest that the accumulation rate of

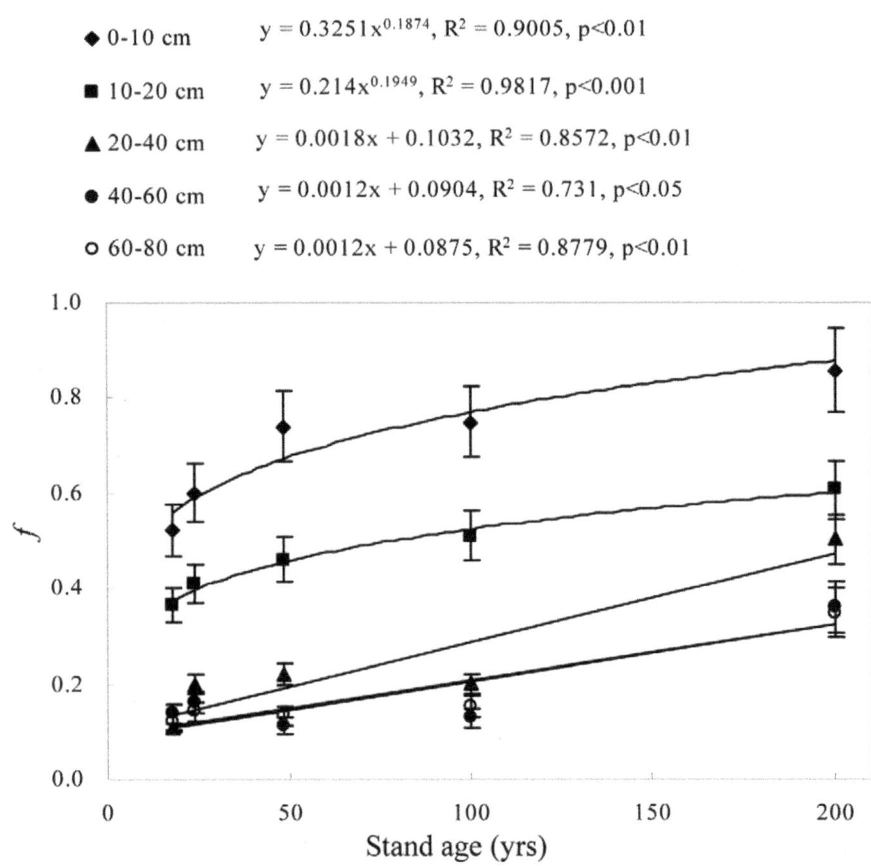

Figure 5. Change in the proportion (*f*) of forest-derived organic carbon to total organic carbon with different stand ages, by soil depth (the error bars are standard deviation of the means).

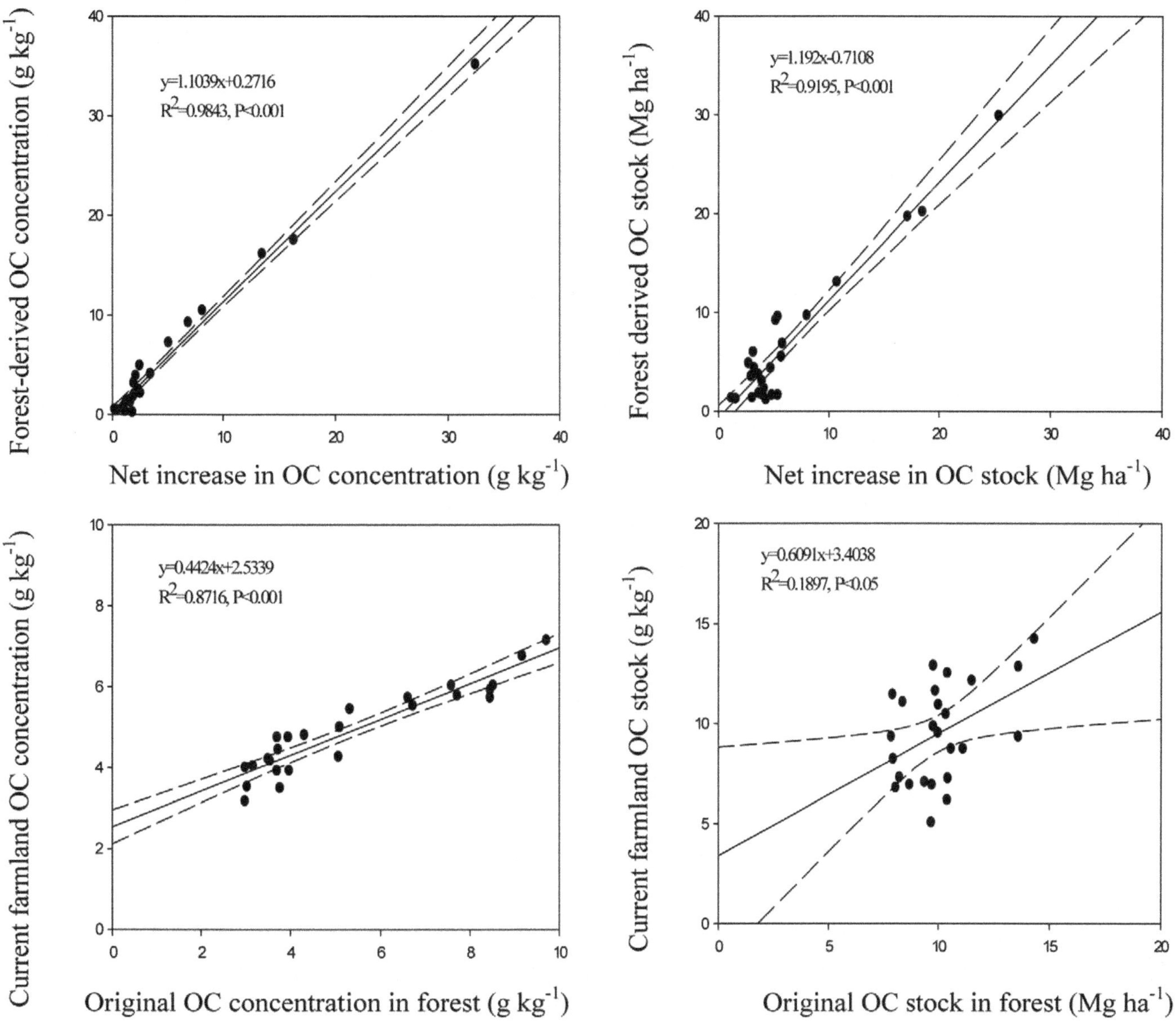

Figure 6. Relationships between forest-derived OC (C_3-C) and the net increase in OC in afforested soil, and between original OC in afforested soil (C_4-C) and the OC in current farmland (the inner bound represents the 95% confidence interval).

forest-derived OC will reach a constant value for each depth as the forests age.

Slight decrease in original, crop-derived OC

The concentrations of original crop-derived OC (C_4-C) in the topsoil of 200 yr stands were not significantly different than those of other stand ages. However, crop-derived OC stocks were lower in the 200 yr stands than those of other stand ages because the soil bulk density in the 200 yr stands was less than those of the other stand ages (Fig. 8). There was no significant difference in original OC for different stand ages in the 0–80 cm depth and small decrease of original OC in 0–20 cm depth (Fig. 9), indicating a slight decline in original OC after the afforestation of farmland. Based on our previous assumption that OC in current farmland is equal to that when the forests were established, we calculated that the original OC concentrations in afforested soils decreased across the five forest ages at the rate of 0.064 g kg^{-1} yr^{-1} in the 0–10 cm depth and 0.044 g kg^{-1} yr^{-1} in the 10–20 cm depth. The original OC stocks decreased at the rate of 0.057 Mg ha^{-1} yr^{-1} in the 0–10 cm depth and 0.048 Mg ha^{-1} yr^{-1} in the 10–20 cm depth. The original OC in the subsoil (20–80 cm depth) of the afforested sites was slightly greater than that of current farmland (with slopes less than 1 in Fig. 6). This suggests that afforestation protects the original OC in the subsoil more than modern agricultural practices. The MRTs for original farmland C after afforestation were 330 yrs in the 0–10 cm depth, 342 yrs in the 10–20 cm depth, and 476 yrs in the 0–80 cm depth. The corresponding $T_{1/2}$ values were 229 yrs in the 0–10 cm depth, 237 yrs in the 10–20 cm depth, and 330 yrs in the 0–80 cm depths. These values indicate a slow decrease of original farmland OC and suggest that the original farmland C is relatively stable in forests.

Discussion

Afforestation of cultivated soil increased total soil OC concentrations in the 0–40 cm depth by 88% and soil OC stocks

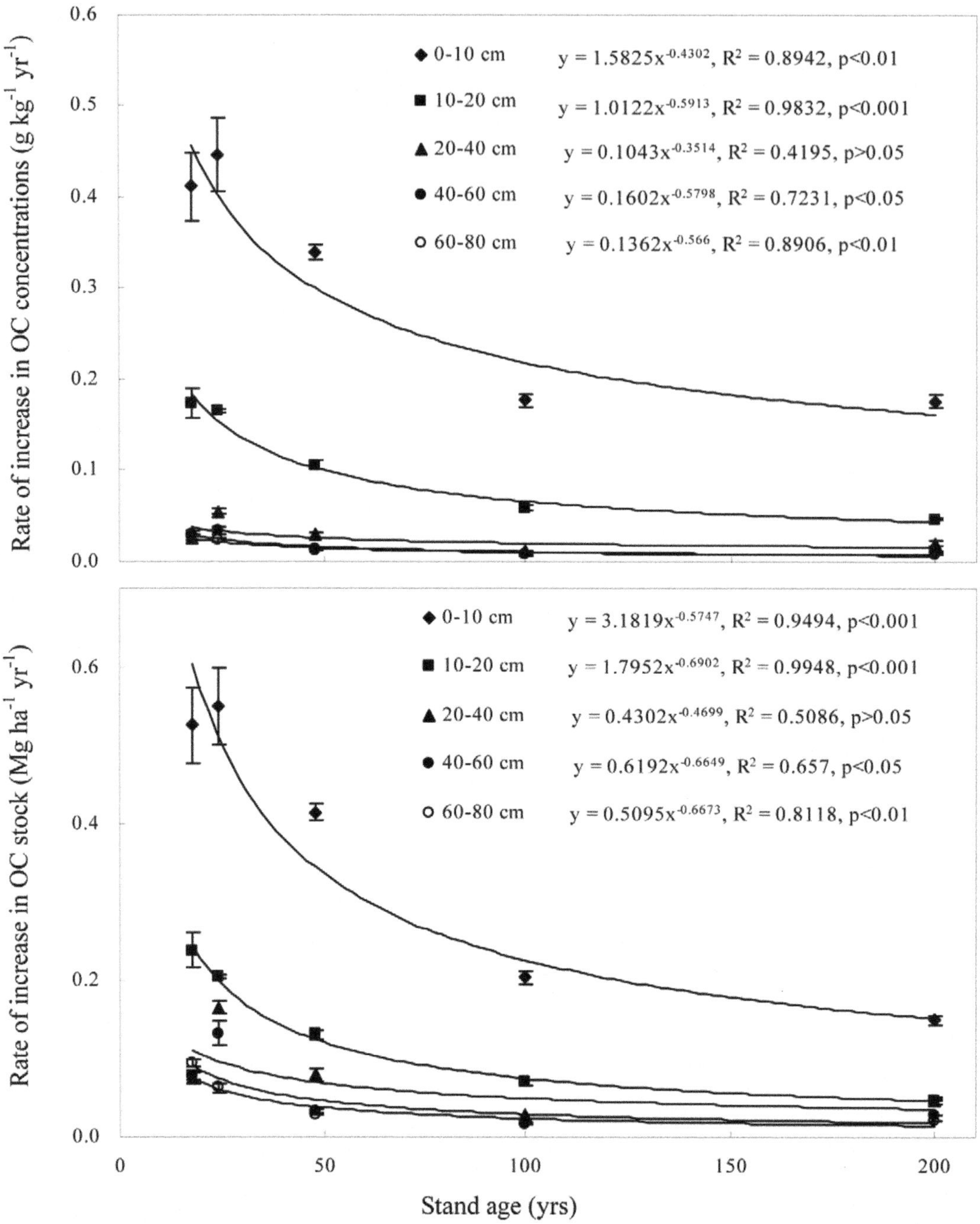

Figure 7. Effect of stand age on the rate of increase in forest-derived OC (the error bars are standard deviation of the means).

by 81% when averaged across all stand ages. In comparison, other studies report that afforestation increased soil OC in the 0–40 cm depth by 26–53% [5,11,18]. The differences between our study and others might be due to differences in stand age. When the data from the 18, 24, and 48 yr stands were averaged, we observed a 52% increase in soil OC concentration and a 60% increase in OC stock. Soil OC concentrations in the 100 and 200 yr stands averaged 143% higher than that in farmland soil. Soil OC stocks averaged 112% higher in the 100 and 200 yr stands than in farmland soil. This demonstrates the large potential for the accumulation of OC in mineral soil due to afforestation.

There was a nonlinear increase in total soil OC with stand age. The rate of increase slowed across time and eventually leveled out 100 yrs after afforestation, consistent with findings that soils continuously accumulate carbon in old-growth stands that have non-steady carbon balances [30,31]. This result is also consistent with the statement by Luyssaert *et al* (2008) that old-growth forests can act as global carbon sinks [32]. The rates of accumulation in our study decreased with soil depth and were lower than those observed by Zhou *et al* (2006) (0.61 Mg ha^{-1} yr^{-1}) for the 0–20 cm depth at a warmer site (i.e., Dinghushan, China) but higher than those by Tang *et al* (2009) (0.04 Mg ha^{-1} yr^{-1}) for the 0–60 cm depth at a colder site (i.e., northern Wisconsin and the

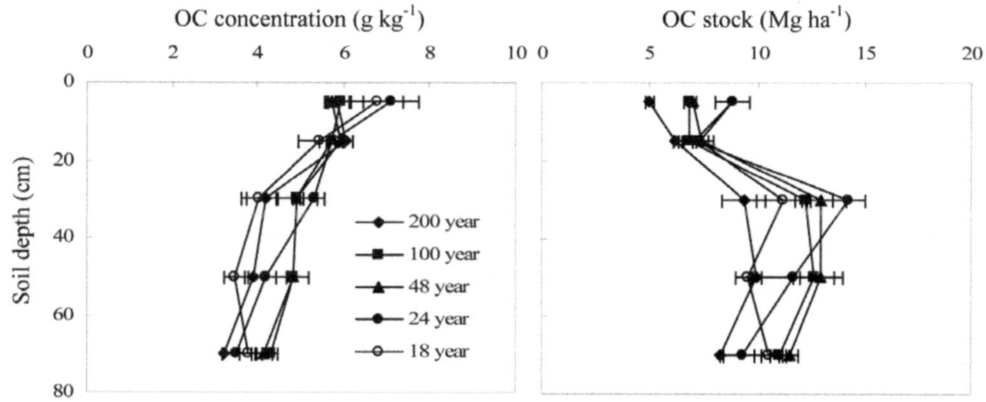

Figure 8. Vertical distribution of crop-derived OC (C_4-C) concentrations and OC stocks in afforested farmland (the error bars are standard deviation of the means).

upper peninsula of Michigan) [30,31]. The large differences in the stable accumulation rate may be due to different climatic conditions at the study sites [8,11] as well as the differences in the total soil OC concentration at the time of afforestation. The OC accumulation rate and the fate of original carbon will probably depend on the initial OC concentrations and OC stocks. Soil OC concentrations were low at the time of afforestation in our study. In this case, we might expect rapid initial increases in soil OC concentrations after afforestation.

Our results indicated that total soil OC stocks were still increasing 100 and 200 yr after afforestation, although the rate of increase slowed across time. This observation is different from the report that soil OC stocks increased steadily at a rate of 0.13 Mg ha^{-1} yr^{-1} in the 0–30 cm depth of a California soil for 82 yr after afforestation and then leveled out [33]. Furthermore, the net increase in the total soil OC of the California soil was less than the net increase in our study (10 Mg ha^{-1} vs 23.8 Mg ha^{-1}).

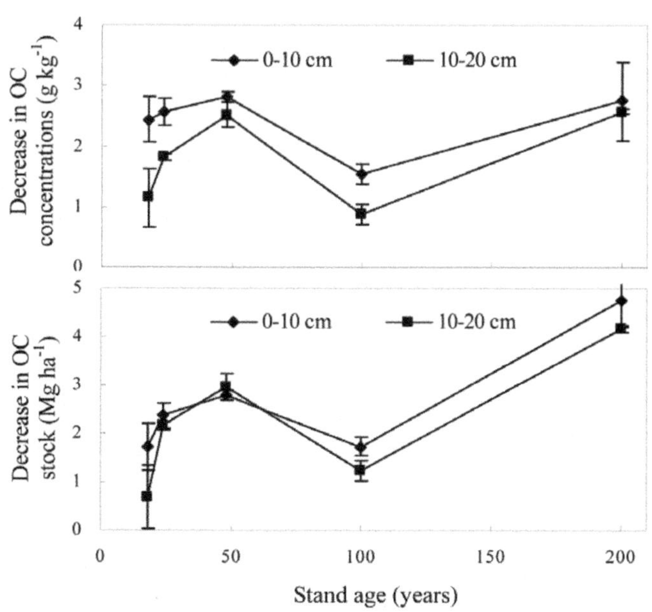

Figure 9. The decrease in OC concentration and OC stock in soils after afforestation of farmland (the error bars are standard deviation of the means).

One explanation is that climate conditions or soil properties limited the ability of the California soil to sequester C.

Many researchers have studied the changes in soil OC after afforestation of farmland [5,11,18], but less attention has been given to the dynamics of the original OC in afforested soils. Such research is essential for understanding how soil OC responds to the afforestation of abandoned farmland. We used the natural ^{13}C abundance of soil organic matter to distinguish between crop- and forest-derived C in the afforested soils. The natural abundance ^{13}C technique has been used in long-term studies about soil OC dynamics [34–36]. An important assumption of this technique is that the isotopic composition of SOM closely resembles the isotopic composition of the vegetation from which it was derived [37,38]. Isotope fractionation during primary plant production or organic matter decomposition [39,40] could affect the interpretation of our results. For example, if the δ^{13} value of the soil OC increased during mineralization, then our calculations of forest-derived OC concentrations and stocks would be too low and our calculations of crop-derived OC would be too high. Our results could also be affected by the observation that the δ^{13} value of organic matter increases (i.e. becomes more positive) with soil depth relative to the forest litter layer [41]. In our study, a 1‰ increase in the δ^{13} value of OC at the 60–80 cm depth would underestimate forest-derived OC concentrations by 11% in the 18 yr stand and 12% in the 200 yr stand.

For the 0–20 cm depth, we observed a 17% decrease in crop-derived OC concentrations 100 yr after afforestation and a 31% decrease in crop-derived OC after 200 yrs of afforestation. For the 20–80 cm depth, crop-derived OC concentrations did not change after afforestation. The changes in original farmland OC after afforestation in our study were relatively small compared with changes in original C after conversion from forest to farmland or grassland. For example, Zach *et al* (2006) reported a 33–57% decrease in original soil OC concentrations after 12–18 yrs of continuous cultivation in central Argentina [24]. Osher *et al* (2003) observed significant decreases in the original OC concentration when a natural forest in Hawaii was converted to pasture or sugarcane cultivation [22]. Powers & Veldkamp (2005) reported that the original OC concentration of the 0–10 cm depth decreased when forest was converted to pasture in northeastern Costa Rica [23]. Similarly, the MRT and $T_{1/2}$ of original crop-derived OC in our study were also larger than those of original OC in other studies [28]. One explanation for the relatively small changes in crop-derived OC after afforestation in our study is that the OC was physically protected by soil structure that developed after afforestation. Alternatively, soil OC concentrations in the farmland were low. The crop-derived OC may have been protected through sorption on to mineral particles.

Our results show that forest-derived OC increased nonlinearly in afforested farmland. Furthermore, the accumulation rate of forest-derived OC in the soil decreased as stand age increased. These results indicate a stand age after which the forest-derived C increases at a low but stable rate. Additionally, our observations of large increases in forest-derived OC and slow losses of original crop-derived OC in afforested farmland suggest that the accumulation of OC in mineral soils is mainly influenced by the input of forest-derived OC rather than the loss of original OC, especially in the subsoil.

Acknowledgments

We thank Dr. Robert Horton from Iowa State University and Dr. Holm Tiessen from Inter American Institute for Global Change for reviewing the manuscript and Gao J, Qiang B, She D, Liu Z, Li X, Jia X, and the Huanglongshan Forest Bureau for their help in field and laboratory experiments.

Author Contributions

Conceived and designed the experiments: XW LQ MS XZ. Performed the experiments: XW LQ. Analyzed the data: XW LQ MS XZ. Wrote the paper: XW WG.

References

1. Kalnay E, Cai M (2003) Impact of urbanization and land-use change on climate. Nature 423: 528–531.
2. Jackson RB, Banner JL, Jobbágy EG, William T, Pockman WT, et al. (2002) Ecosystem carbon loss with woody plant invasion of grasslands. Nature 418: 623–626.
3. Ficetola GF, Maiorano L, Falcucci A, Dendoncker N, Boitani L, et al. (2010) Knowing the past to predict the future: land-use change and the distribution of invasive bullfrogs. Global Change Biol 16: 528–537.
4. Houghton RA (1995) Land-use change and the carbon cycle. Global Change Biol 1: 275–287.
5. Guo LB, Gifford RM (2002) Soil carbon stocks and land use change: a meta analysis. Global Change Biol 8: 345–360.
6. Degryze S, Six J, Paustian K, Morris SJ, Paul EA, et al. (2004) Soil organic carbon pool changes following land-use conversions. Global Change Biol 10: 1120–1132.
7. Piao SL, Fang JY, Ciais P, Peylin P, Huang Y, et al. (2009) The Carbon balance of terrestrial ecosystems in China. Nature 458: 1009–1013.
8. Wei X, Shao M, Fu X, Horton R, Li Y, et al. (2009) Distribution of soil organic C, N and P in three adjacent land use patterns in the northern Loess Plateau, China. Biogeochemistry 96: 149–162.
9. Wei X, Shao M, Fu X, Horton R (2010) Changes in soil organic carbon and total nitrogen after 28 years grassland afforestation: Effects of tree species, slope position, and soil order. Plant Soil 331: 165–179.
10. Arora VK, Boer GJ (2010) Uncertainties in the 20th century carbon budget associated with land use change. Global Change Biol 16: 3327–3348.
11. Laganière J, Angers DA, Paré D (2010) Carbon accumulation in agricultural soils after afforestation: a meta-analysis. Global Change Biol 16: 439–453.
12. Maia SF, Pgle SM, Cerri CP, Cerri CC (2010) Soil organic carbon stock change due to land use activity along the agricultural frontier of the southwestern Amazon, Brazil, between 1970 and 2002. Global Change Biol 16: 2775–2788.
13. IPCC (2000) Land-use, Land-use Change, and Forestry Cambridge University Press, Cambridge.
14. Rudel TK, Coomes OT, Moran E, Achard F, Angelsen A, et al. (2005) Forest transitions: towards a global understanding of land use change. Global Environ Change 15: 23–31.
15. Qiu L (2007) Change of soil quality and its regulation in re-vegetation ecosystem of the Loess Plateau Disseration. Northwest A & F University, Yangling, China.
16. MfE (2008) New Zealand's greenhouse gas inventory 1990–2006 Ministry for the Environment, Wellington, New Zealand.
17. Liu JG, Li SX, Ouyang ZY, Tam C, Chen XD (2008) Ecological and socioeconomic effects of China's policies for ecosystem services. Proc Nat Acad Sci U S A 105: 9477–9482.
18. Paul KI, Polglase PJ, Nyakuengama JG, Khanna PK (2002) Change in soil carbon following afforestation. Forest Ecol Manag 168: 241–257.
19. Smal H, Olszewska M (2008) The effect of afforestation with Scots pine (Pinus silvestris L.) of sandy post-arable soils on their selected properties. II. Reaction, carbon, nitrogen and phosphorus. Plant Soil 305: 171–187.
20. Paul KI, Polglase PJ, Richards GP (2003) Sensitivity analysis of predicted change in soil carbon following afforestation. Ecol Model 164: 137–152.
21. Neill C, Fry B, Melillo JM, Steudler RA (1996) Forest- and pasture-derived carbon contributions to carbon stocks and microbial respiration of tropical pasture soils. Oecologia 107: 113–119.
22. Osher LJ, Matson PA, Amundson R (2003) Effect of land use change on soil carbon in Hawaii. Biogeochemistry 65: 213–232.
23. Powers JS, Veldkamp E (2005) Regional variation in soil carbon and d13C in forests and pastures of northeastern Costa Rica. Biogeochemistry 72: 315–336.
24. Zach A, Tiessen H, Noellemeyer E (2006) Carbon turnover and carbon-13 natural abundance under land use change in semiarid savanna soils of La Pampa, Argentina. Soil Sci Soc Am J 70: 1541–1546.
25. Marin-Spiotta E, Silver WL, Swanston CW, Ostertag R (2009) Soil organic matter dynamics during 80 years of reforestation of tropical pastures. Global Change Biol 15: 1584–1597.
26. Walker LR, Wardle DA, Bardgett RD, Clarkson BD (2010) The use of chronosequences in studies of ecological succession and soil development. J Ecology 98: 725–736.
27. Yang HJ, Hou YJ (2005) The policy effect for huanglong hilly area reclamation during Qing Dynasty. J Chin Hist Geog 20: 125–131.
28. Paul S, Flessa H, Veldkamp E, López-Ulloa M (2008) Stabilization of recent soil carbon in the humid tropics following land use changes: evidence from aggregate fractionation and stable isotope analyses. Biogeochemistry 323: 309–322.
29. SAS Institute Inc (1999) SAS user's guide. Version 8, Cary NC.
30. Zhou G, Liu S, Li Z, Zhang D, Tang X, et al. (2006) Old-growth forests can accumulate carbon in soils. Science 314: 1417–1417.
31. Tang J, Bolstad PV, Martin JG (2009) Soil carbon fluxes and stocks in a Great Lakes forest chronosequence. Global Change Biol 15: 145–155.
32. Luyssaert S, Schulze E-D, Börner A, Knohl A, Hessenmöller D, et al. (2008) Old-growth forests as global carbon sinks. Nature 455: 213–215.
33. Turk JK, Graham RC (2009) Soil carbon and nitrogen accumulation in a forested debris flow chronosequence, California. Soil Science Society of America Journal, 73: 1504–1509.
34. Balesdent J, Mariotti A, Guillet B (1986) Natural ^{13}C abundance as a tracer for studies of soil organic matter dynamics. Soil Biol Biochem 19: 25–30.
35. Wittmer MHOM, Auerswald K, Bai YF, Schaufele R, Schnyder H (2010) Changes in the abundance of C3/C4 species of Inner Mongolia grassland: evidence from isotopic composition of soil and vegetation. Global Change Biol 16: 605–616.
36. Moyes AB, Gaines SJ, Siegwolf RTW, Bowling DR (2010) Diffusive fractionation complicates isotopic partitioning of autotrophic and heterotrophic sources of soil respiration. Plant Cell Environ 33: 1804–1819.
37. Nadelhoffer KJ, Fry B (1988) Controls on natural nitrogen-15 and carbon-13 abundances in forest soil organic matter. Soil Sci Soc Am J 52: 1633–1640.
38. Peterson BJ, Fry B (1987) Stable isotopes in ecosystem studies. Annu Rev Ecol Syst 18: 293–320.
39. Benner R, Fogel M, Sprague EK, Hodson RE (1987) Depletion of 13C in lignin and its implications for stable carbon isotope studies. Nature 329: 708–710.
40. Mary B, Mariotti A, Morel JL (1992) Use of ^{13}C variations at natural abundance for studying the biodegradation of root mucilage, roots and glucose in soil. Soil Biol Biochem 24: 1065–1072.
41. Böstrom B, Comstedt D, Ekblad A (2007) Isotope fractionation and ^{13}C enrichment in soil profiles during the decomposition of soil organic matter. Oecologia 153: 89–98.

15

Intercropping of Green Garlic (*Allium sativum* L.) Induces Nutrient Concentration Changes in the Soil and Plants in Continuously Cropped Cucumber (*Cucumis sativus* L.) in a Plastic Tunnel

Xuemei Xiao, Zhihui Cheng*, Huanwen Meng, Lihong Liu, Hezi Li, Yinxin Dong

College of Horticulture, Northwest A&F University, Yangling, Shaanxi, China

Abstract

A pot-based experiment was conducted to investigate nutrient concentrations in cucumber plants intercropped with various amounts of green garlic. In addition, the soil nutrient contents were studied over two consecutive growing seasons. The results revealed that the accumulation of biomass and the nutritional elements nitrogen (N), phosphorus (P), potassium (K), calcium (Ca) and manganese (Mn) in cucumber plants were significantly increased for intercropping treatments during the two growing seasons compared to monoculture. Conversely, magnesium (Mg) concentrations were decreased in the cucumber plants. Shoot iron (Fe) concentrations decreased whereas root Fe concentrations increased in the intercropping system. Shoot and root zinc (Zn) concentrations decreased during the fall of 2011 but increased during the spring of 2012. Soil organic matter and available N, P and K were significantly increased as the proportion of intercropped green garlic increasing. Medium levels of intercropping green garlic improved cucumber nutrient concentrations the most. The regression analysis showed that the concentrations of most elements were significantly related to the amounts of garlic bulbs, especially the microelements in the spring 2011. The available soil N and organic matter were linearly related to the amounts of garlic bulbs. The results indicate that the nutritional status of the soil and plants of continuously cropped cucumber could be improved by intercropping with green garlic.

Editor: Randall P. Niedz, United States Department of Agriculture, United States of America

Funding: This research was supported by the project of the State Natural Science Foundation (No. 31171949) and the project of the State Commonwealth (Agriculture) Scientific Research (No. 200903018).The funders had no role in study design, data collection and analysis, decision to publish, or preparation of the manuscript.

Competing Interests: The authors have declared that no competing interests exist.

* E-mail: chengzh@nwsuaf.edu.cn

Introduction

Cucumber (*Cucumis sativus* L.) is commonly cultivated in continuous cropping monoculture under protected cultivation, resulting in poor growth, reduced crop yield and "soil sickness" [1–3]. Soil sickness may involve nutrient uptake disorders and changes in the availability of soil nutrients [4–5], adverse effects on soil structure [6–7] and the build-up of soil-borne diseases and autotoxicity [8–9].

Intercropping is becoming more important in improving the utilization of land resources and increasing crop productivity [10–11]. Some studies have demonstrated that intercropping can relieve soil sickness by improving soil quality [10] [12] and the ecological microclimate [13]. In addition, intercropping can effectively improve nutrient mobilization in the rhizosphere and nutrient acquisition based on inter-specific root interactions [14–15]. Most studies on nutrient uptake and transfer have focused on legume-cereal intercropping systems [16–19]. Nitrogen (N) input and transfer occurs in this intercropping system based on the specific biological nitrogen fixation of leguminous crops [16] [18]. Intercropping with legumes also facilitates phosphorus (P) uptake and utilization in cereals [20–22]. The use of a P source can improve calcium (Ca) and magnesium (Mg) uptake due to rhizosphere acidification [23]. The uptake and mobilization of micronutrients is also influenced by intercropping, especially Fe in peanut and maize intercropping systems [15] [24]. Except for legume-cereal intercropping systems, other studies reported that the turmeric, maize and onion intercropping systems improved nutrient uptake compared to that of either of the sole crops [25]. Moreover, Shanmugham [26] reported that intercropping of onion improved P uptake of cotton based on chelating compounds exuded by onion increasing the available P to cotton plants (cited from [11]).

Garlic (*Allium sativum* L.) has effectively been incorporated into an intercropping system as a companion crop due to its allelopathic and antimicrobial effects. Some studies have reported that garlic intercropping can prevent insect attack [27–28] and weed invasion [29]. Recent research demonstrated that intercropping with garlic, green garlic (planting whole bulbs and cutting green garlic more than once) [30] and onion [31] can alleviate problems associated with continuous cropping (soil sickness) of cucumbers under protected cultivation. However, most attention has been directed to the yield and growth parameters, as well as soil biological characteristics. The objective of this study was to ascertain the characteristics of nutrient uptake and utilization in a continuously cropped cucumber crop intercropped with green garlic. It has been reported that garlic (shoots, roots and bulbs) exhibits a "hormesis effect" at increased concentrations [32–33]. For this reason, various levels of intercropping with green garlic were tested in this study to select the best concentration and investigate their effects on the N, P, K, Ca, Mg, Fe, Mn, Zn concentrations of cucumber plants and changes in the levels of soil organic carbon and available N, P and K in this intercropping system.

Materials and Methods

Growth Medium and Experimental Design

The pot-based experiment was conducted from fall 2011 to spring 2012 in a plastic tunnel at the experimental station (Northwest China, N 34° 16', E 108° 4') of the College of Horticulture, Northwest A&F University, in Yangling, Shaanxi province, China. Pot soil was collected from another plastic tunnel, which had been previously planted with cucumber for five years. The chemical characteristics of the soil were as follows: electrolytic conductivity (1:5 soil: water) 340 $\mu S \cdot cm^{-1}$; pH (1:1 soil: water) 7.59; organic matter 25.82 $g \cdot kg^{-1}$; total N 2.15 $g \cdot kg^{-1}$; available N 110.13 $mg \cdot kg^{-1}$; total P 1.34 $g \cdot kg^{-1}$; available K 216.32 $mg \cdot kg^{-1}$ and available P 191.61 $mg \cdot kg^{-1}$. Seventeen kilograms of soil were put into each plastic pot (46 cm diameter ×40 cm depth). The pots were arranged in a complete randomized block design with three replicates for each treatment. Ten pots were used for each treatment in one replication.

The experiment was designed to test one factor (the amount of intercropped green garlic) at the following six levels: 0, 150, 300, 450, 600 and 750 g of garlic bulbs (which were planted with the intention of repeatedly harvesting green garlic). Monocropped cucumber was used as a control.

Experimental Arrangement

In northern China, cucumber is often planted two cropping in plastic tunnel, namely spring cucumber and fall cucumber, respectively. In our experiment green garlic is interplanted with fall cucumber in early fall, cut harvested for about three times during fall and early winter, then totally removed in spring of next year before cucumber is transplanted.

In fall 2011, before the fall cucumbers were transplanted, experimental pots were uniformly tilled by hand and fertilized with 100 g of organic fertilizer, 10 g of calcium superphosphate and 10 g of compound fertilizer (N-P_2O_5-K_2O:18-18-18) per pot. Cucumber seedlings (Xintiandi No. 1) were purchased from the Xintiandi Agriculture Science and Technology Demonstration Garden at Yangling and transplanted on August 13. Each cucumber seedling was planted in the centre of a pot. Twenty days later, various amounts of garlic bulbs were uniformly planted around the cucumber plants. The cucumbers and the garlic bulbs were planted 10–12 cm apart. After 45 days of co-growth, the fall cucumber plants were totally removed from pots at the full bearing stage. Green garlic was growing until the spring of 2012 and removed on March 19. All pots were fertilized according to the previous growing season. On March 21, spring cucumber seedlings were transplanted and grown for one and a half months and then harvested at the early fruiting stage. During the experiment, soil moisture was maintained at approximately 70% of the field water-holding capacity. All pots were irrigated well manually and hand-weeded during crop growth.

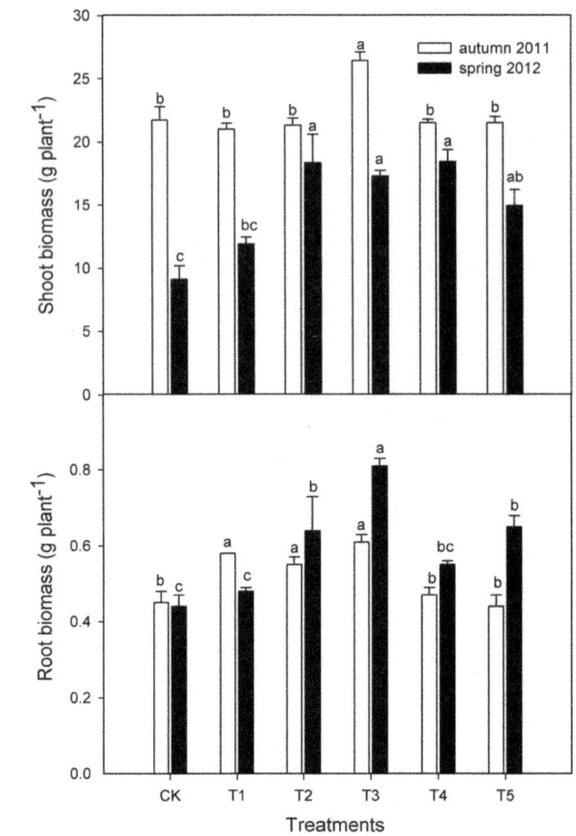

Figure 1. Cucumber shoot and root biomasses with various amounts of intercropped green garlic in the fall of 2011 and the spring of 2012. Samples were taken 45 days after cucumber intercropping with green garlic in the fall of 2011 and the transplantation of cucumber in the spring of 2012. CK represents cucumber monocropping. T1, T2, T3, T4 and T5 represent 150, 300, 450, 600 and 750 g of garlic bulbs intercropped with cucumber, respectively. Different letters above bars indicate significant differences at the $P<0.05$ level, $n=10$.

Biomass and Plant Nutrient Element Analysis

At harvest, cucumber plant samples were divided into shoots and roots and weighed for fresh weight determinations. The samples were then rinsed with tap water and distilled water. The dry weights of the shoots and roots were measured after drying at 65°C for 72 h to a constant weight after killing enzymatic activity at 105°C for 15 min. The samples (shoots and roots) were pulverized to pass through 0.149 and 0.250 mm mesh screens and analyzed to determine the total N, P, K, Ca, Mg, Fe, Zn and Mn contents. To analyze N, P and K, the samples were digested a mixture of concentrated H_2SO_4 and H_2O_2. The nitrogen (N) content was analyzed using a semi-micro Kjeldahl apparatus [34]. The phosphorus (P) content was evaluated spectrophotometrically, and the potassium (K) content was measured by an atomic absorption spectrophotometer (Hitachi Z-2000, Tokyo). Prior to measuring of the remaining nutrients, the samples were subjected to dry ashing at 550°C for 6 h and then dissolved (1:1, v/v) in 5 mL of hydrochloric acid (HCl) and finally diluted to 25 mL with deionized water. The extracts were then filtered and stored in plastic vials until analyzed. Ca, Mg, Fe, Zn and Mn were determined using an atomic absorption spectrophotometer (Hitachi Z-2000, Tokyo).

Table 1. The effect of various amounts of intercropped green garlic on the macronutrient concentrations of cucumber plants in the experiment carried out during the fall of 2011.

Treat-ment	Amount of bulbs intercropped (g)	N (%)		P (%)		K (%)		Ca (mg g-1)		Mg (mg g-1)	
		shoot	root	shoot	root	shoot	root	shoot	root	shoot	root
CK	0	4.75±0.08	3.15±0.01	0.36±0.01	0.98±0.01	2.20±0.03	2.84±0.06	20.07±0.17	18.84±0.20	4.58±0.08	10.18±0.16
T1	150	4.58±0.09	3.11±0.01	0.37±0.01	0.96±0.01	2.23±0.06	3.16±0.02**	22.27±0.28**	19.73±0.15	4.50±0.38	8.08±0.30***
T2	300	5.11±0.01	3.51±0.03***	0.45±0.01***	1.06±0.01***	2.09±0.01	2.78±0.07	21.32±0.81	19.57±0.29	4.78±0.14	7.60±0.08***
T3	450	4.53±0.26	3.37±0.01***	0.40±0.01*	0.95±0.01	2.60±0.04***	2.81±0.05	20.77±0.09	19.70±0.21	4.25±0.14	8.75±0.43**
T4	600	4.70±0.09	3.45±0.04***	0.52±0.00***	0.96±0.01	2.50±0.05***	2.91±0.09	17.84±0.72**	20.02±0.12*	4.77±0.27	8.08±0.08***
T5	750	4.60±0.08	3.26±0.01**	0.47±0.01***	0.98±0.00	2.22±0.02	2.96±0.02	18.13±0.26*	20.12±0.59*	4.03±0.12	8.22±0.37***

Values are the means ± standard error of three replicates.
*Significant at the 0.05 probability level.
**Significant at the 0.01 probability level.
***Significant at the 0.001 probability level.

Soil Analysis

Soil taken from a depth of 0–10 cm was sampled from each of the ten pots randomly at cucumber harvest. Cucumber roots were taken from the pots and gently shaken by hand to obtain soil that had adhered to the roots, hereafter referred to as rhizosphere soil. Soil that was more distant from the roots was regarded as bulk soil [35]. After air-drying, the bulk soil samples were pulverized and passed through 1.000 and 0.149 mm sieves for analysis of organic carbon, available N, available P and available K. Rhizosphere soil were pulverized and passed through 1.000 mm sieves for analysis of pH. Rhizosphere soil pH was prepared by a 1:1 water extract and measured using a pH meter. Soil organic carbon was estimated using the dilution heat $K_2Cr_2O_7$ oxidation volumetric method. Available N was analyzed through the alkali-hydrolytic diffusion method [36]. Available P was determined by extraction with 1 M sodium bicarbonate ($NaHCO_3$) according to the method of Olsen et al. [37]. Air-dried soil samples were extracted with 1 M ammonium acetate (NH_4Ac), and the available K concentration of the filtrate was determined using atomic absorption spectrophotometer (Hitachi Z-2000, Tokyo).

Statistical Analysis

Statistical analysis was carried out using SAS 8.1 software. Data were analyzed by ANOVA for a completely randomized block design. The difference of the treatments was compared using the least significant difference (LSD) test at $P \leq 0.05$, $P \leq 0.01$ and $P \leq 0.001$. Simple regression (REG) analysis was used to examine relationships between the amounts of garlic bulbs and biomass, nutrient concentrations in cucumber plants, available soil N, P, K and organic carbon content.

Results

The Effect of Intercropping with Green Garlic on the Biomass of Continuously Cropped Cucumber in an Experiment Carried out in Fall 2011 and Spring 2012

The shoot and root biomass of cucumber was significantly increased by intercropping with green garlic during two growing seasons (Figure 1). Only treatment T3 (intercropping with 450 g of bulbs for green garlic production) exhibited a significant positive effect on shoot biomass during the fall of 2011. With the exception of treatment T1 (intercropping with 150 g of bulbs for green garlic

Table 2. The effect of various amounts of intercropped green garlic on the micronutrient concentrations of cucumber plants in the experiment carried out during the fall of 2011.

Treatment	Amount of bulbs intercropped (g)	Fe (µg g-1)		Mn (µg g-1)		Zn (µg g-1)	
		shoot	root	shoot	root	shoot	root
CK	0	186.67±2.85	1390.55±7.83	32.83±1.36	48.92±0.71	43.00±0.87	49.79±0.63
T1	150	166.00±8.72	1501.67±56.45	34.67±0.44	48.50±2.60	37.17±1.01**	47.00±1.26
T2	300	203.28±10.22	1288.22±77.37	34.46±0.84	47.12±1.42	34.18±1.41***	46.82±1.19
T3	450	128.67±4.63***	1285.00±29.30	32.17±0.17	46.83±1.69	43.50±0.58	45.83±1.48*
T4	600	146.19±4.21**	1350.63±4.72	30.62±0.64	48.11±2.40	44.31±0.65	39.73±0.87***
T5	750	162.97±8.48*	1993.53±5.96***	33.33±0.93	68.01±1.45***	41.83±1.30	40.20±1.46***

Values are the means ± standard error of three replicates.
*Significant at the 0.05 probability level.
**Significant at the 0.01 probability level.
***Significant at the 0.001 probability level.

Table 3. The effect of various amounts of intercropped green garlic on the macronutrient concentrations of cucumber plants in the experiment carried out during the spring of 2012.

Treatment	Amount of bulbs intercropped (g)	N (%)		P (%)		K (%)		Ca (mg g-1)		Mg (mg g-1)	
		shoot	root	shoot	root	shoot	root	shoot	root	shoot	root
CK	0	3.71±0.06	2.90±0.04	0.50±0.01	0.55±0.01	4.11±0.06	5.26±0.02	54.55±4.02	21.10±0.97	13.50±0.58	9.08±0.22
T1	150	4.44±0.01***	3.47±0.05***	0.59±0.01***	0.59±0.01*	4.90±0.09***	5.25±0.04	65.42±6.32*	22.90±1.06	12.00±1.04	8.67±0.58
T2	300	4.26±0.06***	3.12±0.03**	0.61±0.00***	0.67±0.01***	4.68±0.02***	5.26±0.04	55.33±2.32	22.73±0.38	10.33±0.73*	8.83±0.36
T3	450	4.11±0.02**	3.55±0.10***	0.61±0.01***	0.74±0.01***	4.12±0.06	5.49±0.02***	56.33±0.98	21.67±0.52	10.33±0.44*	9.17±0.46
T4	600	4.27±0.02***	3.33±0.02***	0.61±0.00***	0.70±0.01***	4.06±0.06	5.44±0.04**	53.42±2.32	24.13±0.85*	10.00±1.00**	8.75±0.14
T5	750	4.03±0.17*	3.52±0.03***	0.54±0.01*	0.64±0.00***	4.14±0.05	5.85±0.00***	53.58±1.21	24.80±0.97**	9.67±0.73**	8.58±0.44

Values are the means ± standard error of three replicates.
*Significant at the 0.05 probability level.
**Significant at the 0.01 probability level.
***Significant at the 0.001 probability level.

production), intercropping with green garlic significantly increased shoot biomass until the spring of 2012. And shoot biomass was significantly related to the amounts of garlic bulbs ($Y = 8.4832 + 0.04046X - 0.00004185X^2$, $R^2 = 0.9024$). Otherwise, cucumber root biomass was more susceptible to intercropped green garlic; cucumber root biomass was significantly promoted by relatively small amounts (150, 300 and 450 g) of bulbs grown for green garlic production during the first growing season. In contrast, relatively large amounts (300, 450 and 750 g) exhibited the same effect on root biomass during the spring of 2012.

The Effect of Intercropping with Green Garlic on Plant Nutrient Concentrations in Continuously Cropped Cucumber during the Experiment Carried out in Fall 2011

The co-growth stage of cucumber intercropped with green garlic lasted for 45 days in fall 2011, and the concentrations of nutrient elements in the cucumber plants were significantly affected to varying degrees by the intercropping of green garlic at different bulb-planting rates (Table 1 and 2). Except for the shoot concentration of nitrogen (N), the macroelement contents of the cucumber plants were significantly influenced (Table 1). Intercropping with any amount of green garlic except for 150 g (T1) significantly increased the N concentration of cucumber roots and the P concentration in the shoots. P concentrations of shoot were significantly linearly related to the amounts of garlic bulbs ($Y = 0.36048 + 0.0001809X$, $r^2 = 0.6617$). Only treatment T2 (intercropping with 300 g of bulbs for green garlic production) exhibited a significantly stimulative effect on root P concentrations. The concentrations of K in the shoots and roots were significantly promoted by treatment T4 (intercropping with 600 g of bulbs for green garlic production). Shoot calcium (Ca) concentrations exhibited an up and down trend with increasing amounts of green garlic. Treatments T4 and T5 significantly increased root calcium (Ca) concentrations, which were significantly linearly related to the amounts of garlic bulbs ($Y = 19.1348 + 0.001410X$, $r^2 = 0.7609$). No significant difference in shoot magnesium (Mg) concentrations was observed among the treatments. Surprisingly, root Mg concentrations were significantly reduced by any amount of intercropping with green garlic.

The microelement contents were also significantly affected by intercropping with green garlic (Table 2). Intercropping with relatively large amounts (450 g, 600 g and 750 g) of bulbs (for green garlic production) significantly decreased shoot iron (Fe) concentrations in cucumber. In contrast, root concentrations of Fe were significantly increased using treatment T4, as did root manganese (Mn) concentrations. Intercropping with relatively small amounts (150 and 300 g) and large amounts (450, 600 and 750 g) of bulbs showed significantly negative effects on shoot and root concentrations of Zn, respectively. Root Zn concentrations were significantly linearly related to the amounts of garlic bulbs ($Y = 49.9486 - 0.01348X$, $r^2 = 0.8753$).

The Effect of Intercropping with Green Garlic on Plant Nutrient Concentrations in Continuously Cropped Cucumber during the Experiment Carried out in Spring 2012

Although the green garlic was harvested before the cucumbers were transplanted in spring, the effect of the garlic on nutrient uptake existed until 45 days after cucumber transplantation in spring 2012. Intercropping with various amounts of green garlic displayed significantly different effects on the nutrient element contents (Table 3 and 4). The shoot and root concentrations of N and P in the cucumbers were significantly increased in all

Table 4. The effect of various amounts of intercropped green garlic on the micronutrient concentrations of cucumber plants in the experiment carried out during the spring of 2012.

Treatment	Amount of bulbs intercropped (g)	Fe (μg g-1)		Mn (μg g-1)		Zn (μg g-1)	
		shoot	root	shoot	root	shoot	root
CK	0	370.00±15.28	538.33±30.05	61.67±0.60	27.00±1.04	44.67±0.33	60.33±1.76
T1	150	313.67±4.70***	1351.67±59.47***	61.50±0.76	35.50±1.89***	48.00±0.58	64.33±2.96
T2	300	329.67±3.18**	1380.00±39.69***	64.83±0.73*	48.67±0.73***	50.00±2.65	70.00±0.58**
T3	450	338.33±6.17**	1401.67±47.64***	70.33±0.44***	51.33±1.88***	51.00±2.00*	77.00±2.00***
T4	600	289.00±3.21***	943.33±36.55***	61.33±0.33	42.50±1.50***	50.33±1.86*	77.33±1.20***
T5	750	299.33±2.33***	648.33±6.01	61.33±1.59	32.67±0.17*	51.33±1.76*	75.33±1.76***

Values are the means ± standard error of three replicates.
*Significant at the 0.05 probability level.
**Significant at the 0.01 probability level.
***Significant at the 0.001 probability level.

treatments compared with monocropping (Table 3). Shoot P concentrations were significantly related to the amounts of garlic bulbs ($Y = 0.5057+0.0005733X-0.00000006984X^2$, $R^2 = 0.9489$), and root P showed the same trend ($Y = 0.5293+0.0007452X-0.00000007778X^2$, $R^2 = 0.8814$). Intercropping with relatively small amounts (150 and 300 g) of bulbs significantly promoted shoot whereas large amounts (450, 600 and 750 g) did the same for root K concentrations, which were significantly related to the amounts of garlic bulbs ($Y = 5.2679-0.0003929X+0.000001480X^2$, $R^2 = 0.8961$). Shoot and root Ca concentrations were significantly enhanced by intercropping with 150 g (T1) and 600 g and 750 g (T4 and T5) of bulbs, respectively. Except for T1, shoot Mg concentrations were significantly decreased. The change of Mg concentrations in shoot was significantly related to the amounts of garlic bulbs ($Y = 13.4354-0.01146X+0.000008900X^2$, $R^2 = 0.9652$). No significant difference was found in root Mg concentrations among the treatments.

As is shown in Table 4, intercropped green garlic significantly lowered Fe concentrations in shoot, while facilitated in root. Root Fe concentrations were significantly related to the amounts of garlic bulbs ($Y = 644.8193+4.3329X-0.005940X^2$, $R^2 = 0.8889$). Compared to monocropping cucumber, intercropping with medium amount (300 g, 450 g) of bulbs significantly increased the Mn concentrations of whole plant. Root Mn concentrations were significantly related to the amounts of garlic bulbs ($Y = 25.2032+0.1168X-0.0001426X^2$, $R^2 = 0.9358$). As the amount increasing, intercropping with green garlic enhanced significantly Zn concentrations in shoot and root. The regression analysis showed that both shoot and root Zn concentrations were significantly related to the amounts of garlic bulbs ($Y = 44.9432+0.02116X-0.00001772X^2$, $R^2 = 0.9570$; $Y = 59.0200+0.05362X-0.00004076X^2$, $R^2 = 0.9483$).

The Effect of Intercropping with Green Garlic on Soil Nutrient Characters in Continuously Cropped Cucumber during the Experiments Carried out during Fall 2011 and Spring 2012

The soil nutrients and cucumber rhizosphere pH values were significantly affected by intercropping with green garlic (Tables 5 and 6). Levels of available N, P and K in the soil were higher for the intercropped treatments, especially when relatively large amounts of green garlic were intercropped, compared to the monocropped cucumber cops in both growing seasons. Moreover, intercropping with green garlic increased the available levels of K as reflected in the N: P: K ratio. Available soil N was significantly linearly related to the amounts of garlic bulbs in both fall 2011 and spring 2012 ($Y = 109.4048+0.04008X$, $r^2 = 0.7780$;

Table 5. The effect of various amounts of intercropped green garlic on available soil N, P and K, organic carbon and rhizosphere pH in the experiment carried out during the fall of 2011.

Treatment	Amount of intercropped bulbs (g)	Available soil N (mg kg^{-1})	Available soil P (mg kg^{-1})	Available soil K (mg kg^{-1})	Soil organic carbon (g kg^{-1})	Rhizosphere pH (1:1 soil: water)
CK	0	110.8±5.8	185.8±2.8	285.1±5.9	13.95±0.09	7.30±0.01
T1	150	110.8±7.7	224.0±9.4**	420.8±2.4***	14.45±0.07	7.28±0.03
T2	300	128.3±5.8	237.1±9.8***	399.0±5.3***	15.20±0.28**	7.25±0.03
T3	450	119.6±7.7	221.9±7.9**	427.9±4.6***	16.01±0.27***	7.24±0.02
T4	600	140.0±5.1**	234.0±8.4**	520.8±2.9***	15.48±0.29**	7.29±0.03
T5	750	137.1±2.9**	212.3±7.7*	437.2±3.3***	17.71±0.41***	7.35±0.02

Values are the means ± standard error of three replicates.
*Significant at the 0.05 probability level.
**Significant at the 0.01 probability level.
***Significant at the 0.001 probability level.

Table 6. The effect of various amounts of intercropped green garlic on available soil N, P and K, organic carbon and rhizosphere pH in the experiment carried out during the spring of 2012.

Treatment	Amount of intercropped bulbs (g)	Available soil N (mg kg^{-1})	Available soil P (mg kg^{-1})	Available soil K (mg kg^{-1})	Soil organic carbon (g kg^{-1})	Rhizosphere pH (1:1 soil: water)
CK	0	115.7±1.7	186.2±9.4	287.2±5.9	14.77±0.08	7.66±0.02
T1	150	116.2±1.1	190.6±1.1	388.9±0.3***	16.91±0.07***	7.32±0.02***
T2	300	122.5±0.4*	216.8±5.8*	348.5±1.4***	16.96±0.25***	7.53±0.02**
T3	450	128.1±2.5**	190.6±5.3	329.6±1.1***	16.99±0.09***	7.37±0.02***
T4	600	126.2±2.0**	226.8±11.9**	375.6±0.9***	17.34±0.09***	7.27±0.04***
T5	750	126.9±3.6**	214.7±11.9*	407.8±1.5***	17.51±0.09***	7.43±0.01***

Values are the means ± standard error of three replicates.
*Significant at the 0.05 probability level.
**Significant at the 0.01 probability level.
***Significant at the 0.001 probability level.

$Y = 114.0571+0.01745X$, $r^2 = 0.7976$). During both growing seasons, intercropping with green garlic significantly increased the soil organic carbon content. Moreover, the most amount of green garlic exhibited the highest content of soil organic carbon. Only in the fall of 2011, soil organic carbon content was significantly related to amounts of garlic bulbs ($Y = 13.8452+0.004320X$, $r2 = 0.8429$). The pH of cucumber rhizosphere soil was not significantly affected by intercropping with green garlic during the fall of 2011 (Table 5). However, in spring 2012, the pH of cucumber rhizosphere soil was lower in the intercropping system compared that in monocropped cucumber (Table 6).

Discussion

This study demonstrated that biomass and the concentrations of N, P, K, Ca and Mn in cucumber plants were increased by intercropping with green garlic during the two studied growing seasons (Figure 1 and Tables 1, 2, 3, 4). The increased concentrations of these nutrients could be related to the enhancement of soil organic matter and available nutrients (Table 5 and 6). There might be two reasons for the increase in soil organic matter in the intercropping system. Garlic bulbs (used to grow green garlic) with their abundant nutrients increased the input of nutrient compared to the cucumber monocrop. Plant residues as one of major fractions of soil organic matter may be present in various stages of decomposition [38]. So on the other hand, the green garlic produced numerous roots, which may be remained as a carbon source in the soil after harvesting in the spring of 2012. Some studies have shown that intercropping with garlic was able to improve soil enzyme activities, and the effects lasted to the second growing season [30–31]. Soil enzyme activities can be used as suitable indicators of soil quality [39] and play a crucial role in the decomposition of organic residues and nutrient cycling in the soil [40]. In the present study, the observed increase in available N, P and K in the soil could be attributed to higher soil enzyme activities in the green garlic and cucumber intercropping system [30]. The level of available K (as reflected in the N: P: K ratio) was increased in the intercropping systems, which better reflects the nutritional requirements of cucumbers during the fruiting period, when they were sampled. This finding indicates that the soil nutrient balance in the continuously cropped cucumber system was enhanced by intercropping with green garlic. From fall 2011 to spring 2012, available N, P and K levels in the soil increased in the control treatment but decreased in the intercropped treatments, indicating that intercropping with green garlic improved nutrient uptake in the continuously cropped cucumbers even though the garlic was harvested before the cucumbers were transplanted.

Large ion antagonistic effects have been reported between K, Ca and Mg in some plants [41–42]. In our study, Mg concentrations were decreased by intercropping with green garlic whereas K and Ca contents were increased. In the fall of 2011, the lower Zn concentrations in cucumber plants may be partly due to an increase in P availability [43–44]. However, Zn concentrations were significantly increased by intercropping with green garlic in the spring of 2012, which was related to the lower pH of cucumber rhizosphere soil (Table 6). This is supported by the findings of a report [45] that a decrease in soil pH could enhance the availability of some nutrients in soil, particularly Fe, Mn and Zn. Most studies of Fe uptake have focused on graminaceous plants and dicotyledon intercropping systems. Two mechanisms are present, including increases in the ferric reductase activity of roots and rhizosphere acidification as "Strategy I" [46–47] by dicotyledons and exuding phytosiderophores to increase the available Fe by graminaceous plants [48]. Incredibly, the present study showed that Fe concentrations were decreased in the shoots of cucumbers but increased in the roots in the intercropping system. This result indicates that intercropping of green garlic inhibits the transfer of Fe from the roots to the shoots in cucumber plants. However, the available data and literature could not provide an explanation of the mechanisms involved.

In the present study, although the available nutrients and organic matter increased in the soil with increasing amounts of green garlic, the highest nutrient uptake by the cucumber plants did not always result from the highest amounts of green garlic. The result may be ascribed to the allelopathic feature of garlic, which exhibits a positive effect at low concentrations and a negative effect at high concentrations by bioassay [33]. However, further research into the mechanism behind the effect of green garlic on nutrient uptake in cucumbers is needed, particularly with regard to the aspect of root exudates.

Conclusion

This study indicates that intercropping with green garlic significantly affects nutrient concentrations and the soil nutrient character in cucumber crops. The accumulation of biomass and most nutrient elements (N, P, K, Ca and Mn) in intercropped cucumber was significantly greater than that found in monocropped cucumber, and the effect was even sustained to the second growing season. Mg concentrations were decreased by intercropping. Fe concentrations in cucumber shoots were decreased whereas those in the roots increased in the intercropping system. In the fall of 2011, Zn concentrations were decreased, but these increased in the second growing season. The best effects were always found when medium amounts (300 g, 450 g, 600 g) of bulbs were planted for green garlic production. Soil organic matter and available N, P and K were higher in the green garlic-cucumber intercropping system than in a cucumber monocropping system. The regression analysis showed that the concentrations of the most elements were significantly related to the amounts of garlic bulbs, especially the microelements in the experiment of spring 2011. The available soil N and organic matter were linearly related to the amounts of garlic bulbs.

Author Contributions

Conceived and designed the experiments: ZHC. Performed the experiments: XMX. Analyzed the data: HWM. Contributed reagents/materials/analysis tools: LHL HZL YXD. Wrote the paper: XMX.

References

1. Yu JQ, Matsui Y (1994) Phytotoxic substances in root exudates of *Cucumis sativus* L. J Chem Ecol 20: 21–31.
2. Yao HY, Jiao XD, Wu FZ (2008) Effects of continuous cucumber cropping and alternative rotations under protected cultivation on soil microbial community diversity. Plant Soil 284: 195–203.
3. Ye SF, Yu JQ, Peng YH, Zheng JH, Zou LY (2004) Incidence of fusarium wilt in *Cucumis sativus* L. is promoted by cinnamic acid, an autotoxin in root exudates. Plant Soil 263: 143–150.
4. Yu JQ, Shou SY, Qian YR, Zhu ZZ, Hu WH (2000) Autotoxic potential of cucurbit crops. Plant Soil 223: 147–151.
5. Lin S, Dittert K, Tao HB, Kreye C, Xu YC, et al. (2002) The ground-cover rice production system (GCRPS): a successful new approach to save water and increase nitrogen fertilizer efficiency? In: Bouman BAM, Hengsdijk H, Hardy B, Bindraban PS, Tuong TP, Ladha JK (Eds.), Water-wise rice production. International Rice Research Institute, Philippines, 187–195.
6. Ventura W, Watanabe I, Komada H, Nishio M, dela Cruz A, et al. (1984) Soil sickness caused by continuous cropping of upland rice, mungbean, and other crops. IRRI Res Paper Ser 99: 13.
7. Huang HC, Chou CH, Erickson RS (2006) Soil sickness and its control. Allelopathy J 18: 1–21.
8. Nishizawa T, Ohshima Y, Kurihara H (1971) Survey of the nematode population in the experimental fields of successive or rotative plantation. Proc Kanto-Tosan Plant Prot Soc 18: 121–122.
9. Nishio M, Kusano S (1975) Effect of root residues on the growth of upland rice. Soil Sci Plant Nutr 21: 391–395.
10. Li L, Yang SC, Li XL, Zhang FS, Christie P (1999) Interspecific complementary and competitive interactions between intercropped maize and faba bean. Plant Soil 212: 105–114.
11. Li L, Sun JH, Zhang FS, Li XL, Yang SC, et al. (2001) Wheat/maize or wheat/soybean strip intercropping. I. Yield advantage and interspecific interactions on nutrients. Field Crops Res 71: 123–137.
12. Latif MA, Mehuys GR, Mackenzie AF, Alli I, Faris MA (1992) Effects of legumes on soil physical quality in a maize crop. Plant Soil 140: 15–23.
13. Olasantan FO, Ezumah HC, Lucas EO (1996) Effects of intercropping with maize on the micro-environment, growth and yield of cassava. Agr Ecosyst Environ 57: 149–158.
14. Wasaki J, Yamamura T, Shinano T, Osaki M (2003) Secreted acid phosphatase is expressed in cluster lupin in response to phosphorus deficiency. Plant Soil 248: 129–136.
15. Inal A, Gunes A, Zhang F, Cakmak I (2007) Peanut/maize intercropping induced changes in rhizosphere and nutrient concentrations in shoots. Plant Physiol Bioch 45: 350–356.
16. Adu-Gyamfi JJ, Myaka FA, Sakala WD, Odgaard R, Vesterager JM, et al. (2007) Biological nitrogen fixation and nitrogen and phosphorus budgets in farmer-managed intercrops of maize–pigeonpea in semi-arid southern and eastern Africa. Plant Soil 295: 127–136.
17. Li YF, Ran W, Zhang RP, Sun SB, Xu GH (2009) Facilitated legume nodulation, phosphate uptake and nitrogen transfer by arbuscular inoculation in an upland rice and mung bean intercropping system. Plant Soil 315: 285–296.
18. Jensen ES (1996) Barley uptake of N deposited in the rhizosphere of associated field pea. Soil Biol Biochem 28: 159–162.
19. Martin RC, Voldeng HD, Smith DL (1991) Nitrogen transfer from nodulating soybean to maize or to non-nodulating soybean in intercrop: The 15N dilution methods. Plant Soil 132: 53–63.
20. Ae N, Arihara J, Okada TK, Yoshihara CJ (1990) Phosphorus uptake by pigeon pea and its role in cropping systems of the Indian subcontinent. Science 248: 477–480.
21. Horst WJ, Waschkies C (1987) Phosphorus nutrition of spring wheat *(Triticum aestivum* L.) in mixed culture with white lupin (*Lupinus albus* L.). Z. Pflanzenernähr. Bodenkd 150: 1–8.
22. Gardner WK, Boundy KA (1983) The acquisition of phosphorus by *Lupinus albus* L. IV. The effect of interplanting wheat and white lupin on the growth and mineral composition of the two species. Plant Soil 70: 391–402.
23. Li L, Tang C, Rengel Z, Zhang FS (2004) Calcium, magnesium and microelement uptake as affected by phosphorus sources and interspecific root interactions between wheat and chickpea. Plant Soil 261: 29–37.
24. Zheng Y, Zhang F, Li L (2003) Iron availability as affected by soil moisture in intercropped peanut and maize. J Plant Nutr 26: 2425–2437.
25. Sivaraman K, Palaniappan SP (1996) Turmeric - maize and onion intercropping systems. III. Nutrient uptake. Journal of Spices and Aromatic Crops 5(1): 49–57.
26. Shanmugham K (1988) Effect of onion and greengram intercrops on phosphorous release and its uptake by cotton. Curr Sci 57: 1128–1130.
27. Mogahed MI (2003) Influence of intercropoing on population dynamics of major insect-pests of potato (*Solanum tuberosum*) in North Sinai Governorate, Egypt. Indian J Agr Sci 73: 546–549.
28. Lai RQ, You MS, Lotz LAP, Vasseur L (2011) Response of green peach aphids and other arthropods to garlic intercropped with tobacco. Agron J 103: 856–863.
29. Mueller S, Durigan JC, Banzatto DA, Kreuz CL (1998) Benefits to yield and profits of garlic and beet intercropping under three weed management epochs. Pesqui Agropecu Bras Pesquisa 33: 1361–1373.
30. Xiao XM, Cheng ZH, Meng HW, Khan MA, Li HZ (2012) Intercropping with garlic alleviated continuous cropping obstacle of cucumber in plastic tunnel. Acta Agr Scand B-SP 62 (8): 696–705.
31. Zhou XG, Yu GB, Wu FZ (2011) Effects of intercropping cucumber with onion or garlic on soil enzyme activities, microbial communities and cucumber yield. Eur J Soil Biol 47: 279–287.
32. Cheng ZH, Wang CH, Xiao XM, Khan MA (2011) Allelopathic effects of decomposing garlic stalk on some vegetable crops. Afr J Biotechnol 10(69): 15514–15520.
33. Wang CH, Cheng ZH, Niu Q, Liang JN, Xue SH (2009) Allelopathy of ultrasonic extract of garlic plant on different receiver vegetable crops. Journal of Northwest A& F University 37(7): 103–109. (In Chinese).
34. Bremner JM (1965) Nitrogen availability indexes. In: Black, C. A. (Ed), Methods of soil analysis, part 2. Madison, Wisc., ASA, 1324–1345.
35. Ren LX, Su SM, Yang XM, Xu YC, Huang QW, et al. (2008) Intercropping with aerobic rice suppressed *Fusarium* wilt in watermelon. Soil Biol Biochem 40: 834–844.
36. Keeney DR, Nelson DW (1982) Nitrogen inorganic forms. In: Page AL, Miller RH, Keeney DR (Eds), Methods of soilanalysis, Agronomy monograph 9 part 2, 2nd edn. American Society of Agronomy, Madison Wisconsin, 643–698.
37. Olsen SR, Cole CV, Watanabe FS, Dean LA (1954) Estimation of available phosphorus in soils by extraction with sodium bicarbonate. US Dep of Agric Circ 939.
38. Liang BC, MacKenzie AF, Schnitzer M, Monreal CM, Voroney PR, et al. (1998) Management-induced change in labile soil organic matter under continuous corn in eastern Canadian soils. Biol Fertil Soils 26: 88–94.39.
39. Acosta-Martínez V, Cruz L, Sotomayor-Ramírez D, Pérez-Alegría L (2007) Enzyme activities as affected by soil properties and land use in a tropical watershed. Appl Soil Ecol 35: 35–45.
40. Lalande R, Gagnon B, Simard RR, Cote D (2000) Soil microbial biomass and enzyme activity following liquid hog manure application in a long-term field trial. Can J Soil Sci 80: 263–269.
41. Ding Y, Luo W, Xu G (2006) Characterisation of magnesium nutrition and interaction of magnesium and potassium in rice. Ann Appl Biol 149: 111–123.
42. Ohno T, Grunes DL (1985) Potassium-Magnesium Interactions Affecting Nutrient Uptake by Wheat Forage. Soil Sci Soc Am J 49: 685–690.
43. Zhu YG, Smith FA, Smith SE (2002) Phosphorus efficiencies and their effects on Zn, Cu, and Mn nutrition of different barley (*Hordeum vulgare*) cultivars grown in sand culture. Aust J Agric Res 53: 211–216.
44. Warnock RE (1970) Micronutrient uptake and mobility within corn plants *(Zea mays* L.) in relation to phosphorus-induced zinc deficiency. Soil Sci Soc Am J 34: 765–769.
45. Marschner H (1995) Mineral Nutrition of Higher Plants. 2nd edn. Academic Press, London.
46. RÖmheld V, Marschner H (1986) Mobilization of iron in the rhizosphere of different plant species. Adv Plant Nutr 2: 155–204.
47. Zuo YM, Li X, Cao YP, Zhang FS, Christie P (2003) Iron nutrition of peanut enhanced by mixed cropping with maize: possible role of root morphology and rhizosphere microflora. J Plant Nutr 26: 2093–2110.
48. Marschner H (1998) Role of root growth, arbuscular mycorrhiza, and root exudates for the efficiency in nutrient acquisition. Field Crop Res 56: 203–207.

16

Watershed Sediment Losses to Lakes Accelerating Despite Agricultural Soil Conservation Efforts

Adam J. Heathcote*, Christopher T. Filstrup, John A. Downing

Department of Ecology, Evolution, and Organismal Biology, Iowa State University, Ames, Iowa, United States of America

Abstract

Agricultural soil loss and deposition in aquatic ecosystems is a problem that impairs water quality worldwide and is costly to agriculture and food supplies. In the US, for example, billions of dollars have subsidized soil and water conservation practices in agricultural landscapes over the past decades. We used paleolimnological methods to reconstruct trends in sedimentation related to human-induced landscape change in 32 lakes in the intensively agricultural region of the Midwestern United States. Despite erosion control efforts, we found accelerating increases in sediment deposition from erosion; median erosion loss since 1800 has been 15.4 tons ha^{-1}. Sediment deposition from erosion increased >6-fold, from 149 g m^{-2} yr^{-1} in 1850 to 986 g m^{-2} yr^{-1} by 2010. Average time to accumulate one mm of sediment decreased from 631 days before European settlement (ca. 1850) to 59 days mm^{-1} at present. Most of this sediment was deposited in the last 50 years and is related to agricultural intensification rather than land clearance or predominance of agricultural lands. In the face of these intensive agricultural practices, traditional soil conservation programs have not decelerated downstream losses. Despite large erosion control subsidies, erosion and declining water quality continue, thus new approaches are needed to mitigate erosion and water degradation.

Editor: Han Y.H. Chen, Lakehead University, Canada

Funding: This work was funded by a research grant from the Iowa Department of Natural Resources (Contract #: 08-39HA-18-05) (http://www.iowadnr.gov/). The funders had no role in study design, data collection and analysis, decision to publish, or preparation of the manuscript.

Competing Interests: The authors have declared that no competing interests exist.

* E-mail: aheathco@iastate.edu

Introduction

Soil erosion and nutrient loss from agricultural and urban lands are important problems facing inland, coastal, and marine waters [1]. Despite the US spending $5 billion annually to limit soil and nutrient losses from fields [2], intensive agricultural practices still threaten clean water resources [3].

Historical agricultural development in the Midwestern United States primarily consisted of large-scale landscape change and agricultural intensification. Historical change occurred in three phases (Fig. 1A) that are consistent across the agricultural regions of North America. (I) Land clearing of the native wet tallgrass prairie vegetation for agriculture began around 1850 and was 95% complete by 1910. (II) Land drainage by stream channelization and subsurface drains began around 1850, draining 70% of wetlands by 1920 [4]. (III) 1950s intensification of agriculture via increased average farm size, mechanization, fertilizer, and biocide application increased yields almost 4-fold [5].

Soil erosion mitigation programs began in 1935 and the US federal government has spent $294 billion (inflation adjusted) since then to support these programs nationwide (Figure 1A). Programs commonly paid farmers to plant cover crops, but retiring farmlands to native vegetation has been more common since 1985 [6]. According to the U.S. Department of Agriculture National Resource Conservation Service (USDA NRCS; http://www.ia.nrcs.usda.gov), in the state of Iowa alone, $53 million was spent on conservation programs, almost half of which went to installing practices such as grassed waterways, terraces, and other sediment and nutrient management strategies. The success of these programs in reducing sedimentation in some regions has been demonstrated [7], but other studies have questioned whether current soil conservation methods are enough to curb these problems, especially in areas where agricultural land use is most intense [8]. It has been suggested that the heart of this controversy is based on the prevalent application of generalized models to estimate soil loss changes, in lieu of actual measurements. There has been a call for more direct measurements of erosion in order to better inform models and more effectively implement conservation practices [9].

We measured the flux of erosional sediment to lakes over the last 180 years and quantified in-lake production of organic matter caused by accompanying nutrient runoff. The lakes in this study are located in an intensively agricultural region where erosion and nutrient mitigation efforts have been intense. Previous studies have used sediment to examine the correlation of agricultural practices with erosion rates with sometimes conflicting results, but these studies have often only considered a small number of lakes [10,11,12]. Here we examine a regional lake data set (32 lakes) and show that, on average, sediment accumulation rates have increased exponentially since the intensification of agriculture in the 1950s and that up to 75% of deposited sediment derives from erosion, despite efforts at control. This suggests the need for new approaches to permit the coexistence of productive agriculture and healthy water.

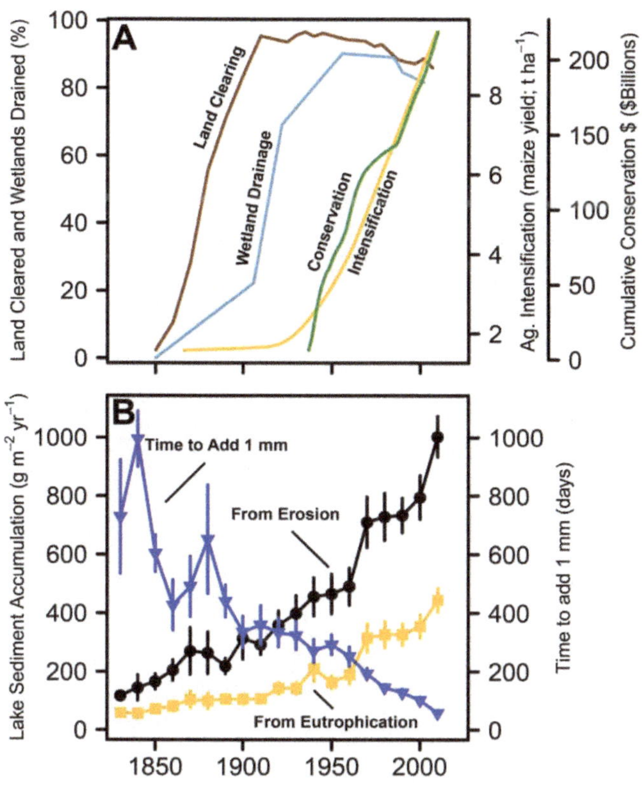

Figure 1. Changing agricultural practices and regional lake sedimentation rates since European settlement, shown as decadal averages across all 32 lakes. (A) Percent land in farms (brown), percent of wetlands drained (light blue), maize yield (t ha^{-1}) (yellow), and cumulative USDA financial assistance (inflation adjusted) for soil and water conservation programs in the USA (green). (B) Average regional lake mass accumulation rates for erosional (black) and in-lake (fueled by nutrient enrichment; yellow) derived sediment (g m^{-2} yr^{-1}). The time for lakes to add one mm of sediment is also shown (dark blue; days). Error bars represent ±1 standard error. Agricultural data were summarized from the United States Department of Agriculture's Census of Agriculture (1850–2007) and the National Agricultural Statistics Service. Annual maize yield data were fitted to a LOWESS model.

Materials and Methods

Thirty-two glacially formed (ca. 10,000–13,000 years old) lakes in the intensively agricultural Midwestern state of Iowa, USA, were selected to characterize the history of eroded soil transport and sediment delivery in the region (Table 1). Estimates from 2002 land cover data indicate that some form of conservation practices was present in the watersheds of all lakes (average: 6% of area) and agriculture was the dominant land use (average: 50% of area) (Iowa Geological and Water Survey, http://www.igsb.uiowa.edu). A sediment core was collected from a representative location in each lake, near the average depth, and sediment mass accumulation rates (MARs) were calculated based on ^{210}Pb dated core intervals [13]. Two lakes (Black Hawk Lake and Storm Lake) had an additional core taken from a secondary basin. Loss-on-ignition analysis [14] was performed on all dated sections to estimate the proportion of sediment derived from erosional inputs versus in-lake processes. Lake locations and basic water quality information are presented in Table S1. All lakes sampled in this study were publically owned and managed by the State of Iowa. No specific permits were required for the described field studies. Permissions to collect sediment cores from public lakes in this study were given by the Iowa Department of Natural Resources.

Previous studies in this region have conservatively estimated that 42% of the sediment organic matter is derived from terrestrial sources [4], so this proportion was added to the terrestrially derived inorganic fraction [15] to determine total erosional inputs. The remaining organic matter plus the fraction of calcium carbonate were assumed to be derived from in-lake processes stimulated by nutrient enrichment. Average MARs were calculated based on the average across all lakes for each decade from 1830–2010 (n = 459). Very few dated sections were available prior to 1830 (n = 8), so sections from 1760–1839 were combined.

Cumulative sediment delivery from the watershed was calculated by estimating average annual erosional deposition to lakes from a fitted locally weighted regression model (LOWESS; span = 0.67) [16] and multiplying by total lake area to estimate total erosional deposition. The total erosional deposition was then divided by total watershed area to determine the cumulative erosional loss to lakes from the watershed per unit area.

Results and Discussion

Sediment cores, dated with the radio-isotope ^{210}Pb, revealed that sediment deposition from erosion and internal production has increased continuously since settlement and that sediment deposition and agricultural changes followed similar chronologies (Figure 1B). Total mass accumulation rates (MARs) of sedimentation in lakes prior to European settlement and at present are given in Table 1. The sediment deposition from erosion increased from 149 g m^{-2} yr^{-1} in 1850 to 986 g m^{-2} yr^{-1} by 2010 and the average deposition rates for internally-produced sediment increased from 58 g m^{-2} yr^{-1} to 434 g m^{-2} yr^{-1}. Erosional sediment delivery and in-lake produced deposition both increased by approximately 7-fold since initial settlement of the region. The largest increases in sediment deposition occurred after 1950, concurrent with agricultural intensification. Average time to accumulate one mm of sediment decreased from 631 days before European settlement (ca. 1850) to 59 days mm^{-1} at present (Fig. 1B).

Sediment deposition from erosion represents a median loss of 15.4 t ha^{-1} (mean: 27.6 t ha^{-1}) delivered to lakes from the watersheds over the length of time covered in this study (~180 years) and was the source of up to 75% of total sediment deposited in these lakes. Cumulative erosion calculated from deposition in lakes ranged from 0.8 to 132.8 t ha^{-1}. A LOWESS smooth fit to erosion data across all lakes indicates an average increase from 0.1 t ha^{-1} yr^{-1} in 1850 to 0.4 t ha^{-1} yr^{-1} in 2010 (Figure 2). Individual erosion losses from watersheds for each of the 32 lakes are presented in Figure S1.

The eroded soil that leaves watersheds can be a small percentage of total upland erosion in some systems [9]; however, increases in delivery seen in these lakes may reflect increased erosion of topsoil as well as destabilization of upland soil previously deposited as unstable alluvium (e.g., stream bank and bed erosion) [17]. The estimates from this study, therefore, should be considered a measurement of annual net loss of erosional materials from the watersheds rather than a direct measurement of annual topsoil loss.

Eroded sediment and run-off from watersheds with intense agriculture also delivers particulate and dissolved phosphorus (P) [18], which led to increased in-lake production through nutrient enrichment of these systems (Figure 1B). Monitoring data indicate that many of the lakes in this study were eutrophic (Total P>30 μg/l) and highly productive by the mid-1970's [19] and

Table 1. Lake and watershed sizes, maximum ^{210}Pb ages at the bottom of cores (± SE), historic and modern sediment total (erosional+in-lake) mass accumulation rates (MAR; ±SE) for the sediment cores taken from 32 lakes in this study.

Lake Name	Lake Area (km^2)	Watershed Area (km^2)	Max ^{210}Pb Date (± Years)	Historic MAR (g cm^{-2} yr^{-1})	Modern MAR (g cm^{-2} yr^{-1})
Black Hawk Lake	3.1	53.2	1895±20	0.065±0.036	0.120±0.006
Black Hawk Lake (Slough)	0.7	53.2	1881±64	0.081±0.140	0.418±0.034
Burt Lake	0.8	22.4	1829±37	0.033±0.033	0.338±0.020
Center Lake	0.9	1.9	1835±31	0.055±0.052	0.130±0.006
Clear Lake	14.9	38.8	1929±38	0.005±0.004	0.005±0.001
Crystal Lake	1.0	7.4	1857±22	0.053±0.035	0.256±0.014
Diamond	0.6	6.6	1831±34	0.019±0.016	0.201±0.012
East Lake Okoboji	7.4	47.5	1799±28	0.035±0.028	0.142±0.007
Five Island Lake	3.9	31.5	1863±36	0.052±0.053	0.230±0.014
High Lake	1.9	6.3	1889±67	0.040±0.076	0.186±0.016
Ingham Lake	1.4	3.7	1813±39	0.010±0.009	0.098±0.005
Iowa Lake	3.2	36.3	1866±35	0.046±0.048	0.218±0.011
Lake Cornelia	1.0	3.0	1755±96	0.006±0.009	0.101±0.004
Lake Minnewashta	0.5	1.2	1825±21	0.044±0.025	0.260±0.011
Little Spirit Lake	2.4	5.8	1820±45	0.027±0.030	0.193±0.008
Little Wall Lake	1.0	0.8	1792±44	0.013±0.016	0.063±0.004
Lost Island Lake	4.7	20.9	1814±33	0.006±0.005	0.088±0.004
Lower Gar Lake	1.0	40.5	1829±23	0.024±0.015	0.184±0.007
Morse Lake	0.4	1.2	1794±62	0.012±0.016	0.173±0.010
North Twin Lake	1.8	8.4	1849±39	0.032±0.035	0.116±0.007
Pickerel Lake	0.7	6.6	1843±41	0.028±0.032	0.160±0.009
Rice Lake	4.0	61.9	1873±9	0.012±0.002	0.037±0.002
Silver Lake (Dickinson Co.)	4.2	60.0	1823±42	0.013±0.012	0.106±0.004
Silver Lake (Palo Alto Co.)	2.6	33.6	1828±29	0.021±0.016	0.207±0.013
Silver Lake (Worth Co.)	1.3	7.0	1863±32	0.004±0.002	0.005±0.001
Storm Lake	12.3	55.7	1929±21	0.038±0.020	0.161±0.014
Storm Lake (Inlet)	0.4	55.7	1836±128	0.0374±0.110	0.449±0.043
Trumbull Lake	4.8	191.5	1894±41	0.134±0.169	0.152±0.011
Tuttle Lake	9.2	496.7	1822±56	0.016±0.021	0.180±0.010
Upper Gar Lake	0.1	0.8	1869±29	0.037±0.030	0.071±0.003
Virgin Lake	0.9	4.3	1838±21	0.017±0.009	0.135±0.007
West Lake Okoboji	15.6	61.0	1813±30	0.029±0.024	0.099±0.003
West Swan Lake	1.5	35.0	1829±22	0.013±0.006	0.095±0.005
West Twin Lake	0.4	0.5	1863±37	0.059±0.066	0.205±0.012

data from our study show that the largest increases of internally-produced material began around 1950. The efficiency with which herbivores consume additional organic matter decreases with eutrophication because increasingly large and inedible algae may be favored under these conditions [20,21], which has led to increased deposition of internally-produced material to the sediments. On average, this material accounts for up to 31% of the sediment in these systems and the rate of deposition has nearly tripled since 1950.

Despite well-intentioned efforts, recent changes in farming have increased the potential for erosional deposition of sediment and nutrients. As demand for farm products increased, farms aggregated and expanded onto previously unfarmed slopes and riparian buffer zones [8]. As a result of this consolidation, average farm size in this region increased from 68 ha in 1950 to 134 ha in 2007, despite a 10% decrease in the total area in farmland. Maize yields rose geometrically after 1950 (Figure 1A) due in part to increasing use of chemical fertilizers and intensive row-cropping practices. Both practices have been linked to increasing organic matter transport and burial in lakes and are strongly correlated with agricultural intensification rather than initial land clearance [4].

The complex conveyance processes that transport eroded topsoil out of watersheds, specifically when temporarily stored as colluvium or alluvium, may cause a lag time between changes in land-use (i.e., source erosion) and sediment delivery [22]. Sediment delivery in regions where agricultural influences have not been as intensive in modern years have shown peaks in

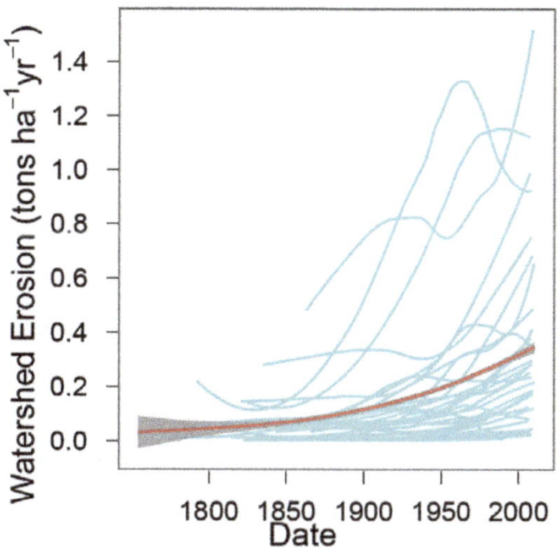

Figure 2. Watershed erosion (tons ha^{-1} yr^{-1}) versus time for the 32 lakes in this study since 1800. Red line represents average rate of erosion (LOWESS smoothed fit to the data) from the watershed across all 32 lakes in this study, surrounded by the 95% confidence region. Blue lines represent LOWESS smoothed fits for each of the individual lakes and are included to visualize variability among lakes.

sediment delivery by 1960 [23] and others have found lake sediments to reflect landscape changes in watersheds within as few as 12 years [11]. The absence of any discernible peak in the rate of sediment delivery across our study sites and the long timescale of our analysis (180 years) suggest that lags in transport of stored sediment alone cannot sufficiently explain the exponential increases in sediment delivery seen in these lakes.

Possibly, erosion and sediment deposition would have been more severe without erosion control subsidies. Soil erosion mitigation programs were created in response to excessive losses of soil from farmers' fields and water quality concerns. Despite these efforts, however, sediment deposition downstream has not decelerated in one of the most intensively agricultural regions. Our results suggest that decreasing downstream sediment accumulation and water impairment will require more than voluntary programs of subsidized farmland easement. Projected increased global demands for food, energy, and fiber [5] imply that even current erosion rates will be exceeded without better land management practices. Practices aimed at decreasing hydrologic connectivity, surgically stabilizing the most erosion-prone parcels, slowing water transport, restoring riparian corridors, and converting to low intensity agricultural practices may be more efficient options.

Supporting Information

Figure S1 Watershed erosion (tons ha^{-1} yr^{-1}) versus time for each of the lakes in this study. Black lines represent a LOWESS smoothed fit to the data.

Acknowledgments

We would like to acknowledge the Limnological Research Center at the University of Minnesota-Minneapolis for assistance in sediment core processing, Daniel Engstrom and the St. Croix Watershed Research Station for their work on the ^{210}Pb chronologies, and the members of the Iowa State University Limnology Laboratory for assistance in the field and laboratory.

Author Contributions

Conceived and designed the experiments: AJH JAD. Performed the experiments: AJH. Analyzed the data: AJH. Contributed reagents/materials/analysis tools: JAD AJH. Wrote the paper: AJH JAD CTF.

References

1. Carpenter SR, Caraco NF, Correll DL, Howarth RW, Sharpley AN, et al. (1998) Nonpoint pollution of surface waters with phosphorus and nitrogen. Ecological Applications 8: 559–568.
2. Winsten JR, Hunter M (2011) Using pay-for-performance conservation to address the challenges of the next farm bill. Journal of Soil and Water Conservation 66: 111A–117A.
3. Foley JA, DeFries R, Asner GP, Barford C, Bonan G, et al. (2005) Global Consequences of Land Use. Science 309: 570–574.
4. Heathcote AJ, Downing JA (2012) Impacts of Eutrophication on Carbon Burial in Freshwater Lakes in an Intensively Agricultural Landscape. Ecosystems 15: 60–70.
5. Tilman D, Cassman KG, Matson PA, Naylor R, Polasky S (2002) Agricultural sustainability and intensive production practices. Nature 418: 671–677.
6. Pavelis GA, Helms D, Stalcup S (2011) Soil and Water Conservation Expenditures by USDA Agencies, 1935–2010. United States Department of Agriculture.
7. Trimble SW, Lund SW (1982) Soil conservation and the reduction of erosion and sedimentation in the Coon Creek Basin, Wisconsin. Available from Br of Distr, USGS 604 S Pickett St, Alexandria, VA 22304, Price$ 2 75 Geological Survey Professional Paper 1234, 1982 35 p, 22 Fig, 7 Tab, 44 Ref, 3 Append.
8. Pimentel D, Harvey C, Resosudarmo P, Sinclair K, Kurz D, et al. (1995) Environmental and economic costs of soil erosion and conservation benefits. Science 267: 1117–1123.
9. Trimble SW, Crosson P (2000) US Soil Erosion Rates–Myth and Reality. Science 289: 248–250.
10. Wren DG, Davidson GR, Walker WG, Galicki SJ (2008) The evolution of an oxbow take in the Mississippi alluvial floodplain. Journal of Soil and Water Conservation 63: 129–135.
11. Wren DG, Davidson GR (2011) Using lake sedimentation rates to quantify the effectiveness of erosion control in watersheds. Journal of Soil and Water Conservation 66: 313–322.
12. Renwick WH, Andereck ZD (2006) Reservoir sedimentation trends in Ohio, USA: sediment delivery and response to land-use change. In: Rowan JSDRWWA, editor. Sediment Dynamics and the Hydromorphology of Fluvial Systems. 341–347.
13. Appleby PG, Oldfield F (1978) The calculation of lead-210 dates assuming a constant rate of supply of unsupported 210 Pb to the sediment. Catena 5: 1–8.
14. Dean WE (1974) Determination of carbonate and organic-matter in calcareous sediments and sedimentary-rocks by loss on ignition - comparison with other methods. Journal of Sedimentary Petrology 44: 242–248.
15. Davis MB (1976) Erosion rates and land-use history in southern Michigan. Environmental Conservation 3: 139–148.
16. Cleveland WS, Grosse E, Shyu WM (1992) Local regression models. Statistical models in S: 309–376.
17. Beach T (1994) The fate of eroded soil: Sediment sinks and sediment budgets of agrarian landscapes in southern Minnesota, 1851–1988. Annals of the Association of American Geographers 84: 5–28.
18. Sharpley AN, Chapra SC, Wedepohl R, Sims JT, Daniel TC, et al. (1994) Managing agricultural phosphorus for protection of surface waters: Issues and options. Journal of Environmental Quality 23: 437–451.
19. Jones JR, Bachmann RW (1978) Trophic status of Iowa lakes in relation to origin and glacial geology. Hydrobiologia 57: 267–273.
20. McCauley E, Kalff J (1981) Empirical relationships between phytoplankton and zooplankton biomass in lakes. Canadian Journal of Fisheries and Aquatic Sciences 38: 458–463.
21. Watson S, McCauley E (1988) Contrasting Patterns of Net- and Nanoplankton Production and Biomass among Lakes. Canadian Journal of Fisheries and Aquatic Sciences 45: 915–920.
22. Trimble SW (1981) Changes in sediment storage in the Coon Creek Basin, Driftless Area, Wisconsin, 1853 to 1975 Science 214: 181–183.
23. Engstrom DR, Thommes K, Balogh SJ, Swain EB, Post HA (1999) Trends in atmospheric mercury deposition across Minnesota: evidence from dated sediment cores from 50 Minnesota lakes. Legislative Commission on Minnesota Resources.

Quantifying Spatial Variability of Selected Soil Trace Elements and Their Scaling Relationships Using Multifractal Techniques

Fasheng Zhang[1,2]*, Guanghua Yin[1], Zhenying Wang[1], Neil McLaughlin[2], Xiaoyuan Geng[2], Zuoxin Liu[1]*

1 Institute of Applied Ecology, Chinese Academy of Sciences, Shenyang, China, **2** Eastern Cereal and Oilseed Research Centre, Agriculture and Agri-Food Canada, Ottawa, Canada

Abstract

Multifractal techniques were utilized to quantify the spatial variability of selected soil trace elements and their scaling relationships in a 10.24-ha agricultural field in northeast China. 1024 soil samples were collected from the field and available Fe, Mn, Cu and Zn were measured in each sample. Descriptive results showed that Mn deficiencies were widespread throughout the field while Fe and Zn deficiencies tended to occur in patches. By estimating single multifractal spectra, we found that available Fe, Cu and Zn in the study soils exhibited high spatial variability and the existence of anomalies ($[\alpha(q)_{max}-\alpha(q)_{min}]\geq 0.54$), whereas available Mn had a relatively uniform distribution ($[\alpha(q)_{max}-\alpha(q)_{min}]\approx 0.10$). The joint multifractal spectra revealed that the strong positive relationships ($r\geq 0.86$, $P<0.001$) among available Fe, Cu and Zn were all valid across a wider range of scales and over the full range of data values, whereas available Mn was weakly related to available Fe and Zn ($r\geq 0.18$, $P<0.01$) but not related to available Cu ($r=-0.03$, $P=0.40$). These results show that the variability and singularities of selected soil trace elements as well as their scaling relationships can be characterized by single and joint multifractal parameters. The findings presented in this study could be extended to predict selected soil trace elements at larger regional scales with the aid of geographic information systems.

Editor: Helmut Ahammer, Medical University of Graz, Austria

Funding: This work was funded by the National Natural Science Foundation of China (41201212), the National Food Production Project of China (2011BAD16B12) and the National Major Technology Program of China (2012BAD09B02). The funders had no role in study design, data collection and analysis, decision to publish, or preparation of the manuscript.

Competing Interests: The authors have declared that no competing interests exist.

* E-mail: fasheng.zhang@yahoo.com (FSZ); liuzuoxin@iae.ac.cn (ZXL)

Introduction

Although trace elements are needed by plants in only very small quantities [1,2], their deficiencies in soils may result in severe plant malnutrition as well as animal and human health problems [3,4]. Quantifying the spatial variability of soil trace elements enables prediction of trace elements at intermediate locations where measurements have not been made [5]. The thematic map produced by the prediction is critical to understanding, assessing and managing trace element-deficient soils [6].

Plant availability of trace elements in soils have been shown to have a high degree of spatial variability [7,8]. This arises mainly from the existence of anomalies in their distributions, which is a consequence of local extreme soil and weather conditions [3] such as alkalinity, soil infertility, erosion, drought, etc. From a practical standpoint, anomalies may carry valuable information for identifying specific trace element deficiencies. Including anomalies is essential when developing models that characterize the spatial variability of soil trace elements.

Under an assumption of intrinsic stationarity, the semivariogram has been the most common method for quantifying the spatial variability of soil trace elements [7−11]. In modern geostatistics, anomalies in a trace element dataset are usually transformed or eliminated in advance to avoid erratic behavior of a semivariogram [12]. As a second-order moment statistic, semivariogram cannot differentiate subtle variations in a trace element dataset. These facts reveal that current semivariogram models may not be able to fully characterize the spatial variability of soil trace elements, especially the singularities of anomalies.

Over a given area, plant availability of soil trace elements is commonly associated with similar soil and environmental conditions. An understanding of the relationships among soil trace elements can help in the analysis of the underlying processes controlling their spatial variability [7,8]. Thus, there is a need to examine the spatial variability of soil trace elements in a way that takes account of the relationships among them.

Factors and processes affecting soil properties including plant availability of trace elements operate at different spatial scales and influence these properties in different ways [9]. Relationships among soil trace elements may be heterogeneous across nested spatial scales. Although correlation analysis can quantify the relationship among soil trace elements at the measured scale, this relationship may not be valid at other spatial scales. As such, relationships among soil trace elements across a range of spatial scales, i.e., scaling relationships, are needed.

Over the past three decades, single multifractal spectrum has been applied in soil science to quantify the spatial variability of soil properties [13−18] including pH, organic matter content, cation exchange capacity, macro- and moderate-nutrient content, texture, structure, hydraulic properties, etc. The relationships among these properties and their associations with other environmental variables across spatial scales have also been investigated by researchers using a joint multifractal spectrum [18–20]. To date, however, studies examining spatial variability and singularity of soil trace elements and their scaling relationships using multifractal techniques are still rare [21]. Usually, the spatial distributions of pH, organic matter content, macro-nutrient content, texture, and other basic soil properties are relatively uniform. Geostatistical methods are able to characterize their spatial variability well and predict them with high accuracy. However, the distribution of soil trace elements is quite different from that of other soil properties, and is often characterized by high spatial variability and anomalies that are accompanied by singularities. As previously mentioned, semivariogram analysis and correlation analysis may not be able to completely characterize the singularity of soil trace elements and their scaling relationships. Panahi and Cheng pointed out that an exploratory dataset with anomalies may be more likely to follow a multifractal rather than a normal or lognormal distribution [22]. Meanwhile, the multifractal techniques have long been known to be able to characterize singularities and scaling relationships. Thus, the application of multifractal techniques in the analysis of soil spatial variability and their scaling relationships may be more appropriate for spatial analysis of soil trace elements rather than for other soil properties.

The objective of this study was therefore to examine whether the spatial distribution of selected soil trace elements in an agricultural field in northeast China exhibits a multifractal property, and if so, to quantify their spatial variability and scaling relationships using multifractal techniques. We also discuss the practical significance of the findings for prediction of selected soil trace elements.

Theory of Multifractal Analysis

Single Multifractal Spectrum

For a spatial dataset, there are several methods for estimating a single multifractal spectrum (e.g., the moment method, Rényi method). Here we used the moment method discussed by Agterberg [23], which is much easier to apply yet still provides accurate results.

Using this method, a normalized variable, $P_i(\varepsilon)$, is defined first according to:

$$P_i(\varepsilon) = \frac{\varphi_i(\varepsilon)}{\sum_{i=1}^{n} \varphi_i(\varepsilon)}$$

where $\varphi_i(\varepsilon)$ is the measured value of variable φ in the ith segment of scale ε in a plot, and n is the number of segments of scale ε over the whole plot. A partition function, $\chi_q(\varepsilon)$, which varies with the moment order q and scale ε, is then calculated as:

$$\chi_q(\varepsilon) = \sum_{i=1}^{n} [P_i(\varepsilon)]^q$$

For a spatial dataset, the power-law relationship between the partition function and scale is used to determine whether or not it exhibits a multifractal property. This power-law relationship can be confirmed if the logarithm of the partition function varies linearly against a logarithm of scale with a slope of the mass exponent, $\tau(q)$:

$$\log \chi_q(\varepsilon) \propto \tau(q) \cdot \log(\varepsilon)$$

By plotting $\log\chi_q(\varepsilon)$ against $\log\varepsilon$ at many moment orders q, we compute the generalized fractal dimension, $D(q)$:

$$D(q) = \frac{\tau(q)}{q-1} = \frac{1}{q-1} \times \lim_{\varepsilon \to 0} \frac{\log \chi_q(\varepsilon)}{\log \varepsilon}, \quad q \neq 1$$

$$D(q) \approx \lim_{\varepsilon \to 0} \frac{\sum_{i=1}^{n} [P_i(\varepsilon) \cdot \log P_i(\varepsilon)]}{\log \varepsilon}, \quad q = 1$$

Then the Hölder exponent (i.e., singularity index), $\alpha(q)$, with respect to variable φ, is computed by Legendre transformation:

$$\alpha(q) = \frac{d\tau(q)}{dq}$$

The fractal dimension, $f(q)$, with respect to variable φ, is given as:

$$f(q) = q\alpha(q) - \tau(q)$$

Plotting $f(q)$ against $\alpha(q)$ produces the so-called single multifractal spectrum. $\alpha(q)$ can be interpreted as an index of crowding of variable φ at the moment order q when ε approaches zero. $f(q)$ is the fractal dimension of the subsets of variable φ having specific $\alpha(q)$. Thus, the abundance of $\alpha(q)$ reflects the extent of spatial variability in variable φ. More $\alpha(q)$ (i.e., wider spectrum) represent greater spatial variability in distribution of variable φ and vice versa. In addition, with the amplification of high positive or negative q values, the spatial variability controlled by the higher or lower values of variable φ is separated from each other and represented by the $\alpha(q)$ in the left and right parts of the single multifractal spectrum, respectively. In particular, the $\alpha(q)_{max}$ and $\alpha(q)_{min}$ correspond to the singularity strength of the lowest and highest values of variable φ, respectively. The single multifractal spectrum therefore can detect a spatial pattern and separate anomalous variation from the background variation [13,23].

Chhabra and Jensen [24] proposed a method to avoid the difficult task of estimating $\alpha(q)$ by Legendre transformation. In the Chhabra and Jensen method, we define another normalized variable, $\delta_i(q,\varepsilon)$:

$$\delta_i(q,\varepsilon) = \frac{[P_i(\varepsilon)]^q}{\sum_{i=1}^{n} [P_i(\varepsilon)]^q}$$

Then the singularity index and fractal dimension are computed respectively as:

$$\alpha(q) = \lim_{\varepsilon \to 0} \frac{\sum_{i=1}^{n} [\delta_i(q,\varepsilon) \cdot \log P_i(\varepsilon)]}{\log \varepsilon}$$

$$f(q) = \lim_{\varepsilon \to 0} \frac{\sum_{i=1}^{n} [\delta_i(q,\varepsilon) \cdot \log \delta_i(q,\varepsilon)]}{\log \varepsilon}$$

Joint Multifractal Spectrum

For coexisting variables, we need to extend the single multifractal spectrum of each variable to a joint multifractal spectrum for two variables [25]. Two normalized variables, $P_i(\varepsilon)$ and $Q_i(\varepsilon)$, are first defined respectively for variables φ and μ. Then the joint normalized variable for variables φ and μ, $\delta_i(q,t,\varepsilon)$, is calculated as:

$$\delta_i(q,t,\varepsilon) = \frac{[P_i(\varepsilon)]^q [Q_i(\varepsilon)]^t}{\sum_{i=1}^{n} [P_i(\varepsilon)]^q [Q_i(\varepsilon)]^t}$$

where q and t are both moment orders and ε is the scale. The singularity indices, $\alpha(q,t)$ and $\beta(q,t)$, with respect to variables φ and μ, are computed respectively as:

$$\alpha(q,t) = \lim_{\varepsilon \to 0} \frac{\sum_{i=1}^{n} [\delta_i(q,t,\varepsilon) \cdot \log P_i(\varepsilon)]}{\log \varepsilon}$$

$$\beta(q,t) = \lim_{\varepsilon \to 0} \frac{\sum_{i=1}^{n} [\delta_i(q,t,\varepsilon) \cdot \log Q_i(\varepsilon)]}{\log \varepsilon}$$

Then the joint fractal dimension, $f(q,t)$, with respect to variables φ and μ, is given as:

$$f(q,t) = \lim_{\varepsilon \to 0} \frac{\sum_{i=1}^{n} [\delta_i(q,t,\varepsilon) \cdot \log \delta_i(q,t,\varepsilon)]}{\log \varepsilon}$$

Plotting $f(q,t)$ against $\alpha(q,t)$ and $\beta(q,t)$ produces the so-called joint multifractal spectrum, contours of which describe the scaling relationships between variables φ and μ. For positive scaling relationships, the joint distribution of higher $\alpha(q,t)$ and $\beta(q,t)$ values at the upper right part of the joint multifractal spectrum indicates the positive correlation between lower values of variables φ and μ, while the joint distribution of lower $\alpha(q,t)$ and $\beta(q,t)$ values at the lower left part indicates the positive correlation between higher values of variables φ and μ. For negative scaling relationships, higher $\alpha(q,t)$ values correspond to lower $\beta(q,t)$ values and vice versa at the diagonal parts of the joint multifractal spectrum, which indicates that lower and higher values of variables φ and μ are distributed together. If variables φ and μ are not correlated, there will be no correlation between $\alpha(q,t)$ and $\beta(q,t)$ [25,26].

Materials and Methods

Site and Soil

The study site is located in Fuxin County in northeast China (42°08′14″ N, 121°44′21″ E). The climate is typical of a warm temperate zone with continental monsoons, hot summers and cold winters. The mean annual temperature is 7.2 degrees Celsius with an average of 2868 hrs of annual sunshine. This region is a typical dryland farming area with an average annual precipitation of 480 mm, over 60% of which occurs between June and August (SWCP I soil moisture sensor & WITU-WS weather station, WITU Technologies, Inc., China).

The main agricultural soil in the region is cinnamon soil which develops through a combination of calcium carbonate leaching, illuviation and humification. It is characterized by a thin humus layer and medium or thick solum. Soil pH (H_2O) is 7.5−8.5 and soil organic matter content is 10.2 g·kg^{-1}. The cation exchange capacity is 19.2 meq/100 g, and mean concentrations of available Fe, Mn, Cu, and Zn are 9.9 mg·kg^{-1}, 36 mg·kg^{-1}, 1.32 mg·kg^{-1}, and 1 mg·kg^{-1}, respectively (The Second National Soil Survey of China, 1985).

Data Collection and Analysis

A total of 1024 soil samples were collected from a flat agricultural field of 10.24-ha (320 m by 320 m) in Matiyingzi Village, 10 km from the center of Fuxin County, in April 2008 (before annual planting). The soils were sampled over the 20-cm depth within a 10-m grid. The samples were taken with a stainless steel sampler and brought to the laboratory. After removing visible plant debris and air drying, soil samples were ground to pass through a 2-mm plastic sieve. Available Fe, Mn, Cu and Zn in each sample were measured by a DTPA (diethylene triamine pentaacetic acid) soil test [27]. The extractant consisted of 0.005 M DTPA, 0.1 M TEA (triethanolamine), and 0.01 M $CaCl_2$, with a pH of 7.3. Hooda and Alloway reported that DTPA method could provide accurate information about plant availability of trace elements in soils [28]. All trace element concentrations were determined by inductively coupled plasma mass spectrometry (Perkin Elmer Optima 3000). All necessary permits were obtained for the described field studies. The land user and owner, Mingfu Zhang, approved the field-work activities at each sampling location. The field examined in this study is not protected in any way, and the study did not involve any endangered or protected species.

A series of five scales were superimposed on the sample plot: 10-m pixel, 20-m pixel, 40-m pixel, 80-m pixel, and 160-m pixel with 1,024, 256, 64, 16 and 4 samples, respectively. The dataset for the minimum scale was the measured trace element data in the initial 10-m pixel grid, while the dataset for the other scales was the average of the measured trace element data of all initial 10-m pixel grids included within the larger pixel in the respective grid.

The single and joint multifractal spectra parameters for selected soil trace element data were calculated using the equations in the part of theory of multifractal analysis by an Office VBA project and were presented in graphs by SigmaPlot 10.0. The correlation analysis between $\alpha(q,t)$ and $\beta(q,t)$ was conducted using SPSS 16.0.

Results

Descriptive Statistics

A summary of descriptive statistics for selected trace elements in the study soils is presented in Table 1. Available Fe, Cu and Zn varied to a high degree with coefficient of variation (CV) of 82.7%, 41.7% and 89.5%, respectively. The CV is very sensitive to the anomalies in a dataset and high CV's for available Fe, Cu and Zn data suggest that there must be some significant anomalous values in their distributions. Available Mn had a relatively uniform distribution with a CV of only 9.5%. Based on the thresholds recommended by Liu [29], we found that Mn deficiency was widely present throughout the field. Although Fe and Zn deficiencies were not widespread, 25.5% of soil samples were deficient in available Fe, and 13.3% were deficient in available Zn. This indicated that Fe and Zn deficiencies tended to occur in

Table 1. Descriptive statistics for selected soil trace elements in the study agricultural field of northeast China.

Trace elements	Min† ($mg \cdot kg^{-1}$)	Max ($mg \cdot kg^{-1}$)	Mean ($mg \cdot kg^{-1}$)	CV	Deficiency‡
Fe	0.30	73.73	18.93	82.7%	25.5%
Mn	25.53	91.11	69.27	9.5%	100%
Cu	0.18	5.27	1.34	41.7%	1.4%
Zn	0.02	14.81	2.48	89.5%	13.3%

†Min, Max and CV are abbreviations for minimum, maximum and coefficient of variation.
‡Critical levels for available Fe, Mn, Cu and Zn recommended by Liu [29] are 4.5 $mg \cdot kg^{-1}$, 100 $mg \cdot kg^{-1}$, 0.3 $mg \cdot kg^{-1}$ and 1.0 $mg \cdot kg^{-1}$ (DTPA extracted). The concentrations of available Fe, Mn, Cu and Zn under the critical levels are recognized as deficiencies.

patches rather than throughout the field. Only a very limited percentage of soil samples were deficient in available Cu.

Empirical Tests of Multifractality

Scatter plots and their fitted relationships between the logarithm of partition function and the logarithm of superimposed scale for selected soil trace elements are shown in Figure 1. The coefficients of determination R^2 for all linear regression lines were larger than 0.99 (F-test, $P<0.001$) which confirms the assumption of the existence of the power-law relationship between partition function and scale. This finding revealed that the spatial distribution of available Fe, Mn, Cu and Zn in the study soils exhibited a multifractal property. Hence, multifractal techniques can be used to characterize the spatial variability of selected soil trace elements and their scaling relationships.

Variables showing a multifractal property usually have a $D(q)$ that demonstrates a monotonic and nonlinear trend with q values. This pattern can be observed when examining the trend of $D(q)$ for available Fe, Cu and Zn in Figure 2. The $D(q)$ values for available Fe, Cu and Zn decreased markedly as q values increased. This is a typical characteristic of a multifractal system, and also indicated that the multifractal property observed here was strong. In contrast, the $D(q)$ values for available Mn decreased very slowly against q values and showed small deviations from the capacity dimension of the sample plot (i.e., 2). This revealed that the distribution of available Mn can be represented by a monofractal or a weak multifractal. Note that a monofractal is a special case of a multifractal.

Variability and Singularity of Soil Trace Elements

To examine the spatial variability of selected soil trace elements, their single multifractal spectra were constructed and shown in Figure 3. The single multifractal spectra for available Fe, Cu and Zn all had large widths ($[\alpha(q)_{max}-\alpha(q)_{min}]\geq 0.54$). This indicated that available Fe, Cu and Zn had quite high spatial variability in studied soils. To characterize such high spatial variability of available Fe, Cu and Zn, a large number of singularity indices ranging from 1.69 to 2.83, 1.81 to 2.35, and 1.60 to 2.38

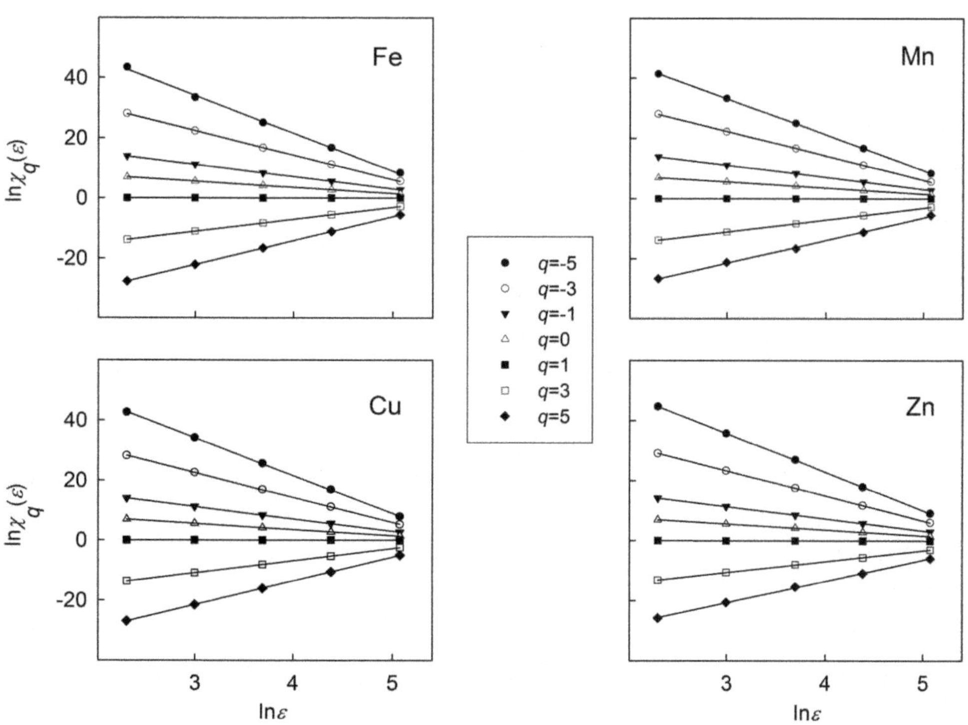

Figure 1. Logarithms of partition functions of selected soil trace elements plotted against scales. The R^2 values of the linear fit for all q considered were always greater than 0.99.

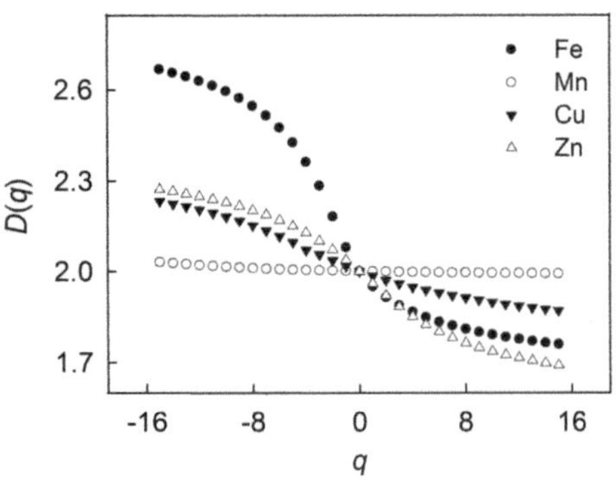

Figure 2. General fractal dimension $D(q)$ for selected soil trace elements at −15 to 15 range of q evaluated at 1 increment.

respectively were needed. In contrast, the single multifractal spectrum of available Mn had a very narrow width ($[\alpha(q)_{max} - \alpha(q)_{min}] \approx 0.10$). This indicated that the spatial variability of available Mn was relatively low. Such low spatial variability can be characterized by a very small set of singularity indices, ranging from 1.99 to 2.09.

The single multifractal spectra of available Fe, Mn, Cu and Zn all had asymmetric tails on either the right or left part (Figure 3), which is another typical characteristic of a multifractal property. The left and right parts of a single multifractal spectrum respectively represent the higher and lower values of a dataset, emphasized by high positive and negative q values. The long left and right tails of the single multifractal spectra for available Fe, Cu and Zn were therefore the result of both extremely high and low values in their distributions. The singularities of these anomalies were characterized by the corresponding singularity indices in the single multifractal spectra of available Fe, Cu and Zn. The single multifractal spectrum of available Mn had a relatively long tail only on the right part. This indicated that there were no anomalies but just some relatively low values in the distribution of available Mn in the study soils.

Scaling Relationships among Soil Trace Elements

The relationships among selected soil trace elements across the spatial scales superimposed on the sample plot were analyzed using the joint multifractal spectra (Figure 4). In general, there were strong positive scaling relationships among available Fe, Cu and Zn, which were evident from the compact joint distribution of their singularity indices at the upper right and lower left parts of

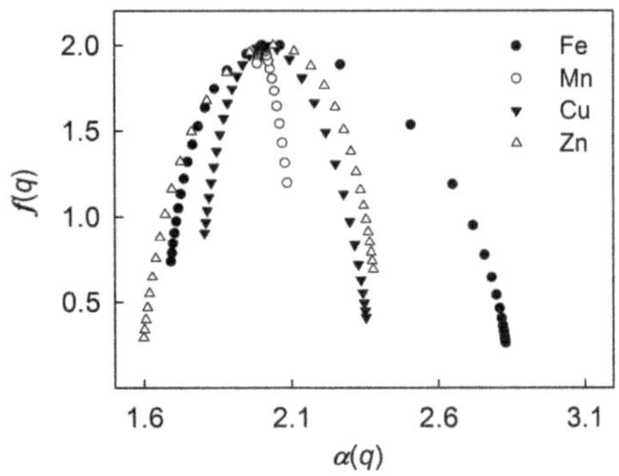

Figure 3. Single multifractal spectrum for selected soil trace elements at −15 to 15 range of q evaluated at 1 increment.

their joint multifractal spectra (Figures 4A, 4B and 4C). The contours in Figure 4B were tightly concentrated in one bundle. This indicated that there were strong positive relationships between available Fe and Zn for both higher and lower values across the superimposed spatial scales. With a small diagonal stretch at the upper right, Figure 4A and Figure 4C had a similar pattern of joint distribution of singularity indices as Figure 4B. This indicated that strong positive relationships also existed between available Cu and available Fe and Zn over a wider spatial scale and for the entire range of data values. In Figures 4D, 4E and 4F, the wide stretch of diagonal contours showed that the singularity indices of available Mn was not correlated tightly with that of available Fe, Cu and Zn. This suggested that available Mn may not be associated, or may be weakly associated, with available Fe, Cu and Zn over the range of measured concentrations and superimposed spatial scales.

To further confirm the scaling relationships among selected soil trace elements, the Pearson correlation coefficients (r) between their singularity indices were analyzed (Table 2). The singularity indices between available Fe and Zn had the highest positive correlation coefficient ($r = 0.98$, $P < 0.001$). The positive correlation coefficients between the singularity indices of available Fe versus available Cu and available Zn versus available Cu were relatively high as well ($r = 0.87$ and 0.86, $P < 0.001$). These results indicated that the positive correlation relationships among available Fe, Cu and Zn were all very strong across superimposed spatial scales. Although the associations between singularity indices of available Fe versus available Mn and available Zn versus available Mn were also significant, their correlation coefficients ($r = 0.19$ and 0.18, $P < 0.01$) were relatively low compared to those among available Fe, Cu and Zn. This suggested the positive correlations between available Mn and available Fe and Zn were weak. The correlation coefficient between singularity indices of available Cu and Mn was not significant ($r = -0.03$, $P = 0.40$), which implied that available Cu and Mn were not correlated across superimposed spatial scales. The strong scaling relationships among available Fe, Cu and Zn were likely due to a common controlling process, which was quite different from that for available Mn.

Discussion

The influences of factors in the CLORPT function or the SCORPAN model on the variations of selected soil trace elements should be relatively consistent over a flat 10.24-ha level field [30,31]. In most soils, available Fe, Mn, Cu and Zn are not directly related to their total contents, but are more related to other soil characteristics such as pH, redox potential, organic matter content, cation exchange capacity, and texture [32]. In agricultural soils, long-term intensive agricultural practices also add another layer of complexity on the distribution of available Fe, Mn, Cu and Zn. Kravchenko et al. [13], Perrier and Bird [14], Caniego et al. [15], Guan et al. [16] and Perrier et al. [33] have reported that pH, redox potential, organic matter content, cation exchange capacity, texture and structure interact with each other and are prone to lead to a nonlinear soil system in which multifractal properties are expressed. The multifractal nature of available Fe, Mn, Cu and Zn observed in this study can be attributed to the complexity of this system.

The spatial variability of selected soil trace elements characterized by single multifractal spectrum was partly consistent with the result of CV analysis. However, the results based on the single multifractal spectrum were obtained over a wider range of spatial scales and moment orders and better reflected the true spatial variability of selected soil trace elements. The single multifractal spectrum also captured the detailed singularity strength of anomalies in available Fe, Mn, Cu and Zn data while the joint multifractal spectrum characterized the scaling relationships among them. These two points demonstrate that multifractal techniques present unique advantages over CV and semivariogram analyses.

Knowledge of soil spatial variability is a prerequisite needed to predict soil properties at unsampled locations. In this study, the spatial variability and singularity of selected soil trace elements quantified by the parameters of single multifractal spectra would have practical significance for prediction of these trace elements. Appropriate prediction procedures for variables with high spatial variability and a large amount of singularity indices are different than that for those with low spatial variability and limited singularity indices. This implication has been summarized in the study by Cheng [34] and justified by Cheng [35] and Xie et al. [36]. In our study, available Mn had a narrow multifractal spectrum with a limited number of singularity indices. Its spatial variability was very low and there were no anomalies within its spatial distribution. For a variable with those distribution characteritics and spatial autocorrelation, a simple prediction procedure such as an ordinary Kriging interpolation may be appropriate. In contrast, available Fe, Cu and Zn in this study's soils had wide multifractal spectra with many singularity indices both at $\alpha(q)>2$ and $\alpha(q)<2$. Their spatial variability was high with both extremely high and extremely low values in their spatial distributions. Under these circumstances, it is possible, but difficult, to apply a prediction procedure. For example, $\alpha(q)>2$ implied that there were some specific locations with extremely low concentrations of available Fe, Cu and Zn. An ordinary interpolation will smooth out the extremely low values of available Fe, Cu and Zn, resulting in the loss of valuable information for identifying Fe, Cu and Zn deficiencies. In this case, incorporating singularity indices $\alpha(q)>2$ into the prediction procedure can help preserve the extremely low values of available Fe, Cu and Zn. Similarly, $\alpha(q)<2$ is also needed when predicting available Fe, Cu and Zn with extremely high values at specific locations.

Table 2. Pearson correlation coefficients (r) between singularity indices of joint multifractal spectra for selected soil trace elements.

	Fe	Mn	Cu
Mn	0.19 ($P<0.01$)		
Cu	0.87 ($P<0.001$)	−0.03 ($P=0.40$)	
Zn	0.98 ($P<0.001$)	0.18 ($P<0.01$)	0.86 ($P<0.001$)

Recent studies have delineated the spatial distribution of selected trace elements in soils using their relationships with environmental data, but those relationships were observed at only a single scale and may not be valid at other scales. Across a regional soil landscape, factors and processes impacting the plant availability of soil trace elements are expected to be much more complicated and heterogeneous than that of the 10.24-ha agricultural field observed in this study. The major controlling processes for different soil trace elements may differ at specific spatial scales, which would further impact the spatial variability of soil trace elements. The results of joint multifractal analysis in this study had demonstrated the ability of joint multifractal spectrum in characterizing the scaling relationships among selected soil trace elements at the field scale. It is well-known that fractals and multifractals are ubiquitous in complex ecosystems [37]. Further study is needed to evaluate the technique in modeling the scaling relationships of soil trace elements at larger regional scales where their variability in relation to the local environments would be expected. Using scale-specific relationships future researchers should be able to improve models when predicting soil trace elements with auxiliary environmental data.

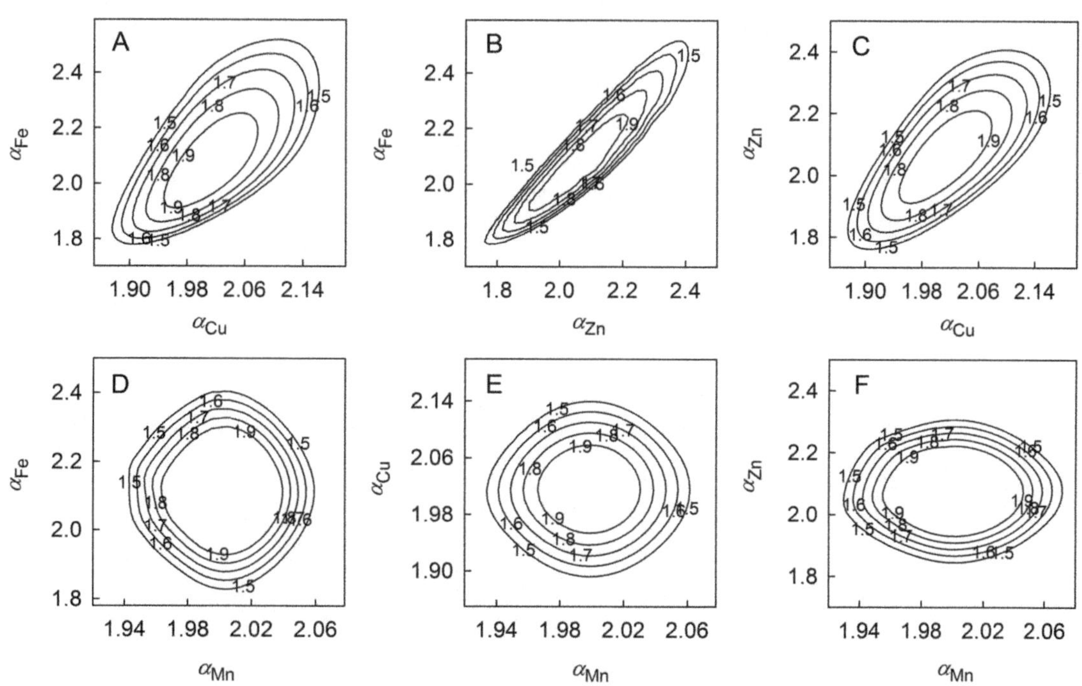

Figure 4. Joint multifractal spectrum for selected soil trace elements at −15 to 15 range of q evaluated at 1 increment.

In recent years, geographic information systems (GIS) have facilitated the process of spatial analysis and computation. If further research demonstrates that the multifractal phenomenon exists over larger regional scales, the findings in this study could be incorporated into GIS to improve the prediction of selected trace elements over a range of scales for optimizing soil use and management.

Conclusions

Spatial variability of available Fe, Mn, Cu and Zn and their scaling relationships were examined using multifractal techniques for a 10.24-ha agricultural field in northeast China. The distribution of selected soil trace elements exhibited a multifractal property. The variability and singularities of available Fe, Cu and Zn in the study soils needed to be quantified by multifractal spectra, whereas that of available Mn can be characterized by a monofractal or a weak multifractal spectrum. The joint multifractal spectra and the Pearson correlation analysis of their singularity indices captured the relationships among selected soil trace elements across a wider range of spatial scale and over the full range of data values.

The use of single and joint multifractal parameters has potential to enhance the prediction of selected soil trace elements in future research. We are currently examining the multifractality of soil trace elements at the county scale and are building a model that incorporates the scaling relationships between selected soil trace elements and environments into a prediction procedure. Such studies are essential to validate the applicability and extension of the findings presented here.

Acknowledgments

We thank Guixian Li for providing assistance in carrying out the field experiment.

Author Contributions

Conceived and designed the experiments: FSZ ZXL. Performed the experiments: FSZ GHY ZYW. Analyzed the data: FSZ ZYW. Contributed reagents/materials/analysis tools: FSZ. Wrote the paper: FSZ NM XYG ZXL.

References

1. Havlin JL, Beaton JD, Tisdale SL, Nelson WL (2005) Soil fertility and fertilizers: An introduction to nutrient management. 7th Edition. Upper Saddle River, NJ:Prentice Hall. 515 p.
2. Hood PS (2010) Trace elements in soils. Chichester, UK:Wiley-Blackwell. 616 p.
3. Alloway BJ (2008) Micronutrient deficiencies in global crop production. Dordrecht, The Netherlands:Springer. 353 p.
4. Gupta UC, Wu KN, Liang SY (2008) Micronutrients in soils, crops, and livestock. Earth Science Frontiers 15: 110–125.
5. Heuvelink GBM, Webster R (2001) Modelling soil variation: Past, present and future. Geoderma 100: 269–301.
6. White JG, Zasoski RJ (1999) Mapping soil micronutrients. Field Crops Research 60: 11–26.
7. Wu CF, Luo YM, Zhang LM (2010) Variability of copper availability in paddy fields in relation to selected soil properties in southeast China. Geoderma 156: 200–206.
8. Behera SK, Singh MV, Singh KN, Todwal S (2011) Distribution variability of total and extractable zinc in cultivated acid soils of India and their relationship with some selected soil properties. Geoderma 162: 242–250.
9. Goovaerts P (1999) Geostatistics in soil science: State-of-the-art and perspectives. Geoderma 89: 1–45.
10. Shi J, Xu J, Huang P (2008) Spatial variability and evaluation of status of micronutrients in selected soils around Taihu Lake, China. Journal of Soils and Sediments 8: 415–423.
11. Liu XM, Xu JM, Zhang MH, Si BC, Zhao KL (2008) Spatial variability of soil available Zn and Cu in paddy rice fields of China. Environmental Geology 55: 1569–1576.
12. Gringarten E, Deutsch CV (2001) Teacher's aide: Variogram interpretation and modeling. Mathematical Geology 33: 507–534.
13. Kravchenko AN, Boast CW, Bullock DG (1999) Multifractal analysis of soil spatial variability. Agronomy Journal 91: 1033–1041.
14. Perrier EMA, Bird NRA (2002) Modelling soil fragmentation: the pore solid fractal approach. Soil and Tillage Research 64: 91–99.
15. Caniego FJ, Espejo R, Martín MA, José FS (2005) Multifractal scaling of soil spatial variability. Ecological Modelling 182: 291–303.
16. Guan XY, Yang PL, Ren SM, Li YK (2007) Multifractal analysis of soil structure under long-term wastewater irrigation based on digital image technology. New Zealand Journal of Agricultural Research 50: 789–796.
17. Wang ZY, Shu QS, Liu ZX, Si BC (2009) Scaling analysis of soil water retention parameters and physical properties of a Chinese agricultural soil. Australian Journal of Soil Research 47: 821–827.
18. Biswas A, Cresswell HP, Si BC (2012) Application of multifractal and joint multifractal analysis in examining soil spatial variation: A Review. In: Ouadfeul SA, editor. Fractal analysis and chaos in geosciences. InTech. pp. 109–138.
19. Zeleke TB, Si BC (2005) Scaling relationships between saturated hydraulic conductivity and soil physical properties. Soil Science Society of America Journal 69: 1691–1702.
20. Wang ZY, Shu QS, Xie LY, Liu ZX, Si BC (2011) Joint multifractal analysis of scaling relationships between soil water-retention parameters and soil texture. Pedosphere 21: 373–379.
21. Jesus AP, Mateus A, Gonçalves MA, Munhá J (2013) Multi-fractal modelling and spatial Cu-soil anomaly analysis along the southern border of the Iberian Terrane in Portugal. Journal of Geochemical Exploration 126–127: 23–44.
22. Panahi A, Cheng Q (2004) Multifractality as a measure of spatial distribution of geochemical patterns. Mathematical Geology 36: 827–846.
23. Agterberg FP (1995) Multifractal modeling of the sizes and grades of giant and supergiant deposits. International Geology Review 37: 1–8.
24. Chhabra A, Jensen RV (1989) Direct determination of the $f(\alpha)$ singularity spectrum. Physical Review Letters 62: 1327–1330.
25. Meneveau C, Sreenivasan KR, Kailasnath P, Fan MS (1990) Joint multifractal measures: Theory and applications to turbulence. Physical Review A 41: 894–913.
26. Kravchenko AN, Bullock DG, Boast CW (2000) Joint multifractal analysis of crop yield and terrain slope. Agronomy Journal 92: 1279–1290.
27. Lindsay WL, Norvell WA (1978) Development of a DTPA soil test for zinc, iron, manganese, and copper. Soil Science Society of America Journal 42: 421–428.
28. Hooda PS, Alloway BJ (1994) The plant availability and DTPA extractability of trace metals in sludge-amended soils. Science of the Total Environment 149: 39–51.
29. Liu Z (1996) Microelements in soils of China. Nanjing:Jiangsu Sci. and Technol. Publishing House. 396 p. (in Chinese).
30. Jenny H (1941) Factors of soil formation: A system of quantitative pedology. New York: McGraw-Hill. 281 p.
31. McBratney AB, Mendonça-Santos ML, Minasny B (2003) On digital soil mapping. Geoderma 117: 3–52.
32. Fageria NK, Baligar C, Clark RB (2002) Micronutrients in crop production. Advances in Agronomy 77: 185–268.
33. Perrier E, Bird N, Rieutord T (2010) Percolation properties of 3-D multiscale pore networks: how connectivity controls soil filtration processes. Biogeosciences Discussion 7: 2997–3018.
34. Cheng Q (2005) A new model for incorporating spatial association and singularity in interpolation of exploratory data. In: Leuangthong O, Deutsch CV, editors. Geostatistics Banff, 2004, Quantitative geology and geostatistics. Dordrecht, The Netherlands:Springer. pp. 1017–1025.
35. Cheng Q (2008) Modeling local scaling properties for multiscale mapping. Vadose Zone Journal 7: 525–532.
36. Xie SY, Cheng Q, Xing XT, Bao ZY, Chen ZJ (2010) Geochemical multifractal distribution patterns in sediments from ordered streams. Geoderma 160: 36–46.
37. Solé RV, Bascompte J (2006) Self-organization in complex ecosystems. Princeton:Princeton University Press. 392 p.

18

A Drought Resistance-Promoting Microbiome Is Selected by Root System under Desert Farming

Ramona Marasco[1], Eleonora Rolli[1], Besma Ettoumi[2], Gianpiero Vigani[3], Francesca Mapelli[1], Sara Borin[1], Ayman F. Abou-Hadid[4], Usama A. El-Behairy[4], Claudia Sorlini[1], Ameur Cherif[2], Graziano Zocchi[3], Daniele Daffonchio[1]*

1 Dipartimento di Scienze per gli Alimenti, la Nutrizione e l'Ambiente, Università degli Studi di Milano, Milan, Italy, **2** Laboratoire Microorganismes et Biomolécules Actives, Université Tunis El Manar, Tunis, Tunisia and Laboratoire Biotechnologie et Valorisation des Bio-Géo Ressources, Institut Supérieur de Biotechnologie, Université de La Manouba, Sidi Thabet, Ariana, Tunisia, **3** Dipartimento di Scienze Agrarie e Alimentari- Produzione, Territorio, Agroenergia; Università degli Studi di Milano, Milan, Italy, **4** Department of Horticulture, Ain Shams University, Cairo, Egypt

Abstract

Background: Traditional agro-systems in arid areas are a bulwark for preserving soil stability and fertility, in the sight of "reverse desertification". Nevertheless, the impact of desert farming practices on the diversity and abundance of the plant associated microbiome is poorly characterized, including its functional role in supporting plant development under drought stress.

Methodology/Principal Findings: We assessed the structure of the microbiome associated to the drought-sensitive pepper plant (*Capsicum annuum* L.) cultivated in a traditional Egyptian farm, focusing on microbe contribution to a crucial ecosystem service, i.e. plant growth under water deficit. The root system was dissected by sampling root/soil with a different degree of association to the plant: the endosphere, the rhizosphere and the root surrounding soil that were compared to the uncultivated soil. Bacterial community structure and diversity, determined by using Denaturing Gradient Gel Electrophoresis, differed according to the microhabitat, indicating a selective pressure determined by the plant activity. Similarly, culturable bacteria genera showed different distribution in the three root system fractions. *Bacillus* spp. (68% of the isolates) were mainly recovered from the endosphere, while rhizosphere and the root surrounding soil fractions were dominated by *Klebsiella* spp. (61% and 44% respectively). Most of the isolates (95%) presented *in vitro* multiple plant growth promoting (PGP) activities and stress resistance capabilities, but their distribution was different among the root system fractions analyzed, with enhanced abilities for *Bacillus* and the rhizobacteria strains. We show that the *C. annuum* rhizosphere under desert farming enriched populations of PGP bacteria capable of enhancing plant photosynthetic activity and biomass synthesis (up to 40%) under drought stress.

Conclusions/Significance: Crop cultivation provides critical ecosystem services in arid lands with the plant root system acting as a "resource island" able to attract and select microbial communities endowed with multiple PGP traits that sustain plant development under water limiting conditions.

Editor: Jack Anthony Gilbert, Argonne National Laboratory, United States of America

Funding: This work was financially supported by the FIRB project n° RBIN047MBH "Strategy to improve crop productivity under water stress" of the Italian Ministry of University and Research, the European Union in the ambit of project BIODESERT (European Community's Seventh Framework Programme CSA-SA REGPOT-2008-2 under grant agreement n° 245746). ER and FM are supported by Università degli Studi di Milano, European Social Found and Regione Lombardia (contract "Dote Ricerca"). The funders had no role in study design, data collection and analysis, decision to publish, or preparation of the manuscript.

Competing Interests: The authors have declared that no competing interests exist.

* E-mail: daniele.daffonchio@unimi.it

Introduction

The "reverse desertification" includes a series of interventions aimed to sustain soil stability and productivity in arid lands, providing tools and strategies to support crop production for human feeding while preserving biodiversity and counteracting climate changes. Desert farming represents a strategy to protect soil fertility and aims at gaining arable land at expenses of desert soil, subjected to low resources landscape [1]. Traditional and more technologically efficient desert farming systems are well established in North Africa and their spread represents an impellent necessity to provide food for the increasing world population that will rapidly reach 9 billion people in few decades [2]. Desert farming primarily relies on irrigation in an ecosystem where water is a limiting and often polluted resource. Water stress is a primary cause of crop losses, reducing average yields by more than 50% [3]. Such a decrease in productivity is attributable to a direct negative effect of water scarcity on plant physiology. Despite the recognized importance of root associated microorganisms for plant growth and health, few studies are available on how desert farming affects the diversity of the crop associated-microbiome and whether the selected microorganisms still retain plant growth abilities to sustain plant development under water limiting conditions [4]. In particular, it is poorly explored whether desert farming may promote the selection of microbes capable of enhancing a key primary ecosystem service like plant tolerance to drought.

In the family *Solanaceae*, *Capsicum annum* L. is one of the horticulture plants most sensitive to water stress [5,6]. Pepper has great economic, agricultural and food relevance, and despite it is largely cultivated where climatic conditions are generally characterized by high temperatures and scarce water availability [7], it requires a relatively high water supply during the whole crop life cycle to obtain high yield productivity [5,8,9,10,11]. Pepper has gained the role of a model plant in physiology studies, like those conducted on the effects that plant growth promoting (PGP) bacteria have in increasing the plant resistance to stress conditions such as salinity [12,13,14,15,16]. Nevertheless, little information is available either about the distribution and diversity of the autochthonous PGP microbiome of pepper cultivated in arid lands, or the potential of the associated PGP bacteria in directly promoting plant development through a stimulation of plant drought tolerance.

Therefore, this study is aimed to assess the impact of desert farming on plant-microbe association in pepper cultivated in arid conditions. We aimed to assess the diversity and topological repartition of bacteria in the pepper root system grown under desert farming and investigate whether under such a crop management practice the root system enriches bacteria capable of supporting the plant resistance to drought and water stress.

With this aim we adopted both culture-independent and -dependent approaches. Cluster analysis was applied to DGGE (Denaturing Gradient Gel Electrophoresis) to dissect the structure and the composition of the microbiome associated to pepper endosphere, rhizosphere and root surrounding soil in comparison to unvegetated soil (bulk). A large collection of isolates from different fractions of the plant root system was established and screened *in vitro* for PGP activities. The rhizo-competence of the bacterial strains was evaluated through an adhesion assay on both *Arabidopsis thaliana* and pepper rhizoplane. Finally we assessed the capacity of selected strains to support plant growth under water deficiency.

We demonstrated that the application of desert greening techniques in arid lands generate hotspots of microbial diversity in the rhizosphere of plants. These techniques include a virtuous use of water for irrigation, field fertilization with organic fertilizers originating from residues of crops and animal manure and other similar traditional agricultural management practises. Furthermore we documented that plant rhizosphere and endosphere are repository for selected and specialized microbial populations, able to promote plant growth under drought. Thus, desert farming hampers desertification by establishing fertility islands and allows to achieve crop yields despite the adverse environmental conditions.

Results

Variability of the Bacterial Community Structure as Revealed by Community Fingerprinting

A 16S rRNA gene PCR-DGGE analysis was performed to explore the structure of the microbial communities associated to the pepper root system. The rhizosphere (R), composed of the soil particles tightly adhering to the rhizoplane, the root surrounding soil (S), composed of the soil particles not attached to the root system, and surface sterilized root tissues (E, endosphere) were compared to the non cultivated soil (B, bulk soil) (Fig. S1 and 1). While all soil fractions resulted inhabited by a complex microbiome, represented by a multiple band pattern, the pepper endosphere was represented by a restricted community (Fig. S1). Cluster analysis of the DGGE band profiles revealed a sharp difference in the microbial community structure associated to the different fractions (Fig. 1). The composition of the microbiome associated to the soil fractions R and S hosting the plant clearly differed from the arid root-free soil, indicating that farming practices profoundly affect soil microbiome structure (Fig. 1). A rhizosphere effect could be also observed since the closeness of the root tissues determined a change of bacterial community structure in the R samples respect to the S samples. The pepper endosphere resulted rather different from the soil-borne fractions by approximately 50% of the detected bands, indicating a strong selection pressure determined by the plant tissues (Fig. 1).

The dominant taxa associated with the PCR-DGGE profiles were identified by partial 16S rRNA band sequencing and their prevalence in the pepper root system and the non-cultivated arid soil was determined (Table 1). The major taxa associated to the pepper root system were affiliated to *Actinobacteria*, *Bacilli*, *Alpha*, *Beta* and *Gammaproteobacteria*. A certain taxa specificity was associated to the different fractions of the root system (Table 1). R and S fractions were dominated by *Proteobacteria* and spore forming bacteria of the genus *Bacillus* and related genera. *Actinobacteria* were retrieved only associated to plant root tissues and uncultivated root-free arid soil (Table 1). A differential repartition between the fractions was also observed for some *Proteobacteria*: *Thiobacillus* sp. was found only in the bulk soil, while some *Pseudoxanthomonas* sp. were typical of the endosphere fraction (Table 1).

Quantitative Analysis of Bacterial Abundance

Statistically higher microbial counts were recorded for the culturable bacteria associated to R fraction in both R2A and KB media [$(5.13\pm3.44)\times10^9$ and $(1.28)\times10^8$ CFU g^{-1} fresh weight, respectively] in comparison to the non-cultivated arid soil [$(1.28\pm0.72)\times10^8$ and $(3.74\pm2.64)\times10^7$ CFU g^{-1} fresh weight,

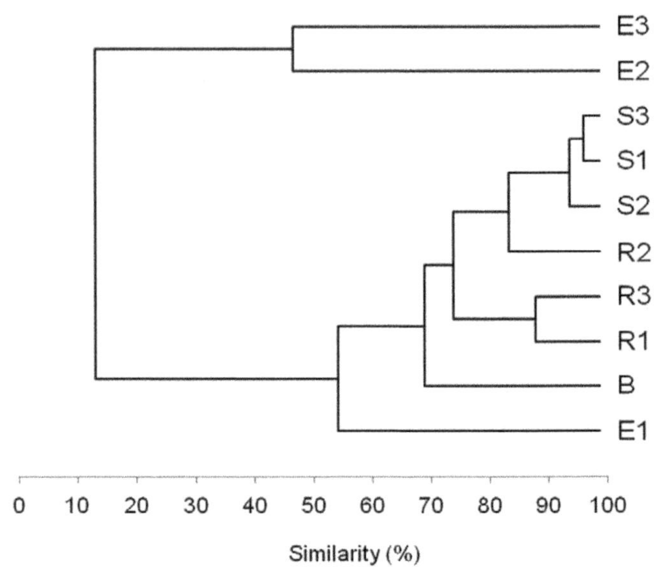

Figure 1. Cluster analysis of total microbial communities according to 16S rRNA DGGE profiles. The cluster analysis of the plot line was obtained from 16S rRNA PCR-DGGE bacterial community profiles, according to Pearson correlation. The analyzed fractions were root tissues (E), rhizosphere (R), root-surrounding soil (S) and bulk soil (B) of three replicate plants of pepper.

Table 1. Phylogenetic identification and distribution of bacteria excised and sequenced from DGGE bands.

Band	Class (RDP)	Closest Relative (accession number)	%	Closest type strain or described cultivable strain (accession number)	%	Fraction of the pepper root system									
						E1	E2	E3	R1	R2	R3	S1	S2	S3	B
4	*Actinobacteria*	*Microbacterium phyllosphaerae* (HM355641)	100	*Microbacterium phyllosphaerae* (HM355641)	100	**X**									
22	*Actinobacteria*	Uncultured Bacterium (DQ129271)	97	*Nocardioides mesophilus* (EF466117)	97										**X**
1	*Bacilli*	Uncultured Bacterium (FN563205)	98	*Paenibacillus taichungensis* (FJ944633)	95	**X**	**X**		X			X	X	X	
8				*Paenibacillus gansuensis* (NR_043219)	95										
3		Uncultured Bacterium (FN563205)	97	*Paenibacillus motobuensis* (EU982886)	94			X	X	X	**X**	X	X	X	X
16	*Bacilli*	Uncultured *Paenibacillus* sp. (JX501979)	99	*Paenibacillus chitinolyticus* (AB680938)	95					**X**	X	X	X	**X**	X
20			98	*Paenibacillus chitinolyticus* (FJ944683)	96										
21	*Bacilli*	*Bacillus megaterium* (JX312633)	98	*Bacillus megaterium* (JX312633)	98	X			X	X	X	X	X	X	**X**
2	*Alphaproteobacteria*	Unclassified *Alphaproteobacterium* (AY162055)	98	*Sphingobium ummariense* (NR_044171)	98	**X**			X	X	X	X	X	X	
11	*Betaproteobacteria*	Uncultured Bacterium (HQ272664)	98	*Thiobacillus thioparus* (HM173633)	97	X	X		**X**	X	X	X	X	X	X
19	*Betaproteobacteria*	*Thiobacillus denitrificano* (EU546130)	98	*Thiobacillus denitrificano* (EU546130)	98				X	X	X	X	X	**X**	X
5	*Gammaproteobacteria*	*Rhodanobacter lindaniclasticus* (AB245366)	100	*Rhodanobacter lindaniclasticus* (AB245366)	100	**X**									
13	*Gammaproteobacteria*	*Rhodanobacter lindaniclasticus* (AB245366)	97	*Rhodanobacter lindaniclasticus* (AB245366)	97				**X**	X	X	**X**	X		
17	*Gammaproteobacteria*		99		99										
6	*Gammaproteobacteria*	*Pseudoxanthomonas ginseng isoli* (JF778717)	99	*Pseudoxanthomonas ginseng isoli* (JF778717)	99	**X**	X								
10		*Pseudoxanthomonas ginseng isoli* (JF778717)	99	*Pseudoxanthomonas ginseng isoli* (JF778717)	99				**X**	**X**	X		X	X	
15			99		99										
7	*Gammaproteobacteria*	*Pseudoxanthomonas ginseng isoli* (JF778717)	99	*Pseudoxanthomonas ginseng isoli* (JF778717)	99	**X**	X								
9	*Gammaproteobacteria*	*Dyella yeojuensis* (FN796854)	99	*Dyella yeojuensis* (FN796854)	99			**X**							
12	*Gammaproteobacteria*	Uncultured *Thermomonas* sp. (EF072902)	99	*Lysobacter pocheonensis* (EU273938)	98				**X**	X	X	X	X		
14	*Gammaproteobacteria*	Uncultured *Thermomonas* sp. (EF072902)	99	*Lysobacter pocheonensis* (EU273938)	98				**X**	X	X	X	**X**	X	
18			98		97										

Identification of the dominant bands in the PCR-DGGE fingerprinting profiles (marked in Fig. S1) and their distribution in the different fractions of the pepper root system. The codes of the different fractions of the pepper root systems are as follow: E, Endosphere; R, rhizosphere; S, root-surrounding soil; B, non-cultivated root-free arid soil. The numbers following the codes indicate the different replicates.
X: presence of the band in the DGGE profile of the indicated fraction; in bold are indicated the bands that were actually sequenced.
Sequences of bands with the same mobility in the DGGE gel are reported in the same white/grey boxes. In some cases the different bands showed slightly different sequences whit few nucleotide variations. When the variation resulted within the 3% divergence on the 16S rRNA sequence, the bands where assumed to belong to the same OTU at the 97% identity threshold, as evaluated using DOTUR [68].

respectively] as shown in Table 2. While the culturable microbiome associated to S fraction showed viable counts with intermediate values between R and B fractions, significantly lower CFU [$(9.62\pm4.53)\times10^6$ and $(1.92\pm1.07)\times10^5$ CFU g^{-1} fresh weight, respectively] were observed in the pepper endosphere (Table 2). In contrast, the abundance of culturable ACC-deaminase (ACCd) bacteria showed a dramatic reduction in the non cultivated soil [$(9.81\pm2.64)\times10^4$ CFU g^{-1} fresh weight] in comparison to plant associated fractions, where bacterial counts were detected at least four order of magnitude higher (Table 2).

Phylogenetic Analysis of Cultivable Bacteria Associated to Pepper Root System

The generated microbial collection from the root system of pepper included a total of 299 bacterial strains (Table 3). Phylogenetic affiliation was performed by 16S rRNA partial sequencing; prior to this procedure, ACCd bacteria were de-replicated by strain typing through ribosomal spacers fingerprinting in order to define the different haplotypes. (Table 2).

Isolates were assigned to four phyla, namely *Firmicutes, Beta* and *Gamma*-subgroups of *Proteobacteria* and *Actinobacteria*, similarly to what observed by the cultivation-independent approach (Table 3). A differential distribution pattern of the major bacterial taxa among the different fractions of the pepper root system was observed (Table 3). According to cluster analysis, the composition of the cultivable community associated to R and S fractions shared a high similarity (83%), whereas that associated to the non-cultivated arid bulk soil differed significantly (Fig. 2). Despite being rather different under DGGE analysis (Fig. S2), the pepper root endosphere and the non-cultivated arid root-free soil resulted less distant according to cluster analysis (Fig. 2), presumably because of the abundance of the *Bacillus* isolates in both fractions (68% and 39%, respectively). *Bacillus, Klebsiella* and *Cellulosimicrobium* represented the most abundant genera in the bacterial collection (41%, 26% and 14%, respectively). In more detail, the pepper endosphere was dominated by the *Firmicutes* phylum and the strains were assigned to 3 genera: *Bacillus, Paenibacillus* and *Lysinibacillus,* which accounted for 68%, 30% and 3% of the isolates, respectively (Table 3 and Fig. S2). Thus E fraction was colonized by a restricted and peculiar community, as reflected by Shannon and Evenness indices (Table 4). In contrast, the R fraction showed the greatest biodiversity in terms of community structure (Table 3–4 and Fig. S2). The strains isolated from R were grouped within the *Proteobacteria* phylum (71%), comprising mainly *Gammaproteobacteria* (70%) and *Betaproteobacteria* (1%). Members of the *Gammaproteobacteria* group belonged to the genera *Klebsiella* (61%), *Pseudomonas* (4%), *Citrobacter* (4%) and *Acinetobacter* (1%). The *Betaproteobacteria* were represented by a single genus, *Achromobacter.* Members of the phylum *Firmicutes* were the second most abundant group in the rhizosphere (R fraction) and all the isolates belonged to the genus *Bacillus* (Table 3 and Fig. S2).

Similarly to the rhizosphere, in the S fraction two dominant phyla were detected: *Gammaproteobacteria* (71%) and *Firmicutes* (28%), with 4 genera in total: *Klebsiella* (44%), *Bacillus* (27%), *Citrobacter* (16%) and *Raoultella* (7%). The non-cultivated arid root-free soil was affected by the lowest Shannon and Evenness indices, pointing to a highly stable microbial community. The isolates from the B fraction were affiliated to three phyla: *Actinobacteria* (60%), *Firmicutes* (35%) and *Gammaproteobacteria* (5%). The genus *Cellulosimicrobium* was the major taxon (57%), followed by the genera *Bacillus* (39%), *Rhodococcus* (3%) and *Klebsiella* (1%) (Table 3 and Fig. S2).

A comparative analysis highlighted that strains of *Paenibacillus* (30%) were isolated only from fraction E. While members of *Gammaproteobacteria* were retrieved only in soil fractions, some genera showed a specific distribution: *Pseudomonas* was found only in R and S fractions; *Acinetobacter* only in R, strains of the *Raoultella* genus only in S and bacteria affiliated to *Cellulosimicrobium* and *Rhodococcus* only in B (Table 3).

Plant Growth Promoting Activities and Tolerances to Abiotic Stress of the Isolates

The potential functionality of pepper associated isolates to sustain plant growth under drought was assessed by a large screening for PGP abilities in relation to drought tolerance, and the resistance to abiotic stresses occurring in arid soils (Table 5). We assessed whether PGP abilities are differentially distributed in the different microhabitats of the pepper root system. All the fractions demonstrated to be colonised with a similar frequency by potential beneficial strains, even though in the non-cultivated arid soil PGP traits were less abundant (Table 5). While none of the isolates showed all the assayed PGP activities, 31,7% and 22,5% of strains presented respectively four and five PGP activities (Fig. S1). All the isolates presented the potential to adapt to unfavourable environmental conditions of arid soils, showing a certain halotolerance, resistance to low water availability and to variable temperature range (Table 5). Similarly, bacteria isolated from the E, R and S plant-associated fractions exhibited a large number of PGP traits compared to isolates from arid non-cultivated root-free soil (B fraction) (Table 5). Nevertheless, some abilities like nutrient supply (phosphate solubilisation, siderophore release), are more frequent in soil bacteria, while auxin synthesis, directly affecting plant hormone homeostasis, was primarily presented by endophytes (Table 5). PGP traits distribution among the different bacterial genera revealed that the *Bacillus* and *Klebsiella* showed a predominant role, even though other genera less frequently isolated, like *Pseudomonas, Raoultella* and *Paenibacillus*, exhibited a higher number of PGP potential activities (Table 6 and Table S2).

Table 2. Abundance of culturable bacteria associated to the different fractions of the pepper root system.

Fraction	Bacterial Count (CFU g^{-1} fresh weight)			N° isolates			ACCd haplotypes
	R2A	KB	ACC	R2A	KB	ACC	
E	$(9.62\pm4.53)\ 10^6$	$(1.92\pm1.07)\ 10^5$	$(1.60\pm4.53)\ 10^8$	12	12	53	5
R	$(5.13\pm3.44)\ 10^9$	$(1.28\pm0.00)\ 10^8$	$(2.24\pm4.53)\ 10^9$	12	12	50	8
S	$(5.83\pm4.06)\ 10^8$	$(2.47\pm1.81)\ 10^7$	$(2.48\pm2.28)\ 10^9$	12	12	49	6
B	$(1.28\pm0.72)\ 10^8$	$(3.74\pm2.64)\ 10^7$	$(9.81\pm2.64)\ 10^4$	12	12	51	5

The isolation was performed on different cultivation media. In the table it is reported the amount of bacterial isolates composing the strain collection associated to pepper endosphere and root-associated soil fractions. E, Endosphere; R, rhizosphere; S, root-surrounding soil; B, non-cultivated arid soil.

Table 3. Distribution of microbial taxa in the collection of culturable bacterial isolates associated to pepper plants.

Phylogenetic group	Genus	E (77)	R (74)	S (73)	B (75)	Species	E (77)	R (74)	S (73)	B (75)
Firmicutes	*Bacillus*	52	21	20	25	*B. acidiceler*			1	
						B. altitudinis				5
						B. amyloliquefaciens	1		3	2
						B. antraci	1			
						B. aquimaris				1
						B. cereus	3	6	2	3
						B. endophyticus	3	1		1
						B. firmus	1			
						B. funiculus			1	
						B. massiliensis			1	
						B. megaterium	13	7	7	13
						B. pumilus			1	2
						B. subtilis	30	5	4	1
						B. thuringiensis		2	1	1
	Lysinibacillus	2		1		*L. fusiformis*	2			
	Paenibacillus	23				*P. illinoisensis*	23			
Betaproteobacteria	*Achromobacter*		1			*A. xylosoxidans*		1		
Gammaproteobacteria	*Acinetobacter*		1			*A. calcoaceticus*		1		
	Citrobacter		3	12		*C. freundii*		3	12	
	Klebsiella		45	32	1	*K. oxytoca*		12		
						K. pneumoniae		33	32	1
	Pseudomonas		3	3		*P. fluorescens*		1		
						P. koreensis		1		
						P. mediterranea			2	
						P. plecoglossicida			1	
						P. putida		1		
	Raoultella			5		*R. planticola*			5	
Actinobacteria	*Cellulosimicrobium*				43	*C. cellulans*				43
	Rhodococcus				2	*R. fascians*				2

Numbers indicate the number of strains assigned to each genus or species, respectively.
The numbers in parentheses are the total number of isolates for each fraction.

In vitro Rhizoplane Colonization

To assess the ability of soil bacteria to adhere and colonize the rhizoplane, an adhesion assay was performed *in vitro* on *Arabidopsis thaliana* roots by taking advantage of a *gfp*-labelled bacterium. Root colonization is a key requirement to ensure an intimate association with the plant and thus a support against water stress. Of the different strains assayed for transformation with plasmids carrying a *gfp* (Green Fluorescent Protein) cassette, we succeeded in transforming a *Klebsiella pneumoniae* strain. The *gfp*-tagged isolate was used to track the bacterial adhesion on *Arabidopsis* and pepper root system. After 15 h of exposure to the *gfp*-tagged bacterial suspension, confocal microscopy analysis revealed that *Arabidopsis* primary root and root hairs were massively colonized by *gfp*-tagged cells. The *gfp*-labelled bacterium completely enwrapped root hairs, with an adherence profile that was adapted to the root hair morphology (Fig. 3A–B). In pepper the strain was massively detected on the rhizoplane but only few cells were found on root hairs (Fig. 3C–D), suggesting a differential colonization profile according to the model plant.

Selection of Rhizobacteria for Plant Growth Promotion under Drought Stress

Rhizobacteria were evaluated for the capability of promoting plant growth under water stress. A cluster analysis performed by combining the rhizobacteria PGP phenotypic traits (Fig.4) grouped the strains in three major clusters. Cluster I is the largest and summed *Bacillus* spp., *Klebsiella* spp. and *Pseudomonas* spp. The great majority of bacteria exhibiting ACCd activity were in this cluster that, moreover, included the strains with the highest number of potential PGP abilities. Clusters II and III displayed only one strain, respectively an ACCd-producing *Achromobacter xylosoxidans* and an *Acinetobacter calcoaceticus*. Both isolates exhibited just one PGP trait (Fig. 4). Consistent with ACCd activity in lowering plant ethylene under abiotic stress conditions, ACCd-producing rhizobacteria from the three clusters were selected to be further assayed in their ability to sustain plant growth *in vivo* under drought. These isolates were affiliated to genera *Citrobacter* (R16ACCd), *Klebsiella* (R01ACCd, R05ACCd, R08ACCd and R15ACCd), *Achromobacter* (R10ACCd) and *Acinetobacter*

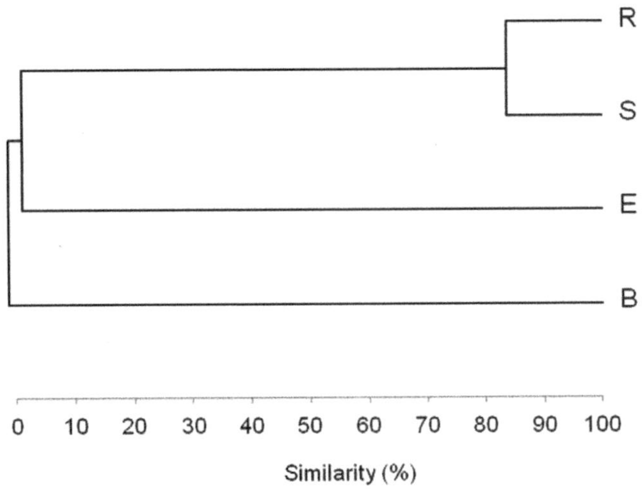

Figure 2. Cluster analysis of the cultivable bacteria associated to pepper fractions. The cultivable fraction of pepper-associated bacteria was compared to uncultivated soil, by performing a cluster analysis according to Pearson correlation.

Table 4. Diversity indexes of the microbial collection.

Diversity Index (OTUs 99%)	E	R	S	B
Taxa	9	12	15	10
Individuals	77	74	73	75
Dominance	0.262	0.252	0.240	0.372
Shannon	1.586	1.799	1.930	1.429
Simpson	0.738	0.748	0.760	0.628
Evenness	0.543	0.504	0.459	0.417

The indexes were calculated for the sequences of bacterial strains isolated from the different fractions of the pepper root system and the non-cultivated arid soil. Sequences have been grouped in OTUs based on nucleotide similarity at 99%.

(R04ACCd), with *Klebsiella* spp. as the most frequent, as showed in the phylogenetic tree (Fig. 4 and S3).

Plant Growth Promotion of Rhizobacteria Associated to Pepper Plants under Water Stress

Well irrigated pepper seedlings inoculated or not with the rhizobacterial suspensions were suddenly exposed to a twelve days period of water stress. After eight days of water stress, control plants were severely affected, whereas plants exposed to ACCd-producing rhizobacteria exhibited a higher shoot turgor (Fig. 5A). Pepper plants inoculated with ACCd rhizobacteria R4, R10 and R16 showed net photosynthesis (Pn), evaporation/transpiration (E), stomatal conductance (Gs) significantly higher than untreated plants (NC), while R1, R5 and R15 strains positively affected water-stressed plants only at a photosynthetic level (Fig. 5B). At the end of the twelve days drought period, three days of re-watering were applied and plants were carefully harvested for biomass and length measure analysis. All plants exposed to the selected bacteria exhibited a more robust root system with a quantitative effect depending on the strain (Fig. 5C). A similar increase of about 20% in root length was observed both in non-stressed plants and in those inoculated with rhizobacteria respect to the plants exposed to drought (Fig. 5C). Root fresh weight in the inoculated plants showed a 40–60% increase depending on the bacterial strain, compared to the non inoculated stressed control plants (Fig. 5C).

Discussion

The traditional management of agriculture in arid ecosystems is essential to preserve land from soil degradation and maintain food production ensuring a sustainability and preserving soil biodiversity [17]. A signature feature of arid and semi-arid lands is plant patchiness with scattered plant clumps dispersed in a bare landscape [18]. While the structure of the microbiota under and inter desert shrubs and canopies has been largely investigated [19], little attention was paid to the effect of desert farming on the structure and functionality of the microbiomes associated to plant root system. Recently, in a farm located at north-east Cairo, Egypt, Köberl et al. [4] reported higher biodiversity indices in cultivated fields than in the desert soil and the enrichment in bacteria with antagonistic activity against plant pathogens. Similarly, in cultivated fields at north-west Cairo, we found dramatic changes in the structure and activity of the bacteria associated to pepper root system compared to non-cultivated soil. A strong rhizosphere effect in terms of higher bacterial densities and species richness was observed in the soil fractions more closely associated to the root system, the R and S fractions, compared to the bulk root-free soil, whereas the endophytic fraction showed the lowest values, presumably because root tissues selected specific bacterial colonizers [20]. A certain variability was detected among endosphere replicates that could originate from multiple factors, including: microvariability in the soil field [20,21,22,23], plant physiological condition [24,25], growth stage [22], extent of root exudation [26], bacteria inter-species interactions and even random events [20]. Despite the sampled plants were coeval, the variability of the field conditions may have influenced the plant physiological state preventing to exclude a certain effect on the endosphere composition. Which combination of driving forces has determined the differences in the three replicates remains unresolved, however, the difference of the endosphere microbiome from the microbiome of the rest of the root system was clearly evident. This differential distribution is presumably triggered by the burst of microbial biomass that can use the organics rhizodeposed by the root determining the realization of a "resource island" effect typical of desert ecosystems where plant growth and interaction with the soil microbiome locally improves soil properties that in turn sustain the overall soil activity and biotic diversity [19]. Such a repartition of bacteria abundance and diversity observed in the pepper root system found analogies in other plant cultivated in arid soils, such as sugarcane [27], bamboo [28], chick pea [29], and olive tree (Marasco et al., unpublished data).

The distribution of the bacterial genera reflects adaptation to the different microhabitats. The *Bacillus* genus was isolated in all the pepper fractions, with higher prevalence in the endosphere. Garbeva et al. [30] showed that the majority of Gram-positive bacteria in soils under different types of management regimes (permanent grassland, grassland turned into arable land and arable land), were putative *Bacillus* species. *Bacillus* spp. are also commonly found in arid land as a consequence of their ability to form endospores that allow bacterial survival for extended time periods under adverse environmental conditions [31]. *Bacillus* and related genera have been already reported to be associated to and promote the growth of a wide range of plants [32].

In our study *Paenibacillus* and *Lysinibacillus* genera were isolated only from pepper root endosphere. *Paenibacillus* is a common soil

Table 5. PGP potential of the microbial collection.

Isolation media	Plant fraction (N° of isolates)	PGP activities and Tolerance to Abiotic stress (%)													
		Aux.	P sol.	EPS	Sid.	NH_3	Prot	5% NaCl	8% NaCl	10% NaCl	10% PEG	20% PEG	4°C	42°C	50°C
ACCd (Haplotypes)	E (5)	80	20	20	20	60	20	20	20	20	100	60	0	100	20
	R (8)	38	88	63	75	88	0	75	0	0	100	100	88	100	0
	S (6)	83	100	83	100	100	0	100	0	0	100	100	67	100	0
	B (5)	20	100	40	20	100	40	100	0	0	100	100	20	100	0
R2A/KB	E (24)	92	58	8	29	100	83	88	75	17	100	100	4	92	0
	R (24)	79	50	17	33	100	92	96	79	42	100	100	8	83	29
	S (24)	92	42	29	17	92	79	88	58	33	100	100	13	96	46
	B (24)	92	50	13	8	100	63	83	67	42	100	100	0	96	25

Percentage distribution of plant growth promoting activities and tolerance to abiotic stress among the isolates of the bacterial collection obtained from the different fractions of the pepper root system and the non-cultivated arid soil.
Auxin = auxin production; P Sol. = inorganic phosphate solubilization; EPS = exopolysaccharide release; Sid. = siderophores production; NH_3 = ammonia production; Prot. = protease activity; PEG = poly-ethylen-glycol.

bacterium that has been described to present PGP properties. In particular, *P. polymyxa* has multiple plant beneficial activities, such as nitrogen fixation, soil phosphorus solubilisation and production of exopolysaccharides, hydrolytic enzymes, antibiotics and cytokinin [33]. Inoculation of *Arabidopsis* and wheat with a *P. polymyxa* strain, isolated from rhizosphere of wild barley in northern Israel, resulted in enhanced drought tolerance [34]. The presence in the pepper root tissue of *Lysinibacillus* spp., a poorly studied genus isolated also from rather different plants such as bamboo [28], citrus [35], tomato [36], medicinal plants [37] and halophytes [38] needs a clarification of its role in the microbe-plant interaction.

The pepper root systems in the arid Egyptian soil showed to host endophytes only within the *Firmicutes* class, while previous studies for endophytes in both herbaceous and arboreal plants reported a diverse array of bacterial species, including members of *Acetobacter*, *Arthrobacter*, *Bacillus*, *Burkholderia*, *Enterobacter*, *Herbaspirillum*, *Serratia* and *Pseudomonas* [16,39,40,41,42]. In sweet pepper, culturable endophytes were assigned to high-G+C Gram-positive *Microbacterium*, *Micrococcus* and *Rhodococcus* but also to *Firmicutes* of the *Bacillus* and *Staphylococcus* genera. In other studies, the variability in the diversity of culturable endophytic bacteria has been associated to different selective pressures determined by the different pepper cultivars [43].

Table 6. Bacterial genera distribution of the PGP potential.

Genus	N° of Isolates	PGP activities and Tolerance to Abiotic stress (%)													
		Auxin	P sol	EPS	Sid	NH_3	Prot	5% NaCl	8% NaCl	10% NaCl	10% PEG	20% PEG	4°C	42°C	50°C
Achromobacter	1	0	0	0	0	100	0	0	0	0	100	100	100	100	0
Acinetobacter	1	0	100	0	0	0	0	0	0	0	100	100	100	100	0
Bacillus	92	88	48	18	18	98	79	89	70	35	100	100	43	94	27
Cellulosimicrobium	2	0	100	50	0	100	0	100	0	0	100	100	0	100	0
Citrobacter	2	0	100	0	100	100	0	100	0	0	100	100	50	100	0
Klebsiella	8	75	100	100	100	100	0	100	0	0	100	100	62	100	0
Lysinibacillus	1	100	0	0	0	1	0	0	0	0	100	100	0	100	0
Paenibacillus	3	100	33	0	0	33	0	0	0	0	100	33	0	100	0
Pseudomonas	6	100	83	0	50	100	83	83	67	17	100	100	33	50	0
Raoultella	3	100	100	100	100	100	0	100	0	0	100	100	100	100	0
Rhodococcus	1	0	100	0	0	100	100	100	0	0	100	100	100	100	0

Percentage distribution of plant growth promoting activities and tolerance to abiotic stress according to the genera of the bacterial collection obtained from the pepper root system and the non-cultivated arid soil.
Auxin = auxin production; P Sol. = inorganic phosphate solubilization; EPS = exopolysaccharide release; Sid. = siderophores production; NH_3 = ammonia production; Prot. = protease activity; PEG = poly-ethylen-glycol.

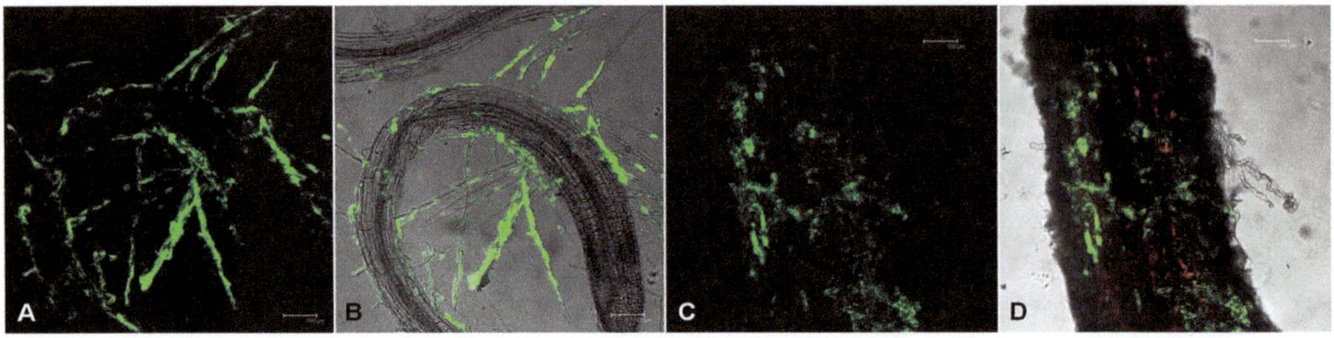

Figure 3. Rhizocompetence of *gfp*-labelled bacteria on different plant models. Plant root colonization experiments performed with a *Klebsiella pneumoniae* strain isolated from the pepper rhizosphere genetically labeled with a *gfp*. (A) and (B) colonization of *Arabidospis thaliana* rhizoplane; (C) and (D) colonization of the pepper rhizoplane. Red spots represent root autofluorescence as acquired through the TRICT filter. The scale bars of the different images in the figure correspond to 100 μm.

In all the soil fractions strains belonging to *Gammaproteobacteria* were predominant with many of the isolates assigned to the *Enterobacteriaceae* family. It comprises many species with enteric habitat, which origin could be attributable to the low hygienic quality of the irrigation water. The decline in the availability of pristine freshwater for irrigation due to allocation to urban and/or industrial supply, often results, especially in arid and semi-arid regions, in the intensive use of low-quality water to satisfy the increasing demand for irrigation. Representative species of *Enterobacteriaceae* genera, especially *Klebsiella*, *Enterobacter*, *Citrobacter*, have been isolated from different plant species grown in arid lands [44,45,46,47]. In non-cultivated soil not subjected to irrigation and soil amendment, *Enterobacteriaceae* decreased in favour of *Actinobacteria*, with the prevailing genera *Cellulosomicrobium* and *Rhodococcus*. Together with *Bacillus* spp., *Actinobacteria* can survive as spores under adverse environmental conditions, hence making them typical desert taxa [4,48].

The PGP features of bacteria associated to the pepper root system indicated that arid soils are excellent reservoir of bacteria responsible for the efficient functioning of the plant-soil ecosystem services. Twenty three percent of the assayed isolates exhibited multiple PGP activities, which may promote plant growth directly, indirectly or synergistically. Moreover, *gfp* labelling of a rhizobacterium demonstrated a versatile colonization capabilities being capable of colonizing the roots of two different plant models, as previously described for other bacteria [49].

A relative large range of PGP activities was recorded for bacteria isolated from non-cultivated soil, with 38% of isolates displaying more than 4 PGP activities, compared to 58% for isolates from R and S fractions. As the boundary between the non-cultivated soil and the desert areas around the farm was labile, we can perceive the still unexplored biotechnological potential of arid lands. Chanal and colleagues [50] found new radiotolerant bacterial species in Tataouine desert and recently *Ramlibacter tataouinensis* genome annotation revealed unexpected adaptation mechanisms to hot and dry environments, including sensitivity to light and to water availability at the dew time [50,51]. A survey of PGP bacteria associated to *Hordeum spontaneum* in the "Evolution canyon" in Israel reported a significant higher population of osmotic tolerant, phosphate solubiliser, EPS producer and ACCd bacteria in the stressful sunny site than in the shadowed site [34]. According to these data it was assumed that the foundations for the adaptability to the harsh conditions of agriculture in arid lands are based on the co-evolution of the association between plant and microbes under harsh environmental conditions [34,52]. In our bacterial collection from the pepper root system 88% of isolates showed multiple PGP activities and were able to grow at high temperature and at low water potential indicating that they can be active and hence express their PGP features *in vivo* under water stress conditions.

Drought is responsible for the weakening of ecosystem services, even at temperate latitudes. In 2003, a summer heat wave along with a prolonged drought event in Europe caused a reduction of 36% in the net productivity of maize in the Po valley in Italy and dramatically compromised agricultural production in France (−17%) and Eastern Europe (−20%) [53]. In Egypt agriculture strongly relies on the exploitation of the water from the Nile river, considering the limited availability of groundwater. In such general condition of water limitation it is supposed that other factors, like the root-bacteria association are selected for contributing to alleviate plant water stress. A candidate group of PGP bacteria that can have a potential protecting effect against water stress are ACCd rhizobacteria. ACCd bacteria are capable of lowering the concentration of ethylene that is overproduced in response to stressful conditions [54]. ACCd bacteria have been shown to recover plants from different stresses [55]. Different plant models have been successfully recovered from a variety of stressful conditions such as salinity [56] drought [57] and heavy metals [58] following the exposure to ACCd bacteria [59]. Hence, we have selected the collection of ACCd bacteria isolated from the pepper root system for assessing the capability of protecting the plant from drought and water stress.

Early responses to water stress include a decrease in photosynthesis efficiency [60]. Pepper plants treated with ACCd rhizobacteria recorded higher values for the photosynthesis processes and even a higher tissue turgor. These beneficial effects result in the increase of root biomass and length, up to 50% respect to non-inoculated plants. Although the rhizobacterial strains exhibited a variable extent in the improvement of plant drought tolerance, the most pronounced protection against drought was obtained with strains of the genera *Achromobacter*, *Klebsiella* and *Citrobacter*. Considering the root-colonization capacity of these genera it is conceivable that such protecting activity can be performed also in field conditions.

Desert bloom remains a general vision, although the real efficacy of PGP treatment of plants for desert restoration remains contradictory. A three-years field trial in the Sonoran desert with different tree species exposed to AM fungi and *Azospirillum brasilense* to restore degraded lands showed that the treatments were only partially successful. Positive results were obtained only with

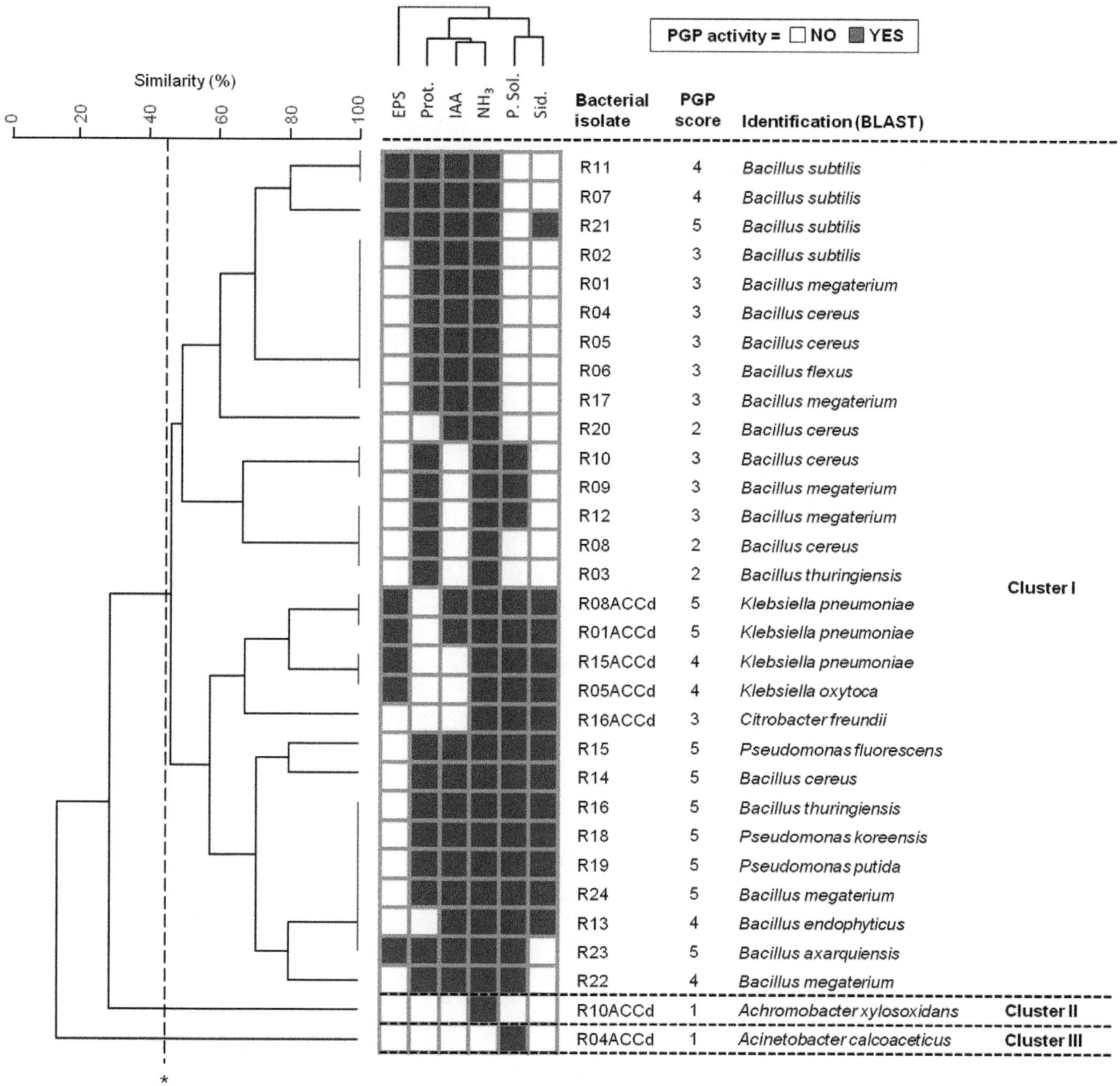

Figure 4. Analysis of the PGP potential of pepper associated rhizobacteria. Cluster analysis of the distribution of PGP activities in the rhizobacterial collection, according to Pearson correlation coefficient. Total PGP potential is indicated as a score value resulting from the sum of the number of the different PGP abilities exhibited by each strain. Cluster group were defined based on a cluster cutoff value of 42% of similarity.

autochthonous leguminous trees while other combination of tree-inoculant-amendment resulted in small negative or no effect at all [1]. Our data indicate that consolidated traditional desert farms represents "resource islands" were topsoil is preserved from destruction by the wind or other soil erosion agents, contributing to act as a sink for organic matter and beneficial microbes. Desert farming remains a bulwark for protecting soil fertility in desert ecosystems and an effective strategy for enriching plant growth promoting microorganisms capable of directly protecting plants from drought stress.

Materials and Methods

Site Description and Sampling

Plant and soil samples were collected in a cultivated field in a private traditional farm located in the north-western desert region in Egypt, near El-Tawheed Village. The permission for sample collection was obtained by the Department of Horticulture of the University of Ain Shams, Egypt. Crop irrigation was performed using the water from the Nile river and groundwater. Four different fractions were collected in triplicates: E (endosphere), R (rhizosphere) and S (root surrounding soil) of *Capsicum annum* L. plants and B (bulk soil) as control. Intact roots were collected after

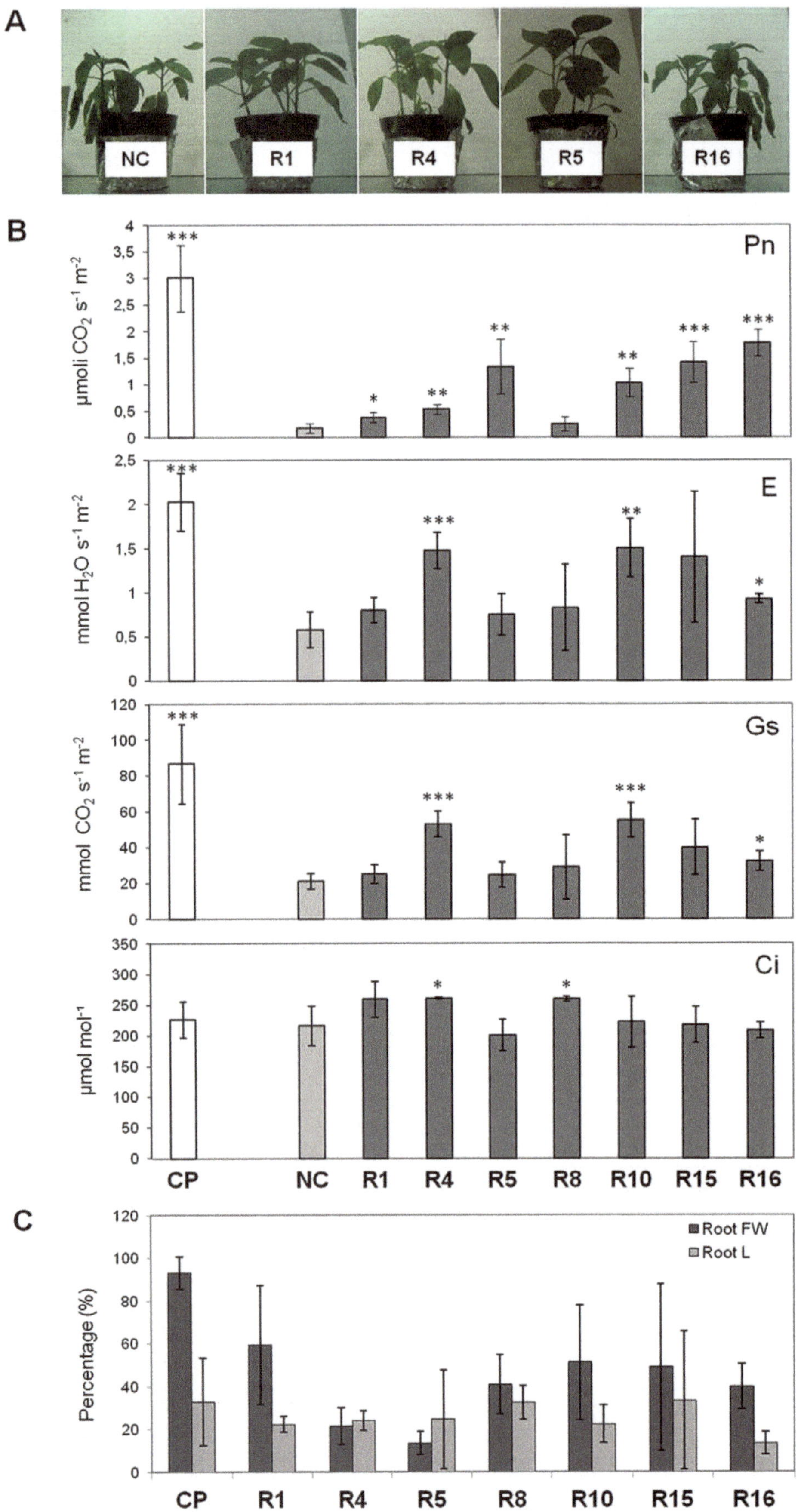
A
NC
R1
R4
R5
R16
B
Pn
µmoli CO2 s-1 m-2
E
mmol H2O s-1 m-2
Gs
mmol CO2 s-1 m-2
Ci
µmol mol-1
CP
NC
R1
R4
R5
R8
R10
R15
R16
C
Root FW
Root L
Percentage (%)
CP
R1
R4
R5
R8
R10
R15
R16

Figure 5. Rhizobacteria increased plant resistance to drought stress. Abbreviations for the figure: CP, (positive) abiotic control, irrigated at the water holding capacity of the soil along all the experiment; NC, (negative) abiotic control, subjected to drought by interrupting water supply for 12 days. (A) Representative images of plants exposed to rhizobacteria compared to untreated plants eight days after the induction of drought. (B) Leaf physiological parameters in treated and untreated plants eight days after the induction of drought.Abbreviations: Pn, net photosynthesis; E, evapo-transpiration; Gs, stomatal conductance; Ci, internal carbon dioxide (CO_2). Student t-test was adopted to statistically analyse the data. *:p≤0,05; **:p≤0,01; ***:p≤0,001. The data reported in the graphs are representative of one replicate experiment. (C) Percentage increase in root fresh weight (FW) and root length (L) of water stressed plants, compared to the abiotic stressed control, set as 0%.

plant eradication, with soil particles still adhering on the rhizoplane (E+R fractions). Soil around the collected roots and not attached to plant root system was sampled (S fraction). Uncultivated soil (B fraction) was kept as control at 4 m far from the cultivated field, in an area not subjected to irrigation and that was not cultivated in the last years. All soils and roots samples were collected under sterile condition using sterile tools. Recovered samples were stored at 4°C for microbiological isolation or stored at −20°C for molecular analysis.

PCR-DGGE Analysis of Pepper Associated Bacterial Communities

Primers 907R and 357F with a GC-clamp were used in this study for the amplification of bacterial 16S rDNA genes [61]. PCR reaction was performed in 0.2 ml tubes using 50 µl reaction volume. The reaction mixture contained the diluted buffer 1 X, 1,5 mM $MgCl_2$, 5% of DMSO, 0,12 mM of a mixture of dNTPs, 0,3 µM of each primer, 1 U Taq polymerase, and 10 ng of template. If necessary, DNA was properly diluted. Cycling conditions used to amplify the 16S rDNA gene fragment were 94°C for 4 min, followed by 10 cycles of 94°C for 0.5 min, 61°C for 1 min, and 72°C for 1 min; followed by further 20 cycles of 94°C for 0.5 min, 56°C for 1 min, and 72°C for 1 min; and a final extension at 72°C for 7 min. 2 µl of the PCR products were visualized by electrophoresis in 1.5% agarose gels stained with ethidium bromide prior to DGGE. For DGGE analysis, 100–150 ng of the PCR products generated from each sample were separated using polyacrylamide gel (8% of a 37:1 acrylamide–bisacrylamide mixture in a Tris acetate EDTA (TAE) 1X buffer, 0.75 mm thick, 16×10 cm) with a 40–60% denaturant gradient). Gel was run overnight at 90 V in TAE 1X buffer at 60°C in DCode apparatus (Bio-Rad, Milan, Italy). The gel was stained with 1X Syber Green (Life Technologies) in TAE buffer and the gel was scanned with gel photo GS-800 system.

The DGGE bands were excised from the gel using a sterile cutter and eluted in 50 µl water at 37°C for 6 hours. The reamplification of DNA eluted from DGGE bands was performed using 907R and 357F primers without the GC-clamp, using the following protocol: 95°C for 5 min, 30 cycles of 95°C for 1 min, 61°C for 1 min, and 72°C for 1 min and a final extension at 72°C for 7 min. PCR products were checked by electrophoresis in 1% agarose gel. The sequencing service was performed by Macrogen Inc. (Korea). The band profile of fragments in the DGGE gel was converted in line plots with ImageJ software [62], and the x/y values obtained were imported into an Excel file. The matrix of x/y values of rRNA 16S line profiles was subjected to cluster analysis using the Pearson correlation coefficient. The multivariate analysis were conducted using XLSTAT software (vers. 7.5.2 Addinsoft, France).

Isolation of Bacteria, Media and Culture Condition

R fraction, the soil particles tightly adhering to the rhizoplane, were separated from the root tissue (E) by applying the "pull and shake method". Root surface was sterilized as described by Sun et al. [63] and the efficacy of the sterilization method was verified by plating the last wash water on King's medium [64]. One gram of smashed E, R, S and B were suspended in 9 ml of sterile physiological solution (9 g/L NaCl) and shaken for 15 min at 200 rpm at room temperature. Suspension were diluted in 10-fold series and plated in triplicate onto KB medium and on R2A medium (Oxoid). After 3 days at 30°C, Colony Forming Units (CFU) per gram were determined. 12 colonies per medium per fraction were randomly selected and spread on the original medium for three times to avoid contamination risks. Moreover, 1 g of sample from each fraction was used as inoculum for ACC-deaminase enrichments as described by Penrose and Glick [65]. 50 colonies were randomly picked and propagated three times on PAF medium (10 g/L proteose peptone, 10 g/L hydrolyzed casein, 3 g/L $MgSO_4$, 1,5 g/L K_2HPO_4, 10 mL/L glycerol and 15 g/L agar for solid medium). Pure strains were frozen in 25% glycerol at −80°C. A total of 299 isolates were collected and further characterized in this study.

Phylogenetic Affiliation of Bacterial Strains

DNA was extracted from isolates by boiling lysis. The bacterial cells were resuspended in 50 µl of sterile TE (10 mM Tris/HCl, pH 8, 1 mM EDTA) in 1.5 ml tubes and incubated at 100°C for 8 min. After centrifugation (13000 g, 10 min), the supernatant containing the released DNA was stored at −20°C and used as template for PCR amplification. The bacterial collection originated from ACC enrichments was de-replicated by fingerprinting analysis of the rRNA 16S-23S Intergenic Transcribed Spacer (ITS) region. The ITS-PCR protocol was performed as described by Cardinale et al. [66]. The PCR products were separated by gel electrophoresis in 2% agarose gel and the fingerprinting profiles were visualized using Gel Doc system (Bio-Rad, Milan, Italy). Isolates which showed the same banding pattern were grouped in haplotypes, and for each haplotype a representative strain was selected for further analysis. Phylogenetic identification of isolates was performed by partially sequencing of the 16S rRNA gene, using universal primers 27F and 1492R. PCR products were checked by electrophoresis in 1% agarose gel. The sequencing service was performed by Macrogen Inc. (Korea). The sequences were compared with those deposited in the GenBank database, using the online software BLAST.

Diversity and Phylogenetic Analyses

16S rRNA gene sequences were aligned using the ClustalX software [67] and the output file was used to define operational taxonomic units (OTUs) using DOTUR [68]. A quantitative matrix was created basing on the absence/presence of each polymorphic OTU calculated at 99% nucleotide similarity. Cluster analysis has been performed with the XLSTAT software using the Pearson correlation coefficient.

Number of Taxa, Shannon, Evenness, Simpson and Dominance indexes of the OTUs, defined at 99% of similarity, have been calculated using the PAST software [69].

The alignment of ACCd rhizobacteria sequences and the construction of the phylogenetic tree were performed using the neighbor-joining method [70] of MEGA version 4 [71].

Nucleotide Sequence Accession Numbers

The partial 16S rRNA gene sequences (800–900 bp) from the isolates and the partial 16S rRNA gene sequences (500 bp) from the DGGE bands have been deposited in the GeneBank database from the accession numbers HE610774 to HE610892 and from HE856290 to HE856311 respectively.

Evaluation of Direct and Indirect Plant Growth Promoting Activity and Tolerance to Abiotic Stresses

Indolacetic acid production was estimated following the protocol described by Brick et al. [72]. The mineral P-solubilizing ability of the strains was determined on Pikovskaya's liquid medium amended with 0.5% [$Ca_3(PO_4)_2$] as described by Mehta and Nautiyal [73]. Siderophore release was determined as described by Schwyn and Neilands [74]. Exopolysaccharides (EPS) production was estimated as described by Santaella et al. [75], using modified Weaver mineral media enriched with sucrose.

Ammonia production was evaluated as described by Cappuccino and Sherman [76]; protease production was determined in 5% agar skimmed milk [77]. Resistance to salt was assessed by adding 5–8–10% NaCl to the culture media and incubating the plates at 30°C for 5 days. Tolerance to osmotic stress was evaluated by adding to liquid media 10–20% of Poly-Ethylen-Glycol (PEG). The ability to growth at 4°, 42° and 50°C was verified in solid media by incubation at the indicated temperatures and the growth was qualitatively scored after 5 days of incubation.

In vitro Bacterial Rhizocompetence Assay

The plasmid pHM2-*gfp* [78] was used to label R1-ACCd strain, affiliated to *Klebsiella* spp. Overnight culture of R1-ACCd was re-inoculated in fresh KB medium and the growth was monitored spectrofotometrically. When the culture reached 0.3 OD, 1 ml aliquot of cells were centrifuged (4000 rpm, 4°C) and washed twice with MilliQ water prior to be resuspended in 50 µl of MilliQ water and 10% glycerol. 30 µl of cells were used to be transformed by electroporation (Eppendorf 2510) with 50 ng of pHM2-*gfp* plasmid. Successful transformation was checked by growth on a selective medium (KB+50 µg/ml of kanamicin). To evaluate R1-*gfp* colonization ability, three-days *Arabidopsis thaliana* seedlings or seven-days *Capsicum annuum* L. seedlings were exposed to 10^8 cells/mL. Seedlings dipped in sterilized water were used as negative control. After 15h, plants were rapidly washed to remove weakly bound bacteria and observed under a confocal laser scanning microscope (Leica TCSNT). Images were acquired using Leica Confocal Software, using BP530/30 GFP filter (exitation at 488 nm) and LP590 TRITC filter (excitation length at 568 nm). For pepper rhizocompetence analysis, images were acquired also using the TRICT filter to observe root architecture by exploiting root autofluorescence in this channel. The acquired images were analyzed by using the MBF ImageJ software.

Plant Growth Promotion under Water Atress in Soil

Pepper seeds were sown in trays in wet agriperlite. After 1 week, uniform-sized seedlings were selected and planted in soil, three plants per 14-cm plastic pot. The seedlings were maintained in a growth chamber at a day/night temperature of 25/20°C with ~100 µmol photons m^{-2} s^{-1} of light supplied for 12 h during the daytime. During the second week, the seedlings were fertilized once with a bacterial suspension at the concentration of 10^8 cells/g of soils, while uninoculated plants were watered with tap water. One week after bacteria treatment, water was withhold for 12 days. A (positive) abiotic control, PC, was included and was properly irrigated all the experiment long. Seven-eight days after drought induction, physiological measures have been performed. To characterize photosynthesis performance, gas exchange measurements were taken with a portable photosynthesis system (CIRAS-2, PP System, USA). Measurements were taken on young, fully expanded, intact leaves of capsicum plants. Net CO_2 assimilation rate, stomatal conductance and transpiration were assessed at a CO_2 concentration of 400 µmol mol^{-1}, 50% relative humidity, 28°C chamber temperature, 500 ml min^{-1} airflow and a photon flux density of 1500 µmol m^{-2} s^{-1}. The instrument was stabilized according to manufacturer guidelines. After drought, water irrigation was resumed for three days and plants were harvested for biomass and length measures. Three independent experiments were performed with three replicate plants each. The statistical analysis was performed by analysing data by the T student test with ($p<0,05$).

Supporting Information

Figure S1 16S rRNA PCR-DGGE analysis of the bacterial communities in soil and endosphere of pepper plants. (A) 16S rRNA gene PCR-DGGE profiles in different plant fractions (E, R, S and B) obtained from three replicate plants (indicated as 1, 2 and 3). Circles on the bands indicate the DNA fragments that were excised from the gel and successfully amplified and sequenced (see also Table 1). (B) Plot line conversion for each DGGE fingerprinting profile obtained using Image-J software.

Figure S2 Diversity of culturable bacteria in pepper plant fractions. Distribution of bacterial isolate genera associated to different fractions of the pepper root system compared to non-cultivated root free arid soil.

Figure S3 Phylogenetic affiliation of pepper ACCd rhizobacteria. Neighbour-joining phylogenetic tree based on 16S rRNA gene sequences of ACCd rhizospheric bacteria and their closest phylogenetic neighbours. Bootstrap values are indicated at nodes. Scale bar represents observed number of changes per nucleotide position.

Table S1 Percentages of bacteria displaying PGP activities in different fractions of the pepper root system. Isolates recovered from the pepper root system and its different fractions, presenting different numbers (from 0 to 6) of PGP activities.

Table S2 Distribution of the PGP potential according to the microbial genera. The percentage of isolates displaying different numbers (from 0 to 6) of PGP activities are classified according to genus level, considering the whole microbial collection.

Acknowledgments

The authors would like to thank Dr. Umberto Fascio (CIMA, Centro Interdipartimentale di Microscopia avanzata of the University of Milan) for technical support at the confocal microscope, Dr. Isabella Tamagnini for the contribution in sampling activities, Dr. Gian Attilio Sacchi and BIOGESTECA project (n° 15083/RCC "Fondo per la promozione di accordi istituzionali") for helpful discussion on the biology of plant microbe interaction under water stress and Mr. Andrea Fiorentini for his availability for the recovery of soil, seeds and pots for the microcosm experiments.

Author Contributions

Conceived and designed the experiments: RM ER SB DD AC AAH UEB. Performed the experiments: RM ER BE GV FM. Analyzed the data: RM ER GV. Contributed reagents/materials/analysis tools: DD SB CS GZ. Wrote the paper: ER RM GV SB DD. Collected the samples: AAH UEB GV GZ DD. Critically revised the manuscript: AAH UEB CS AC GZ.

References

1. Bashan Y, Salazar BG, Moreno M, Lopez BR, Linderman RG (2012) Restoration of eroded soil in the Sonoran Desert with native leguminous trees using plant growth-promoting microorganisms and limited amounts of compost and water. J Environ Manage 102: 26–36.
2. FAO website. Available: http://www.fao.org/news/story/en/item/35571/icode/. Accessed 2012 May 15.
3. Boyer JS, Westgate ME (2004) Grain yields with limited water. J Exp Bot 55: 2385–2394.
4. Köberl M, Müller H, Ramadan EM, Berg G (2011) Desert farming benefits from microbial potential in arid soils and promotes diversity and plant health. PLoS ONE 6: e24452–e24452.
5. Gonzalez-Dugo V, Orgaz F, Fereres E (2007) Responses of pepper to deficit irrigation for paprika production. Sci Hortic 114: 77–82.
6. Showemimo FA, Olarewaju JD (2007) Drought tolerance indices in sweet pepper (*Capsicum annuum* L.). Int J Plant Breed Genet 1: 29–33.
7. FAOSTAT website. Available: http://faostat.fao.org/site/339/default.aspx. Accessed 2012 May 17.
8. Dalla Costa L, Gianquinto G (2002) Water stress and water table depth influence yield, water use efficiency and nitrogen recovery in bell pepper, lysimeter studies. Aust J Agric Res 53: 201–210.
9. Delfine S, Loreto F, Alvino A (2001) Drought-stress effects on physiology, growth and biomass production of rainfed and irrigated bell pepper plants in the mediterranean region. J Am Soc Hortic Sci 126: 297–304.
10. Dorji K, Behboudian MH, Zegbe-Dominguez JA (2005) Water relations, growth, yield and fruit quality of hot pepper under deficit irrigation and partial rootzone drying. Scientia Hortic 104: 137–149.
11. Sezen SM, Yazar A, Eker S (2006) Effect of drip irrigation regimes on yield and quality of field grown bell pepper. Agric Water Manage 81: 115–131.
12. Mayak S, Tirosh T, Glick B (2004a) Plant growth-promoting bacteria confer resistance in tomato plants to salt stress. Plant Physiol Biochem 42: 565–572.
13. Mayak S, Tirosh T, Glick B (2004b) Plant growth-promoting bacteria that confer resistance to water stress in tomatoes and peppers. Plant Sci 166: 525–530.
14. Kokalis-Burelle N, Vavrina CS, Rosskopf EN, Shelby RA (2002) Field evaluation of plant growth-promoting rhizobacteria amended transplant mixes and soil solarization for tomato and pepper production in Florida. Plant Soil 238: 257–266.
15. Kang S, Zhang L, Hu X, Li Z, Jerie P (2001) An improved water use efficiency for hot pepper grown under controlled alternate drip irrigation on partial roots. Sci Hortic 89: 257–267.
16. Sziderics AH, Rasche F, Trognitz F, Sessitsch A, Wilhelm E (2007) Bacterial endophytes contribute to abiotic stress adaptation in pepper plants (Capsicum annum L.). Can J Microbiol 53: 1195–2102.
17. Mäder P, Fliessbach A, Dubois D, Gunst L, Fried P, et al. (2002) Soil fertility and biodiversity in organic farming. Science 296: 1694–1697.
18. Goberna M, Sànchez J, Pascual G, Garcìa C (2007) *Pinus halepensis* Mill. plantations did not restore organic carbon, microbial biomass and activity levels in a semi-arid Mediterranean soil. Appl Soil Ecol 36; 107–115.
19. Bachar A, Soares MIM, Gillor O (2012) The effect of resource islands on abundance and diversity of bacteria in arid soils. Microb Ecol 63: 694–700.
20. Hardoim PR, van Overbeek LS, van Elsas JD (2008) Properties of bacterial endophytes and their proposed role in plant growth. Trends Microbiol 16: 463–471.
21. Ryan RP, Germaine K, Franks A, Ryan DJ, et al. (2008). Bacterial endophytes: recent developments and applications. FEMS Microbiol. Lett. 278: 1–9.
22. Lundberg DS, Lebeis SL, Peredes SH, Yourstone S, et al. (2012) Defining the core *Arabidopsis thaliana* root microbiome. NATURE 488: 86–90.
23. Bulgarelli D, Rott M, Schaleppi K, Ver Loren van Themaat E, et al. (2012) Revealing structure and assembly cues for arabidopsis root-inhabiting bacterial microbiota. NATURE 488: 91–95.
24. Kniskern JM, Traw MB, Bergelson J (2007) Salycilic acid and jasmonic acid signalling defense pathway reduce natural bacterial diversity on *Arabidopsis thaliana*. MPMI 20: 1512–1522.
25. Long HH, Sonntag DG, Schmidt DD, Baldwin IT (2010) The structure of the culturable root bacterial endophyte community of *Nicotiana attenuata* is organized by soil composition and host plant ethylene production and perception. New Phytol 185: 554–567.
26. Balachandar D, Sandhiya GS, Sugitha TKC, Kumar K (2006) Flavonoids and growth hormones influence endophytic colonization and in planta nitrogen fixation by a diazotrophic Serratia sp in rice. World Journal of Microbiology and Biotechnology 22: 707–712.
27. Mendes R, Pizzirani-Kleiner AA, Araújo WL, Raaijmakers JM (2007) Diversity of cultivated endophytic bacteria from sugarcane: genetic and biochemical characterization of *Burkholderia cepacia* complex isolates. Appl Environ Microbiol 73: 7259–7267.
28. Han J, Xia D, Li L, Sun L, Yang K, et al. (2009) Diversity of Culturable Bacteria Isolated from Root Domains of Moso Bamboo (*Phyllostachys edulis*). Microbial Ecology 58: 363–373.
29. Joseph B, Ranjan Patra, Lawrence R (2007) Characterization of plant growth promoting rhizobacteria associated with chickpea (*Cicer arietinum* L.). Int J Plant Production 1: 141–152.
30. Garbeva P, van Veen JA, van Elsas JD (2003) Predominant Bacillus spp. in Agricultural Soil under Different Management Regimes Detected via PCR-DGGE. Microbial Ecology 45: 302–316.
31. Benardini JN, Sawyer J, Venkateswaran K, Nicholson WL (2003) Spore UV and acceleration resistance of endolithic *Bacillus pumilus* and *Bacillus subtilis* isolates obtained from Sonoran desert basalt: implications for lithopanspermia. Astrobiology 3: 709–717.
32. Francis I, Holsters M, Vereecke D (2010) The Gram-positive side of plant–microbe interactions. Environ Microbiol 12: 1–12.
33. Lal S, Tabacchioni S (2009) Ecology and biotechnological potential of *Paenibacillus polymyxa*: a minireview. Indian Journal of Microbiology 49: 2–10.
34. Timmusk S, Paalme V, Pavlicek T, Bergquist J, Vangala A, et al. (2011) Bacterial Distribution in the Rhizosphere of Wild Barley under Contrasting Microclimates. PLoS ONE 6: e17968–e17977.
35. Trivedi P, Spann T, Wang N (2011) Isolation and Characterization of Beneficial Bacteria Associated with Citrus Roots in Florida. Microbial Ecology 62: 324–336.
36. Kavroulakis N, Ntougias S, Besi MI, Katsou P, Damaskinou A, et al. (2010) Antagonistic bacteria of composted agro-industrial residues exhibit antibiosis against soil-borne fungal plant pathogens and protection of tomato plants from *Fusarium oxysporum* f.sp. *radicis-lycopersici*. Plant Soil 333: 233–247.
37. Kumar G, Kanaujia N, Bafana A (2012) Functional and phylogenetic diversity of root-associated bacteria of *Ajuga bracteosa* in Kangra valley. Microbiol Res 167: 220–225.
38. Sgroy V, Cassán F, Masciarelli O, Florencia Del Papa M, Lagares A, et al. (2009) Isolation and characterization of endophytic plant growth-promoting (PGPB) or stress homeostasis-regulating (PSHB) bacteria associated to the halophyte *Prosopis strombulifera*. Appl Microbiol Biotechnol 85: 371–381.
39. Aravind R, Kumar A, Eapen SJ, Ramana KV (2009) Endophytic bacterial flora in root and stem tissues of black pepper (*Piper nigrum* L.) genotype: isolation, identification and evaluation against *Phytophthora capsici*. Lett Appl Microbiol 48: 58–64.
40. Khan Z, Doty SL (2009) Characterization of bacterial endophytes of sweet potato plants. Plant Soil 322: 197–207.
41. Lodewyckx C, Vangronsveld J, Porteous F, Moore ERB, Taghavi S, et al. (2002) Endophytic bacteria and their potential applications. Plant Sci 21: 583–606.
42. Sturz AV, Christie BR, Nowak J (2000) Bacterial endophytes: potential role in developing sustainable systems of crop production. Plant Sci 19: 1–30.
43. Rasche F, Trondl R, Naglreiter C, Reichenauer TG, Sessitsch A (2006) Chilling and cultivar type affect the diversity of bacterial endophytes colonizing sweet pepper (*Capsicum anuum* L.). Can J Microbiol 52: 1036–1045.
44. Ambrosini A, Beneduzi A, Stefanski T, Pinheiro FG, Vargas LK, et al. (2012) Screening of plant growth promoting Rhizobacteria isolated from sunflower (*Helianthus annuus* L.). Plant Soil 356: 245–264.
45. Hayat R, Ali S, Amara U, Khalid R, Ahmed I (2010) Soil beneficial bacteria and their role in plant growth promotion: a review. Annal Microbiol 60: 579–598.
46. Park MS, Jung SR, Lee MS, Kim KO, Do JO, et al. (2005) Isolation and characterization of bacteria associated with two sand dune plant species, *Calystegia soldanella* and *Elymus mollis*. The Journal of Microbiology 43: 219–227.
47. Shankar M, Ponraj P, Ilakkiam D, Gunasekaran P (2011) Root colonization of a rice growth promoting strain of *Enterobacter cloacae*. J Basic Microbiol 51: 523–530.
48. Madigan MT, Martinko JM, Dunlap PV, Clark DP (2009) Brock Biology of Microorganisms. San Francisco: Pearson Benjamin Cummings. 12th ed.
49. Fan B, Borriss R, Bleiss W, Wu X (2012) Gram-positive rhizobacterium *Bacillus amyloliquefaciens* FZB42 colonizes three types of plants in different patterns. J Microbiol 50: 38–44.
50. Chanal A, Chapon V, Benzerara K, Barakat M, Christen R, et al. (2006) The desert of Tataouine: an extreme environment that hosts a wide diversity of microorganisms and radiotolerant bacteria. *Environ Microbiol* 8: 514–525.
51. De Luca G, Barakat M, Ortet P, Fochesato S, Jourlin-Castelli C, et al. (2011) The Cyst-dividing bacterium *Ramlibacter tataouinensis* TTB310 genome reveals a well-stocked toolbox for adaptation to a desert environment. PLoS ONE 6: e23784–e23784.
52. Redman RS, Kim YO, Woodward CJDA, Greer C, Espino L, et al. (2011) Increased fitness of rice plants to abiotic stress via habitat adapted symbiosis: a strategy for mitigating impacts of climate change. PLoS ONE 6: e14823–e14823.

53. Ciais P, Reichstein M, Viovy N, Granier A, Ogée J, et al. (2005) Europe-wide reduction in primary productivity caused by the heat drought in 2003. Nature 437: 529–533.
54. Glick BR, Cheng Z, Czamy J, Duan J (2007) Promotion of plant growth by ACC deaminase-containing soil bacteria. Eur J Plant Pathol 3: 329–339.
55. Penrose DM, Glick BR (2003) Methods for isolating and characterizing ACC deaminase-containing plant growth-promoting rhizobacteria. Physiologia Plantarum 118: 10–15.
56. Siddikee MA, Chauhan PS, Anandham R, Han GH, Sa T (2010) Isolation, characterization, and use for plant growth promotion under salt stress, of ACC deaminase-producing halotolerant bacteria derived from coastal soil. Journal of Microbiology and Biotechnology 20: 1577–1584.
57. Belimov AA, Dodd IC, Hontzeas N, Theobald JC, Safronova VI (2009) Rhizosphere bacteria containing 1-aminocyclopropane-1-carboxylate deaminase increase yield of plants grown in drying soil via both local and systemic hormone signaling. New Phytologist 181: 413–423.
58. Rajkumar M, Sandhya S, Prasad MNV, Freitas H (2012) Perspectives of plant-associated microbes in heavy metal phytoremediation. Biotechnology Advances doi:10.1016/j.biotechadv.2012.04.011.
59. Balloi A, Rolli E, Marasco R, Mapelli F, Tamagnini I, et al. (2010) The role of microorganisms in bioremediation and phytoremediation of polluted and stressed soils. Agrochimica 54: 353–369.
60. Bartels D, Sunkar R (2005) Drought and salt tolerance in plants. CRC Crit Rev Plant Sci 24: 23–58.
61. Muyzer G, De Waal EC, Uitterlinden AG (1993) Profiling of complex microbial populations by denaturing gradient gel electrophoresis analysis of polymerase chain reaction-amplified genes coding for 16S rRNA. Appl Environ Microbiol 59: 695–700.
62. Rasband WS 1997–2012 ImageJ. US National Institutes of Health. Bethesda, Maryland, USA.
63. Sun L, Qiu F, Zhang X, Dai X, Dong X, et al. (2008) Endophytic bacterial diversity in rice (*Oryza sativa* L.) roots estimated by 16S rDNA sequence analysis. Microb Ecol 55: 415–424.
64. King EO, Ward MK, Raney DE (1954) Two simple media for the demonstration of phycocyanin and fluorescin. J Lab Clin Med 44: 301–307.
65. Penrose DM, Glick BR (2003) Methods for isolating and characterizing ACC deaminase-containing plant growth-promoting rhizobacteria. Physiol Plant 118: 10–15.
66. Cardinale M, Brusetti L, Quatrini P, Borin S, Puglia AM, et al. (2004) Comparison of different primer sets for use in automated ribosomal intergenic spacer analysis of complex bacterial communities. Appl Environ Microbiol 70: 6147–6156.
67. Thompson JD, Gibson TJ, Plewniak F, Jeanmougin F, Higgins DG (1997) The CLUSTAL_X windows interface: flexible strategies for multiple sequence alignment aided by quality analysis tools. Nucleic Acids Res 25: 4876–4882.
68. Schloss PD, Westcott SL, Ryabin T, Hall JR, Hartmann M, et al. (2009) Introducing mothur: open-source, platform-independent, community-supported software for describing and comparing microbial communities. Appl Environ Microbiol 75: 7537–7541.
69. Hammer Ø, Harper DAT, Ryan PD (2001) PAST: paleontological statistics software package for education and data analysis. Palaeontol Electronica 4: 1.
70. Saitou N, Nei M (1987) The neighbor-joining method: A new method for reconstructing phylogenetic trees. Mol Biol Evol 4: 406–425.
71. Tamura K, Dudley J, Nei M, Kumar S (2008) MEGA4: molecular evolutionary genetics analysis (MEGA) software version 4.0. Mol Biol Evol 24: 1596–1599.
72. Brick JM, Bostock RM, Silverstone SE (1991) Rapid in situ assay for indoleacetic acid production by bacteria immobilized on nitrocellulose membrane. Appl Environ Microbiol 57: 535–538.
73. Mehta S, Nautiyal SC (2001) An efficient method for qualitative screening of phosphate-solubilizing bacteria. Curr Microbiol 43: 51–56.
74. Schwyn B, Neilands JB (1997) Universal chemical assay for the detection and determination of siderophores. Ana Biochem 160: 46–56.
75. Santaella C, Schue M, Berge O, Heulin T, Achouak W (2008) The exopolysaccharide of *Rhizobium* sp. YAS34 is not necessary for biofilm formation on *Arabidopsis thaliana* and *Brassica napus* roots but contributes to root colonization. Environ Microbiol 10: 2150–2163.
76. Cappuccino JC, Sherman N (1992) In:Microbiology: A Laboratory Manual. New York: third ed. Benjamin/cummings Pub. Co. 125–179.
77. Nielsen P, Sørensen J (1997) Multi-target and medium-independent fungal antagonism by hydrolytic enzymes in *Paenibacillus polymyxa* and *Bacillus pumilus* strains from barley rhizosphere. FEMS Microbiol Ecol 22: 183–192.
78. Favia G, Ricci I, Marzorati M, Negri I, Alma A, et al. (2008) Bacteria of the genus *Asaia*: a potential paratransgenic weapon against malaria. Adv Exp Med Biol 627: 49–59, 2008.

19

Emissions of NO and NH_3 from a Typical Vegetable-Land Soil after the Application of Chemical N Fertilizers in the Pearl River Delta

Dejun Li*

Department of Microbiology and Plant Biology, University of Oklahoma, Norman, Oklahoma, United States of America

Abstract

Cropland soil is an important source of atmospheric nitric oxide (NO) and ammonia (NH_3). Chinese croplands are characterized by intensive management, but limited information is available with regard to NO emissions from croplands in China and NH_3 emissions in south China. In this study, a mesocosm experiment was conducted to measure NO and NH_3 emissions from a typical vegetable-land soil in the Pearl River Delta following the applications of 150 kg N ha^{-1} as urea, ammonium nitrate (AN) and ammonium bicarbonate (ABC), respectively. Over the sampling period after fertilization (72 days for NO and 39 days for NH_3), mean NO fluxes ($\pm$ standard error of three replicates) in the control and urea, AN and ABC fertilized mesocosms were 10.9±0.9, 73.1±2.9, 63.9±1.8 and 66.0±4.0 ng N m^{-2} s^{-1}, respectively; mean NH_3 fluxes were 8.9±0.2, 493.6±4.4, 144.8±0.1 and 684.7±8.4 ng N m^{-2} s^{-1}, respectively. The fertilizer-induced NO emission factors for urea, AN and ABC were 2.6±0.1%, 2.2±0.1% and 2.3±0.2%, respectively. The fertilizer-induced NH_3 emission factors for the three fertilizers were 10.9±0.2%, 3.1±0.1% and 15.2±0.4%, respectively. From the perspective of air quality protection, it would be better to increase the proportion of AN application due to its lower emission factors for both NO and NH_3.

Editor: Caroline P. Slomp, Utrecht University, The Netherlands

Funding: This work was financially supported by an innovative project of Guangzhou Institute of Geochemistry (54O7342101). The funders had no role in study design, data collection and analysis, decision to publish, or preparation of the manuscript.

Competing Interests: The author has declared that no competing interests exist.

* E-mail: dejunl@gmail.com

Introduction

Increases of atmospheric trace gases are of increasing concern due to their roles in detrimental health and environmental effects. Nitrogen oxides ($NO_x = NO+NO_2$) and ammonia (NH_3) are among these gases. NO_x catalyze the photochemical formation of ground-level ozone [1], which is a potent greenhouse gas [2] and poses a threat to human health and vegetation on regional scales [3]. The photochemical end product, HNO_3, is a major component of acid rain [1]. NH_3 enhances aerosol formation [4], and hence influences regional air quality.

Soil is a major source for atmospheric NO_x. According to Davidson and Kingerlee (1997), the source strength of soils is just inferior to fossil fuel combustion globally, while cropland is the second largest contributor to atmospheric NO_x among all types of land uses. Current estimates of soil NO emissions from cropland range from 2.4–5.4 Tg N yr^{-1} [5,6,7]. The large uncertainty is largely caused by limited data availability, especially in tropical and subtropical regions. Furthermore, fertilized cropland (12.6 Tg N yr^{-1}) ranks second among all the sources of atmospheric NH_3 and is responsible for 23.5% of the global annual emission [8]. Of the emitted NO and NH_3 from cropland, the major part is induced by fertilizer application [9].

Chinese croplands have been intensively managed. With the cropland area only occupying 7% of the global total, the annual consumption of synthetic N fertilizer in China accounted for ca. 30% of the total global consumption in 2004 [10]. The annual fertilizer N consumption in China has been increasing by about 6.5%, which is higher than the mean rate of 4.8% for other Asian countries [11]. Urea ($CO(NH_2)_2$) and ammonium bicarbonate (NH_4HCO_3, ABC) are the two most widely used synthetic N fertilizers in China, with the latter only used in China [9,12].

Most of the previous studies on NH_3 volatilization from Chinese croplands were carried out in North China Plain and eastern China's Yangtze River Delta [12], but, to our knowledge, little information is available for south China. NO emissions from Chinese croplands have been reported by several studies [11,13,14,15,16,17,18]. However, no study has been conducted to investigate NO emission following ABC application, although ABC is a main N fertilizer in China (Figure 1), Due the lack of a NO emission factor for ABC, one study used the emission factor for ammonium sulphate as an estimate of that for ABC [19].

Vegetable land forms an important part of China's croplands. On a national scale, vegetable land accounts for ca. 7% of the total cropland, and the ratios are typically higher in developed regions [15]. For example, in south China's Guangdong province, the ratio is 24.2% (1.16×10^6 ha) in 2005 [15]; in the Yangtze Delta of east China, vegetable land (ca. 6×10^6 ha) also occupies one fifth of the total cropland area [14]. Compared with other cultivated lands, vegetable lands receive more fertilizer-N per unit of area. For example, N application rates in individual growing season for vegetables range from 300 to 700 kg N ha^{-1}, compared to 150–300 kg N ha^{-1} for non-vegetable crops [20]. These higher N fertilizer application rates to vegetable lands would inevitably

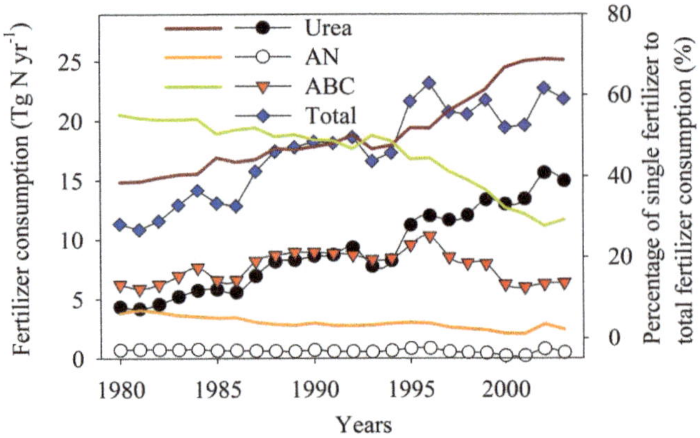

Figure 1. Changes of fertilizer consumption (line and scatter) and percentage (bold line) of single fertilizer consumption to total consumption of urea, ammonium nitrate (AN) and ammonium bicarbonate (ABC) during 1980 to 2003 in China [10].

induce larger amount of N loss to the environment, including leaching of nitrate to surface/ground water, and emission of nitrogen-containing gases to the atmosphere.

In the present study, we conducted a mesocosm experiment in which we measured emissions of NO and NH_3 from a subtropical vegetable-land soil following the application of urea, ABC, and NH_4NO_3 (AN), respectively. The main objective was to quantify the emission factors for the three synthetic fertilizers.

Materials and Methods

Ethics Statement

The vegetable field for arable soil sampling was privately owned. We collected soil under the permission of the land owner Mrs Cheng. All other necessary permits were obtained for the described field studies.

Site Description

The experiment was conducted at a suburban site (23°10'N, 113°23'E) of Guangzhou, the capital of Guangdong province. This area enjoys a subtropical monsoon climate, with an annual mean rainfall of 1938.2 mm and annual mean air temperature of 22.5°C. Monthly mean air temperature is lowest in January (13.3°C) and highest in July (29.5°C) [21].

Flowering Chinese cabbage (*Brassica campestris* L. ssp. *Chinensis* var. *utilis* Tsen et Lee) is the most widely cultivated vegetable type in Guangdong Province [15]. Vegetables can grow under field conditions all year round in this region. Typically, a new batch of vegetables is planted approximately five days after a batch of vegetables is harvested, and thus no long fallow period exists. At the study site, a typical cultivation cycle for flowering Chinese Cabbage takes about 47 days and mainly involves five management events: sowing, harrowing and transplanting, first fertilization, second fertilization, harvesting and harrowing. The rate of fertilizer application is about 320 kg N ha^{-1} yr^{-1} with about 5 batches of vegetables grown in a year according to our previous field study [15].3

Experimental Design

The arable soil (top layer of 20 cm) used in this study was collected from a vegetable field which has been cultivated for flowering Chinese cabbage for more than 10 years (Personal communication with the farmer). The soil type is lateritic red earth, which is typical in subtropical China. Soil properties (mean ± SE, n = 5) were analyzed after the soil was well mixed. Soil pH (extracted with KCl solution), contents of total C and total N were

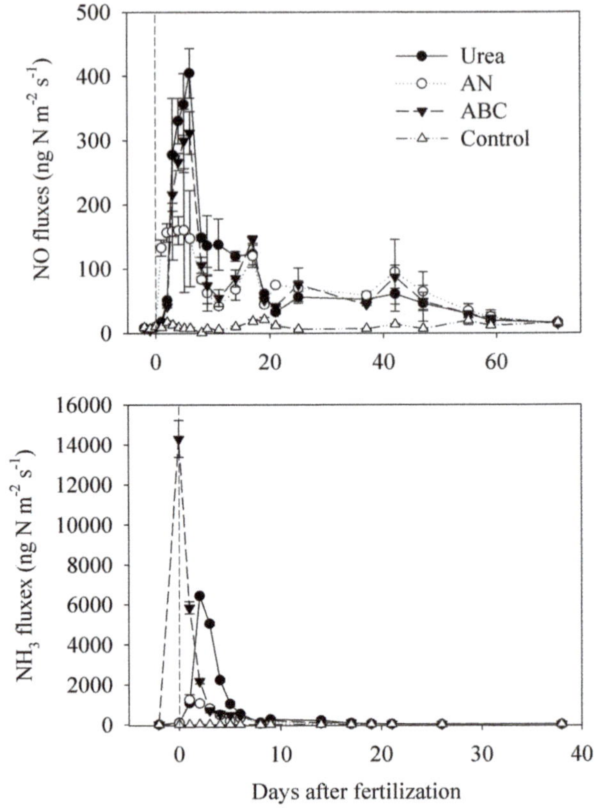

Figure 2. NO and NH_3 fluxes during the sampling period. Each value is the mean of three replicates, and error bars represent standard errors. Vertical dash line indicates the day when fertilizers were applied.

Table 1. Fluxes of NO and NH_3, total emissions (E_{total}), net emissions (E_{net}), and emission factors (EF) for the three fertilizers.

	Flux	E_{total}	E_{net}	EF
	ng N m^{-2} s^{-1}	mg N m^{-2}	mg N m^{-2}	%
NO				
Urea	73.1±2.9 a	454.6±17.9 a	386.7±18.7 a	2.6±0.1 a
AN	63.9±1.8 b	397.5±10.9 b	329.6±12.2 b	2.2±0.1 b
ABC	66.0±4.0 b	410.3±25.1 b	342.4±25.7 b	2.3±0.2 b
Control	10.9±0.9 c	67.9±5.5 c		
NH_3				
Urea	493.6±4.4 b	1663.1±27.3 b	1633.2±27.3 b	10.9±0.2 b
AN	144.8±0.1 c	488.0±0.5 c	458.1±1.1 c	3.1±0.1 c
ABC	684.7±8.4 a	2307.1±52.2 a	2277.2±52.2 a	15.2±0.4 a
Control	8.9±0.2 d	29.9±1.0 d		

Values are presented as mean ± standard error of three replicates. Different letters in a column denote that the mean values are significantly different at *P*<0.05 level.

Table 2. NO emissions from fertilized upland arable fields in China.

Location	Land-use type	Fertilizer type	Duration	N rates	NO flux	Emission factors %
				kg N ha^{-1}	ng N m^{-2} s^{-1}	
39°57′N; 116°18′E[a]	Corn	OM[j]	N/A	88.5	0.14	0.04
39°57′N; 116°18′E[a]	Corn	Urea	N/A	150	77.5	1.6
39°57′N; 116°18′E[a]	Corn	Urea	N/A	300	76.4	0.75
39°57′N; 116°18′E[a]	Corn	OM+urea	N/A	238.5	66.4	0.78
39°57′N; 116°18′E[a]	Corn	OM+urea	N/A	238.5	150.8	1.3
34°56′N; 110°43′E[b]	Cotton	Urea+DP+PS	Jan-Dec	66.3	2.4	0.24
32°35′N; 119°42′E[c]	Bare soil	Urea	Jan-Dec	153	70.4	0.67
32°35′N; 119°42′E[c]	Vegetables*		Jan-Dec	118–548	2.5–142.7	0.05–1.24
31°16′N; 120°38′E[d]	Wheat	M+CF+urea	Nov-Jun	191	21.3	1.75
31°16′N; 120°38′E[d]	Wheat	CF+urea	Nov-Jun	191	28.5	2.5
31°16′N; 120°38′E[d]	Wheat	CF+urea	Nov-Jun	191	22.4	1.87
30°50′N; 120°42′E[e]	Cabbage	CF	Mar -Jun	45	11.5	0.6
30°50′N; 120°42′E[e]	Potato	CF+M	Mar -Jun	45.6	34.2	3.6
30°50′N; 120°42′E[f]	Cabbage	CF	Mar -Jun	135	20.9	1.05
30°50′N; 120°42′E[f]	Potato	CF+ M	Mar -Jun	108	27.4	1.75
30°50′N; 120°42′E[f]	Soybean	CF+urea	Mar -Jun	81	21.4	1.83
30°50′N; 120°42′E[g]	Cabbage	CF+M+urea	Aug-Dec	271.2	33.8	1.2
30°50′N; 120°42′E[g]	Garlic	CF+M+urea	Aug-Dec	267.3	360	11.56
30°50′N; 120°42′E[g]	Radish	CF+M+urea	Aug-Dec	263.6	76	2.56
23°10′N; 113°23′E[h]	FCC	CF+M+urea	Sep-Oct	63.5	47.5	2.4
23°10′N; 113°23′E[i]	Bare soil	Urea	Dec-Mar	150	73.1	2.6
23°10′N; 113°23′E[i]	Bare soil	AN	Dec-Mar	150	63.9	2.2
23°10′N; 113°23′E[i]	Bare soil	ABC	Dec-Mar	150	66	2.3

[a]Walsh [37];
[b]Liu et al. [18];
[c]Mei et al., [16];
[d]Zheng et al. [11];
[e]Fang and Mu [14];
[f]Fang and Mu [38];
[g]Pang et al. [17];
[h]Li and Wang [15];
[i]This study.
[j]OM–organic matter; CF–compound fertilizer; M–Manure; FCC– Flowering Chinese cabbage; DP-diammonium phosphate; PS-potassium sulphate;
*Different vegetables were cultivated from 2004 to 2008 and various fertilizers were used.
N/A: not available.

6.99±0.02, 1.69±0.07% and 0.14±0.01%. Soil texture was sandy clay loam with a bulk density of 1.18±0.18 g cm^{-3} at the mid of a cultivation cycle of the vegetable [22].

On December 10, 2005, mixed soil (after excluding coarse roots and residuals) was put into twelve plastic containers, each with a height of 40 cm and an area of 1 m^2. There was a 10-cm layer of gravel and 10 holes (each with a diameter of about 0.5 cm), which were used for drainage, at the bottom of each container. Each container was filled with 362 kg soil (with gravimetric soil water content of about 18.5%) to a depth of 25 cm to achieve the field bulk density measured at the mid of a cultivation cycle. Therefore, our study can be regarded as a mesocosm experiment. The mesocosms were placed on an open field in our institute, at a distance of approximately 2 km from the vegetable fields.

Urea, AN and ABC were applied at a rate of 150 kg N ha^{-1}. Each treatment had three replications. On December 29, 2005 (the temperature of this period was similar to the annual average, details are shown in the following section), the fertilizers were dissolved in 1 L water and applied to the soil surface with a handhold sprayer. Three mesocosms were used as control and was only treated with 1 L water. The treatment process completed within 30 min. Deionized water was sprayed to the soil to maintain the soil moisture within a range similar to that of the field (with gravimetric water content of 16.6%–22.5%, which are equivalent to volumetric soil water content of 19.6–26.6% based on the relationship between gravimetric water content and volumetric water content [23]) [15] by using moisture probe meter (MP-508B, China) to monitor the change of volumetric soil water content at the depth of 10 cm.

Measurement of NO Fluxes

Measurement of NO fluxes started on December 25, 2005 and lasted until March 10, 2006. Measurements were conducted every one day before fertilization. Upon fertilization, NO fluxes were measured once a day for a week and then once every two days for the following two weeks. Thereafter, measurement was conducted

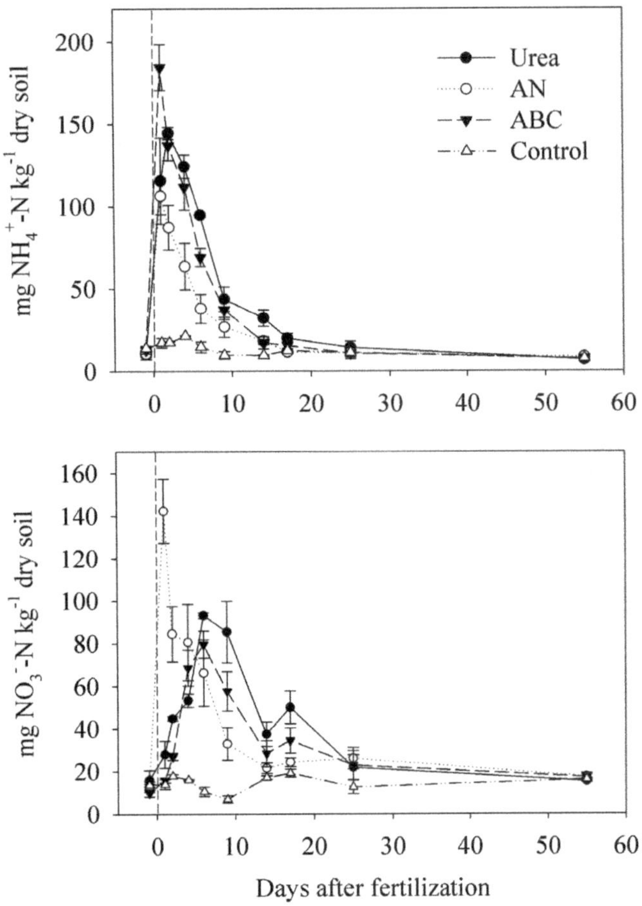

Figure 3. Soil inorganic N pools during the sampling period. Each value is the mean of three replicates, and error bars represent standard errors.

once every 4–12 days. On each sampling day, flux measurements were conducted in the morning (9:30–10:30), at noon (12:00–13:00), and in the afternoon (3:00–4:00).

A dynamic flow-through chamber technique was used to measure NO fluxes. The chamber system was described in detail previously [24,25]. The chambers are made of stainless steel (inner walls coated with Teflon films), each covering an area of 30 cm ×30 cm with a total volume of 9 L. Each chamber has one inlet port, one exhaust port and one outlet port for sampling. Inside each chamber, a thermo-sensor is fixed to measure air temperature, and a fan is attached to ensure sufficient mixing of air within the chamber. Different from the chamber system described previously [24,25], for this study each chamber was coupled with a steel pedestal (with a depth of 10 cm), at the top of which was a water-filled channel. One pedestal was put in the center of each container and inserted into the soil with a depth of 10 cm. During sampling, water was added to the channel of the pedestal so that the chamber and pedestal were sealed by the water. An additional reference chamber, closed at the bottom with Teflon sheet, was employed for in situ quantification of chemical reactions and chamber wall deposition effects. Ambient air was pumped into the chambers at a rate of 4 L min^{-1} through 10 m long Teflon tubes with inner diameters of 4.8 mm, and the sample air was taken in through tubes of the same dimension. The residence time of air in the chambers was about 2.25 min. After about 15 min (over 5 cycles of residence time) when a steady state was reached inside the chambers, NO was analyzed by a model 42C chemiluminescence NO-NO_2-NO_x analyzer (Thermo Environmental Instruments Inc., USA) for 3 minutes. By the difference of sampling chambers and the reference chamber, net fluxes from the soils could be calculated from equation (1) [26]:

$$F = (C_{eq} - C_0) \times \frac{M}{Vm} \times \frac{Q}{A} \qquad (1)$$

where F is the net flux in ng m^{-2} s^{-1}, M is the atomic weight of the element (N = 14.008 g mol^{-1}, C = 12.011 g mol^{-1}), V_m is the standard gaseous molar volume (24.055×10^{-3} m^{-3} mol^{-1}), C_{eq} is the mixing ratio (ppbv = 10^{-9} m^{-3} m^{-3}) of the gas when the chamber under consideration has reached steady state, C_0 is the mixing ratio of the gas in the reference chamber, Q is the mass flow rate of air through the chamber (0.667×10^{-5} m^3 s^{-1}), and A is the soil surface area (0.09 m^2) covered by the chamber.

Measurement of NH_3 Fluxes

Measurement of NH_3 fluxes started on December 25, 2005 and lasted until February 4 2006. After sampling of NO fluxes, NH_3 was sampled with an acid trap method, which is commonly used for NH_3 measurement [27]. During sampling, the flow rate was also 4 L min^{-1}, and sample air was drawn at a rate of 1 L min^{-1} for 30 min to the bubblers containing 20 ml of 0.7 M H_2SO_4. The NH_4^+ concentration trapped in the acid was determined by the indophenol blue method. Results gained with the acid trap method have been proven to be comparable with those by chemiluminescence NH_3-NO_x analyzer [27]. Net fluxes were calculated by the difference of sampling chambers and the reference chamber using the equation (2):

$$F = \frac{C_s - C_r}{A \times T} \qquad (2)$$

where C_s and C_r represent the amount of N (ng) trapped in the bubblers connected to the sampling chamber and reference chamber, respectively. A is the soil surface area (0.09 m^2) covered by the chamber, and T is the sampling time (1800 s).

Auxiliary Measurements

Along with measurement of fluxes, air temperature at a height of 1.5 m, temperature inside the chambers, and soil temperature at 0–5 cm depth was measured with soil temperature probes (TES Ltd., China).

At each sampling date, three composite soil samples were collected from 0–5 cm depth. These samples were used for determinations of gravimetric soil moisture and KCl-extractable inorganic nitrogen pools. Briefly, 40 g fresh soil samples were extracted with 200 ml KCl solution (1 mol L^{-1}) for one hour. NH_4^+-N was measured by indophenol blue method and determined spectrophotometrically at 630 nm; NO_3^-N was determined by reduction to nitrite (NO_2-N) via a cadmium reactor, diazotized with sulfanilamide and is coupled to N-(1-Napthyl)-ethylenediamine dihydrochloride to form an azochromophore measured spectrophotometrically at 543 nm [22].

Statistics and Data Analysis

Daily averaged flux of NO or NH_3 for each mesocosm was the average of the three measurements on each sampling day. Linear interpolation approach was used to estimate the daily fluxes for dates on which measurements were not made based on the fluxes from the dates when measurements were conducted immediately preceding and immediately following the dates that were not measured. Mean fluxes (ng N m^{-2} s^{-1}) over the sampling period

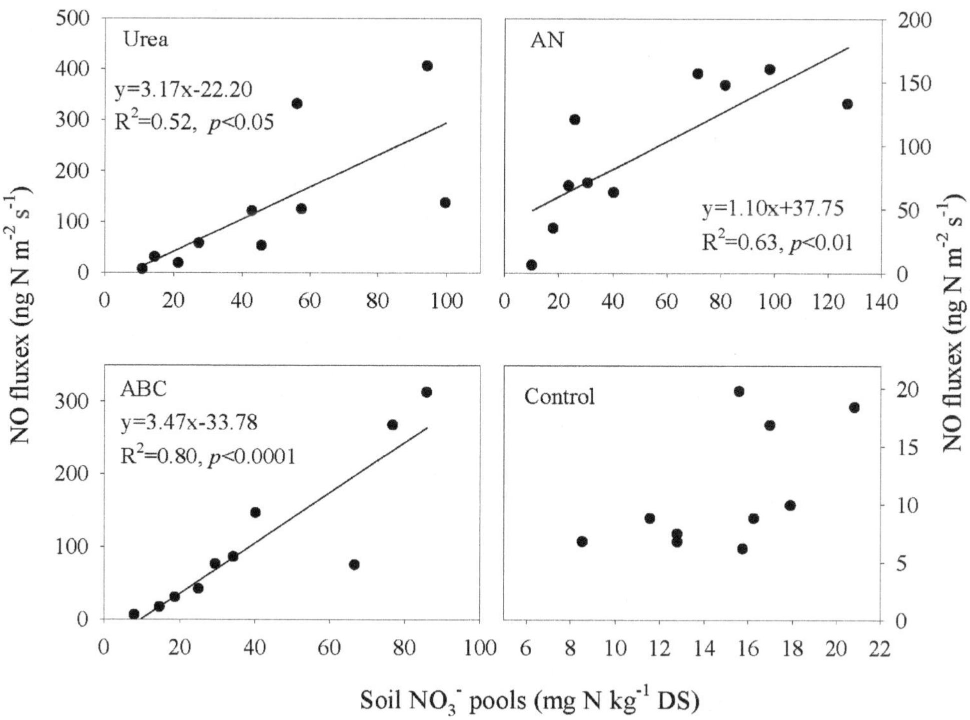

Figure 4. Correlation between NO fluxes and soil NO_3^- pools (n = 10). Each value is the mean of three replicates.

after fertilization (72 days for NO and 39 days for NH_3) were calculated after data interpolation. Total emissions (E_{total}, mg N m^{-2}) were calculated by multiplying mean fluxes by sampling days after fertilization. Net emission (E_{net}, mg N m^{-2}) was obtained by subtracting E_{total} of the control from that of the corresponding fertilized mesocosms. The emission factor (%) of NO (F_{NO}) or NH_3 (F_{NH3}) was calculated based on N atoms according to equation (3):

$$emission\,factor = \frac{total\ N\ emission_{fertilizer\ treatment} - total\ N\ emission_{control}}{applied\ N} \times 100 \quad (3)$$

Reported data of each treatment were the mean of the three replicate mesocosms on a daily basis. ANOVA analyses with post hoc LSD tests were performed using SPSS 10.0 (SPSS Ltd., USA) to identify differences among treatments. In this paper, analyses with *P* values <0.05 were considered significant.

Results and Discussion

NO Fluxes

Before fertilization, the mean NO flux was 6.9±0.6 (mean ± standard error of three replicates) ng N m^{-2} s^{-1} (Figure 2). The NO flux increased quickly in the AN fertilized mesocosms and reached 132.8±12.4 ng N m^{-2} s^{-1} one day after fertilization. The peaks of NO fluxes for different fertilizers occurred at different times. For AN fertilized mesocosms, the flux peaked 5 days after fertilization. For both the urea and ABC fertilized mesocosms, flux peaks were observed 6 days after fertilization. However, the peak fluxes for different fertilizers varied greatly. The averaged peak flux for urea was the greatest (404.9±38.3 ng N m^{-2} s^{-1}), much higher (P<0.05) than that for AN (160.4±96.5 ng N m^{-2} s^{-1}) or for ABC (312.1±33.0 ng N m^{-2} s^{-1}) (Figure 2). The enhancement effects of fertilization on NO emissions lasted nearly two months.

Over the experimental period after fertilization, mean NO fluxes were 10.9±0.9, 73.1±2.9, 63.9±1.8 and 66.0±4.0 ng N m^{-2} s^{-1}, respectively, for the control, urea, AN and ABC treatments (Table 1). The averaged NO flux for urea was significantly greater than those for AN and ABC (P<0.05). We previously reported that NO fluxes in a typical vegetable field (planted with flowering Chinese cabbage), where the soil for the current study was collected, varied from 20.0 to 122.1 ng N m^{-2} s^{-1} with an average of 47.5 ng N m^{-2} s^{-1} [15], which was relatively small compared with the fluxes in the present study. However, considering that only 63.5 kg N ha^{-1} was applied in the field study (also see Table 2), a lower averaged NO flux was reasonable. According to a review paper, NO emissions ranged from 1.19 to 44 ng N m^{-2} s^{-1} from recently fertilized agricultural soils in temperate regions [5]. However, surprisingly high NO fluxes were also recorded in temperate croplands [28]. For example, Boeckx and Van Cleemput [28] observed that NO fluxes in autumn of 1997 (fertilized with 144 kg N ha^{-1}) and spring of 1998 (fertilized with 88 kg N ha^{-1} as cattle slurry) and summer of 1998 (fertilized with 122 kg N ha^{-1} as chicken manure) were19, 68, and 581 ng N m^{-2} s^{-1}, respectively, from sandy arable soil in Belgium. Measurements of NO fluxes in tropical croplands were rather limited, but the available fluxes from tropical croplands were not significantly higher than those from temperate croplands [5], suggesting that the effect of latitude was small.

Chinese croplands are famous for relatively high N fertilization. Table 2 compiles the available data of NO fluxes from Chinese cropland soils. Averaged NO fluxes over the sampling periods varied from 0.14 to 360 ng N m^{-2} s^{-1}. There is a strong

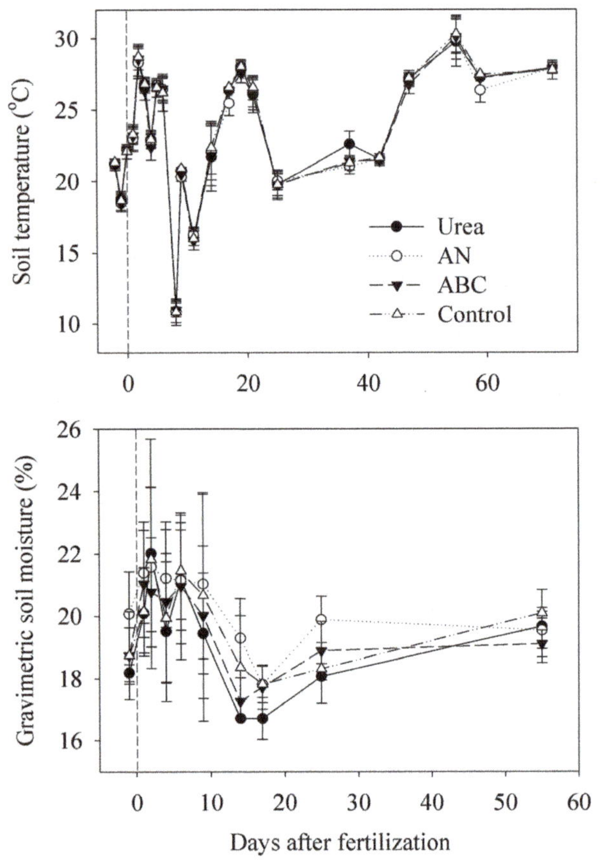

Figure 5. Variation of soil temperature and soil moisture in different treatment mesocosms during the sampling period. Each value is the mean of three replicates, and error bars represent standard errors. Vertical dash line indicates the day when fertilizers were applied.

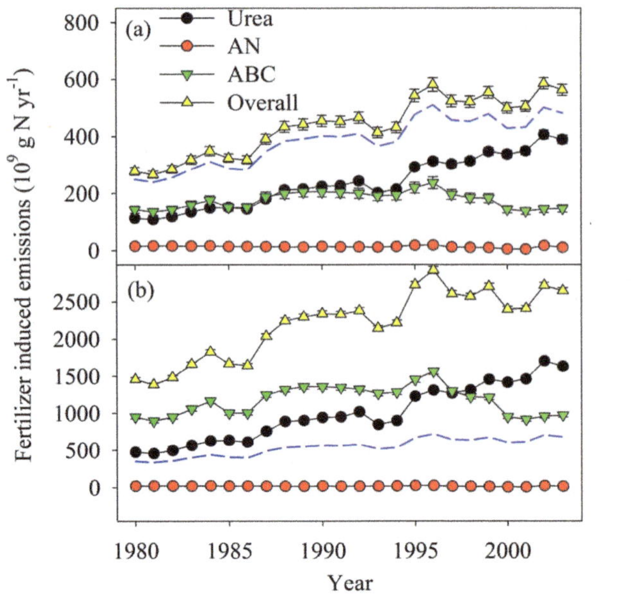

Figure 6. Changes of fertilizer induced emissions of (a) NO and (b) NH_3 by consumption of urea, ammonium nitrate (AN) and ammonium bicarbonate (ABC) in China between 1980 and 2003. Fertilizer induced emissions of NO and NH_3 were calculated by multiplying annual consumption of each fertilizer (Figure 1) by the corresponding emission factor obtained in the current study. The blue line in each panel represents the total emission of NO or NH_3 if the other two fertilizers were replaced by AN.

correlation between N fertilizer rates and NO fluxes in an exponential way ($y = 6.7861e^{0.0094x}$, $R^2 = 0.27$, $P<0.05$, $n = 23$). However, there is no significant correlation between latitude and NO fluxes across these studies ($P>0.05$, Table 2). Our results confirm that there is no strong effect of latitude on NO fluxes.

Influencing Factors for NO Fluxes

Many factors influence soil NO emissions, including N availability, soil moisture, soil temperature, atmospheric NO concentration, and other environmental factors that regulate the underlying processes of NO production and consumption in soils [29]. In well-drained agricultural soils, application of N fertilizers is often observed to cause large NO emissions [30]. This is because N fertilization supplies substrates (NH_4^+ and NO_3^-) to nitrifying or denitrifying bacteria, which are responsible for soil biogenic NO production [29]. In the current study, soil NO_3^- pools increase after NH_4^+-based fertilizer application, indicating the occurrence of nitrification (Figure 3). Soil NH_4^+ pools were found to be significantly correlated with NO emissions only in AN fertilized mesocosms ($R^2 = 0.60$, $p<0.01$), but soil NO_3^- pools were significantly correlated with NO emissions in all fertilized mesocosms (Figure 4). Since NO_3^- is the product of nitrification and substrate of denitrification, it is difficult to judge which process is mainly responsible for NO production from the correlation of NO_3^- and NO fluxes in the current study. Nevertheless, our field study indicated that nitrification was the dominant process [15]. Since the soil used in the current study was collected from the same vegetable field where our field study was conducted, nitrification was probably also the main process responsible for NO production in the current study [15].

Soil temperature and moisture as important regulators on NO emissions have often been reported [11,15,24,29]. In the present study, soil temperature at 0–5 cm depth ranged between 10–30°C (Figure 5). A strong correlation was only found in the control ($F = 0.46\exp^{0.12T}$, $R^2 = 0.62$, $P<0.0001$, $n = 23$), probably because the effects of fertilization masked the effects of soil temperature in the fertilized mesocosms. Soil moisture (Figure 5) was maintained within a narrow range (16–25%) during the experiment, which was similar to the field soil moisture (16.6–22.5%) of typical vegetable land in this region [15]. No clear correlation was found between NO fluxes and soil moisture, probably because during the experimental period of this study, soil moisture varied within a narrow range.

NH_3 Fluxes

Before fertilization, the average NH_3 flux was 10.7±3.8 ng N m^{-2} s^{-1}. NH_3 volatilization from the ABC fertilized mesocosms increased sharply immediately after fertilization, and reached a maximum on the treatment day, then decreased rapidly (Figure 2). NH_3 flux from urea fertilized mesocosms peaked 3 days after fertilization (Figure 2). Peak values of NH_3 fluxes from AN fertilized mesocosms were observed 2 days after fertilization (Figure 2). The enhancement effect of fertilization on NH_3 fluxes lasted nearly three weeks. Our results were consistent with others in the pattern of changes of NH_3 volatilization following fertilization, i.e., ABC usually led to large NH_3 volatilization immediately after application, then reduced dramatically [31]. Urea is not volatile itself, but will become volatile after it is hydrolyzed into NH_4^+ and CO_2 by ureases in the soil. This process was found to be completed within 2–3 days, and NH_3 volatilization increases with increasing pH and temperature [8].

NH_3 flux measurements lasted 39 days after fertilization. During this period, mean NH_3 fluxes from urea, AN and ABC fertilized mesocosms were 493.6±4.4, 144.8±0.1 and 684.7±8.4 ng N m^{-2} s^{-1}, respectively, compared to 8.9±0.2 ng N m^{-2} s^{-1} from the control (Table 1). The difference of averaged fluxes between any two fertilizers were significantly different ($P<0.5$).

N loss as NO

Total NO emissions from urea, AN and ABC fertilized mesocosms were 454.6±17.9, 397.5±10.9, 410.3±25.1 mg N m^{-2}, respectively, significantly greater than that from the control ($P<0.05$) (Table 1).

With regard to fertilizer-induced NO emission factor (EF), there was no significant difference between AN and ABC ($P>0.05$), but EF for urea (2.6±0.1%) was significantly greater ($P<0.05$) than the other two fertilizers (Table 1). Our data supported that urea had the highest EF [32], but was in conflict with some studies which reported that a mixed form fertilizer such as AN had the greatest emission in tropical savanna soil [33]. FAO/IFA [19] used EFs for urea, AN and ABC of 0.6%, 0.5% and 0.8%, respectively. However, they used the EF for ammonium sulphate as an estimate of EF for ABC since no study was conducted to determine EF for ABC at that time.

The fertilizer-induced NO emission factors based on studies in fertilized upland arable fields in China were presented in Table 2. The field measured EFs varied from 0.04 to 11.56 with an average of 1.93 (excluding EFs from the current study). The EFs observed in the present study were within the reported range but greater than the average of all the filed observed values. However, the EFs in the current study are quite similar to those derived from the field study (EF = 2.4%) conducted at a nearby site [15]. Nevertheless, it should be noted that the EFs in the current study ware based on a mesocosm experiment, therefore limitations may be involved when making comparison with field based studies.

N Loss as NH_3

Total emission of NH_3 from ABC fertilized mesocosms was the highest (2.31 g N m^{-2}), which was 1.4 and 4.7 times that from urea and AN fertilized mesocosms, respectively, and 77 times that from the control ($P<0.05$, Table 1). Total emission of NH_3 was significantly greater from urea than from AN fertilized mesocosms ($P<0.05$).

Fertilizer-induced NH_3 emission factors varied greatly for the three fertilizers (Table 1). EF for ABC was 15.2±0.4%, which was 1.4 times that for urea (10.9±0.2%), and 5 times that for AN (3.1±0.1%). Similarly, some studies in China reported that EF for ABC was about 1.5 times that for urea [34,35]. Yan et al. (2003) estimated NH_3 emission from dry croplands in East, South East and South Asia by adopting emission factors of 13.7% for urea and 20.5% for ABC. As no emission data for AN were available from Asian croplands, Yan et al. (2003) adopted an EF of 2.0%, which was proposed by European Environment Agency. According to the present study, the adoption of the European proposed EF for AN likely only slightly underestimated NH_3 emission due to AN fertilization in China. Nevertheless, more studies are needed since only one cropland soil was investigated in the current study and since our study was conducted as a mesocosm experiment.

It must be stressed that the chamber, though widely used, may lead to NH_3 volatilization rates somewhat different from those under field conditions. Wind speed exerts positive effects on NH_3 volatilization loss [36], but air currents inside the chamber by flowing through of air in addition to the mixing by fans, can not well simulate the wind above the soil under natural conditions. Thus as an approximation, NH_3 volatilization measured in this study might not be always consistent with that under field conditions.

Implications for National Emissions of NO and NH_3

In China, urea and ABC are two main N fertilizers, while AN only accounts for a small portion in total fertilizer N consumption. During 1980 to 2003 (Figure 1), urea consumption increased continuously; ABC consumption increased until 1996 and decreased thereafter; AN consumption tended to decrease continuously [10]. The total consumption of fertilizer N (only for the three fertilizers) peaked in 1996 and then levelled off (Figure 1). The portion of urea consumption increased from 38.6% to 68.6% of the total consumption of the three fertilizers, but that of ABC and AN decreased accordingly from 55.2% to 29.2% and from 6.3% to 2.2%, respectively (Figure 1). Since the emission factors of both NO and NH_3 are different for the three fertilizers, changes in fertilizer consumption and proportion would have a substantial impact on national budgets of NO and NH_3. Since the current study compars the emission factors for the three fertilizers from a typical cropland soil in China, the impact of changes in fertilizer consumption and proportion on national budgets of NO and NH_3 can be roughly estimated. However, since only one soil type was studied using a mesocosm experiment design, there are limitations involved in these estimates.

Total emissions of both NO and NH_3 from the three fertilizers peaked in 1996 and then levelled off (Figure 6). But for different fertilizers, the changes of emission during 1980–2003 are different. Emissions from urea application increased continuously for both gases. Emissions from ABC application peaked in 1996 and then decreased. Emissions from AN application were much smaller compared to the other two fertilizers and decreased over the period. Urea application has been the dominant source since 1994 for NO and since 1998 for NH_3. Since EFs of both NO and NH_3 emissions were the lowest for AN among the three studied fertilizers, the complete replacement of the other two fertilizers by AN would greatly decrease NO and NH_3 emissions, especially NH_3 (Figure 6). Therefore, in order to reduce air pollution, it would be better to increase the proportion of AN application.

Conclusions

In the present study, a mesocosm experiment was conducted to measure NO and NH_3 emissions from bare soil, which was collected from vegetable land, following the application of urea, AN and ABC. Emission factors of NO were 2.6%, 2.2% and 2.3% for the above three fertilizers, respectively; and emission factors of NH_3 were 10.9%, 3.1% and 15.2%, respectively. From the perspective of air quality protection, it would be better to increase the proportion of AN application due to its lower emission factors for both NO and NH_3.

Author Contributions

Conceived and designed the experiments: DL. Performed the experiments: DL. Analyzed the data: DL. Contributed reagents/materials/analysis tools: DL. Wrote the paper: DL.

References

1. Crutzen PJ (1979) The role of NO and NO_2 in the chemistry of the troposphere and stratosphere. Annual Reviews of Earth and Planetary Sciences 7: 443–472.
2. IPCC (2001) Climate Change 2001: The Scientific Basis. Cambridge, UK: Cambridge University Press.
3. Fowler D, Flechard C, Skiba U, Coyle M, Cape JN (1998) The atmospheric budget of oxidized nitrogen and its role in ozone formation and deposition. New Pyhtologist 139: 11–23.
4. Barthelmie RJ, Pryor SC (1998) Implications of ammonia emissions for fine aerosol formation and visibility impairment-a case study from the Lower Fraser Valley, British Columbia. Atmospheric Environment 32: 345–352.
5. Davidson EA, Kingerlee W (1997) A global inventory of nitric oxide emissions from soils. Nutrient Cycling in Agroecosystems 48: 37–50.
6. Yan XY, Ohara T, Akimoto H (2005) Statistical modeling of global soil NOx emissions. Global Biogeochemical Cycles 19: GB3019, doi:3010.1029/2004GB002276.
7. Yienger JJ, II HL (1995) Empirical model of global soil biogenic NOx emissions. Journal of Geophysical Research 100: 11447–11464.
8. Bouwman AF, Lee DS, Asman WAH, Dentener FJ, Hoek KWVD, et al. (1997) A global high-resolution emission inventory for ammonia. Global Biogeochemical Cycles 11: 561–587.
9. Yan XY, Akimoto H, Ohara T (2003) Estimation of nitrous oxide, nitric oxide and ammonia emissions from croplands in East, Southeast and South Asia. Global Change Biology 9: 1080–1096.
10. IFA (2007) Nitrogen, phosphate and potash statistics from 1973–1973/74 to 2004–2004/05. IFADATA statistics online, updated: 7 Feb. 2007. (http://www.fertilizer.org/ifa/statistics/IFADATA/dataline.asp).
11. Zheng X, Huang Y, Wang Y, Wang M (2003) Seasonal characteristics of nitric oxide emission from a typical Chinese rice-wheat rotation during the non-waterlogged period. Global Change Biology 9: 219–227.
12. Xing GX, Zhu ZL (2000) An assessment of N loss from agricultural fields to the environment in China. Nutrient Cycling in Agroecosystems 57: 67–73.
13. Walsh M (2001) NO_x and N_2O fluxes in an upland agroecosystem of the North China Plain: field measurements, biogeochemical simulation, and climatic sensitivity [PhD dissertation]. Fort Collins, Colorado: Colorado State University.
14. Fang SX, Mu Y (2006) Air/surface exchange of nitric oxide between two typical vegetable lands and the atmosphere in the Yangtze Delta, China. Atmospheric Environment 40: 6329–6337.
15. Li D, Wang X (2007) Nitric oxide emission from a typical vegetable field in the Pearl River Delta, China. Atmospheric Environment 41: 9498–9505.
16. Mei B, Zheng X, Xie B, Dong H, Zhou Z, et al. (2009) Nitric oxide emissions from conventional vegetable fields in southeastern China. Atmospheric Environment 43: 2762–2769.
17. Pang X, Mu Y, Lee X, Fang S, Yuan J, et al. (2009) Nitric oxides and nitrous oxide fluxes from typical vegetables cropland in China: Effects of canopy, soil properties and field management. Atmospheric Environment 43: 2571–2578.
18. Liu C, Zheng X, Zhou Z, Han S, Wang Y, et al. (2010) Nitrous oxide and nitric oxide emissions from an irrigated cotton field in Northern China. Plant and Soil 332: 123–134.
19. FAO/IFA (2001) Global estimates of gaseous emissions of NH_3, NO and N_2O from agricultural land. Rome, Italy: Food and Agriculture Organization (FAO) of the United Nations.
20. Zheng X, Han S, Huang Y, Wang Y, Wang M (2004) Re-quantifying the emission factors based on field measurements and estimating the direct N2O emission from Chinese croplands. Global Biogeochemical Cycles 18: GB2018, doi:2010.1029/2003GB002167.
21. Guangzhou-Municipal-Statistics-Bureau (2006) Guangzhou Statistical Yearboo. Beijing: China Statistics Press.
22. National-Standard-Bureau-of-China (1987) Analytical Methods for Forest Soils (Third fascicule). Beijing: National Standard Bureau Press.
23. WMO (2008) Guide to Meteorological Instruments and Methods of Observation, WMO-No. 8, Seventh edition. Geneva: World Meteorological Organization.
24. Li DJ, Wang XM, Mo JM, Sheng GY, Fu JM (2007) Soil nitric oxide emissions from two subtropical humid forests in south China. J Geophys Res 112: D23302, doi:23310.21029/22007JD008680.
25. Li D, Wang X, Sheng G, Mo J, Fu J (2008) Soil nitric oxide emissions after nitrogen and phosphorus additions in two subtropical humid forests. J Geophys Res 113: D16301, doi:16310.11029/12007JD009375.
26. Pilegaard K, Hummelshøj P, Jensen NO (1999) Nitric oxide emission from a Norway spruce forest floor. J Geophys Res 104: 3433–3445.
27. Akiyama H, Mctaggart IP, Ball B, Scott A (2004) N_2O, NO, and NH_3 emissions from soil after the application of organic fertilizers, urea and water. Water, Air and Soil Pollution 156: 113–129.
28. Boeckx P, Cleemput OV (2006) "Forgotten" terrestrial sources of N-gases. International Congress Series 1293: 363–370.
29. Ludwig J, Meixner FX, Vogel B, Förstner J (2001) Soil-air exchange of nitric oxide: An overview of processes, environmental factors, and modeling studies. Biogeochemistry 52: 225–257.
30. Veldkamp E, Keller M (1997) Fertilizer-induced nitric oxide emissions from agricultural soils. Nutrient Cycling in Agroecosystems 48: 69–77.
31. He ZL, Alva AK, Calvert DV, Banks DJ (1999) Ammonia volatilization from different fertilizer sources and effects of temperature and soil pH. Soil Science 164: 750–758.
32. Bouwman AF, Boumans LJM, Batjes NH (2002) Modeling global annual N_2O and NO emissions from fertilized fields. Global Biogeochemical Cycles 16: 1080, doi:1010.1029/2001GB001812.
33. Sanhueza E (1992) Biogenic emissions of NO and N2O from tropical savanna soils. Proceedings of International Symposium on Global Climate Change. Tokyo: Japan National Committee for the IGBC. 22–34.
34. Cai GX, Zhu ZL, Trevitt ACF, Freney JR, Simpson JR (1986) Nitrogen loss from ammonium bicarbonate and urea fertilizers applied to flooded rice. Fertilizer Research 10: 203–215.
35. Zhu ZL, Cai GX, Simpson JR, Zhang SL, Chen DL, et al. (1989) Processes of nitrogen loss from fertilizers applied to flooded rice fields on a calcareous soil in north-central China. Fertilizer Research 18: 101–115.
36. Sommer SG, Olesen JE (2000) Modelling ammonia volatilization from animal slurry applied with trail-hoses to cereals. Atmospheric Environment 34: 2361–2372.
37. Walsh M (2001) NO_x and N_2O Fluxes in an Upland Agroecosystem of the North China Plain: Field Measurements, Biogeochemical Simulation, and Climatic Sensitivity: Colorado State University.
38. Fang S, Mu Y (2007) NOX fluxes from three kinds of agricultural lands in the Yangtze Delta, China. Atmospheric Environment 41: 4766–4772.

20

Affordable Nutrient Solutions for Improved Food Security as Evidenced by Crop Trials

Marijn van der Velde[1]*, Linda See[1], Liangzhi You[2,3], Juraj Balkovič[1], Steffen Fritz[1], Nikolay Khabarov[1], Michael Obersteiner[1], Stanley Wood[4]

1 International Institute for Applied Systems Analysis (IIASA), Ecosystem Services and Management Program, Laxenburg, Austria, 2 International Food Policy Research Institute (IFPRI), Washington D.C., United States of America, 3 College of Economics and Management, Huazhong Agricultural University, Wuhan, China, 4 Global Development Program, Bill & Melinda Gates Foundation, Seattle, Washington, United States of America

Abstract

The continuing depletion of nutrients from agricultural soils in Sub-Saharan African is accompanied by a lack of substantial progress in crop yield improvement. In this paper we investigate yield gaps for corn under two scenarios: a micro-dosing scenario with marginal increases in nitrogen (N) and phosphorus (P) of 10 kg ha^{-1} and a larger yet still conservative scenario with proposed N and P applications of 80 and 20 kg ha^{-1} respectively. The yield gaps are calculated from a database of historical FAO crop fertilizer trials at 1358 locations for Sub-Saharan Africa and South America. Our approach allows connecting experimental field scale data with continental policy recommendations. Two critical findings emerged from the analysis. The first is the degree to which P limits increases in corn yields. For example, under a micro-dosing scenario, in Africa, the addition of small amounts of N alone resulted in mean yield increases of 8% while the addition of only P increased mean yields by 26%, with implications for designing better balanced fertilizer distribution schemes. The second finding was the relatively large amount of yield increase possible for a small, yet affordable amount of fertilizer application. Using African and South American fertilizer prices we show that the level of investment needed to achieve these results is considerably less than 1% of Agricultural GDP for both a micro-dosing scenario and for the scenario involving higher yet still conservative fertilizer application rates. In the latter scenario realistic mean yield increases ranged between 28 to 85% in South America and 71 to 190% in Africa (mean plus one standard deviation). External investment in this low technology solution has the potential to kick start development and could complement other interventions such as better crop varieties and improved economic instruments to support farmers.

Editor: Luis Herrera-Estrella, Centro de Investigación y de Estudios Avanzados del IPN, Mexico

Funding: The Austrian Research Funding Agency (FFG) funded projects. The funders had no role in study design, data collection and analysis, decision to publish, or preparation of the manuscript. This research was supported by the Austrian Research Funding Agency (FFG) for the Project FarmSupport (No. 833421).

Competing Interests: The authors have declared that no competing interests exist.

* E-mail: velde@iiasa.ac.at

Introduction

Farming looks mighty easy when your plow is a pencil and you're a thousand miles from the corn field. –Dwight D. Eisenhower, 1956.

The increases in global population and food demand clearly indicate that current growth in agricultural productivity is not sufficient to sustain the 9 billion people that will inhabit the Earth by 2050 [1]. Feeding the world is a multifaceted and complex challenge and a number of solutions have been offered, where closing the yield gap is one of the most frequently cited recommendations [2;3;4;5;6]. The FAO [1] suggests that 70% of the required increase in crop production in developing countries should be realized through boosting the productivity of fields already under cultivation. Without this intensification, the inevitable cropland expansion will lead to deforestation, accelerate land degradation and threaten natural habitats and biodiversity [7]. A large part of this augmentation in crop production must therefore come from soils in tropical regions which are often highly weathered, have low levels of chemical soil fertility and will need additional inputs to improve crop productivity [8]. Global fertilizer use has already increased significantly since 1960 [9] and this increase has played an important role in the Green Revolution, benefitting many developing countries in South America and Asia. Yet substantial progress is still lacking in Africa [10]. From 1960 to 2000, yields of staple crops such as wheat, rice and corn increased in South America by over 180% while African yields did not improve substantially (see Figure 1). These contrasting trajectories reflect disparities in infrastructure development, primary crop types grown, agricultural R&D and extension capacities, socio-economic conditions as well as environmental differences [10;11;12].

In many smallholder fields, fertilizer and manure inputs have been too low for too long [8]. Agricultural soils cultivated without adequate nutrient replenishment cannot reach their full crop production potential and are at risk of irreversible degradation [13]. In large parts of Africa, this has led to seemingly perpetual low per capita food production [8]. Without maintaining adequate soil fertility levels, crop yields cannot be sustained, increase over time, or respond to improved agricultural management practices.

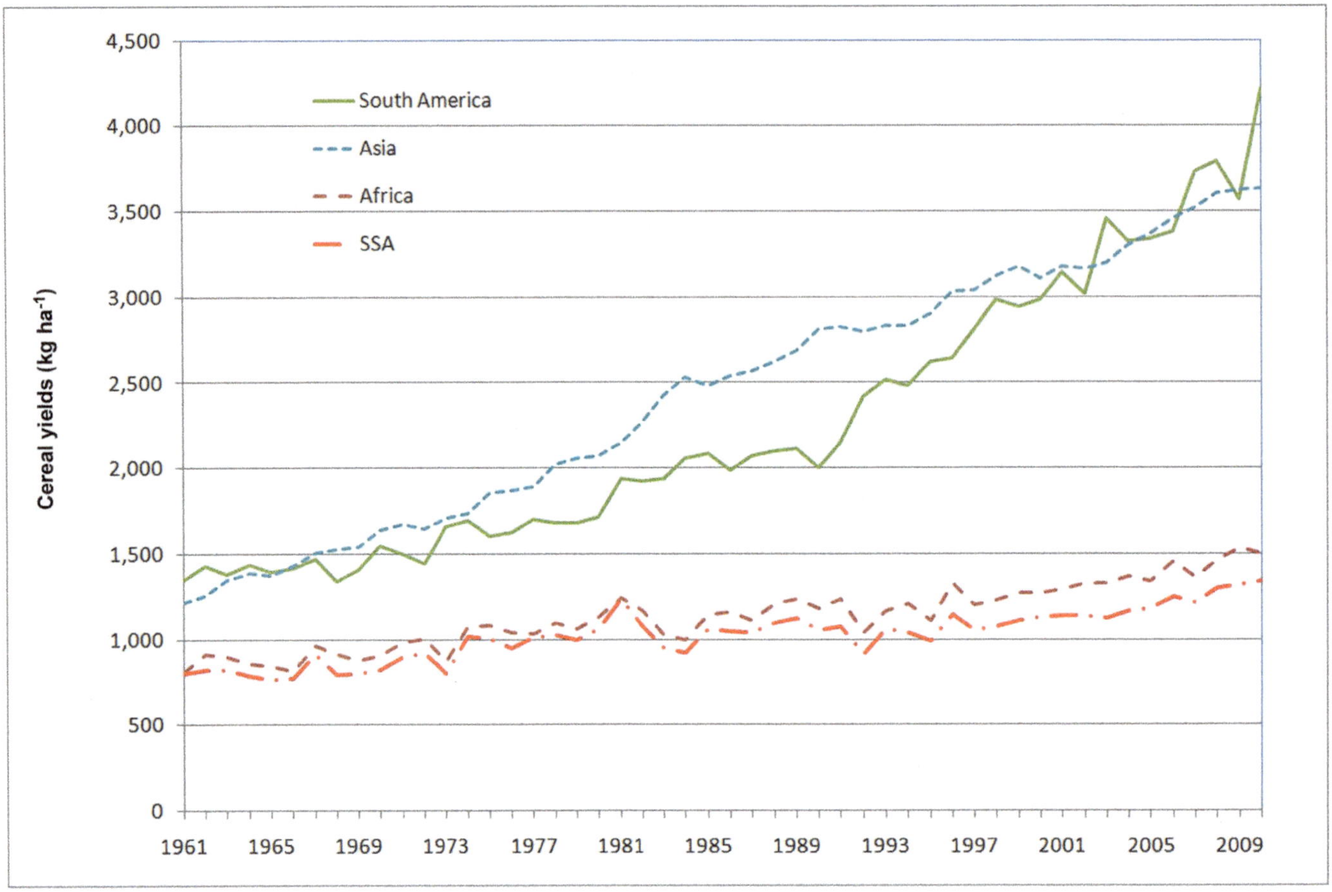

Figure 1. Cereal yield trends since 1960 in Africa, Sub-Sahara Africa (SSA), South America and Asia.

To improve nutrient input, smallholder farmers need actionable strategies such as micro-dosing: applications of small quantities of fertilizers. Field studies have shown that micro-dosing presents an attainable strategy for smallholder farmers in line with their financial means that can result in significant yield gains [14]. Importantly, previous higher fertilizer rate recommendations have ignored the sizeable but unlikely investment that would be required by poor and risk adverse smallholders.

Fertilizer prices in Africa are often higher relative to other developing countries. The small size of the fertilizer market, the high transportation and handling costs and the inefficient supply chain all contribute to the relatively high retail prices for fertilizers in Africa [15]. Fertilizer price also varies across regions, through different years and even among cropping seasons in the same year. In a landlocked country such as Uganda, the prices for urea in 2000 ranged from 600 shillings kg^{-1} (300 US\$ ton^{-1}) in the central districts to over 750 shillings kg^{-1} (375 US\$ ton^{-1}) in the eastern districts while prices for phosphate (Diammonium Phosphate, DAP) ranged from 560 shillings kg^{-1} (280 US\$ ton^{-1}) in the long rainy season to over 700 shillings kg^{-1} (350 US\$ ton^{-1}) in the short rainy season [16]. As a comparison, US farmers in 2000 paid from US\$80 to US\$120/ton for Urea, and US\$140 to US\$170/ton for DAP. Depending on the locations and seasons, the NPK (Nitrogen-Phosphorus-Potassium) compound fertilizer ranged from 700 shillings kg^{-1} (350 US\$ ton^{-1}) to over 1000 shillings kg^{-1} (500 US\$ ton^{-1}) for 1:1:2 NPK, the most common compound fertilizer in Uganda. The transportation cost in Uganda is over one third of the total fertilizer cost while it is about a fifth in Tanzania [17] due to the major sea port of Dar Es Salaam. Moreover, there are global pressures that lead to price volatility in fertilizer prices. For example, there was a fourfold increase in the price of urea from 2000 to 2008, reaching over 500 US\$ ton^{-1}, falling to around 200 US\$ ton^{-1} in 2009 and which is currently at around 400 US\$ ton^{-1}.

To formulate more realistic sustainable intensification pathways, we need better estimates of smallholder yield gaps in tropical countries and to then align these with local fertilizer prices and associated investment costs. There are a number of yield gap approaches that estimate different types of attainable yield potentials across varying spatial and temporal scales [18]. Many global assessments of yield gaps use crop models or data that currently lack sufficiently detailed spatial information on soil characteristics, crop management practices and crop responses to fertilizers [19,20]. Mueller et al. [19] found that large crop production increases are possible, but will require considerable changes in nutrient and water management. Crop trials represent a valuable source of information for yield gap analysis and could be analyzed more comprehensively for this purpose, yet are rarely collected systematically. Furthermore, nutrient specific analysis of the relationships between fertilizers and crop yields has been limited, especially in the tropical and subtropical regions where crop yields are relatively low and must increase the most to meet growing demands [21].

In this paper we analyze historic data from FAO corn fertilizer trials carried out between 1969–1993 at 1358 locations in Africa and South America (Figure 2) where corn is the most commonly cultivated crop [18]. Mitscherlich-Baule crop response functions were fit to the crop trial data by optimizing the factors describing yield responses to elemental nitrogen (N) and phosphorus (P) inputs as well as an initial (residual) soil N and P. Nitrogen and Phosphorus specific fertilizer application rates from [22] were then used as inputs to the crop response functions, and the resulting yields were validated using sub-national yield statistics on corn from the International Food Policy Research Institute (IPFRI) [23] to estimate yield gaps in corn at the continental level. Only water-limited (i.e. rain fed) yield potential as opposed to irrigated yields potentials are considered here. Furthermore, we consider the importance of soil nutrient stoichiometry and the viability of micro-dosing [14,24] in order to further the African crop productivity discourse [25,26]. Yield increases associated with two different scenarios are considered here: 1) depicting a micro-dosing strategy and 2) a topping up to a conservative estimate of average nutrient fertilizer rates in the USA. Finally, the investment costs of scaling up these scenarios are calculated using the latest average fertilizer prices in Africa and South America and used to evaluate whether these approaches can function as part of an actionable development blueprint for Sub-Saharan Africa.

Materials and Methods

Crop Trials and Response Functions

Recently historic FAO crop fertilizer field trials have become publically accessible (http://www.fao.org/ag/agl/agll/nrdb/). These data were collected as part of FAO's Fertilizer Programme [27] that ran from 1969 until 1993. The purpose behind the programme was to undertake trials to determine suitable fertilizer application rates for locally grown crops and to demonstrate to as many farmers as possible, the positive effect of fertilizer application on crop yields and farm income. The information available from the trials includes crop yields (kg ha^{-1}), and application rates of the main nutrients (nitrogen (N), phosphorus (P) and potassium (K) and farmyard manure (kg ha^{-1})). Unfortunately, detailed soil or meteorological information was not recorded. Nitrogen was mostly applied as urea, phosphorus (P) was mostly applied as superphosphate and potassium as part of NPK compound fertilizers. All the applied nutrients were recalculated to elemental application rates (with P calculated from P_2O_5 and K from K_2O). The application rates of farm yard manure were converted to N, P-P_2O_5 and K-K_2O application rates following [28]. Data on corn yields from trials with at least five N and P input combinations were selected for this analysis. The Mitscherlich-Baule crop response function was used to analyze relations between nitrogen (N) fertilizer input, phosphorus-phosphate (P) fertilizer input and corn yields y_{mb}:

$$y_{mb} = a_1[1 - \exp(-a_2(a_3 + N))] * [1 - \exp(-a_4(a_5 + p))] \quad (1)$$

The function allows for growth that plateaus with increasing fertilizer application, and accommodates cases of both near perfect factor substitution and near zero factor substitution, and performs superior to a quadratic and von Liebig type production function [29]. The growth plateau is represented by a_1, which was set equal to the maximum yield obtained in each field trial, while a_3 and a_5 represent the residual available nitrogen and phosphorus in the soil. Taking account of residual soil phosphorus is important; the cumulative cropland P surplus in certain countries in Western Europe has led to a buildup of residual soil P with expected future benefits to crop production, although lower and no effects are generally expected for Latin America and Africa [30]. The coefficients a_2 and a_4 describe the influence of the corresponding N or P fertilization on yield. The parameters a_2, a_3, a_4, and a_5 were obtained by minimizing the sum of squared errors for all applications in each experiment. The Nelder-Mead multidimensional unconstrained nonlinear minimization algorithm was used to minimize the objective function. Only those trials where a crop response function could be fit were used. This resulted in a total of 1358 unique experiments with at least five N and P input combinations; 752 in Africa and 606 in South America.

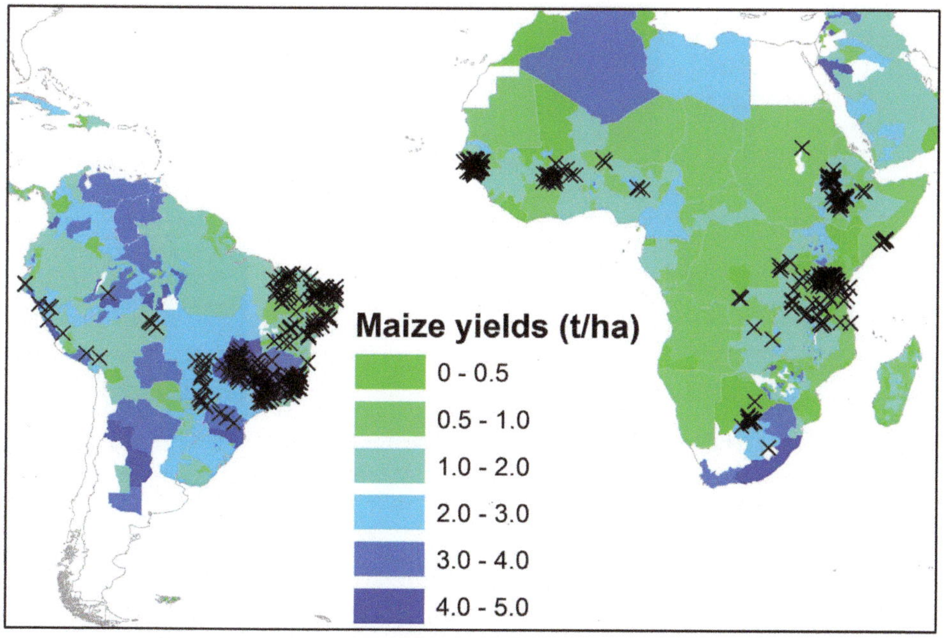

Figure 2. Crosses indicate locations of 1358 historic FAO corn field trials with at least five N and P input combinations in Africa and South America carried out between 1969 and 1993. Colors indicate (subnational) maize (corn) yields (ton/ha) as collected by [23].

Current Fertilizer Application Rates

The fertilizer dataset of [22] containing crop fertilizer rates was used to assign the current N and P inputs from chemical fertilizer (Nfer, Pfer) and manure (Nman, Pman; see Figures S1, S2, S3) at the trial locations and considered representative for corn fertilizer rates [31]. Data on corn yields collected by IFPRI [23] were used as a comparison with the yields obtained from the crop response functions (Figure S4). The average cereal area, production and yield as reported by FAOSTAT for 2008–2010 were calculated for each of the scenarios. In the area scenario a constant production was assumed and an increase in yield would reduce the requirement for cropland. In the people scenario an increase in yield would produce more on the same cropland area. We assume the average cereal calorie content to be 3000 kcal kg^{-1} and the average annual calorie need of a person to be 1 million kcal $year^{-1}$.

Costs

The cost of the two proposed scenarios was calculated using the Agricultural GDP in US dollars for 2009 [32]: South America (SA) $192 billion USD; Sub-Saharan Africa (SSA) $150 billion USD. Then the costs for each region were calculated by taking the total cost, dividing by the Agricultural GDP and multiplying by 100 to arrive at the percentage of Agricultural GDP that would be required to finance the scenario.

Results

An example of crop response trial data and the corresponding modeled Mitscherlich-Baule crop response function is shown in Figure 3. The median r^2 obtained by fitting the individual crop trials equaled 0.81; the 25th percentile equaled 0.66 and the 75th percentile 0.91. The resulting median, 25th and 75th percentiles for the a_1, a_2, a_3 and a_4 parameters obtained across all crop trials are presented in Table 1. Median values corresponded to 0.017 ton kg^{-1}, 68.4 kg N ha^{-1}, 0.29 ton kg^{-1}, 3.18 kg P ha^{-1} for parameters a_1, a_2, a_3 and a_4 respectively. This is in correspondence with the parameter values reported by [29] and [33]. The results from the individual crop trials indicate that out of the 1358 trials, there were 1037 trials (76%) that responded stronger to added phosphorus than nitrogen. Similarly, for 82% of the trials the pool of residual soil N was larger than the accessible residual P. Clearly, these site-specific analyses indicate that overall, phosphorus is the nutrient most limiting crop yield. Nevertheless, at the same time, the range of parameter values obtained highlights the variety of crop yield responses depending on site-specific conditions.

Overall, the yields modeled using Mitscherlich-Baule crop response functions show a good relationship ($r^2 = 0.94$, Figure 4).

Yield Gaps and Potentials

The fertilizer dataset of Potter et al. [22] was used to assign the current N and P inputs from chemical fertilizer (Nfer, Pfer) and manure (Nman, Pman) to the crop response functions; the average of these inputs and their distribution across Africa and South America as well as manure and fertilizer nutrient specific histograms are shown in Figures S1, S2, S3. Corn yield from the crop responses functions was compared to IFPRI reported data in Figure S4; median yields are comparable but there is a much larger variability in the yields derived from the crop response functions. This reflects both the coarser resolution of the IFPRI data [23] and the more realistic representation of the frequency distribution of yields that are attained at individual locations across both continents from the FAO crop trial data.

To indicate the potential for production increase, we calculated the average percentage yield increases resulting from an additional application of 10 kg N ha^{-1}, 10 kg P ha^{-1}, and both. Adding only N will lead to increases in crop production by ~4% and ~8%. Adding only P, on the other hand, will lead to substantially larger increases of ~12% and ~26% for South America and Africa respectively (Figure 5). This highlights the critical importance of P, of which many subsistence farmers may not be aware. The addition of both nutrients leads to increases of ~15% and ~35%, respectively, indicating that the effect of both nutrients is additive once P is applied. Thus P is clearly the limiting nutrient in improving crop yields.

In Africa, adding 10 kg ha^{-1} of N or P will result in mean and median percentage increases of respectively ~5.5 and ~5.7% and ~11.7 and ~16.3%; this would thus bring a significant proportion of farmers with the lowest yields closer towards attaining average yield levels and effectively shift a bulk of smallholders out of current marginal productivity. Since these are indicative for rainfed yields, additional water resources to attain these yield increases would not be required [6]. A final experiment considers the percentage yield increase that would be obtained if 80 kg ha^{-1} of N and 20 kg ha^{-1} of P - a relatively conservative estimate of average rates in the USA - were applied (Figure 6). In South America this would lead to average yield increases of 30%, up to a maximum of 90% while in Africa these increases would be considerably larger, i.e. average yield increases of 70%, up to a maximum of 190%.

Even though these yield increases are considerable, they are lower than the yield potentials generally estimated in other studies. For example, yield gaps of 180 to 540% for maize have been estimated for sub-Saharan Africa [34] while a yield gap of 118% was found by Tittonell et al. [35] for western Kenya. Since most of the crop trials were done more than 20 years ago, our results will provide conservative yield estimates. In soils that have become increasingly depleted, and with new and better crop varieties that have become available since then, the response to fertilizer may be even stronger than predicted here.

Implications

The implications for cropland expansion will be significant. Unless both N and P nutrient inputs are increased considerably and other complementary inputs and rural services such as seed, irrigation, market access and extension are available to improve crop yields, then necessary crop production gains will largely come from cropland expansion. This would have considerable negative impact on forest and grassland habitats and biodiversity [5]. In contrast, improving cereal yield by just 5% globally, over 33 million ha of forest or grassland would be saved. To put the benefits of these higher yields in context, our first scenario of applying an additional 10 kg N ha^{-1} and 10 kg P ha^{-1} (leading to a corn yield increase by 15% and 32% in South America and Africa) would save more than 4 million ha and 25 million ha of cropland conversion in South America and Africa respectively. Alternatively, if such yield improvement occurs in the currently cultivated cereal areas on these two continents, the improved productivity could feed an additional 64 million and 150 million people respectively.

Costs

Such a scenario would require a total investment of US$148 million in sub-Sahara Africa (using an average of fertilizer prices paid by farmers in 2012 of Urea ($620/ton or 0.29$ kg^{-1} N) and DAP ($950/ton or 0.22$ kg^{-1} P) and US$79 million in South America (using an average fertilizer price paid by farmers in 2012

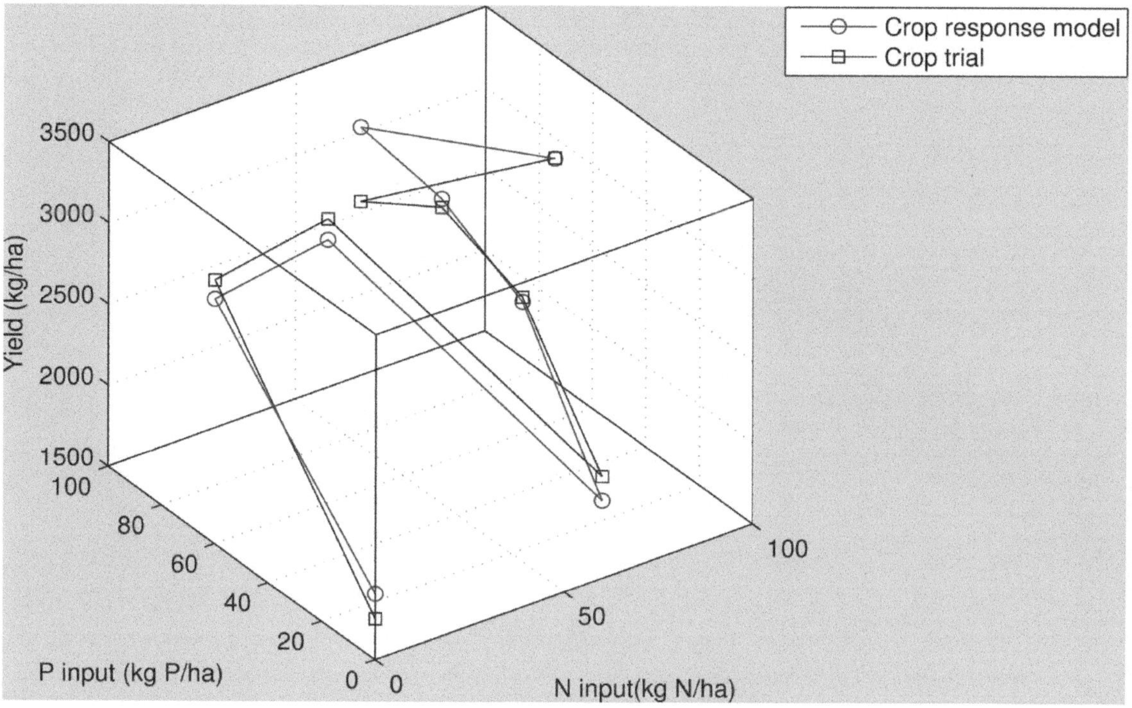

Figure 3. A typical example of a fitted crop response trial with experimental (blue line with circles) and modeled data (black line with squares) with eight N and P input combinations and resulting yields.

of Urea ($460/ton or 0.22$ kg^{-1} N) and DAP ($680/ton or 0.16$ kg^{-1} P)). The second scenario or larger nutrient inputs would amount to investments of US$798 million in sub-Sahara Africa and US$428 million in South America respectively. Maize yield increases range from 15% to over 70%, and such an investment would therefore bring considerable additional revenue to maize farmers. We acknowledge the fact that the prices of both fertilizer and maize vary with location so the actual profitability of fertilizer investment would vary spatially.

However, the direct investment in fertilizers is actually very small and is less than 1% of Agricultural GDP of both Sub-Saharan Africa and South America for the micro-dosing scenario. The calculation of the investment in terms of the percentage of Agricultural GDP that would be needed in SSA for scenario 1 equates to dividing 148 million USD by the Agricultural GDP of $150 billion USD multiplied by 100. The percentage of investment in terms of Agricultural GDP for the other scenario for SSA and the scenarios for SA were calculated in the same way (see Table 2).

Table 1. The median, 25th and 75th percentile values from the distributions of the Mitscherlich-Baule crop response function parameters (a_1, a_2, a_3 and a_4) fitted for the 1358 individual crop trials.

Parameter	Median	25th percentile	75th percentile
a1	0.017321	0.0077335	0.047077
a2	68.4297	21.6344	285.0599
a3	0.29219	0.047732	0.29219
a4	3.1803	0.99095	13.0806

Discussion

In reality a full development blueprint would need to have a broader scope and costs would be compounded with investments in roads, agricultural extension, (local) market access, etc. [5,6]. Nevertheless, external investment in this low technology solution has the potential to kick start development and could complement other interventions such as better crop varieties, improved economic instruments to support farmers as well as new technologies involving mobile phones, crowdsourcing and data mining of internet searches [36,37]. To improve our understanding of how best to target, design and support rural development, it is insightful to compare the costs calculated here with the costs involved in a project such as the Millennium Villages (MV). The MV is an integrated approach to eradicate poverty by involving an entire community in improving their livelihoods and health in a sustainable way (http://www.millenniumvillages.org). The focused investments calculated here are significantly lower per person per year compared to the costs in the MV, which vary between 35 to 100 USD per person per year [38]. However, we clearly acknowledge that the MV has a much broader scope, engages entire communities and contributes to many other aspects of well-being and improved livelihoods as set out in the Millennium Development Goals, which would not be part of the scenarios suggested here.

We have clearly demonstrated the importance of phosphorus for closing yield gaps in Africa. The phosphorus deficiency reflects soil P supply problems that are of widespread concern in highly weathered tropical soils notorious for low levels of available P and exhibit a strong P fixation capacity [39]. For instance, approximately 82% of the land area of the American tropics is deficient in P in its natural state [40]. Combined with the fact that P reserves are likely to become exhausted during the next 30–300 years [41], this paints a bleak picture indeed. Nevertheless, if recycling programs were put in place for animal and

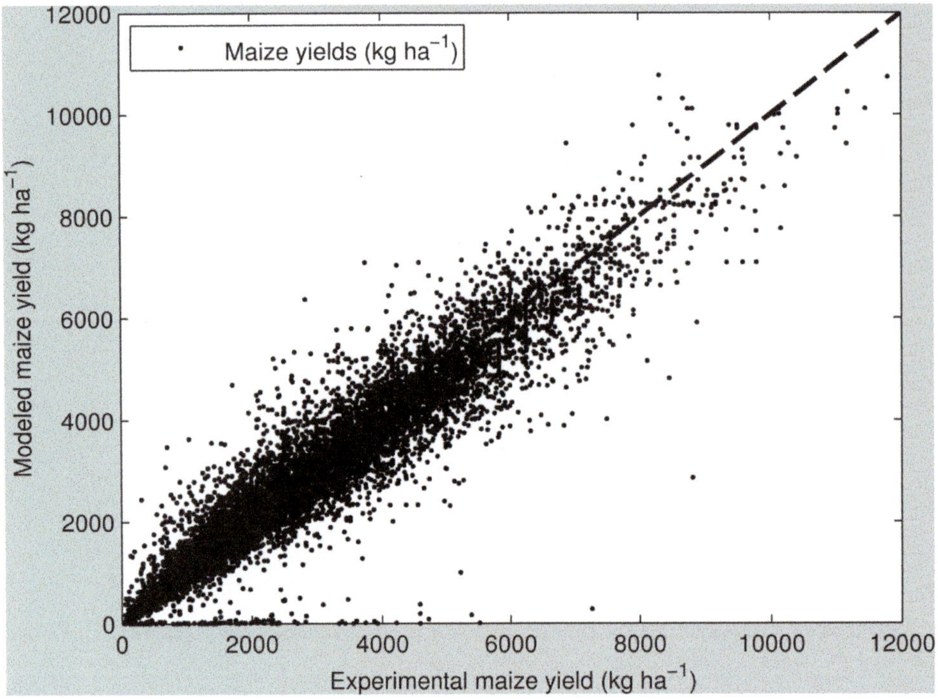

Figure 4. Relationship between historic FAO experimental corn field trials with at least five N and P input combinations and corn yields calculated with the Mitscherlich-Baule crop response function totaling 1358 unique nutrient-yield relations ($r^2 = 0.94$).

human excreta some of these effects might be mitigated, e.g. [42]. Raising awareness of the need to provide a more balanced stoichiometry is also a critical element in improving yields. If farmers continue to add increased supplies of N without P, they will soon reach a saturation point in yields and effectively waste valuable resources as well as contaminating groundwater due to the leaching of nitrogen. This finding also has implications for the current fertilizer subsidy programs in many developing countries. Most of these programs focus mainly on N and do not emphasize the importance of P sufficiently. Our results demonstrate that better balanced subsidy

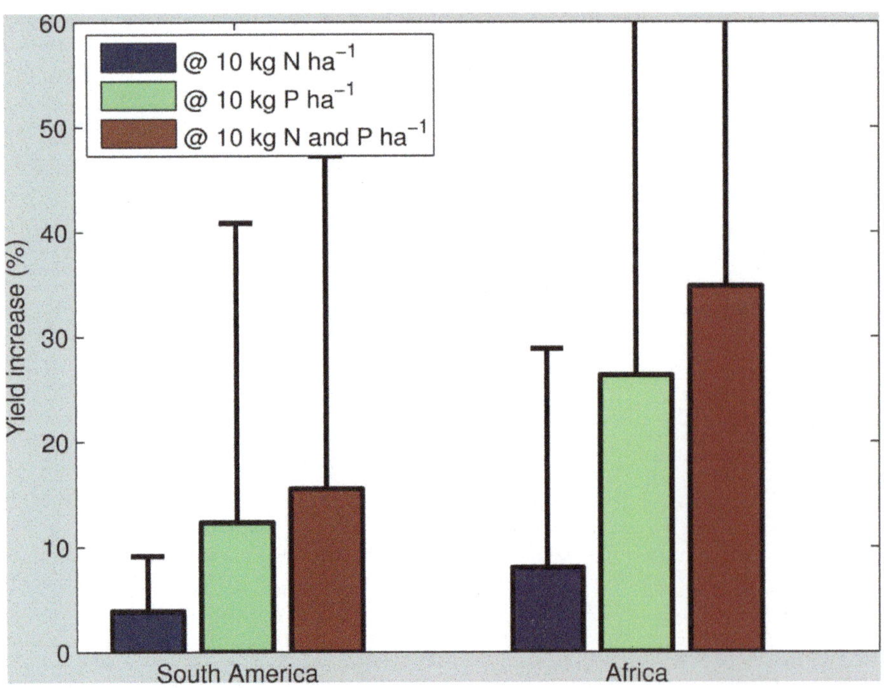

Figure 5. Mean corn yield increase (%) across trial sites at additional applications of 10 kg N ha^{-1}, 10 kg P ha^{-1} or 10 kg N and P ha^{-1} (error bars refer to the standard deviation of the obtained yield increases observed across all trials).

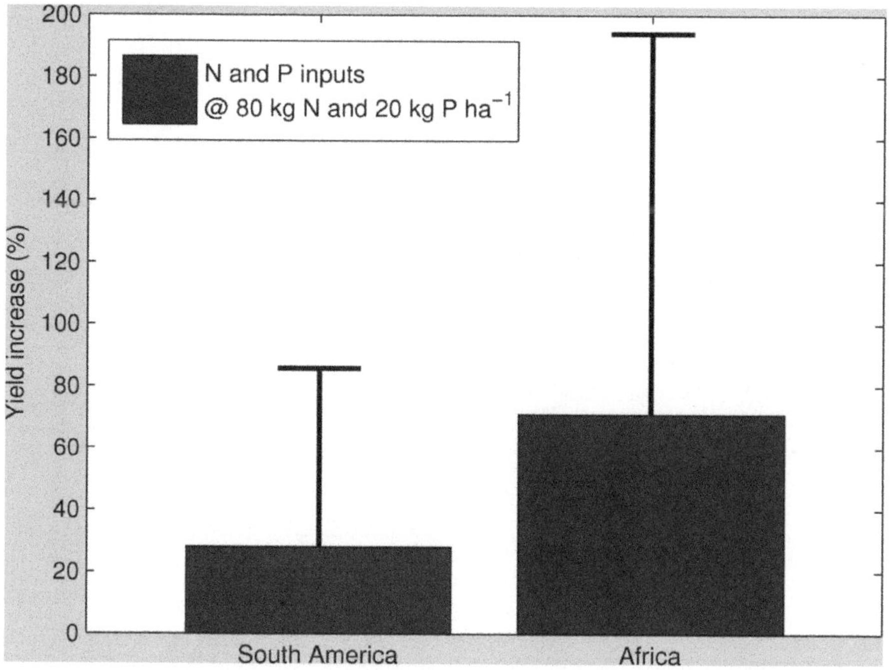

Figure 6. Mean corn yield increase (%) across trial sites at applications of 80 kg N ha^{-1} and 20 kg P ha^{-1} (error bars refer to the standard deviation of the obtained yield increases observed across all trials).

schemes taking account of both N and P would have a larger effect on crop yields.

Climate change is expected to generally have a negative impact on corn in Africa with estimates of lowering yield ranging from 3 to over 12% [43]. Corn will be the most heavily impacted crop in sub-Saharan Africa, where yield losses would occur in 65% of corn growing regions for a 1°C warming, increasing to 100% losses in areas subject to drought stress [44]. Although these pressures are considerable, and will require adaptation and fundamental changes to agricultural management, our results indicate that significant increases in yield are possible by improved nutrient management; especially during growing seasons when soil water availability is not constraining crop yields. Thus, achieving trend-growth in crop productivity through sufficient and balanced nutrient applications coupled with effective storage policies could partly offset negative climate change impacts on food security in Africa. The risk averseness of poor farmers that are prone to drought is one of the reasons why these farmers will be hesitant to invest in higher fertilizer applications. However, we have shown that a relatively small external investment would yield large improvements in both crop production and sustainable use of soils.

In contrast to previous crop productivity assessments for Africa and South America, our study allows farm input management

Table 2. Cost of the proposed scenarios expressed as the percentage of Agricultural GDP.

Region	Cost as a % of Agricultural GDP	
	Scenario 1	Scenario 2
SSA	0.10%	0.53%
SA	0.04%	0.22%

interventions to be directly based on small scale on-the-ground observations accounting for both site-specific conditions, as well as reflecting variability in soil conditions and climates. Furthermore, the study provides yield gaps that are realistically attainable as the assessment is based on farmers' field trials and the costs associated with these interventions amount to less than 1% of Agricultural GDP in both Sub-Saharan Africa and South America. Crop field trials might be considered costly by some but provide essential and hard-won insights, and when analyzed comprehensively, have the potential - through better formulated policies and agreements - to reward global society with improved food security status for many.

Supporting Information

Figure S1 Current N and P inputs from chemical fertilizer (Nfer, Pfer) and manure (Nman, Pman) extracted and averaged from [22] for the 1358 trial locations.

Figure S2 Histograms of the N and P nutrient inputs from chemical fertilizer at the 1358 locations [22].

Figure S3 Histograms of the N and P nutrient inputs from manure at the 1358 locations [22].

Figure S4 Boxplots of regionally reported corn yields collected by IFPRI and corn yields obtained from the 1358 crop response functions (CRFs) with current N and P inputs from chemical fertilizer and manure (Nman, Pman) as reported by [22].

Author Contributions

Analyzed the data: MvdV LS LY. Wrote the paper: MvdV LS LY JB SF NK MO SW.

References

1. FAO (2009). How to Feed the World in 2050. Report from the High-Level Expert Forum. http://www.fao.org/fileadmin/templates/wsfs/docs/expert_paper/How_to_Feed_the_World_in_2050.pdf. Accessed 2013 Feb 28.
2. Rosegrant MW, Cline SA (2003) Global food security: challenges and policies. Science 302: 1917–1918.
3. Godfray HCJ, Beddington JR, Crute IR, Haddad L, Lawrence D, et al. (2010) Food security: the challenge of feeding 9 billion people. Science 327(5967): 812–818. doi: 10.1126/science.1185383.
4. Licker R, Johnston M, Foley JA, Barford C, Kucharik CJ, et al. (2010) Mind the gap: how do climate and agricultural management explain the 'yield gap' of croplands around the world? Global Ecology and Biogeography 19: 769–782.
5. Foley JA, Ramankutty N, Brauman KA, Cassidy ES, Gerber JS, et al. (2011) Solutions for a cultivated planet. Nature 478: 337–342.
6. Mueller ND, Gerber JS, Johnston M, Ray DK, Ramankutty N, et al. (2012) Closing yield gaps through nutrient and water management. Nature. Doi:10.1038/nature11420.
7. Maitima JM, Mugatha SM, Reid RS, Gachimbi LN, Majule A, et al. (2009) The linkages between land use change, land degradation and biodiversity across East Africa. African Journal of Environmental Science and Technology 3(10): 310–325.
8. Sanchez PA (2002) Soil fertility and hunger in Africa. Science 295: 2019–2020.
9. Bumb B, Baanante C (1996) World trends in fertilizer use and projections to 2020. 2020 Brief No. 38, International Food Policy Research Institute, Washington, DC, USA.
10. Ejeta G (2010) African Green Revolution needn't be a mirage. Science 327: 831–832.
11. Thurlow R, Kilman S (2009) Enough: Why the world's poorest starve in an age of plenty. New York: Perseus Books. 302 p.
12. Nin-Pratt A, Johnson M, Magalhaes E, You L, Diao X, et al. (2011) Yield gaps and potential growth in Western and Central Africa. IFPRI Research Monograph. International Food Policy Research Institute (IFPRI), Washington, USA. DOI: 10.2499/9780896291829.
13. Lal R (2010) Managing soils for a warming earth in a food-insecure and energy-starved world. J Plant Nutr Soil Sci 173: 4–15.
14. Tmowlow S, Rohrbach D, Dimes J, Rusike J, Mupangwa W, et al. (2010) Micro-dosing as a pathway to Africa's Green Revolution: evidence from broad-scale on-farm trials. Nutr Cycl Agroecosyst 88: 3–15.
15. The World Bank (2007) Africa fertilizer policy toolkit: Promoting fertilizer use in African agriculture: Lessons learned and good practice guidelines. http://www.worldbank.org/html/extdr/fertilizeruse/about.html. Accessed 2012 August 22.
16. UBOS (2002) Uganda national household survey 1999/2000. Report on the crop survey module. Entebbe, Uganda: Uganda Bureau of Statistics: 4–30.
17. Bumb B, Johnson M, Fuente PA (2011) Policy Options for Improving Regional Fertilizer Markets in West Africa. IFPRI Discussion Paper 01084, International Food Policy Research Institute, Washington, DC, USA.
18. Lobell DB, Cassman KG, Field CB (2009) Crop yield gaps: their magnitudes, and causes. Annu Rev Environ Resourc 34: 179–204.
19. Mueller ND, Gerber JS, Johnston J, Ray DK, Ramankutty N, et al. (2012) Closing yield gaps through nutrient and water management. Nature 490: 254–257.
20. Sanchez PA, Ahamed S, Carré F, Hartemink AE, Hempel J, et al. (2009) Digital soil map of the world. Science 325(5941): 680–681.
21. FAO (2011) The state of the world's land and water resources for food and agriculture. Managing systems at risk. FAO and Earthscan.
22. Potter P, Ramankutty N, Bennett EM, Donner SD (2010) Characterizing the spatial patterns of global fertilizer application and manure production. Earth Interac 14: 1–22.
23. IFPRI (2012) Global subnational crop database. IFPRI, Washington, DC, USA.
24. Van der Velde M, See L, Fritz S (2012) Soil remedies for small-scale farming. Nature 484: 318, doi:10.1038/484318c.
25. Gilbert N (2012) Africa agriculture: dirt poor. Nature 483: 525–527, doi:10.1038/483525a.
26. Bindraban PS, Van der Velde M, Ye L, Van den Berg M, Materechera S, et al. (2012) Assessing the impact of soil degradation on food production. Current Opinion in Environmental Sustainability, 4(5), 478–488, doi:10.1016/j.cosust.2012.09.015.
27. FAO (1989) Fertilizers and food production: Summary review of trial and demonstration results 1961–1986. FAO, Rome.
28. Van Averbeke W, Yoganathan S (1997) Using Kraal Manure as a Fertiliser. Department of Agriculture and Agricultural and Rural Development Research Institute, Fort Hare.
29. Frank MD, Beattie BR, Embleton ME (1990) A comparison of alternative crop response models. Am J Agric Econ 72(3): 597–603.
30. Sattari SZ, Bouwman AF, Giller KE, van Ittersum MK (2012) Residual soil phosphorus as the missing piece in the global phosphorus crisis puzzle. Proc Natl Acad Sci U S A doi: 10.1073/pnas.1113675109.
31. IFA (2002) Fertiliser use by crop. Rome, 2002, 45 pp.
32. The World Bank (2011) Data. http://data.worldbank.org/indicator. Accessed 2013 Feb 28.
33. Finger R, Hediger W (2008) The application of robust regression to a production function comparison. The Open Agriculture Journal 2: 90–98.
34. Pingali PL, Pandey S (2001) World maize needs meeting: technological opportunities and priorities for the public sector. In: PL Pingali (Ed.) CIMMYT 1999–2000 World Maize Facts and Trends. Meeting World Maize Needs: Technological Opportunities and Priorities for the Public Sector, 1–24. Mexico: CIMMYT.
35. Tittonell P, Vanlauwe B, Corbeels M, Giller KE (2008) Yield gaps, nutrient use efficiencies and response to fertilizers by maize across heterogeneous smallholder farms of western Kenya. Plant Soil 313: 19–27.
36. Fritz S, McCallum I, Schill C, Perger C, See L, et al. (2012) Geo-Wiki: An online platform for improving global land cover. Environmental Modelling & Software 31: 110–123, doi: 10.1016/j.envsof.2011.11.015.
37. Van der Velde M, See L, Fritz S, Verheijen FGA, Khabarov N, et al. (2012) Generating crop calendars with Web search data. Environmental Research Letters 7(2): 024022, doi:10.1088/1748-9326/7/2/024022.
38. Pronyk P (2012) The Costs and Benefits of the Millennium Villages: Correcting the Center for Global Development. http://www.millenniumvillages.org/field-notes/archive/2012/4. Accessed 2013 Feb 28.
39. Buresh RJ, Smithson PC, Hellums DT (1997) Building soil phosphorus capital in Africa. In Buresh RJ, Sanchez PA and Calhoun F (eds.) Replenishing soil fertility in Africa. SSSA/ASA. Madison, WI. 111–149.
40. Sanchez PA, Salinas JG (1981) Low-input technology for managing Oxisols and Ultisols in tropical America. Advances in Agronomy 34: 279–406.
41. Cordell D, Drangeert JO, White S (2009) The story of phosphorus: Global food security and food for thought. Glob Environ Change 19: 292–305.
42. Baker LA (2011) Can urban P conservation help to avoid the brown devolution. Chemosphere 84: 779–784.
43. Nelson GC, Rosegrant M, Palazzo A, Gray I, Ingersoll C, et al. (2010) Food security, farming, and climate change to 2050. IFPRI Research Monograph. International Food Policy Research Institute (IFPRI), Washington, USA. DOI: 10.2499/9780896291867.
44. Lobell DB, Banziger M, Magorokosho C, Vivek B (2011) Nonlinear heat effects on African corn as evidenced by historical yield trials. Nat Clim Chang DOI: 10.1038/NCLIMATE1043.Dafadf.

A 3-Year Study Reveals That Plant Growth Stage, Season and Field Site Affect Soil Fungal Communities while Cultivar and GM-Trait Have Minor Effects

Silja Emilia Hannula[1]*, Wietse de Boer[1], Johannes van Veen[1,2]

1 Netherlands Institute of Ecology (NIOO-KNAW), Department of Microbial Ecology, Wageningen, The Netherlands, 2 Insititute of Biology, Leiden University, Leiden, The Netherlands

Abstract

In this three year field study the impact of different potato (*Solanum tuberosum* L.) cultivars including a genetically modified (GM) amylopectin-accumulating potato line on rhizosphere fungal communities are investigated using molecular microbiological methods. The effects of growth stage of a plant, soil type and year on the rhizosphere fungi were included in this study. To compare the effects, one GM cultivar, the parental isoline, and four non-related cultivars were planted in the fields and analysed using T-RFLP on the basis of fungal phylum specific primers combined with multivariate statistical methods. Additionally, fungal biomass and some extracellular fungal enzymes (laccases, Mn-peroxidases and cellulases) were quantified in order to gain insight into the function of the fungal communities. Plant growth stage and year (and agricultural management) had the strongest effect on both diversity and function of the fungal communities while the GM-trait studied was the least explanatory factor. The impact of cultivar and soil type was intermediate. Occasional differences between cultivars, the amylopectin-accumulating potato line, and its parental variety were detected, but these differences were mostly transient in nature and detected either only in one soil, one growth stage or one year.

Editor: Kevin McCluskey, University of Missouri, United States of America

Funding: This research was financed by ERGO grant number 838.06.052 of the Netherlands Organization for Scientific Research. The funders had no role in study design, data collection and analysis, decision to publish, or preparation of the manuscript.

Competing Interests: The authors have declared that no competing interests exist.

* E-mail: e.hannula@nioo.knaw.nl

Introduction

Genetic engineering of plants has been used to improve the quality and quantity of crop production in a cost-effective way (e.g. by enhancing resistance to pests and diseases or introducing tolerance to herbicides) [1]. Despite the great potential of this technology to advance agricultural yields, there are major concerns about the ecological impacts of genetically modified (GM) crops on soil ecosystem functioning. These impacts may be (1) direct (e.g. toxicity of an expressed introduced gene on key non-target species of important functional groups), (2) indirect (e.g. effects via unintended changes in the metabolism of the plant thereby affecting root exudates composition and fluxes) or (3) caused by changes in management regime used with GM crops [2].

The rhizosphere is a hot-spot of microbial abundance and metabolic activity due to the resources released by plants [3,4]. Hence, possible side-effects of GM plants on functioning of soil microbes should be first considered for the rhizosphere. Together with bacteria, fungi in the rhizosphere are very important to functioning of the soil-plant system and their functions range from symbiotic arbuscular mycorrhizal fungi (AMF) and plant pathogens to decomposers [5,6].

The structure and functioning of soil microbial communities is affected by soil type [7–9], plant growth stage [7,10–13], and other abiotic and biotic factors such as agricultural management [14,15]. The magnitude of the effects exerted by these factors compared to possible effects of cultivar and GM-crops is still largely unknown although knowledge of these sources of natural variation is critical for the assessment of the relative effects of specific potential perturbations such as introduced GM-traits.

Most of the studies on soil fungal communities have shown that GM-crops affect soil fungi in a similar way as its isoline [7,13,16–22], and only three studies [23–25] observed significant differences between the GM-variety and its parental isoline which could, however, be explained by factors other than GM-trait. Common to these studies was that the normal variability between cultivars under field conditions was usually very high and that other factors than cultivar-type affected the soil fungal communities more than the cultivar-type did. The aforementioned studies usually focused on one growth stage or one season/year without investigating variability over seasons. Thus, the question remains if different cultivars of potato, including a GM variety, have different effects on diversity or functioning of the soil microbes over multiple years.

Identifying the normal variation in fungal community structure and function in the soil is very important when aiming to evaluate the possible effects of GM-crops on soil communities [26]. In this study we followed the fungal community structure and function in two fields located in the Netherlands during 3 years of growing potatoes (*Solanum tuberosum* L.). Three growth stages of six cultivars (including a GM-variety with modified starch quality and its parental isoline) were included in the study allowing us to determine the long-term (years) and short term (within growth season) effects of the potato cultivars on fungal community dynamics and fungal decomposing activities. This approach facilitated an evaluation of the normal variation in fungal communities between years, growth stages, soils and under different cultivars, thereby providing a necessary baseline for assessing the potential impact of this GM potato variety. Further, we sampled the fields also after the growing seasons as well as in the rhizosphere of the succeeding crop (barley) to learn about possible long term effects of the starch-modified GM-potatoes.

Materials and Methods

Field Set-up and Sampling

Two agricultural sites VMD and BUI were selected for this experiment [19]. They are both located in the northern part of the Netherlands and are 10 km apart. Details on soil type, soil parameters and fertilizer treatments are presented in table S1. Cropping in these sites consists of potato-barley rotation (1 crop per year). Plots with six cultivars of potato were sampled in years 2008, 2009, and 2010 and barley fields were sampled after cultivation with potato in 2009. The fields were fertilized with 180–220 kg ha^{-1} nitrogen (N) in the form of calcium ammonium nitrate, 56–81 kg ha^{-1} phosphorous (P) as P_2O_5 and 145–200 kg ha^{-1} potassium (K) as K_2O or K_2SO_4 in 2008 and 2009. In 2010 organic fertilizer in form of pig manure (14 ton ha^{-1}in field VMD and 25 ton ha^{-1} in field BUI, respectively) was added together with inorganic fertilizers (table S1). Six cultivars of potato; 'Aveka', 'Aventra', 'Désirée', 'Premiere', 'Karnico' and 'Modena' (the modified variety of 'Karnico') were grown each in four replicates on these fields in randomized block design and locations were varied between years. The variety 'Modena' was genetically modified for its starch composition by complete inhibition of the production of amylose via introduction of a RNAi construct of the granule-bound starch synthase gene inhibiting GBSS and amylose formation, which yields pure amylopectin [27]. Cultivars 'Aventra', 'Aveka', 'Karnico' and 'Modena' produced tubers with a relatively high starch content and had a low to medium growth rate, whereas cultivars 'Désirée' and 'Premiere' had lower starch content in the tubers and higher growth rates.

Soil samples were collected from bulk soil before and after harvest whereas both rhizosphere and bulk soil were collected at the growth stages EC30 (seedling/young), EC60 (flowering) and EC90 (senescence) [28]. Bulk soil was collected using 0–15 cm soil corers (diameter 10 cm) and 5 cores per plot were randomly sampled and used to form a composite sample per plot that was further homogenized and sieved (4 mm mesh) to remove possible root fragments and stones. Rhizosphere soil was collected from a combination of 4 plants per plot by brushing roots. Part of the soil sample was subsequently frozen at −80°C for molecular analyses, another part was kept at −20°C prior to enzymatic analyses and ergosterol measurements and another part was used for immediate analyses of soil water content and pH (table S1). Soil water content was determined from fresh material as weight loss after overnight drying at 105°C.

Enzymatic Analyses

Quantification of ergosterol, via the alkaline extraction method, was used as an estimate of fungal biomass [29]. Analyses of activities of enzymes involved in decomposition of lignocellulose-rich organic matter, *i.e.* laccase, cellulase and Mn-peroxidase were performed according to van der Wal *et al.* [30].

Molecular Analyses

DNA was extracted from soil (0.5 g wet weight) with a Power Soil DNA isolation kit (MOBIO Laboratories, Inc. Carlsbad, CA, USA) using a bead beating system. Yields of genomic DNA were checked on 1% agarose gel and visualized under UV after ethidium bromide staining.

Terminal restriction fragment length polymorphism (T-RFLP) combined with the construction of a small library of the most dominant operational taxonomical units (OTUs) was used to determine the fungal community compositions over years. The structures of the three fungal phyla studied, ascomycetes, basidiomycetes and glomeromycetes, were assessed separately. For the analysis of ascomycete and basidiomycete communities, internal transcribed spacer (ITS) regions were used as target regions and the large subunit of ribosomal genes (LSU) was used as a target region for AMF (*Glomeromycota*). PCR conditions, primers and restriction enzymes are given in Hannula *et al.* [19]. Appropriate dilutions based on test runs of terminal restriction fragments (TRFs) were analyzed with an ABI 3130 sequencer using GeneScan^{TM} −500 LIZ (Applied Biosystems) and used as a size standard.

Clone libraries were constructed as described in Hannula *et al.* [19] and partially the same clone libraries were used. The sequenced clones were assigned to OTUs based on comparisons with GenBank using BLAST and considered to belong to a genus or species with similarities of 95% for an order and 97% for a species. These OTUs were related to the orginal peaks and their presence and absence in field samples were evaluated in T-RFLP Analyses Matching Program (TRAMP-R) [31] in the statistical computing environment R. Three out of four of the enzyme/primer combinations within 1.5 bp margin had to be met in a sample for it to be assigned to an OTU.

Data Analyses

Analyses of variance (ANOVA) with a linear mixed effect model was used to compare the ergosterol and enzymatic data as well as number of TRFs using SPSS for windows (Release 17.0.). The assumption of normality was tested with Shapiro-Wilk statistics and homogeneity of variances was assessed with Levene's test. The field site, growth stage, year of sampling, cultivar and GM-variety were used as fixed factors and block was set as the random factor. Differences between treatments were compared by a post hoc Tukey's honestly significant difference (HSD) test. Log transformation was used when data were not normally distributed. To estimate the possible effects of GM variety 'Modena' to its parental variety over years, a mixed model with repeated measure (growth stage) and block as a random factor was built separately for both fields.

The quality of T-RFLP data was first visually inspected in GeneMapper Software v4.1 (Applied Biosystems) and then transferred to T-Rex [32]. True peaks were identified for both labels as those of which the height exceeded the standard deviation (assuming zero mean) computed over all peaks and multiplied by two [33]. Non-Metric Multidimensional Scaling (NMDS) with Jaccard as distance measure were used to assess the similarity of the fungal communities after the harvest and in the rhizosphere of next crop, barley. Principal component analyses (PCA) were used to analyse the communities between years, fields, growth stages and cultivar. The community fingerprints were compared using ANOSIM in PAST [34]. In short, ANOSIM is a non-parametric test of significant differences between groups by comparing distances between groups to distances within groups. We used Jaccard as a distance index and 10000 permutations. Pairwise

Table 1. ANOVA comparisons of several fungal-related parameters between fields, years, growth stages, cultivars and GM-trait and the interaction effects of the cultivar.

	Field (df. 1)		Year (df.2)		Growth stage (df. 3)		Cultivar (df. 5)		GM- parent (df. 1)		Year×cultivar		Field×cultivar		Growth stage×cultivar		Year×Growth stage×field×cultivar	
	F	*P*	*F*	*P*	*F*	*P*	*F*	*P*	*F*	*P*	*F*	*P*	*F*	*P*	*F*	*P*	*F*	*P*
Ergosterol (mg/g)	0.13	0.72	48.17	**<0.001**	19.38	**<0.001**	1.47	0.20	0.12	0.73	1.40	0.18	1.00	0.42	0.97	0.49	1.72	0.071
Laccases (µmol/g)	0.63	0.43	14.39	**<0.001**	21.19	**<0.001**	1.05	0.39	0.36	0.55	1.84	0.052	1.27	0.28	2.39	**0.004**	1.72	0.052
Mn-Peroxidases (µmol/g)	1.06	0.10	1.96	0.14	9.81	**<0.001**	3.31	0.06	0.67	0.42	1.02	0.43	1.07	0.38	1.86	**0.031**	1.69	**0.043**
Cellulases (µmol/g)	17.74	**<0.001**	23.94	**<0.001**	19.01	**<0.001**	1.08	0.37	0.04	0.83	4.03	**<0.001**	3.96	**0.002**	3.12	**<0.001**	1.29	0.35
# of Ascomycetes	0.41	0.52	6.28	**<0.001**	25.15	**<0.001**	1.51	0.19	2.73	0.11	0.72	0.69	0.48	0.79	2.67	**0.001**	1.38	0.16
# of Basidiomycetes	1.65	0.20	51.60	**<0.001**	20.14	**<0.001**	0.72	0.61	0.16	0.69	0.52	0.88	0.08	1.00	1.12	0.34	0.39	0.97
# of AMF	0.61	0.44	15.29	**<0.001**	6.09	**<0.001**	0.66	0.65	0.35	0.55	0.49	0.88	0.89	0.49	0.34	0.98	0.50	0.91

Significant P-values are marked with bold. Only samples from rhizosphere were included in analyses of growth stage, cultivar and GM-parent comparison. # indicates richness of the fungi.

ANOSIMs between field sites, growth stages, years and cultivars are provided.

The diversity was calculated from the matched samples with both Shannon-H' and Simpson diversity indexes and compared with ANOVA as explained above.

Results

Soil Enzymatic Analyses, Fungal Biomass and Fungal Richness

Fungal-related parameters in plots cropped with the GM-variety seemed to fall within normal variation among potato cultivars observed in time (table 1). The largest explaining factor for most of the measured parameters was the plant phenological growth stage, followed by year and the soil type (table 1). Ergosterol analyses indicated that soil fungal biomass strongly dependent on plant growth stage and varied from year to year (table 1, Fig. 1). Although growth stage was affecting the fungal biomass, there were no significant differences between pre- and post-cropping situations or in bulk soils ($F = 1.31$, $p = 0.25$). Hence, no long term effects of cultivation were detected. Cultivar did not affect the fungal biomass in the rhizosphere in general, however, differences between some cultivars were detected in pairwise comparisons: cultivar 'Premiere' had a significantly lower fungal biomass as assayed by the ergosterol method in its rhizosphere than cultivars 'Aveka' and 'Désirée' ($F = 4.131$ and 4.181, $p<0.05$) over the entire period. In field BUI significant effects of cultivar on fungal biomass were detected at the stage of flowering in 2008 and the stage of young plant in 2010 (table 2) while in field VMD there were no effects of cultivar at any stage. Furthermore, there was no consistency in cultivars having the lowest or highest amount of ergosterol in their rhizosphere (Fig.1). The GM cultivar 'Modena' was not significantly different from the other cultivars or the parental variety (table 2) but rather in the middle range of the cultivars in the field BUI. The only significant difference between the GM-variety and its parental variety was the amount of ergosterol in the rhizosphere in the senescent stage (table 2).

Correlations revealed that all the extracellular enzymes measured in this study (laccases, cellulases and Mn-peroxidases) were positively correlated with the fungal biomass indicator ergosterol ($n = 702$, R^2 between 0.23–0.29 and $p<0.001$). Further, there were strong positive correlations among all enzyme activities measured. The richness of both ascomycetes and basidiomycetes was positively correlated with the amount of ergosterol (for basidiomycetes $R^2 = 0.27$ and $P<0.001$ and ascomycetes $R^2 = 0.08$ and $P<0.05$). AMF richness was negatively correlated with the amount of ergosterol ($R^2 = .11$ and $P<0.05$). Furthermore, the amount of Mn-Peroxidases in the soil was positively correlated with the ascomycete diversity ($R^2 = 0.16$, $P<0.001$) while the AMF richness was negatively correlated with production of cellulases ($R^2 = 0.11$ and $P<0.005$).

The measured extracellular enzymes (laccases, Mn-peroxidases and cellulases) were all affected by plant growth stage; highest activities were measured during senescence (table 1, Fig. 1). The amount of laccases and cellulases in the rhizosphere was significantly affected by year and the highest activity of these enzymes was found in 2009. On average the BUI location had higher laccase and cellulase activity than field VMD. The amount of Mn-peroxidases was associated with cultivar, but other enzymes were not. The cultivar 'Modena' had similar amounts of Mn-peroxidase in its rhizosphere as the parental cultivar 'Karnico', but more Mn-peroxidases in its rhizosphere than was found in the rhizospheres of Premiere and Aveka.

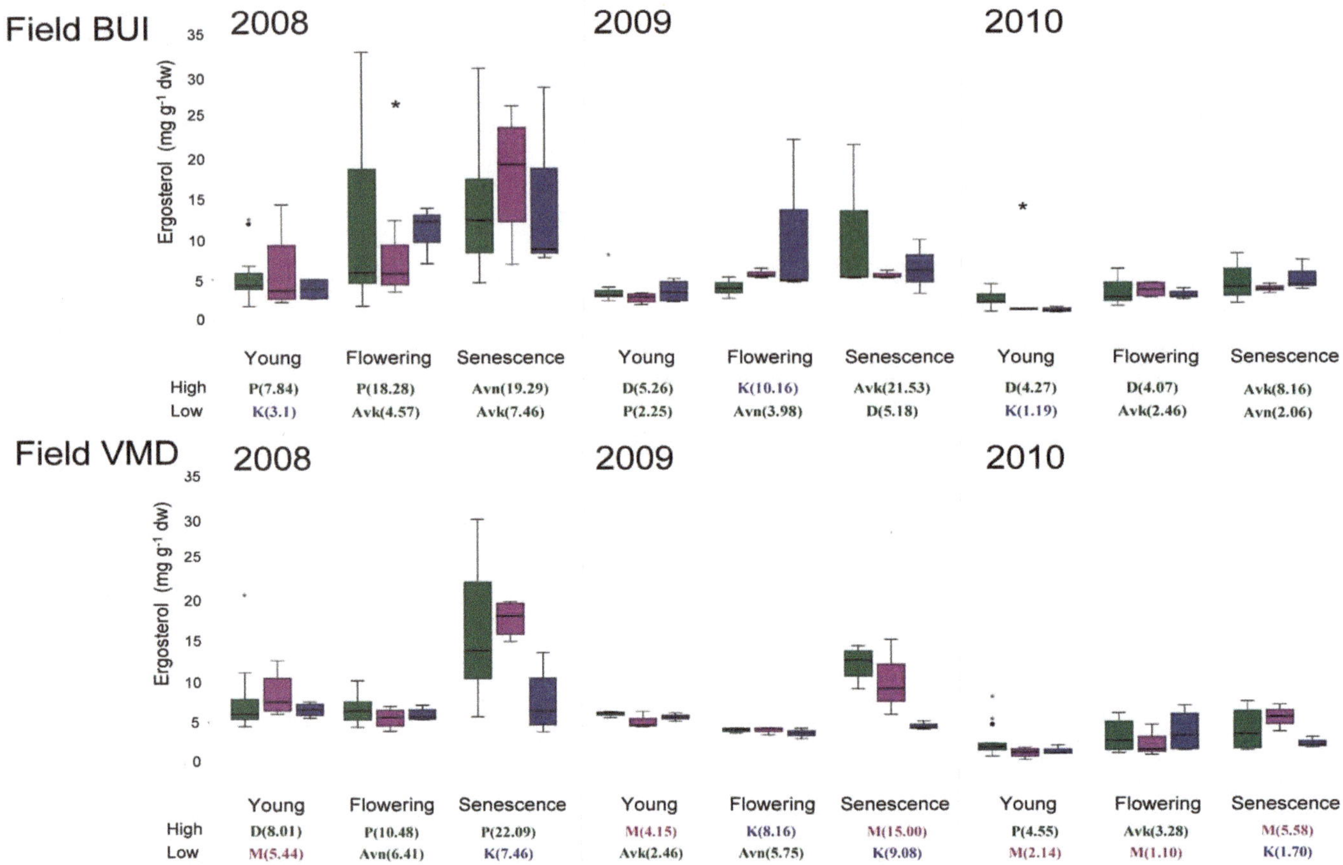

Figure 1. Change in fungal biomass. Boxplots of fungal biomass in the rhizosphere as measured by ergosterol concentrations during 3 years in different growth stages and in both field locations. The baseline (all other cultivars combined, n = 16) is marked with green boxplots, the GM-variety (n = 4) with purple and the parental variety 'Karnico' (n = 4) with blue markers. The star indicates a significant cultivar effect at the indicated time point. The values under the graphs are the cultivars with highest and lowest values (on average) colored the same as in the boxplots where 'D' = 'Désirée', 'Avk' = 'Aveka', 'Avn' = 'Aventra', 'P' = 'Premiere', 'K' = 'Karnico' (parental cultivar) and 'M' = 'Modena' (modified cultivar).

When looking at individual time points and fields the ascomycete, basidiomycete and glomeromycete richness was only once significantly different between cultivars (table 2). The richness of ascomycetes and glomeromycetes in the rhizosphere of GM-cultivar was only once different from the parental cultivar, namely at senescence 2008 and senescence 2010 in field BUI. The basidiomycete richness was at no occassion different between GM- and parental cultivar (table 2).

Data on community function, as based on activities of enzymes involved in decomposition of lignocellulose-rich organic matter, and richness were analysed by principal component analyses (PCA). The PCA analyses revealed that the growth stage was the strongest explanatory factor of differences in the community function (Fig. 2). The stage senescence clearly separated from the other stages along PC1 (ANOVA; $F = 9.57–13.74$, $p<0.001$) which was explained with higher ergosterol and enzymatic activities during senescence. The PC2 was explained by the same factors as PC1 and is thus not used here. The flowering stage separated along PC3 ($F = 4.22–8.28$, $p<0.05$) which is explained by more AMF and ascomycetes and less basidiomycetes during that stage compared to the other stages. Further, the years separated along both axes (PC1: $F = 8.5$, $p<0.001$ and PC3: $F = 124.6$, $p<0.001$) and fields along PC3 ($F = 33.9$, $p<0.001$) (Fig. S1). Cultivar had no significant contribution to explanation of PC1 ($F = 1.83$, $P = 0.15$), PC2 ($F = 1.92$, $P = 0.12$) nor PC3 ($F = 0.88$, $P = 0.47$) and the GM-variety was not significantly different from its parental isoline 'Karnico' (Fig. 2).

Fungal Diversity and Community Structure

According to the ANOSIM, the community fingerprints of all TRF peaks as well as identified OTUs of *Ascomycota*, *Basidiomycota* and *Glomeromycota*, were affected by the growth stage of the plant, field site and year (Fig. 3, Table 3). The fungal community structure was most strongly influenced by year-to-year variation ($R>0.22$) and difference in growth stage ($R>0.09$). The R values for the field site were close to 0 however, due to the size of the data-set a significant difference between fields were found. Plant cultivar did not predict fungal community structure when all growth stages, years and both fields were considered together (Table 3). There were no significant differences in the community structure of ascomycetes, basidiomycetes, glomeromycetes or total fungi between GM-cultivar 'Modena' and its parental variety 'Karnico' in any pairwise comparisons (Fig. 3).

The diversity of all fungal phyla was expressed both by the Shannon-Wiener index (H') and Simpson diversity index. The ascomycete diversity was significantly correlated with ascomycete richness ($R^2 = 0.55$ for total diversity, $R^2 = 0.45$ for orders and $R^2 = 0.36$ for classes, $P<0.001$ for all) and basidiomycete diversity with basidiomycete richness ($R^2 = 0.51$ for total diversity and $R^2 = 0.41$ for orders, $P<0.001$ for both). Further, the ascomycete

Table 2. ANOVA analysis of effects of cultivar (including all cultivars) and GM-cultivar 'Modena' versus parental cultivar 'Karnico'on fungal biomass, enzymatic activities and fungal richness in the rhizosphere in different fields, years and growth stages.

				Ergosterol (mg/g)		Laccases (μmol/g)		Mn-Peroxidases (μmol/g)		Cellulases (μmol/g)		# of Ascomycetes		# of Basidiomycetes		# of AMF	
				Cultivar	GM-Parent	Cultivar	GM-Parent	Cultivar	GM-Parent	Cultivar	GM-Parent	Cultivar	GM-Parent	Cultivar	GM-Parent	Cultivar	GM-Parent
			df.	5	1	5	1	5	1	5	1	5	1	5	1	5	1
Field BUI	**2008**	Young	*F*	0.86	0.24	0.89	2.99	1.02	1.32	0.48	0.92	0.22	0.06	0.90	0.05	1.09	0.10
			P	0.53	0.65	0.51	0.15	0.44	0.30	0.79	0.38	0.95	0.82	0.50	0.83	0.42	0.78
		Flowering	*F*	**12.64**	2.13	1.00	0.59	0.76	0.42	0.79	0.00	0.38	0.25	0.48	0.02	1.89	0.47
			P	**<0.001**	0.20	0.44	0.47	0.59	0.84	0.57	0.96	0.86	0.63	0.79	0.89	0.15	0.53
		Senescence	*F*	2.07	0.15	2.61	0.20	**9.19**	0.85	1.25	0.51	**3.67**	**39.06**	0.66	0.38	1.93	0.28
			P	0.12	0.72	0.06	0.67	**<0.001**	0.40	0.33	0.50	**0.018**	**0.025**	0.66	0.57	0.15	0.62
	2009	Young	*F*	1.22	1.13	1.24	0.50	0.73	0.61	0.94	0.61	0.13	0.07	0.47	0.19	1.12	2.42
			P	0.36	0.33	0.35	0.51	0.59	0.46	0.47	0.46	0.97	0.80	0.75	0.69	0.40	0.17
		Flowering	*F*	1.04	0.72	0.45	2.63	1.00	0.34	0.69	5.53	0.17	3.00	0.87	0.00	0.04	5.44
			P	0.41	0.45	0.77	0.16	0.45	0.58	0.62	0.06	0.85	0.33	0.49	0.96	0.96	0.26
		Senescence	*F*	1.32	0.20	1.68	1.32	1.42	1.36	1.80	0.71	0.32	1.93	1.60	2.10	1.24	1.15
			P	0.32	0.68	0.26	0.33	0.33	0.33	0.23	0.46	0.81	0.21	0.27	0.24	0.37	0.40
	2010	Young	*F*	**9.49**	0.01	1.09	0.11	0.11	0.26	1.32	0.49	0.92	1.94	0.84	1.26	0.34	0.35
			P	**0.001**	0.94	0.41	0.75	0.99	0.63	0.31	0.52	0.49	0.21	0.54	0.31	0.88	0.58
		Flowering	*F*	0.85	0.77	1.06	0.64	0.84	0.69	0.47	0.13	0.19	0.02	2.52	0.06	0.34	1.08
			P	0.54	0.42	0.41	0.46	0.54	0.80	0.79	0.73	0.96	0.89	0.31	0.81	0.88	0.49
		Senescence	*F*	4.73	1.42	0.65	0.22	1.89	0.30	1.55	1.51	0.33	1.04	0.68	0.00	1.53	**6.42**
			P	0.11	0.30	0.63	0.66	0.17	0.61	0.24	0.27	0.86	0.38	0.62	0.95	0.24	**0.044**
Field VMD	**2008**	Young	*F*	1.96	1.49	1.20	1.21	0.76	0.09	1.03	0.34	2.13	2.50	0.34	0.10	0.55	1.54
			P	0.13	0.27	0.35	0.31	0.59	0.78	0.43	0.58	0.11	0.17	0.89	0.77	0.74	0.26
		Flowering	*F*	0.67	0.33	1.43	1.19	0.73	0.65	1.23	1.93	0.56	0.33	0.46	1.47	2.10	1.03
			P	0.65	0.59	0.27	0.32	0.61	0.46	0.35	0.21	0.73	0.59	0.72	0.44	0.12	0.35
		Senescence	*F*	1.43	**17.50**	1.24	0.51	2.43	0.18	0.81	0.10	0.66	1.93	0.56	0.26	0.66	0.93
			P	0.26	**0.006**	0.33	0.50	0.08	0.69	0.56	0.76	0.66	0.21	0.73	0.65	0.66	0.38
	2009	Young	*F*	2.91	0.54	0.66	0.22	0.27	0.06	0.32	0.15	0.57	0.24	0.58	0.36	1.06	0.62
			P	0.17	0.50	0.59	0.66	0.85	0.82	0.81	0.72	0.65	0.64	0.64	0.57	0.40	0.46
		Flowering	*F*	7.03	0.62	0.98	0.18	**4.24**	1.18	1.22	2.83	0.67	2.00	0.25	0.85	1.58	0.29
			P	0.23	0.47	0.44	0.69	**0.03**	0.32	0.35	0.14	0.54	0.22	0.90	0.41	0.29	0.63
		Senescence	*F*	0.77	**14.16**	1.57	1.18	1.35	1.24	0.36	0.01	1.35	1.04	**6.00**	4.09	2.38	5.89
			P	0.68	**0.011**	0.23	0.32	0.30	0.31	0.84	0.92	0.36	0.38	**0.03**	0.11	0.13	0.06

Table 2. Cont.

			Ergosterol (mg/g)		Laccases (μmol/g)		Mn-Peroxidases (μmol/g)		Cellulases (μmol/g)		# of Ascomycetes		# of Basidiomycetes		# of AMF	
			Cultivar	GM-Parent	Cultivar	GM-Parent	Cultivar	GM-Parent	Cultivar	GM-Parent	Cultivar	GM-Parent	Cultivar	GM-Parent	Cultivar	GM-Parent
2010	Young	*F*	0.86	0.30	0.83	1.05	0.65	2.97	0.68	0.93	0.53	0.18	0.83	1.67	**4.01**	4.17
		P	0.53	0.61	0.55	0.34	0.66	0.14	0.65	0.37	0.75	0.68	0.55	0.24	**0.03**	0.11
	Flowering	*F*	0.75	0.99	1.76	1.00	2.13	1.00	0.78	0.86	1.08	0.74	0.25	0.09	2.02	8.17
		P	0.60	0.36	0.17	0.36	0.11	3.56	0.58	0.39	0.41	0.42	0.90	0.78	0.18	0.07
	Senescence	*F*	0.27	**13.59**	1.54	1.17	0.18	0.13	0.30	0.06	0.95	0.46	0.47	1.02	0.12	0.52
		P	0.84	**0.014**	0.30	0.54	0.90	0.74	0.82	0.82	0.45	0.53	0.71	0.35	0.95	0.51

Significant P-values are marked with bold.

diversity was negatively correlated with basidiomycete diversity ($R^2 = -0.15$, $P<0.005$). Ascomycete richness was correlated with the amount of Mn-peroxidases in the soil ($R^2 = 0.15$, $P<0.05$) and basidiomycete richness with ergosterol ($R^2 = 0.18$, $P<0.001$). The AMF diversity was positively correlated with soil moisture content ($R^2 = 0.15$, $P<0.001$), AMF richness ($R^2 = 0.58$, $P<0.001$) and ascomycete diversity ($R^2 = 0.10$, $P<0.05$).

The diversity of ascomycetes or basidiomycetes at the level of OTUs or orders was not significantly affected by field site. However, AMF diversity was. There was no significant difference in diversity of ascomycetes at the level of OTUs and orders from year to year, although diversity between years 2009 and 2010 was significantly different. However, at the level of classes also 2008 and 2009 were different and year was a more pronounced factor explaining the diversity. For basidiomycetes and AMF, year had a strong influence on diversity both at the level of OTUs and orders (table 4). Growth stage, had a strong significant effect on ascomycete and basidiomycete diversities (Fig. 4, table 4) but less effect on the AMF diversity in the rhizosphere.

Cultivar-type had no overall effect on basidiomycete, ascomycete and AMF diversity at the level of OTUs or orders. However, at the level of classes cultivar 'Désirée' had a significantly less diverse community of ascomycetes in its rhizosphere than all the other cultivars causing a general cultivar effect (table 4). When the field sites, growth stages and years were considered separately, cultivar was a weak explanatory factor for the diversity of ascomycetes, basidiomycetes and AMF (Fig. 4, table 5). Both cultivar and GM-variety had an effect on diversity of ascomycetes in the rhizosphere in field BUI 2010 in the young-plant stage where 'Karnico' had a low diversity. The GM-variety had a significantly less diverse community of ascomycetes compared to its parental variety in field VMD 2010 at the stage of flowering plants (table 5). Basidiomycete diversity was different in rhizospheres between cultivars both during flowering and senescence 2009 in field VMD but never between GM and its parental cultivar. For AMF effects of cultivar and GM-variety were observed only at the first sampling moment of rhizosphere field in VMD (young 2008).

Legacy of GM-crops

The fields were sampled after the growth seasons 2008 and 2009 and, in addition, rhizosphere of barley was sampled in June 2009 in the field where potatoes were grown in 2008. There were no significant differences in ergosterol content, enzymatic activities, fungal richness or fungal diversity between soils where 'Modena' and 'Karnico' had been grown (table 6). In the rhizosphere of barley there was no effect of previous genotype detected at all. Furthermore, no effect could be detected of different potato genotypes on the fungal community fingerprints in post-harvest samples and in the rhizosphere of barley (Fig. 5, table 6).

Discussion

The composition and function of fungal communities in the rhizosphere was shown to be highly dynamic and influenced by plant growth stage, soil type, year and, to a smaller extent, also cultivar-type. The largest explaining factor for most of the measured parameters was plant phenological growth stage, followed by year and the soil type. In addition, results confirmed our previous observations that fungal composition and abundance is strongly influenced by the presence of potato roots (i.e. a strong rhizosphere effect) [19].

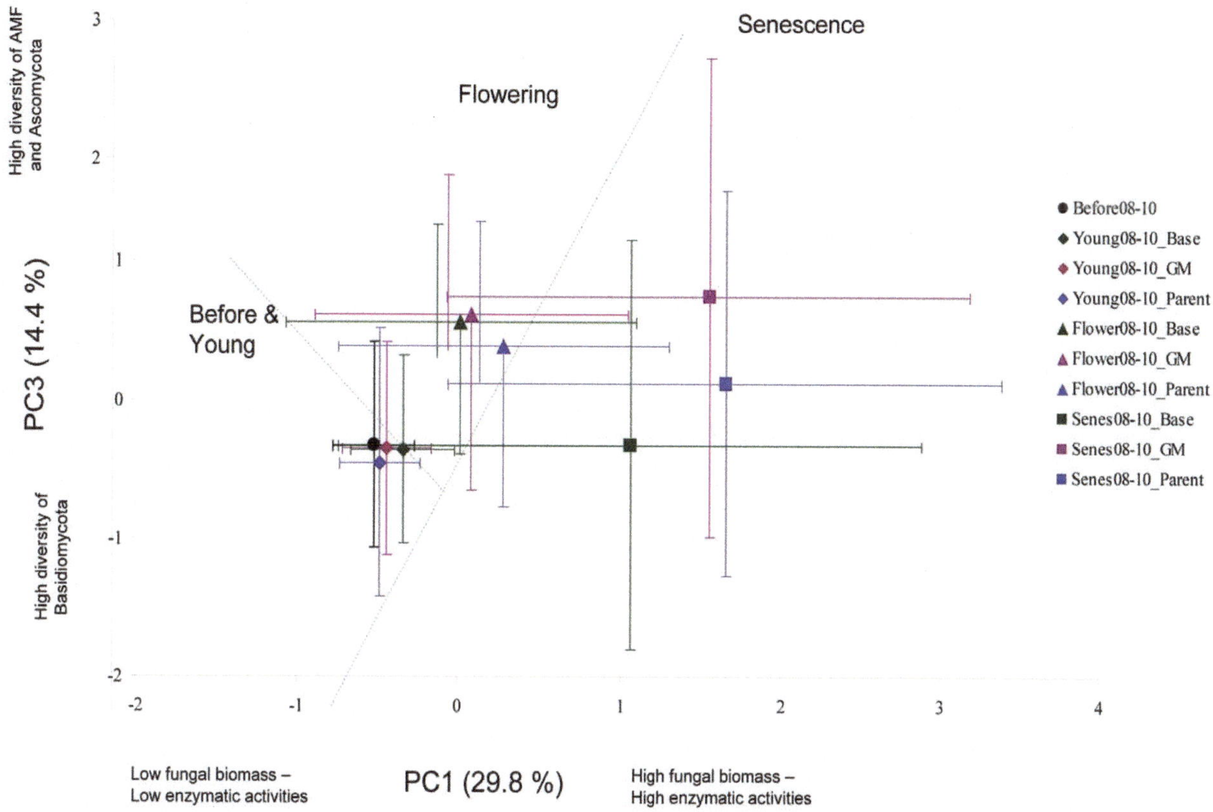

Figure 2. Principal component analysis of functioning and diversity of fungal communities in plots cropped with different potato cultivars. For clarity, the years and field sites are combined. Pre-cropping samples are represented by black circles, young plant stage samples with diamonds, flowering plants stage samples with triangles and senescence stage samples with squares. Green markers and error bars represent baseline cultivars (n = 96), purple markers the GM-variety (n = 24) and blue markers the parental variety 'Karnico' (n = 24). The explanatory parameters are mentioned next to the axis. The enzymes measured as functional parameters were laccases, Mn-peroxidases and cellulases.

The succession of microbial communities during plant growing season can be explained by two possible mechanisms [7]. The first one is related to temporal changes in abiotic conditions such as soil moisture and temperature. However this is not a likely option to explain the fungal community dynamics observed in this study as the three years of study were very contrasting in temperature and moisture. The second, more likely, mechanism is the changes in quality and quantity of root exudates and rhizodeposits with growth stage [35,36] and or changes in root morphology. Although root exudates were not measured in this study, there is evidence of the effect of plant growth stage on root exudate fluxes which in turn affect soil microbial communities [8,37]. Earlier studies indicated that bacterial and fungal communities in the rhizosphere would either decrease [7,13,20,38] or increase [10,11,39–41] during plant maturation. Our results clearly indicate that the plants at the senescence stage (EC90) harbor the most diverse, active and abundant fungal communities. The presence of the highest fungal biomass and diversity at the stage of senescence was expected, as decomposable material (dead roots and leaves) is already available while root exudation still continues thereby broadening the spectrum of substrate availability [42]. Yet, the increase and magnitude of the fungal biomass and its activity in the rhizosphere at that stage is surprising. Until now, the prevailing belief was that the fungal biomass is low in soils under intensive agricultural management. Earlier results with the same cultivars under controlled conditions confirm our observations [43].

Surprisingly, despite the strong differences in soil organic matter content, field location did not affect the community function or diversity of the higher fungi much and results from the two fields could be even combined for baseline purposes. Earlier studies have found soil type as one of the most explanatory factor [7–9,12,18] affecting soil microbial communities. Bacterial communities appeared to differ strongly between the two fields used in this study, both for bulk soil and rhizosphere [44]. In our study, however, only total fungal community structure and diversity of AMF were strongly affected by the field site while fungal biomass and functional parameters such as enzymatic activities seemed to respond to the field type only slightly. The difference in AMF between fields could be probably explained by the higher organic matter content and thus higher AMF diversity in field VMD [14].

We detected interesting differences between the years. In the first years, mineral fertilizer was used and only from the beginning of 2010 pig manure was used as a fertilizer. This might explain differences in fungal communities observed between 2008 and 2010. Previously, it has been shown that different types of fertilizer treatments contribute to different microbial communities [45]. Notably, in our study we detected more ascomycetes and less basidiomycetes and fungi in general in 2010 compared to 2008 in both fields (Fig. 1) which might be an indication of changed community structure due to changed fertilizer treatment. Also the diversity and richness of AMF was higher in 2010.

Community structure and diversity of soil fungi are important determinants of key soil ecosystems functions such as decomposition of organic matter. Indeed, we could detect a correlation

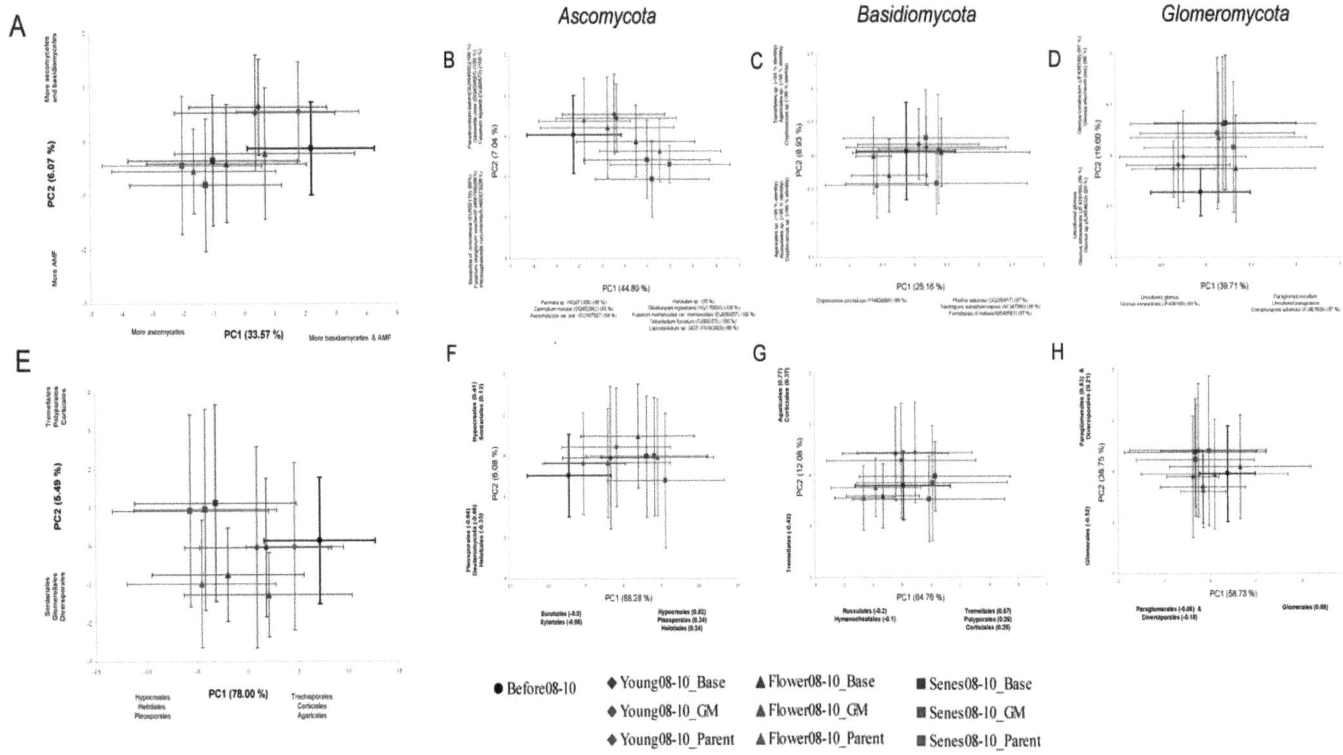

Figure 3. Principal component analysis of community structure of identified fungi. The PCA analysis was done both at the level of individual OTUs and of orders for total fungi (A & E), *Ascomycota* (B & F), *Basidiomycota* (C & G) and *Glomeromycota* (D & H). Figures A–D depict the identified fungal OTUs whereas figures E–H indicate the levels of orders. Orders together with identified OTUs are given in table S1. For clarity, the years and field sites are combined. Pre-cropping soil samples are marked with black circles, young plants stage with diamonds, flowering plant stage with triangles and senescence stage with squares. Green markers and error bars represent baseline cultivars (n = 96), purple markers the GM-variety (n = 24) and blue markers the parental variety 'Karnico' (n = 24). The OTUs (figures A–D) and orders (E–H) that do significantly explain the components are mentioned next to the axis.

between community structure of fungi and decomposition-related enzyme activities. Moreover, the combination of phylogenetic analyses with functional assays proved highly useful, providing a more complete picture of fungal community dynamics. We found a correlation between Mn-peroxidases produced and the ascomycete diversity (and richness). Mn-peroxidases can be produced primarily by basidiomycetes as well as some ascomycetal groups [46]. However, not much is known of the ecology. AM fungi are strongly affected by agricultural practices and changes in soil characteristics [47–49] such as moisture and manure addition. Indeed, we saw an increase of AMF diversity in 2010 when the fertilizer was changed from mineral to pig manure which is in correspondence with results from Verbruggen *et al.* [14] who found organic fertilizers having a positive effect on AMF diversity.

Only few studies have evaluated the potential impacts of GM-plants in the context of impacts of multiple cultivars on fungal rhizosphere communities. Most of them have found some degree of cultivar dependence of soil fungal community composition [13,18,47] while another one [20] found no cultivar dependent alterations in the fungal communities. We found some indications of cultivar dependence, for instance the cultivar 'Premiere' had a lower amount of fungi, as measured by ergosterol, in its rhizosphere than two other cultivars 'Aveka' and 'Désirée' Despite some differences in enzymatic activities, total fungal diversity was not affected by the cultivar-type at the level of OTUs and orders.

Table 3. ANOSIM comparisons between the fields, years, growth stages, cultivars and GM-trait for *Ascomycota*, *Basidiomycota* and *Glomeromycota*.

	Field		Year		Growth stage		Cultivar*		GM-parent*	
	R	P	R	P	R	P	R	P	R	P
Ascomycota	0.07	**<0.001**	0.29	**<0.001**	0.10	**<0.001**	0.013	0.131	−0.006	1
Basidiomycota	0.04	**<0.001**	0.25	**<0.001**	0.19	**<0.001**	0.008	0.188	0.015	0.915
Glomeromycota	0.11	**<0.001**	0.22	**<0.001**	0.09	**<0.001**	−0.005	0.689	−0.011	0.863

*Only samples where plant was present are included in the analyses.
Significant P-values are marked with bold.

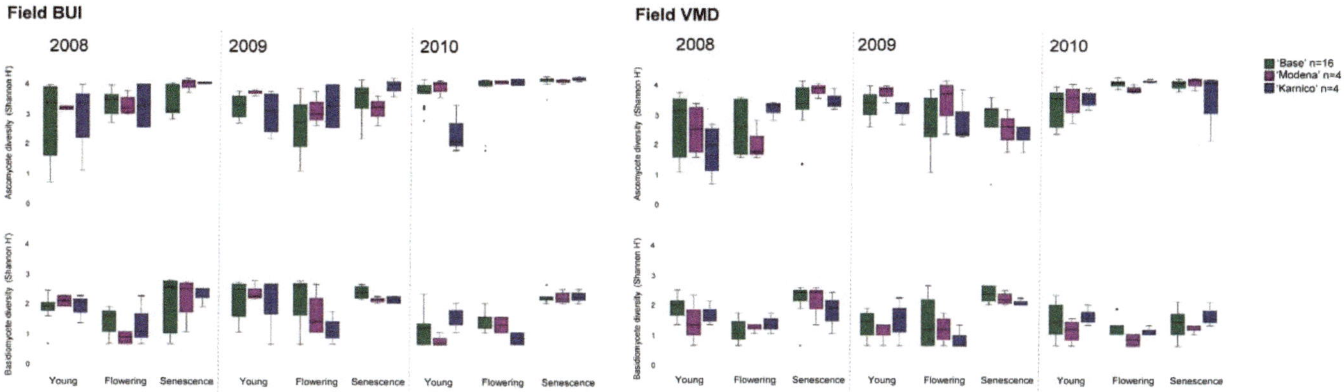

Figure 4. Effect of cultivar, year, growth stage and field on fungal diversity. Boxplots of changes in diversity of *Ascomycota* and *Basidiomycota* between years, growth stages, fields and between baseline, GM and its parental variety. The baseline (all other cultivars combined, n = 16) is marked with green boxplots, the GM-variety (n = 4) with purple and the parental variety 'Karnico' (n = 4) with blue markers. Diversity was calculated using Shannon-Wiener index (H') and statistical comparisons are presented in table 5.

Ascomycetal diversity was affected at the level of classes as one cultivar, 'Désirée', had a less diverse community in its rhizosphere. To conclude, we found some degree of cultivar dependence in measured parameters at some time points, but these differences were mostly not persisting over time and not observed in both fields, similarly as found by Weinert et al. [18].

In this study the GM-variety 'Modena' was not significantly different from its parental variety 'Karnico' in any measured parameter and it seemed that these cultivars had a very similar effect on both the structure and function of soil fungal communities. The only significant effect was the difference in the amount of fungi in the rhizosphere of the two cultivars in the field VMD during senescence, in all years of the study. This was, however, seen only in one of the two soils studied and can, thus, be ruled out as a cultivar-soil interaction effect. There was no overall trend of multiple parameters being consistently changed by any of the cultivars while the other factors (i.e. growth stage and season) had consistent effect on multiple parameters measured.

The growth stage can also affect the outcome of the comparison between the cultivars. Other authors have found differences in microbial communities associated with GM-potatoes mostly at the senescent growth stage [19,33,40,50,51]. The soil micro-organisms have an important role in soil ecosystem functioning such as decomposition of plant residues and nutrient cycling [52]. Thus it is possible that the differences at the stage of senescence as found in this study could lead to changes in function and might, thus, have long lasting effects. In this study, all analyses indicated that when the fungal communities were assessed after removal of the plant or in the rhizosphere of the next crop in rotation, there were no differences between fungal communities from field plots that contained harvested modified potato plants. So, we did not detect any significant connection between the previous cultivar of potato on the fungi in the rhizosphere of the next crop barley. Hence, the changes in the fungal biomass associated with starch modified potato plants detected at certain time points and fields in this study were temporary and did not persist into the next field season. A

Table 4. The effect of field site, year, growth stage and cultivar on soil ascomycete, basidiomycete and glomeromycete diversity for different taxonomic levels.

		Field		Year		Growth stage		Cultivar*		GM-parent*	
		F	P	F	P	F	P	F	P	F	P
Ascomycota	OTUs	0.005/0.004	0.94/0.94	7.80/3.89	**<0.001/ 0.02**	12.76/9.16	**<0.001**	0.65/0.32	0.66/0.91	2.67/0.49	0.11/0.49
	Orders	0.33/0.009	0.57/0.92	7.44/3.56	**<0.005/ 0.03**	10.8/13.22	**<0.001**	0.59/0.52	0.64/0.76	2.74/1.58	0.10/0.21
	Classes	9.30/9.50	**0.03/0.02**	10.80/9.64	**<0.001**	6.78/5.76	**<0.001**	15.58/34.61	**<0.001**	2.97/2.31	0.09/0.31
Basidiomycota	OTUs	1.803/0.523	0.18/0.47	9.49/6.64	**<0.001/ 0.002**	13.84/9.37	**<0.001**	1.24/1.41	0.29/0.23	0.03/0.02	0.87/0.90
	Orders	0.04/0.002	0.85/0.97	21.85/17.86	**<0.001**	8.99/6.48	**<0.001**	1.85/2.08	0.13/0.09	0.19/0.37	0.67/0.54
Glomeromycota	OTUs	14.67/15.04	**<0.001**	24.48/20.72	**<0.001**	3.01/2.76	**0.03/0.04**	1.91/1.63	0.09/0.15	1.91/1.40	0.17/0.24
	Orders	38.22/35.98	**<0.001**	12.50/9.99	**<0.001**	2.29/2.13	0.08/0.09	1.91/1.89	0.09/0.10	1.17/1.59	0.19/0.21

*Only samples where plant was present are included in the analyses.
All diversities were calculated using both Shannon H' and Simpson diversity indexes and presented in the table as Shannon H'/Simpson diversity. If both P-values are the same, only one value is presented. Diversity index for classes was not calculated for basidiomycetes and glomeromycetes due to low numbers or unevenness of classes. Significant P-values are marked with bold.

Table 5. ANOVAs of the effect of cultivar (including all cultivars) and GM-cultivar 'Modena' versus parental cultivar 'Karnico' on diversity of ascomycetes, basidiomycetes and glomeromycetes in the rhizosphere in both fields, all years and growth stages.

				Diversity of Ascomycota (Shannon H)			Diversity of Basidiomycota (Shannon H)			Diversity of Glomeromycota (Shannon H)		
				Cultivar	GM-Parent	Order level	Cultivar	GM-Parent	Order level	Cultivar	GM-Parent	Order level
				df. 5	df. 1		df. 5	df. 1		df. 5	df. 1	
Field BUI	**2008**	Young	*F*	0.12	0.12		0.21	0.60		0.38	0.38	
			P	0.89	0.75	n.s.	0.95	0.48	n.s.	0.86	0.85	n.s.
		Flowering	*F*	0.92	0.00		0.79	0.56		0.62	2.34	
			P	0.91	1.00	n.s.	0.58	0.50	n.s.	0.69	0.22	n.s.
		Senescence	*F*	2.93	0.05		0.99	0.30		0.93	0.60	
			P	0.11	0.83	n.s.	0.47	0.61	n.s.	0.50	0.50	n.s.
	2009	Young	*F*	0.88	1.22		0.23	0.40		0.99	0.88	
			P	0.51	0.52	n.s.	0.92	0.55	n.s.	0.46	0.38	n.s.
		Flowering	*F*	2.86	1.14		3.28	0.21		0.64	0.71	
			P	0.14	0.35	n.s.	0.10	0.63	n.s.	0.56	0.44	n.s.
		Senescence	*F*	1.35	3.29		4.88	1.19		nd	nd	
			P	0.35	0.21	n.s.	0.77	0.29	n.s.	nd	nd	n.s.
	2010	Young	*F*	**6.25**	**13.80**	**3.83/11.27**	0.79	5.98		0.54	0.60	
			P	**0.03**	**0.01**	**0.02/0.001**	0.58	0.07	n.s.	0.74	0.50	n.s.
		Flowering	*F*	0.46	0.00		2.86	2.24		0.43	0.13	
			P	0.80	0.98	n.s.	0.14	0.38	n.s.	0.82	0.74	n.s.
		Senescence	*F*	1.37	1.14		0.31	0.29		4.41	0.82	
			P	0.30	0.35	n.s.	0.86	0.82	n.s.	0.99	0.40	n.s.
Field VMD	**2008**	Young	*F*	1.73	1.89		0.45	0.19		**4.77**	**38.37**	**2.38/40.38**
			P	0.19	0.49	n.s.	0.80	0.69	n.s.	**0.01**	**0.00**	**0.02/0.001**
		Flowering	*F*	0.85	4.35		0.43	0.37		2.23	0.41	
			P	0.52	0.17	n.s.	0.79	0.58	n.s.	0.15	0.59	n.s.
		Senescence	*F*	0.99	0.04		0.57	0.34		0.13	0.32	
			P	0.48	0.86	n.s.	0.73	0.59	n.s.	0.93	0.62	n.s.
	2009	Young	*F*	0.64	1.79		0.45	0.26		0.51	0.20	
			P	0.62	0.41	n.s.	0.73	0.64	n.s.	0.69	0.67	n.s.
		Flowering	*F*	0.64	1.08		**4.95**	0.90		0.47	0.71	
			P	0.61	0.41	n.s.	**0.04**	0.39	n.s.	0.76	0.44	n.s.
		Senescence	*F*	0.35	1.21		**5.54**	0.93	**4.37**/1.61	2.31	23.78	
			P	0.84	0.44	n.s.	**0.02**	0.38	**0.04**/0.26	0.15	0.16	n.s.

Table 5. Cont.

		Diversity of Ascomycota (Shannon H)			Diversity of Basidiomycota (Shannon H)			Diversity of Glomeromycota (Shannon H)		
		Cultivar	GM-Parent	Order level	Cultivar	GM-Parent	Order level	Cultivar	GM-Parent	Order level
		df. 5	df. 1		df. 5	df. 1		df. 5	df. 1	
2010 Young	*F*	0.49	0.42		0.62	2.92		1.12	1.89	
	P	0.78	0.84	n.s.	0.66	0.15	n.s.	0.56	0.49	n.s.
Flowering	*F*	2.66	**7.84**	1.43/**1.17**	2.72	2.34		0.46	0.33	
	P	0.08	**0.05**	0.28/**0.03**	0.13	0.22	n.s.	0.77	0.67	n.s.
Senescence	*F*	0.84	0.91		0.54	2.87		0.49	0.42	
	P	0.50	0.38	n.s.	0.67	0.17	n.s.	0.75	0.86	n.s.

The diversities were estimated using Shannon-H'. The first two columns of each fungal group are performed at the level of OTUs and the third column indicates significance at the level of orders. Significant P-values are marked with bold.

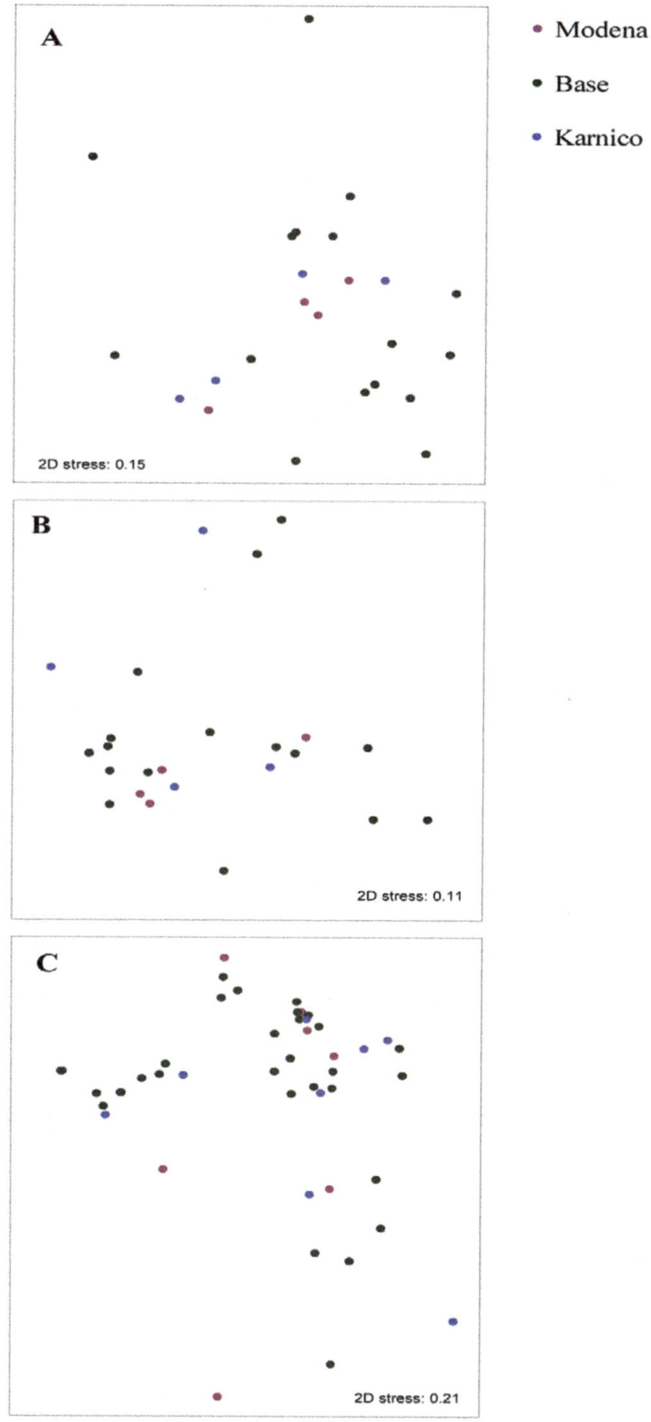

Figure 5. Long term effect of GM-trait on fungal community. NMDS of effects of GM-variety in the next crop (barley) rhizosphere in field BUI on ascomycetes (A), basidiomycetes (B) and in fields BUI and VMD on glomeromycetes (C). The GM-variety 'Modena' is marked with purple markers, the parental cultivar 'Karnico' with blue markers, and baseline (all other cultivars combined) green markers. Details on statistical analysis are given in table 6.

similar observation was made for bacteria after cropping of transgenic canola [53].

In conclusion, plant growth stage, year and field site were the factors contributing most to variation in the potato-associated

Table 6. Statistical analysis of the effects of cultivar and GM-trait on fungal-related parameters in post-harvest soil samples as well as in the rhizosphere of next plant barley.

		Field BUI				Field VMD				Barley rhizosphere			
		Cultivar		GM-parent		Cultivar		GM-parent		Cultivar		GM-parent	
		F/R	P	F/R	P	F/R	P	F/R	P	F/R	P	F/R	P
ANOVA	Ergosterol	0.14	0.89	0.00	0.95	0.81	0.47	1.04	0.35	0.03	0.98	0.00	0.98
	Laccases	0.70	0.63	0.28	0.62	3.63	**0.02**	0.06	0.82	0.56	0.58	1.03	0.36
	Mn-Peroxidases	1.77	0.19	2.76	0.15	1.30	0.31	0.00	0.98	0.00	0.99	0.00	0.99
	Cellulases	1.00	0.43	0.35	0.58	1.56	0.22	5.53	0.06	0.06	0.95	0.05	0.84
	Diversity of basidiomycetes									1.37	0.29	0.06	0.82
	Diversityof ascomycetes	0.87	0.72	0.33	0.64	0.34	0.72	0.58	0.48	0.72	0.51	0.60	0.48
	Diversity of AMF	0.61	0.50	0.02	0.89	0.53	0.68	1.00	0.36	0.55	0.60	1.14	0.35
ANOSIM	Community of ascomycetes	−0.22	0.88	−0.32	1.00	−0.05	0.72	0.07	0.24	0.01	0.42	0.56	0.33
	Community of basidiomycetes	0.03	0.65	−0.12	0.95	−0.13	0.73	−0.38	0.97	−0.24	0.91	−0.58	1.00
	Community of AMF	−0.04	0.67	−0.16	0.80	0.02	0.37	−0.13	0.81	−0.28	1.00	−0.38	1.00

ANOVA was used as a similarity measure for fungal biomass, enzymatic measurements and diversity and F values are presented in the table. ANOSIM was used for the community data derived from T-REX and R-values are presented in the table. Significant P-values for both ANOVA and ANOSIM are marked with bold.

fungal communities. Despite some differences in fungal-related parameters between individual cultivars, there were no directional effects and most of the differences observed were not consistent between fields and years. Even at the level of individual OTUs, there were no consistent significant differences between cultivars in community structure and no differences in community function were found during and after the growth of the plant. However, as was seen from conflicting evidence between different studies, we acknowledge that potential effects of GM-crops on soil fungal communities vary between crop species and types of modifications done to the plant making a case-by–case evaluation strategy advisable. We hypothesized that this modification would have no direct but rather indirect unintended effects of the modification on the plant physiology through production of different exudates. Data presented in this study allowed us to conclude that the modification studied here has no long-lasting effects on soil fungal communities and that the potato plant growth stage, season and field location affect the soil fungal community structure and function more than the cultivar-type or starch modification of tubers.

Supporting Information

Figure S1 Principal component analysis of function and diversity of fungal communities in between growth stages, fields and years. Field BUI is marked with closed symbols and solid lines while field VMD with open symbols and dotted lines. Year 2008 is marked with black markers, year 2009 with red markers and 2010 with blue marker. The explanatory parameters are mentioned next to the axis.

Table S1 Soil characteristics and fertilizers added to the fields. In the fertilizer treatments CAN = Calcium Ammonium Nitrate, NP = nitrogen as ammonium sulphate and phosphorous as P_2O_5 and ORG = organic fertilizer = pig manure.

Acknowledgments

The authors would like to thank Wiecher Smant, Özgül Ineoglu and Nadia Marttin for technical assistance and collaboration during field sampling and Peter Bruinenberg and Paul Heeres from AVEBE/AVERIS seeds for valuable assistance in the field experiment. This is publication 5227 of the Netherlands Institute of Ecology (NIOO-KNAW).

Author Contributions

Conceived and designed the experiments: SEH WdB JvV. Performed the experiments: SEH. Analyzed the data: SEH WdB JvV. Wrote the paper: SEH WdB JvV.

References

1. Wolfenbarger LL, Phifer PR (2000) Biotechnology and ecology - The ecological risks and benefits of genetically engineered plants. Science 290: 2088–2093.
2. Birch ANE, Griffiths BS, Caul S, Thompson J, Heckmann LH, et al. (2007) The role of laboratory, glasshouse and field scale experiments in understanding the interactions between genetically modified crops and soil ecosystems: A review of the ECOGEN project. Pedobiologia 51: 251–260.
3. Lynch JM, Whipps JM (1990) Substrate flow in the rhizosphere. Plant and Soil 129: 1–10.
4. Raaijmakers J, Paulitz T, Steinberg C, Alabouvette C, Moënne-Loccoz Y (2009) The rhizosphere: a playground and battlefield for soilborne pathogens and beneficial microorganisms. Plant and soil 321: 341–361.
5. Carlile MJ, Watkinson SC, Gooday GW (2001) The Fungi. San Diego, CA, USA: Academic Press.
6. Buée M, De Boer W, Martin F, van Overbeek L, Jurkevitch E (2009) The rhizosphere zoo: An overview of plant-associated communities of microorganisms, including phages, bacteria, archaea, and fungi, and of some of their structuring factors. Plant and Soil 321: 189–212.

7. Wang GH, Xu YX, Jin J, Liu JD, Zhang QY, et al. (2009) Effect of soil type and soybean genotype on fungal community in soybean rhizosphere during reproductive growth stages. Plant and Soil 317: 135–144.
8. Garbeva P, van Veen JA, van Elsas JD (2004) Microbial diversity in soil: Selection of microbial populations by plant and soil type and implications for disease suppressiveness. Annual Review of Phytopathology 42: 243–270.
9. Berg G, Smalla K (2009) Plant species and soil type cooperatively shape the structure and function of microbial communities in the rhizosphere. FEMS Microbiology Ecology 68: 1–13.
10. Gomes NCM, Heuer H, Schönfeld J, Costa R, Hagler-Mendonca L, et al. (2001) Bacterial diversity of the rhizosphere of maize (Zea mays) grown in tropical soil studied by temperature gradient gel electrophoresis. Plant and Soil 232: 167–180.
11. Gomes NCM, Fagbola O, Costa R, Rumjanek NG, Buchner A, et al. (2003) Dynamics of fungal communities in bulk and maize rhizosphere soil in the tropics. Applied and Environmental Microbiology 69: 3758–3766.
12. Buyer JS, Roberts DP, Russek-Cohen E (2002) Soil and plant effects on microbial community structure. Canadian Journal of Microbiology 48: 955–964.
13. Gschwendtner S, Reichmann M, Muller M, Radl V, Munch JC, et al. (2010) Effects of genetically modified amylopectin-accumulating potato plants on the abundance of beneficial and pathogenic microorganisms in the rhizosphere. Plant and Soil 335: 413–422.
14. Verbruggen E, Röling WFM, Gamper HA, Kowalchuk GA, Verhoef HA, et al. (2010) Positive effects of organic farming on below-ground mutualists: large-scale comparison of mycorrhizal fungal communities in agricultural soils. New Phytologist 186: 968–979.
15. Griffiths BS, Caul S, Thompson J, Birch ANE, Cortet J, et al. (2007) Microbial and microfaunal community structure in cropping systems with genetically modified plants. Pedobiologia 51: 195–206.
16. Donegan KK, Schaller DL, Stone JK, Ganio LM, Reed G, et al. (1996) Microbial populations, fungal species diversity and plant pathogen levels in field plots of potato plants expressing the Bacillus thuringiensis var tenebrionis endotoxin. Transgenic Research 5: 25–35.
17. Donegan KK, Seidler RJ, Doyle JD, Porteous LA, Digiovanni G, et al. (1999) A field study with genetically engineered alfalfa inoculated with recombinant Sinorhizobium meliloti: effects on the soil ecosystem. Journal of Applied Ecology 36: 920–936.
18. Weinert N, Meincke R, Gottwald C, Heuer H, Gomes NCM, et al. (2009) Rhizosphere communities of genetically modified zeaxanthin-accumulating potato plants and their parent cultivar differ less than those of different potato cultivars. Applied and Environmental Microbiology 75: 3859–3865.
19. Hannula SE, de Boer W, van Veen JA (2010) In situ dynamics of soil fungal communities under different genotypes of potato, including a genetically modified cultivar. Soil Biology & Biochemistry 42: 2211–2223.
20. Milling A, Smalla K, Maidl FX, Schloter M, Munch JC (2004) Effects of transgenic potatoes with an altered starch composition on the diversity of soil and rhizosphere bacteria and fungi. Plant and Soil 266: 23–39.
21. Hart MM, Powell JR, Gulden RH, Dunfield KE, Pauls KP, et al. (2009) Separating the effect of crop from herbicide on soil microbial communities in glyphosate-resistant corn. Pedobiologia 52: 253–262.
22. Lee S-H, Kim C-G, Kang H (2011) Temporal Dynamics of Bacterial and Fungal Communities in a Genetically Modified (GM) Rice Ecosystem. Microbial Ecology 61: 646–659.
23. Donegan KK, Palm CJ, Fieland VJ, Porteous LA, Ganio LM, et al. (1995) Changes in levels, species and DNA fingerprints of soil-microorganisms associated with cotton expressing the *Bacillus-Thuringiensis* var *Kurstaki* endotoxin. Applied Soil Ecology 2: 111–124.
24. Götz M, Nirenberg H, Krause S, Wolters H, Draeger S, et al. (2006) Fungal endophytes in potato roots studied by traditional isolation and cultivation-independent DNA-based methods. FEMS Microbiology Ecology 58: 404–413.
25. Wei XD, Zou HL, Chu LM, Liao B, Ye CM, et al. (2006) Field released transgenic papaya affects microbial communities and enzyme activities in soil. Plant and Soil 285: 347–358.
26. van Overbeek L, van Elsas JD (2008) Effects of plant genotype and growth stage on the structure of bacterial communities associated with potato (*Solanum tuberosum L.). FEMS Microbiology Ecology* 64: 283–296.
27. de Vetten N, Wolters AM, Raemakers K, van der Meer I, ter Stege R, et al. (2003) A transformation method for obtaining marker-free plants of a cross-pollinating and vegetatively propagated crop. Nature Biotechnology 21: 439–442.
28. Hack H, Gal H, Klemke T, Klose R, Meier U, et al. (2001) The BBCH scale for phenological growth stages of potato (*Solanum tuberosum* L.). In: Meier U, ed. Growth Stages of Mono- and Dicotyledonous Plants, BBCH Monograph: Federal Biological Research Centre for Agriculture and Forestry.
29. de Ridder-Duine AS, Smant W, van der Wal A, van Veen JA, de Boer W (2006) Evaluation of a simple, non-alkaline extraction protocol to quantify soil ergosterol. Pedobiologia 50: 293–300.
30. van der Wal A, van Veen JA, Pijl AS, Summerbell RC, de Boer W (2006) Constraints on development of fungal biomass and decomposition processes during restoration of arable sandy soils. Soil Biology & Biochemistry 38: 2890–2902.
31. Fitzjohn RG, Dickie IA (2007) TRAMPR: an R package for analysis and matching of terminal-restriction fragment length polymorphism (TRFLP) profiles. Molecular Ecology Notes 7: 583–587.
32. Culman SW, Bukowski R, Gauch HG, Caliddo-Quiroz H, Buckley DH (2009) T-REX: Software for the processing and analysis of T-RFLP data.BMC Bioinformatics 10: 171.
33. Abdo Z, Schuette UME, Bent SJ, Williams CJ, Forney LJ, et al. (2006) Statistical methods for characterizing diversity of microbial communities by analysis of terminal restriction fragment length polymorphisms of 16S rRNA genes. Environmental Microbiology 8: 929–938.
34. Hammer Ø, Harper DAT, Ryan PD (2001) PAST: Paleontological statistics software package for education and data analysis. Palaeontologia Electronica 4: 9 pp.
35. Jones DL, Hodge A, Kuzyakov Y (2004) Plant and mycorrhizal regulation of rhizodeposition. New Phytologist 163: 459–480.
36. Marschner P, Neumann G, Kania A, Weiskopf L, Lieberei R (2002) Spatial and temporal dynamics of the microbial community structure in the rhizosphere of cluster roots of white lupin (*Lupinus albus*; L.). Plant and Soil 246: 167–174.
37. Duineveld BM, Kowalchuk GA, Keijzer A, van Elsas JD, van Veen JA (2001) Analysis of bacterial communities in the rhizosphere of chrysanthemum via denaturing gradient gel electrophoresis of PCR-amplified 16S rRNA as well as DNA fragments coding for 16S rRNA. Applied and Environmental Microbiology 67: 172–178.
38. Acosta-Martínez V, Dowd S, Sun Y, Allen V (2008) Tag-encoded pyrosequencing analysis of bacterial diversity in a single soil type as affected by management and land use. Soil Biology & Biochemistry 40: 2762–2770.
39. Sessitsch A, Gyamfi S, Tscherko D, Gerzabek MH, Kandeler E (2004) Activity of microorganisms in the rhizosphere of herbicide treated and untreated transgenic glufosinate-tolerant and wildtype oilseed rape grown in containment. Plant and Soil 266: 105–116.
40. Lottmann J, Heuer H, de Vries J, Mahn A, During K, et al. (2000) Establishment of introduced antagonistic bacteria in the rhizosphere of transgenic potatoes and their effect on the bacterial community. FEMS Microbiology Ecology 33: 41–49.
41. Smalla K, Wieland G, Buchner A, Zock A, Parzy J, et al. (2001) Bulk and rhizosphere soil bacterial communities studied by denaturing gradient gel electrophoresis: plant-dependent enrichment and seasonal shifts revealed. Applied and Environmental Microbiology 67: 4742–4751.
42. Broeckling CD, Broz AK, Bergelson J, Manter DK, Vivanco JM (2008) Root exudates regulate soil fungal community composition and diversity. Applied and Environmental Microbiology 74: 738–744.
43. Hannula SE, Boschker HTS, de Boer W, Van Veen JA (2012) ^{13}C pulse-labeling assessment of the community structure of active fungi in the rhizosphere of a genetically starch-modified potato cultivar and its parental isoline. New Phytologist accepted: xx- xx.
44. Inceoglu O, Salles JF, van Overbeek L, van Elsas JD (2010) Effects of plant genotype and growth stage on the betaproteobacterial communities associated with different potato cultivars in two fields. Applied and Environmental Microbiology 76: 3675–3684.
45. Marschner P, Kandeler E, Marschner B (2003) Structure and function of the soil microbial community in a long-term fertilizer experiment. Soil Biology & Biochemistry 35: 453–461.
46. Bödeker ITM, Nygren CMR, Taylor AFS, Olson A, Lindahl BD (2009) ClassII peroxidase-encoding genes are present in a phylogenetically wide range of ectomycorrhizal fungi. Isme Journal 3: 1387–1395.
47. Turrini A, Sbrana C, Nuti MP, Pietrangeli BM, Giovannetti M (2004) Development of a model system to assess the impact of genetically modified corn and aubergine plants on arbuscular mycorrhizal fungi. Plant and Soil 266: 69–75.
48. Helgason T, Daniell TJ, Husband R, Fitter AH, Young JPW (1998) Ploughing up the wood-wide web? Nature 394: 431–431.
49. Giovannetti M, Sbrana C, Turrini A (2005) The impact of genetically modified crops on soil microbial communities. Rivista Di Biologia-Biology Forum 98: 393–417.
50. Lukow T, Dunfield PF, Liesack W (2000) Use of the T-RFLP technique to assess spatial and temporal changes in the bacterial community structure within an agricultural soil planted with transgenic and non-transgenic potato plants. FEMS Microbiology Ecology 32: 241–247.
51. Lottmann J, Heuer H, Smalla K, Berg G (1999) Influence of transgenic T4-lysozyme-producing potato plants on potentially beneficial plant-associated bacteria. FEMS Microbiology Ecology 29: 365–377.
52. Kennedy AC (1999) Bacterial diversity in agroecosystems. Agriculture Ecosystems & Environment 74: 65–76.
53. Dunfield KE, Germida JJ (2003) Seasonal changes in the rhizosphere microbial communities associated with field-grown genetically modified canola (Brassica napus). Applied and Environmental Microbiology 69: 7310–7318.

Land Suitability Assessment on a Watershed of Loess Plateau Using the Analytic Hierarchy Process

Xiaobo Yi[1,2], Li Wang[1,2]*

1 College of Resources and Environment, Northwest A&F University, Yangling, Shaanxi, China, **2** State Key Laboratory of Soil Erosion and Dryland Farming on the Loess Plateau, Northwest A&F University, Yangling, Shannxi, China

Abstract

In order to reduce soil erosion and desertification, the Sloping Land Conversion Program has been conducted in China for more than 15 years, and large areas of farmland have been converted to forest and grassland. However, this large-scale vegetation-restoration project has faced some key problems (e.g. soil drying) that have limited the successful development of the current ecological-recovery policy. Therefore, it is necessary to know about the land use, vegetation, and soil, and their inter-relationships in order to identify the suitability of vegetation restoration. This study was conducted at the watershed level in the ecologically vulnerable region of the Loess Plateau, to evaluate the land suitability using the analytic hierarchy process (AHP). The results showed that (1) the area unsuitable for crops accounted for 73.3% of the watershed, and the main factors restricting cropland development were soil physical properties and soil nutrients; (2) the area suitable for grassland was about 86.7% of the watershed, with the remaining 13.3% being unsuitable; (3) an area of 3.95 km^2, accounting for 66.7% of the watershed, was unsuitable for forest. Overall, the grassland was found to be the most suitable land-use to support the aims of the Sloping Land Conversion Program in the Liudaogou watershed. Under the constraints of soil water shortage and nutrient deficits, crops and forests were considered to be inappropriate land uses in the study area, especially on sloping land. When selecting species for re-vegetation, non-native grass species with high water requirements should be avoided so as to guarantee the sustainable development of grassland and effective ecological functioning. Our study provides local land managers and farmers with valuable information about the inappropriateness of growing trees in the study area along with some information on species selection for planting in the semi-arid area of the Loess Plateau.

Editor: Matteo Convertino, University of Florida, United States of America

Funding: This work was financially supported by the National Natural Science Foundation of China (Nos. 51239009 and 41271239) (http://www.nsfc.gov.cn), and the CAS Action-plan for Western Development (KZCX2-XB3-13) (http://www.cas.cn/). The funders had no role in study design, data collection and analysis, decision to publish, or preparation of the manuscript.

Competing Interests: The authors have declared that no competing interests exist.

* E-mail: wangli5208@nwsuaf.edu.cn

Introduction

Sandy desertification and soil erosion are two of the most serious problems affecting China's water and land resources. The World Bank has suggested that potentially more than 331 million hectares of land are susceptible to desertification (about one third of the area of China) while about 262 million hectares are actually affected [1]. Previous studies have shown that soil erosion affects about 360 million hectares of land in China, which is about 38% of its total area, and this proportion is more than three times the global average [2–3]; land in China is considered to be amongst the most severely eroded in the world. For example, on the Loess Plateau, erosion rates are about 8000–25,000 tonnes km^{-2} $year^{-1}$ in the gully areas of Shanxi and Shaanxi Provinces. Intensive cultivation of the steep hillsides has resulted in the loss of an estimated 1.6 billion tonnes of soil annually to the Yellow River [4]. In addition, dry conditions combined with the fine texture of the loess soil make the area very susceptible to dust storms. Due to the severe soil erosion and sandy desertification, the eco-environment of the Loess Plateau has been severely degraded, which has seriously affected sustainable development in the region [5]. With the environmental goals of reducing soil erosion and desertification, the Chinese Central Government launched the Sloping Land Conversion Program (SLCP, also known as Grain for Green or Grain to Green) in the late 1990s, with the intension of increasing the country's forest and grassland cover by retiring steeply sloping and marginal land from agricultural production [6–9]. This program aimed to restore degraded ecosystems, reduce poverty and assist rural households to move towards more sustainable economic activities [2,10].

However, after more than 10 years, this large-scale ecological-recovery project has faced some key problems that have limited the successful development of the current SLCP policy [11]. Surveys and case studies of SLCP have consistently identified insufficient technical support and arid conditions as being the key constraints to achieving program goals [3]. For example, in the north part of the Loess Plateau, the potential for converting cropland into forests has been over estimated: tree species have been planted in areas that are better suited to growing shrubs or grass, and species with high water requirements have been planted in areas where drought-tolerance is required. The result has been soil desiccation and the development of dry soil layers because the planted trees exploited water stored in the deeper soil layers and prevented them from being recharged by rainwater [12,7]. Consequently, once the stored water was exhausted, the limited precipitation was insufficient to maintain normal growth in the revegetated areas [13], leading to the vegetation dying or the production of stunted trees (colloquially referred to as "little old man trees", which are only about 20% of the normal height for their age) [14–17]. Thus, implementing the SLCP policy on the Loess Plateau requires careful consideration of several factors that have a significant effect on land use change and land suitability associated with carrying out the re-vegetation program. Currently, China is facing increased environmental pressures with shortages of water potentially limiting development, especially in its dryer northern and western regions, including the Loess Plateau [12]. In order to reduce soil water depletion and implement a successful re-vegetation program for environmental improvement on the Loess Plateau it is essential to identify appropriate species to plant. It is important to evaluate the land suitability and to develop land use plans that maximize reduction in soil erosion while at the same time minimizing water yield reduction, thus ensuring the sustainable growth and succession of vegetation. Only the proper implementation of the SLCP and successful re-vegetation can make a real contribution to efforts being made to combat the urgent environmental problems of soil erosion and desertification, as well as of climate change and loss of biodiversity, currently confronting the Loess Plateau, in particular, and China as a whole. In this study, the analytic hierarchy process (AHP) was used to assess the suitability of cropland, grassland and forestland in the Liudaogou watershed, and the main limiting factors for different land use options were quantitatively analyzed. The AHP has been widely used by decision-makers and researchers. It is a mathematical method, developed by Saaty in 1977 [18] and improved by the same author in 1980 [19], for analyzing complex decisions involving many criteria [20]. It has been widely used in site selection, suitability analysis, regional planning, and land consolidation analysis [21–22].The success of the AHP as a practical and reliable method is highlighted by its extensive application in the past two decades [23–24]. Furthermore, its simplicity in relation to its power was a significant factor in the choice for its use in the presented study, and all the mentioned factors ensured that the study objectives would be successfully achieved. The objectives of the study are to determine which land use is best suited to the re-vegetation program by assessing the land suitability, to identify the constraints for future land conservation, and to provide a scientific basis for decision-making in the successful implementation of the Sloping Land Conversion Program, not only for the study's watershed but across the whole Loess Plateau.

Materials and Methods

Study area

This study was conducted in the Liudaogou watershed, located in Shenmu County, Shaanxi Province, China (110°21′–110°23′E, 38°46′–38°51′N; Fig. 1). The Liudaogou watershed covers 6.9 km^2 and is located at the center of the wind-water erosion crisscross region in the north part of the Loess Plateau. This area suffers its most serious water erosion in summer and autumn and its most serious wind erosion in winter and spring [25–26]. The watershed is representative of the ecotone between the grass–pastoral and the agricultural areas, in the transitional zone between the desert aeolian deflation zone (the Mu Us Desert) and the loess hilly area (Loess Plateau), as well as being between the arid and the semi-arid regions. For these reasons, the Shenmu Erosion and Environmental Research Station (SEERS) of the Chinese Academy of Sciences was built within the watershed [27]. The study area is located at altitudes ranging from 1080 to 1270 m above mean sea level and has a semiarid continental monsoon climate, with a mean annual temperature of 8.4°C. The monthly mean temperature ranges from 9.7°C in January to 23.7°C in July. The mean annual precipitation is 437 mm, 77% of which occurs from June to September. Mean annual potential evapotranspiration can be as high as 1800 mm, which would result in a water deficit of 1350 mm year^{-1}. The area has a deep (up to 100 m) loess layer, which originated during the Quaternary period. The dominant soil type (cultivated loessial soil), is a Ust-Sandic Entisol, which is loess-derived and is consequently easily eroded by both water and wind.

Analytical hierarchy process (AHP)

The application of AHP for making a decision about land suitability in this study involved four main steps, as follows:

The first step was to decompose the decision problem into a hierarchical structure where the attributes and plans were present as inter-related elements. Based on a qualitative analysis of the environment in the study area, the final hierarchical structures were separated into four levels (Fig. 2). The first level was the overall land suitability. The second level was composed of subsystems: geological and topographical conditions; nutrient status; and soil physical properties. The third level consisted of the specific factors that affected the land suitability. The fourth level comprised each assessment unit (cell).

Elevation (C_1) and slope gradient (C_2) were selected to represent geological and topographical conditions (B_1). Slope is a crucial factor affecting vegetation structure and soil erosion on the Loess Plateau. Variation in elevation has an impact on soils, microclimatic effects, and other processes that could affect land suitability. Soil organic matter (C_3), total nitrogen (C_4), available phosphorous (C_5), and available potassium (C_6) were selected as factors that represented nutrient status (B_2). These four nutrition factors are closely related to land use and cover change, and understanding the effects of land use change on soil organic matter and nitrogen is important to sustainable management of land resources and associated watershed processes, as well as regional responses to global climatic change [28]. Three factors were selected to represent the soil physical properties. Soil bulk density (C_7), soil texture (C_8) and soil water content (C_9) strongly influence plant growth and land use, and soil properties and plant recovery processes have distinct characteristics in the typical desertified sandy land of the Loess Plateau [17]. Socio-economic factors were not considered in this study because it was limited to a small watershed where such factors are generally less important than the physical ones.

The second step involved a pair-wise comparison of the elements based on a nine point weighting scale; this generated the input data (Table 1). The comparison was carried out for each decision element at 1–(n–1) levels, where n was the matrix size. A matrix was generated as a result of the pair-wise comparisons and weights for the criteria were derived from these calculations.

A matrix of scores could be developed from the comparisons, given by

$$R=\begin{vmatrix} a_{11} & a_{12} & \cdots & a_{1n} \\ a_{21} & a_{22} & \cdots & a_{2n} \\ \cdots & \cdots & \cdots & \cdots \\ a_{n1} & a_{n2} & \cdots & a_{nn} \end{vmatrix}$$

where a_{ij} indicated how much more important the *ith* objective was than the *jth* objective, while making a suitable material handling/equipment selection decision. For all i and j, it was necessary that $a_{ii}=1$ and $a_{ij}=1/a_{ji}$. The possible assessment

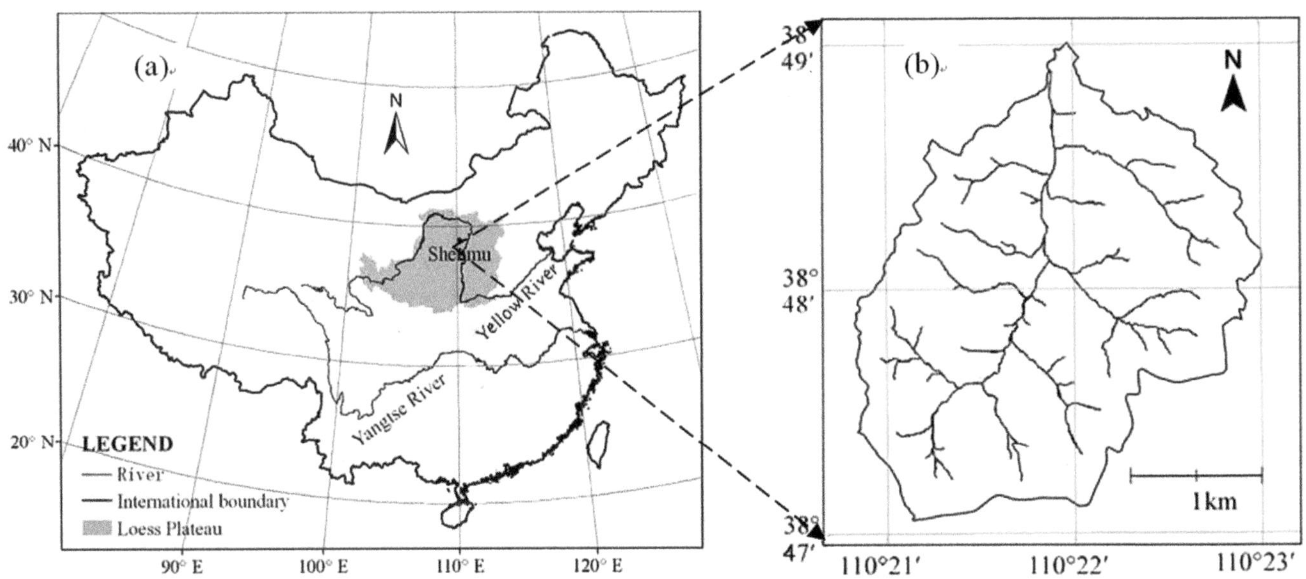

Figure 1. Location of the study site. (a) China; (b) the Liudaogou watershed.

values of a_{ij} in the pair-wise comparison matrix, along with their corresponding interpretations, are shown in Table 1.

The scores were normalized by dividing each element within the matrix by the sum of the column j, to create a normalized matrix, Rw:

$$Rw = \begin{vmatrix} \frac{a_{11}}{\sum a_{i1}} & \frac{a_{12}}{\sum a_{i2}} & \cdots & \frac{a_{1n}}{\sum a_{in}} \\ \frac{a_{21}}{\sum a_{i1}} & \frac{a_{22}}{\sum a_{i2}} & \cdots & \frac{a_{2n}}{\sum a_{in}} \\ \cdots & \cdots & \cdots & \cdots \\ \frac{a_{n1}}{\sum a_{i1}} & \frac{a_{n2}}{\sum a_{i2}} & \cdots & \frac{a_{nn}}{\sum a_{in}} \end{vmatrix}$$

The weight associated with each objective (c_i) could be estimated as the mean of the normalized scores in row i of the Rw matrix. Hence, c_i was calculated to give a matrix of weights, C:

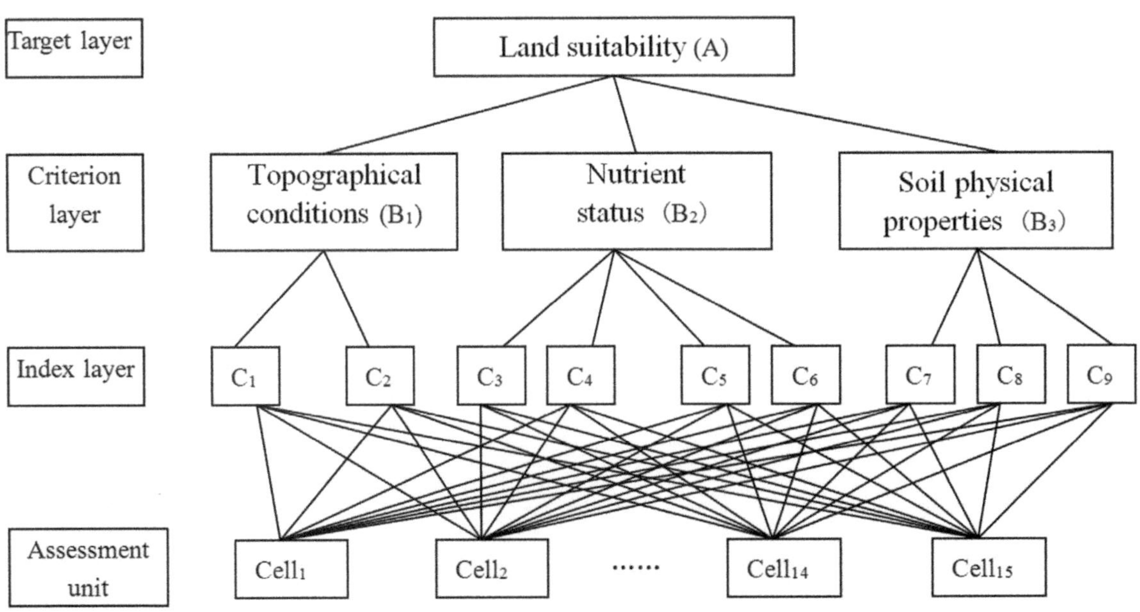

Figure 2. Hierarchical structure of land suitability.

Table 1. AHP pair-wise comparison scale for variables *i and j.*

Intensity of the relative importance	Definition
1	Equal importance of *i* and *j*
3	Moderate importance of *i* over *j*
5	Strong importance of *i* over *j*
7	Very strong importance of *i* over *j*
9	Extremely strong importance of *i* over *j*
2, 4, 6, 8	Intermediate values

$$C=\begin{vmatrix} c_1 \\ c_2 \\ \cdots \\ c_n \end{vmatrix}=\begin{vmatrix} \frac{a_{11}}{\sum a_{i1}}+\frac{a_{12}}{\sum a_{i2}}+\cdots+\frac{a_{1n}}{\sum a_{in}} \\ n \quad\quad n \quad\quad n \\ \frac{a_{21}}{\sum a_{i1}}+\frac{a_{22}}{\sum a_{i2}}+\cdots+\frac{a_{2n}}{\sum a_{in}} \\ n \quad\quad n \quad\quad n \\ \cdots \quad \cdots \quad \cdots \quad \cdots \\ \frac{a_{n1}}{\sum a_{i1}}+\frac{a_{n1}}{\sum a_{i2}}+\cdots+\frac{a_{nn}}{\sum a_{in}} \\ n \quad\quad n \quad\quad n \end{vmatrix}$$

where n is the number of objectives being compared, and the c_i value indicates the relative degree of importance (weight) of the *ith* objective.

The third step was to check for the consistency of the weight values (c_i) underlying the theoretical validity of the comparison matrix. In order to determine consistency, the consistency vector ($R\times C$ matrix) and x_i were calculated as:

$$R\times C=\begin{vmatrix} a_{11} & a_{12} & \cdots & a_{1n} \\ a_{21} & a_{22} & \cdots & a_{2n} \\ \cdots & \cdots & \cdots & \cdots \\ a_{n1} & a_{n2} & \cdots & a_{nn} \end{vmatrix}\times\begin{vmatrix} c_1 \\ c_2 \\ \cdots \\ c_n \end{vmatrix}=\begin{vmatrix} x_1 \\ x_2 \\ \cdots \\ x_n \end{vmatrix}$$

The eigenvalue of the pair-wise comparison matrix, λ_{max}, was then estimated using the following equation:

$$\lambda_{max}=\frac{1}{n}\sum_{i=1}^{n}\frac{x_i}{c_i}$$

An approximation to the consistency index (CI) was calculated and the consistency judgment was checked for the appropriate value of n by CR. CI and CR were calculated as [18]:

$$CI=\frac{\lambda_{max}-n}{n-1}$$

$$CR=\frac{CI}{RI}$$

where RI is the random consistency index. The RI values for different numbers of n are listed in Table 2.

The rule commonly used in the AHP was applied, whereby if CR was less than 0.10 (i.e. 10%), the degree of consistency was acceptable; and if it was greater than 0.10 it was considered that there were significant inconsistencies and, in such a case, the AHP would not produce meaningful results [29], and it would be necessary to review and improve the judgments.

The fourth step was a complete evaluation based on the rating of the final weights in the decision plan. Matrixes for layers B and C and for C and D were generated using the same method as that used for layers A and B. By comparing the final values by simple rankings, the weights of all elements in each level of the hierarchy relative to the entire level could be obtained. These in turn were all ranked, and were carried from the upper layer to the lower layer. The combined weight (W_i) of each assessment factor was determined for the integrated assessment of land suitability in the Liudaogou watershed.

Membership value standardization for assessment factors

In the process of land suitability assessment, a primary step was to ensure a standardized measurement system for all the factors considered. Since those factors have different standards of measurements, they had to be standardized to a uniform rating scale; in this study the scale was between 1 and 4 for ease of analysis. Assigning values to specific factors required specific decision rules in the form of thresholds for each factor. Various statistical and empirical guidelines from the related national codes and literature were used to determine the boundary values. As a general guideline, a positive correlation between the value awarded and suitability was employed. The class boundaries and standardized measurements employed for each factor are given for arable land in Section 3.1.1 and for forestland and grassland in Section 3.2.1. The integer numbers ranging from 1 to 4 were assigned to high, moderate, marginal and unsuitable classes, respectively. The next step involved assigning a new value for the degree of membership of every attribute to each index at each level. All these new values ranged from 0 to 1 where 0 indicated a poor fit and 1 indicated a perfect fit for membership of an attribute to a given index.. The membership degree function (F_i) of each assessment factor is given as follows:

The membership degree of soil organic matter, total nitrogen, available phosphorus and available potassium to the soil nutrient status criterion were calculated using the following equations:

$$F_i=\begin{cases} 0.1 & M_i\leq M_{i1} \\ 0.1+0.9\times\dfrac{M_i-M_{i1}}{M_{i2}-M_{i1}} & M_{i1}<M_i<M_i2 \\ 1 & M_i>M_{i2} \end{cases}$$

where M_{i1} and M_{i2} are the lower and upper limits of the value range of a factor.

Table 2. Random consistency index for AHP.

n	1	2	3	4	5	6	7	8	9
RI	0.00	0.00	0.58	0.90	1.12	1.24	1.32	1.41	1.45

The membership degree of slope gradient was given by:

$$F_i = \begin{cases} 1.0 & M_i < M_{i1} \\ 1 - 0.9 \times \dfrac{M_i - M_{i1}}{M_{i2} - M_{i1}} & M_{i1} \le M_i < M_{i2} \\ 0.1 & M_i \ge M_{i2} \end{cases}$$

The membership degree of physical clay content was given by:

$$F_i = \begin{cases} 0 & M_i \le M_{i1}, M_i \ge M_{i2} \\ \dfrac{M_i - M_{i1}}{M_{i0} - M_{i1}} & M_{i1} < M_i < M_{i0} \\ \dfrac{M_{i2} - M_i}{M_{i2} - M_{i0}} & M_{i0} < M_i < M_{i2} \\ 1 & M_i = M_{i0} \end{cases}$$

where M_{i0} means the best physical clay content in the study area.

The membership degree of altitude was given by:

$$F_i = \begin{cases} 1 & M_i \le M_{i1} \\ \dfrac{M_{i2} - M_i}{M_{i2} - M_{i1}} & M_{i1} < M_i < M_{i2} \end{cases}$$

The degree of membership of soil water content and bulk density was given by:

$$F_i = \begin{cases} 1 & M_i \ge M_{i2} \\ \dfrac{M_i}{M_{i2}} & M_i < M_{i2} \end{cases}$$

Assessment of land suitability

The integrated classification of land suitability for the study area was obtained by integrating the combined weight (W_i) and the total degree of membership for different factors (F_i) as follows:

$$E(j) = \sum_{i=1}^{n} F_i W_i$$

The maximum membership principle was adopted to define the comprehensive classification of land suitability. If $E_{i0} = \max E_i (1 \le i \le n)$, the comprehensive classification was E_{i0}.

Data

Field survey and soil sampling were conducted during August and September 2011. The land use status in 1993 and 2011 is presented in Table 3. When compared with 1993, there was significantly less arable land in 2011, whilst there had been increases in the extent of grassland, land that had been build on and land that was not in use. The main reason was the construction of four collieries during this period; this resulted in the transfer of the work force (local farmers) from traditional cultivation to the coal industry [30]. Based on land use and topography, we selected 15 assessment units: three for representing arable land, five for forestland and seven for grassland. Details of the assessment units are presented in Table 4.

Altitude was measured in the field by double frequency RTK-GPS. Slope gradient was determined using a compass incorporating an inclinometer. Sampling transects were placed along the slope in each assessment unit. For each transect, 3 plots (15×15 m) were established for the field investigation and sampling. In each sampling plot, a 1.0 m long×0.7 m wide×0.5 m deep pit was dug to allow measurements of soil bulk density. Samples were collected using a 5.0 cm diameter by 5.0 cm long stainless cutting ring with which samples were collected from the top 20 cm of the soil profile. Soil samples in the ring were extracted and the roots were carefully removed by hand. The soils were dried at 105°C to constant weight and bulk density was calculated by dividing the dry mass by the known volume. Soils were also collected with a soil auger (4 cm-diameter) to estimate profile water content in 10 cm layers to a depth of 40 cm. Soil water content was determined by oven-drying samples at 105°C.

Composite samples of about 1 kg from each assessment unit were collected and then air-dried and ground to pass through 1.00 mm and 0.25 mm nylon screens prior to laboratory analysis. The sieved soil was used to measure soil particle composition (physical clay content) and soil organic matter (SOM) , respectively. Particle composition was measured by the laser diffraction technique using a MasterSizer 2000 (Malvern Instruments, Malvern, England), equipped with a low-power (2 mW) Helium-Neon laser with a wavelength of 633 nm as the light source. Soil organic carbon (SOC) was measured by an Elementar Vario EL element analyzer and soil organic matter (SOM) was obtained by multiplying SOC content values by 1.723.

Soil total nitrogen (TN) was measured following the Kjeldahl digestion method [31]. Total phosphorus (TP) was determined colorimetrically after wet digestion with sulfuric acid and perchloric acid, and available phosphorus (AP) was extracted following the Olsen bicarbonate extractable P method [32]. Available K was determined by the ammonium acetate extraction method [33].

Table 3. Land use in the Liudaogou watershed (%) in 1993 and 2011.

Year	Arable land	Forestland	Grassland	Others[2)]
1993[1)]	31.3	25.7	39.6	3.4
2011	16	26	44	14

Note:
[1)]Data cited are from Yang et al. (1994).
[2)]Refers to built-up and unused land.

Table 4. Detailed information about the assessment units.

Assessment unit	Land use	Slope gradient (°)	Altitude (m)	Soil organic matter (g/kg)	Total nitrogen (g/kg)	Available phosphorus (mg/kg)	Available potassium (mg/kg)	Bulk density (g/cm^3)	Soil texture (physical clay content ≤0.01 mm) (%)	Soil water content (%)	Main species
1	Arable land	7	1170	14.29	0.41	4.64	203.78	1.60	17.52	9.45	*Solanum tuberosum*
2	Arable land	5	1127	22.26	0.39	15.83	420.90	1.57	12.99	12.16	*Setaria italica*
3	Arable land	0	1155	16.02	0.24	7.36	244.80	1.59	11.37	12.09	*Zea mays*
4	Forestland	15	1195	9.54	0.22	5.12	269.09	1.53	6.01	3.71	*Populus simonii*
5	Forestland	14	1201	13.84	0.35	5.87	279.82	1.60	7.07	4.46	*Populus simonii*
6	Forestland	10	1170	12.50	0.26	5.17	176.56	1.54	5.14	2.00	*Pinus tabulaeformis*
7	Forestland	12	1154	5.80	0.07	3.77	72.55	1.55	6.59	3.78	*Populus simonii*
8	Forestland	17	1175	3.08	0.05	3.14	133.91	1.56	6.86	3.91	*Populus simonii*
9	Grassland	17	1160	2.90	0.12	3.06	119.29	1.60	9.21	3.90	*Stipa bungeana*
10	Grassland	15	1196	1.99	0.08	2.46	92.97	1.58	19.13	3.51	*Medicago sativa*
11	Grassland	20	1189	8.14	0.16	4.58	116.34	1.58	2.27	3.64	*Medicago sativa*
12	Grassland	30	1204	6.23	0.13	3.95	297.40	1.61	12.45	8.60	*Lespedeza daurica*
13	Grassland	10	1237	11.02	0.22	7.28	157.24	1.59	39.50	9.54	*Stipa bungeana*
14	Grassland	9	1152	6.38	0.14	4.11	85.04	1.59	18.86	8.17	*Stipa bungeana*
15	Grassland	18	1197	2.46	0.08	2.98	186.94	1.57	20.52	8.50	*Stipa bungeana*

Results and Discussion

Suitability assessment for arable land

Standardized evaluation factors for arable land. Arable land suitability was divided into 4 classes, i.e. highly suitable (S1), moderately suitable (S2), marginally suitable (S3) and unsuitable (N). All factors were standardized to a uniform rating scale for ease of analysis. Standardized values for specific factors required specific decision rules in the form of thresholds for each factor. Various statistical and empirical guidelines from the related national–provincial codes and literature were used to determine the boundary values. In addition, twelve experts were invited to act as the decision makers to guarantee the reliability of boundary values. The experts involved were ecologists, pedologists, geographers, forest and grass managers, land resource experts and environmental protection experts; all have undertaken related studies of vegetation restoration and ecological– environmental issues in the study area and all are familiar with the Loess Plateau. Thus, they were able to suggest reasonable values for the thresholds. As a general guideline, there was a positive correlation between the value assigned and suitability. The class boundaries and standardized measurements applied for each factor are shown in Table 5.

Distribution of suitable cropland and associated statistics. Based on the method described above, suitable cropland classes were calculated and the main limiting factors were identified for each assessment unit (Table 6). Assessment units 1 and 3 in arable land and unit 14 in grassland were considered to be of marginal suitability, assessment unit 2 in arable land was classified as being of moderate suitability, and the other units from forest and grasslands were classified as being unsuitable for crops. Generally, an area of 0.40 km^2, accounting for 6.7% of the total area of the watershed (not including the built-up and unused lands), was classified as being moderately suitable for crops, and 1.19 km^2 (20.0%) as marginally suitable for crops (Table 7). In total, the area suitable for crops comprised about 26.7% of the watershed, so the area unsuitable for crops accounted for 73.3% of the watershed. The main factors restricting cropland development were soil physical properties and soil nutrients. First, the soil organic matter content and total nitrogen were generally low. This was particularly true in forest and grasslands, where the total nitrogen ranged between 0.05 and 0.35 g/kg (on average 0.17±0.09 g/kg), and was significantly lower than the generally reported values for soils (0.48–0.91 g/kg; on average 0.71) in the center of the Loess Plateau [34]. Soil organic matter was in the range 1.99–13.84 g/kg (on average 6.99±4.05 g/kg), and lower than the generally reported values for soils (7.0–15.0 g/kg; on average 11.0) in the center of the Loess Plateau [35]. Secondly, soil texture and soil water content obviously affected the land's suitability to grow crops in the watershed. In forestland in particular, the clay content was significantly lower than that of crop and grasslands, suggesting that planting trees adversely affected the soil structure in the study area. Our data were consistent with those presented in previous studies [26,30] indicating that species such as *Populus simonii* can cause soil degradation. The watershed is located in the ecotone between arid and the semi-arid regions, and soil water is always a limiting factor for plant growth because of the low levels of precipitation (about 437 mm) and high potential evapotranspiration (1800 mm). In forestland, soil water content ranged between 2.00% and 4.46%, which was close to the wilting point. Grasslands also exhibited soil water deficits compared with the croplands, although soil water content was higher than that in forestlands. Previous studies have shown that planting *P. simonii* (a tree species) and *Medicago sativa* (alfalfa, a forage legume) could cause large reductions in soil water content due to their high water consumption, and this could result in the formation of dry soil layers on land where they have been planted [36–37]. Usually, soil drying due to high plant water use adversely affects soil physical properties, for example, reducing aggregate stability and soil surface roughness [38–39]. Thus, soil water deficit and soil physical properties may have a compounding effect.

It is noteworthy that available phosphorus was in the range 2.46–5.87 mg/kg (on average 4.29±1.39 mg/kg) in forest and grass lands, and was, therefore, very much higher than generally reported for soils (0.91–1.76 g/kg; on average 1.10) in the center of the Loess Plateau. Available potassium was in the range 85.04–297.40 mg/kg (on average 165.60±78.49 mg/kg), which is also significantly higher than generally reported for soils (77.2–170.0 mg/kg) in the center of the Loess Plateau [35]. Therefore, these two nutrients were not the limiting factors for cropland development in the watershed.

Suitability assessment for forest and grassland

Standardized evaluation factors for forest and grassland. As for the arable land suitability assessment, forest and grassland suitability was also divided into 4 categories: highly suitable (S1), moderately suitable (S2), marginally suitable (S3) and unsuitable (N). The class boundaries and standardized measure-

Table 5. Land characteristics, thresholds and degree of suitability for arable land.

Land characteristics	Suitability			
	High (S1)	Moderate (S2)	Marginal (S3)	Unsuitable (N)
Slope gradient (°)	≤5	5~15	15~25	>25
Altitude (m)	≤1170	1170~1195	1195~1220	1220~1273.9
Soil organic matter (g/kg)	≥20	13~20	6~13	<6
Total nitrogen (g/kg)	≥0.28	0.16~0.28	0.05~0.16	<0.05
Available phosphorus (mg/kg)	≥15	9~15	3~9	<3
Available potassium (mg/kg)	≥250	150~250	50~150	<50
Bulk density (g/cm^3)	-	-	-	>1.60
Soil texture (physical clay content) (%)	15	-	-	>30 or <10
Soil water content (%)	≥12	10~12	8~10	<8

Table 6. Cropland suitability and limiting factors for each assessment unit.

Assessment unit	Combined weight	Grade	Limiting factors
1	0.709	S3	-
2	0.989	S2	-
3	0.795	S3	-
4	0.466	N	Soil water and texture
5	0.619	N	Soil water and texture
6	0.501	N	Soil water and texture
7	0.259	N	Soil water, texture and organic matter
8	0.244	N	Soil water, texture and organic matter
9	0.287	N	Soil water, texture and organic matter
10	0.260	N	Soil water, organic matter
11	0.347	N	Soil water and texture
12	0.390	N	Bulk density
13	0.607	N	Soil texture and slope gradient
14	0.427	S2	-
15	0.371	N	Soil organic matter and available P

ments applied for each factor are shown in Table 8. Altitude is not a limiting factor for forest and grassland development.

Distribution of forest and grassland suitability. Only assessment unit 6 in the forestland and unit 11 in the grassland were classified as being unsuitable for grassland (Table 9). The limiting factor for unit 6 was the low soil water content resulting from the high water consumption by tree species, and the limiting factor for unit 11 was soil texture because of its very low clay content (Table 4). Overall, an area of 0.79 km^2, accounting for 13.3% of the total area of the watershed, was classified as highly suitable for grassland, 1.18 km^2 (20.0%) was classified as being moderately suitable for grassland, and 3.16 km^2 (53.4%) as being marginally suitable for grassland (Table 10). In total, the area suitable for grassland was about 86.7% of the watershed, with the remaining 13.3% being unsuitable. This result is consistent with a previous study suggesting that development of vegetation in the watershed should focus on grassland [30]. Hou et al. [27] reported that the watershed is located within the boundary of the 400 mm rainfall isoline, representing the demarcation between cropping and pastoral regions. From the perspective of the climate, the watershed is more suitable for the growth and development of herbaceous plants than for trees. It is also notable that the area is not entirely appropriate for grassland, with 53.4% being classified as only marginally suitable. Besides soil water content and soil texture, total nitrogen may be a subsidiary factor limiting the suitability of this land for use as grassland – it is present at low levels in forest and grasslands (Table 11).

Table 7. Areas of land suitable for growing crops in the Liudaogou watershed.

Grade	Area (km^2)	Percentage (%)
S1	0	0
S2	0.40	6.7
S3	1.19	20.0
N	4.34	73.3

With respect to forestland suitability, assessment units 2 and 3 in the arable land were classified as being moderately suitable, assessment unit 1 in the arable land and units 13 and 14 in the grassland were considered marginally suitable, and other units in the forest and grasslands were unsuitable for use as forest. Overall, an area of 0.79 km^2, accounting for 13.3% of the total area of the watershed, was classified as moderately suitable for forestland and 1.19 km^2 (20.0%) as marginally suitable for forestland; the remaining 3.95 km^2, accounting for 66.7% of the watershed was unsuitable for forest (Table 12).

Like cropland, the main factors restricting the establishment of forest were soil water content, organic matter and total nitrogen. This means soil water and nutrition conditions were not suitable for widespread afforestation. The slope gradients of the arable and grassland assessment units that were suitable for forest were all below 10°, indicating that slope gradient may be a subsidiary limiting factor with respect to forests. Usually, steep slopes are associated with low soil water content, with increasing soil erosion and decreasing infiltration [40–42].

Implications for the Sloping Land Conversion Program

On the Loess Plateau, rapid population growth and an economic boom, coupled with severe soil erosion and desertification, have led to deterioration of the natural environment and a reduction in biodiversity [43,4]. The Sloping Land Conversion Program is, therefore, considered to be a necessary step for increasing vegetation cover and restoring ecosystem service functions, thus promoting sustainable environmental and economic development. However, during implementation of the SLCP, tree planting was overemphasized, resulting in inappropriate planting schemes [34]. Lü et al. [8] reported that the large areas converted from farmland to woodland have resulted in decreased regional water yield as the climate warms and dries on the Loess Plateau. Successful ecological rehabilitation programs have, thus, been largely dependent on innovative ecosystem management systems and technical support. Land suitability evaluation indicates that the Liudaogou watershed is suited to grass growth and is unsuitable for forest growth on the sloping land; this is due mainly to water and nitrogen deficits. In fact, there is still 1.79 km^2 (26% of the total area) of forestland in the watershed, with *Populus simonii* and *Pinus tabulaeformis* that were mainly planted in the late 1970s and after 1999 [27,30]. Hou [27] and Yang et al. [44] reported that the forest in the Liudaogou watershed could not function as a stable forest ecosystem. The height of the 20-year-old *Populus simonii* trees averaged about 4–6 m (the smallest just 2 m), the diameter at breast height (DBH) was about 5–6 cm, and the volume of timber was just 0.0031 m^3 per individual. In general, the growth of *Populus simonii* stopped after about 15 years. Under normal circumstances, the height of a 20-year-old *Populus simonii* tree would be more than 20 m with DBH ≥15 cm. Due to limited cover and poor growth, the *Populus simonii* forest is very restricted in its ecological function. It was originally planted as a part of a shelterbelt in the "Three North" Protective Forest Program, aiming to combat desertification and soil erosion in the Northwest of China [45–46]. However, according to a survey conducted by Hou et al. [47], the wind erosion depth in the *Populus simonii* forestland amounts to 1.91–4.68 cm/a, corresponding to 1.9×10^4–4.7×10^4 m^3 soil loss per year. Compared with the natural grass vegetation, it does not prevent wind erosion, but greatly accelerates it. The main reasons for the stunted tree growth are: (1) there is insufficient soil water available to maintain normal

Table 8. Land characteristics, thresholds and degree of suitability for forest and grasslands.

Land characteristics	Forest or grassland	Suitability			
		High (S1)	Moderate (S2)	Marginal (S3)	Unsuitable (N)
Slope gradient (°)	Both	≤15	15~25	25~35	>35
Soil organic matter (g/kg)	Forest	≥7.97	5.5~7.97	3.08~5.5	<3.08
	Grass	≥5.41	3.50~5.41	1.99~3.5	<1.99
Total nitrogen (g/kg)	Forest	≥0.19	0.12~0.19	0.05~0.12	<0.05
	Grass	≥0.11	0.08~0.11	0.05~0.08	<0.05
Available phosphorus (mg/kg)	Forest	≥4.62	3.88~4.62	3.14~3.88	<3.14
	Grass	≥3.66	2.52~3.66	2.46~2.52	<2.46
Available potassium (mg/kg)	Forest	≥250	150~250	50~150	<50
	Grass	≥250	150~250	50~150	<50
Bulk density (g/cm^3)	Both	-	-	-	>1.65
Soil texture (physical clay content) (%)	Both	15	-	-	>40 or <5
Soil water content (%)	Forest	≥15	11.5~15	8~11.5	<8
	Grass	≥4.78	4.14~4.78	3.50~4.14	<3.50

growth rates; (2) there is insufficient soil fertility (e.g. besides low total nitrogen, the Soil Organic Carbon Density is about 1.18–2.81 kg/m^2, significantly lower than the national mean values for soils of 11.52–12.04 kg/m^2) [48]; and (3) there is insufficient management because of the very limited economic value of forest – local farmers are more willing to devote themselves to coal-mining which generates a much higher income.

Although our forestland suitability assessment indicated that about 33.3% of total area is suitable for forest growth (Table 12), such areas are currently arable land and some grassland with gentle gradients (Table 4). Compared to tree species, crop and grass species have shallow roots and consume relatively little water [30], so do not deplete soil water to any great depth; however, the current arable lands are dammed, terraced fields that collect valuable rainfall because of the low soil erosion rates. Therefore, those arable lands and grasslands with gentle gradients were classified as being moderately or marginally suitable for forest development. We predict that planting trees in these areas would cause soil water depletion because of high water consumption combined with the low precipitation rates in the watershed.

The grassland suitability assessment indicated that about 86.7% of the area of the watershed is suitable for grass growth, thus the SLCP in the watershed should focus on conversion of cropland to grassland. Wang et al. [17] reported that soil physical properties such as bulk density, hydraulic conductivity, mean weight diameter, and the stability of >1 mm macro-aggregates have been significantly ameliorated in the 0–20 cm soil layer under secondary natural grasslands, implying that natural grass (*Stipa bungeana* Trin.) restoration is an appropriate and sensible approach to re-vegetation in the wind–water erosion region of the northern Loess Plateau of China. Planting alfalfa (*Medicago sativa*) and korshinsk peashrub (*Caragana korshinskii*) has no effect on the soil physical conditions, and may even reduce bulk density and soil permeability because of their very high water consumption with deep and vigorous root systems. Therefore, care is required when selecting species for re-vegetation to promote shifts from arable land to grasslands in a more environmentally compatible manner. Non-native grassland species, such as the korshinsk peashrub and alfalfa, have high water requirements and should be avoided

Table 9. Grassland suitability and limiting factors for each assessment unit.

Assessment unit	Combined weight	Grade	Limiting factors
1	0.984	S2	-
2	0.995	S1	-
3	0.989	S2	-
4	0.926	S3	-
5	0.965	S1	-
6	0.821	N	Soil water
7	0.751	S3	-
8	0.538	S3	-
9	0.683	S3	-
10	0.441	S3	-
11	0.865	N	Soil texture
12	0.945	S3	-
13	0.948	S2	-
14	0.948	S3	-
15	0.632	S3	-

Table 10. Areas suitable for use as grassland in the Liudaogou watershed.

Grade	Area (km^2)	Percentage (%)
S1	0.79	13.3
S2	1.18	20.0
S3	3.16	53.4
N	0.79	13.3

Table 11. Forestland suitability and limiting factors for each assessment unit.

Assessment unit	Combined weight	Grade	Limiting factors
1	0.900	S3	-
2	0.952	S2	-
3	0.945	S2	-
4	0.806	N	-
5	0.820	N	-
6	0.756	N	Soil water
7	0.423	N	Organic matter and soil water
8	0.249	N	Organic matter and soil water
9	0.507	N	Organic matter, soil water and total nitrogen
10	0.458	N	Organic matter, soil water and total nitrogen
11	0.708	N	Soil water
12	0.603	N	Organic matter and soil texture
13	0.865	S3	-
14	0.658	S3	-
15	0.556	N	Organic matter and total nitrogen

Table 12. Areas of land suitable for forest growth in the Liudaogou watershed.

Grade	Area (km^2)	Percentage (%)
S1	0	0
S2	0.79	13.3
S3	1.18	20.0
N	3.95	66.7

when choosing species to plant in the watershed.

Conclusion

Land use management involves complex decision-making that requires an understanding of many factors. This paper presents the AHP as a decision support tool for use when selecting an appropriate land use, which is an important issue for the sustainable development of the Liudaogou watershed. Based on the AHP, grassland was found to be the most suitable land-use type to support the aims of the Sloping Land Conversion Program in the watershed. Under the constraints of soil water shortage and nutrient deficit, crops and forests were considered to be inappropriate land uses in the study area, especially on sloping land. All forest should be converted to grassland because continued growth of trees will damage the soil water environment and increase desertification problems. When selecting species for re-vegetation, non-native grass species with high water requirements should be avoided so as to guarantee the sustainable development of grassland and effective ecological functioning. In the future, the area of abandoned cropland is likely to increase rapidly due to the government policy embodied in the Sloping Land Conversion Program. Our study provides local land managers and farmers with valuable information about the inappropriateness of growing trees in the study area along with some information on species selection for planting in the semi-arid area of the Loess Plateau.

Acknowledgments

We wish to thank graduate students Wang Jianguo, Shi Zhanfei and Wang Mei for their help in field survey and data analysis in the Laboratory. We also wish to thank Dr. David Warrington for his invaluable comments and suggestions on this paper.

Author Contributions

Conceived and designed the experiments: LW XBY. Performed the experiments: XBY LW. Analyzed the data: XBY. Contributed reagents/materials/analysis tools: XBY LW. Wrote the paper: LW XBY.

References

1. World Bank (2001) China: air, land and water, environmental priorities for a New Millennium. World Bank, Washington, DC.
2. SFA (2003) Sloping land conversion program plan. (2001–2010) (in Chinese).
3. Michael T, Bennett MT (2008) China's sloping land conversion program: Institutional innovation or business as usual? Ecological economics 65: 699–711.
4. Wang L, Shao MA, Wang QJ, Gale WJ (2006) Historical changes in the environment of the Chinese Loess Plateau. Environmental science & policy 9: 675–684.
5. Ding CR (2003) Land policy reform in China: assessment and prospects. Land Use Policy 20: 109–120.
6. Uchida E, Xu JT, Rozelle S (2005) Grain for Green: cost-effectiveness and sustainability of China's conservation set-aside programme. Land Economics 81: 247–264.
7. Wang L, Wang QJ, Wei SP, Shao M, Li Y (2008) Soil desiccation for Loess soils on natural and regrown areas. Forest Ecology and Management 255(7): 2467–2477.
8. Lü Y, Fu B, Feng X, Zeng Y, Liu Y, et al. (2012) A policy-driven large scale ecological restoration: quantifying ecosystem services changes in the Loess Plateau of China. PLoS ONE 7(2): e31782. doi:10.1371/journal.pone.0031782.
9. Wei J, Cheng J, Li W, Liu W (2012) Comparing the effect of naturally restored forest and grassland on carbon sequestration and its vertical distribution in the Chinese Loess Plateau. PLoS ONE 7(7): e40123. doi:10.1371/journal.pone.0040123.
10. Yin RS, Zhao MJ (2012) Ecological restoration programs and payments for ecosystem services as integrated biophysical and socioeconomic processes—China's experience as an example. Ecological Economics 73: 56–65.

11. Wang XH, Lu CH, Fang JF, Shen YC (2007) Implications for development of grain-for-green policy based on cropland suitability evaluation in desertification-affected north China. Land Use Policy 24: 417–424.
12. McVicar TR, Li LT, Van Nie TG, Zhang L, Li R, et al. (2007) Developing a decision support tool for China's revegetation program: Simulating regional impacts of afforestation on average annual streamflow in the Loess Plateau. Forest Ecology and Management 251: 65–81.
13. Yang RJ, Fu BJ, Liu GH, Ma KM (2004) Research on the relationship between water and eco-environment construction in Loess Hilly and Gully Region. Chinese Journal of Environmental Science 25(2): 37–42 (in Chinese with an English abstract).
14. Yang WZ, Tian JL (2004) Essential exploration of soil aridization in the Loess Plateau. Acta Pedologica Sinica 41(1): 1–6 (in Chinese with English abstract).
15. McVicar TR, Van Niel TG, Li LT, Wen ZM, Yang QK, et al. (2010) Parsimoniously modelling perennial vegetation suitability and identifying priority areas to support China's re-vegetation program in the Loess Plateau: Matching model complexity to data availability. Forest Ecology and Management 259: 1277–1290.
16. Wang L, Wei SP, Horton R, Shao M A (2011) Effects of vegetation and slope aspect on water budget in the hill and gully region of the Loess Plateau of China. Catena 87: 90–100.
17. Wang L, Wei SP, Shao HB, Wu YJ, Wang QJ (2012) Simulated water balance of forest and farmland in the hill and gully region of the Loess Plateau of China. Plant Biosystems, dx.doi.org/10.1080/11263504.2012.709198.
18. Saaty TL (1977) A scaling method for priorities in hierarchical structures. Journal of Mathematical Psychology 15: 234–281.
19. Saaty TL (1980) The analytical hierarchy process. McGraw Hill, New York.
20. Kurttila M, Pesonen M, Kangas J, Kajanus M (2000) Utilizing the analytic hierarchy process (AHP) in SWOT analysis-a hybrid method and its application to a forest-certification case. Forest Policy and Economics 1 (1): 41–52.
21. Ayalew L, Yamagishi H, Marui H, Kanno T (2005) Landslides in Sado Island of Japan: Part II. GIS-based susceptibility mapping with comparisons of results from two methods and verifications. Engineering Geology 81: 432–445.
22. Cay T, Uyan M (2012) Evaluation of reallocation criteria in land consolidation studies using the Analytic Hierarchy Process (AHP). Land Use Policy 30: 541–548.
23. Mardle S, Pascoe S (2004) Management objective importance in fisheries: an evaluation using the Analytic Hierarchy Process (AHP). Environmental Management 33(1):1–11.
24. Pascoe S, Dichmont CM, Brooks K, Pears R, Jebreen E (2013) Management objectives of Queensland fisheries: Putting the horse before the cart. Marine Policy 37: 115–122.
25. Li M, Li ZB, Liu PL, Yao WY (2005) Using cesium-137 technique to study the characteristics of different aspect of soil erosion in the Wind–Water Erosion Crisscross Region on Loess Plateau of China. Applied Radiation and Isotopes 62: 109–113.
26. Wang L, Mu Y, Zhang QF, Jia Z K (2012) Effects of vegetation restoration on soil physical properties in the wind-water erosion region of the northern Loess Plateau of China. CLEAN - Soil, Air, Water 40(1): 7–15.
27. Hou QC (1994) Comprehensive analysis on natural conditions and environmental harnessing in the experimental Area. Memoir of NISWC, Academia Sinica and Ministry of Water Resources, 18: 136–143 (in Chinese with English abstract).
28. Fahey B, Jackson R (1997) Hydrological impacts of converting native forests and grasslands to pine plantations, South Island, New Zealand. Agricultural and Forest Meteorology 84: 69–82.
29. Chakraborty S, Banik D (2006) Design of a material handling equipment selection model using analytic hierarchy process. The International Journal of Advanced Manufacturing Technology 28: 1237–1245.
30. Wang L, Zhang QF, Wei SP, Wang QJ (2009) Vegetation restoration model in a watershed of a coal mining area in the water and wind erosion crossing zone of the Loess Plateau. Journal of Beijing Forestry University 31 (2): 36–43.
31. Bremmer JM, Mulvaney CS (1982) Nitrogen—total. In: Page AL, Miller RH, Keeney DR (Eds.), Methods of soil analysis, Part 2—Chemical and microbiological properties. ASA-SSSA, Madison, WI, pp. 595–624.
32. Olsen SR, Sommers LE (1982) Phosphorous. In: Page AL, Miller RH, Keeney DR (eds) Methods of soil analysis Part 2, Chemical and microbial properties. Agronomy Society of America, Agronomy Monograph 9, Madison, Wisconsin, pp 403–430.
33. Pratt PF (1965) Potassium. In: Black, C.A. (Ed.), Methods of Soil Analysis. Part 2. Chemical and Microbiological Properties. Am. Soc. of Agron, Inc., Madison, p. 1022–1030.
34. Wang XL, Guo SL, Ma YH, Huang DY, Wu JS (2007) Effects of land use type on soil organic C and total N in a small watershed in loess hilly-gully region. Chinese Journal of Applied Ecology18 (6): 1281–1285 (in Chinese with English abstract).
35. Bai WJ, Jiao JY, Ma XH, Jiao F (2005) Soil environmental effects of artificial woods in abandoned croplands in the Loess Hilly-gullied region. Journal of Arid Land Resources and Environment 19(7): 135–141 (in Chinese with English abstract).
36. Jiang N, Shao MA, Lei TW (2007) Soil water characteristics of different typical land use patterns in Water–Wind Erosion Interlaced Region. Journal of Beijing Forestry University 29: 134–137 (in Chinese with English abstract).
37. Huo Z, Shao MA, Horton R (2008) Impact of gully on soil moisture of shrubland in Wind–Water Erosion Crisscross Region of the Loess Plateau. Pedosphere 18: 674–680.
38. Reid JB, Goss MJ (1982) Suppression of decomposition of C-labeled plant roots in the presence of living roots of maize and perennial ryegrass. Journal of Soil Science 33: 387–395.
39. Grant CD, Dexter AR (1987) Generation of microcracks in moulded soils by rapid wetting. Australian Journal of Soil Research 27(1): 169–182.
40. Tang KL, Zhang KL, Lei AL (1998) Research on up-limit slope of "grain for green" in the plowlands of Loess Hilly and Gully Region. Chinese Science Bulletin 43(2): 200–203 (in Chinese).
41. Luk SH, Cai Q, Wang GP (1993) Effects of surface crusting and slope gradient on soil and water losses in the hilly Loess region, North China. Catena 24: 29–45.
42. Ng CWW, Shi Q (1997) A numerical investigation of the stability of unsaturated soil slopes subjected to transient seepage. Computers and Geotechnics 22 (1): 1–28.
43. Li WH (2004) Degradation and restoration of forest ecosystems in China. Forest Ecology and Management 201: 33–41.
44. Yang G, Song YX (1993) Analysis on the current situation of forestry and forestry developing strategy in Liudaogou Watershed of Shenmu Experimental Area. Memoir of Northwestern Institute of Soil and Water Conservation, 18: 106–112.
45. Fang JY, Chen AP, Peng CH, Zhao SQ, Ci LJ (2001) Changes in forest biomass carbon storage in China between 1949 and 1998. Science 292: 2320–2322.
46. Wang XM, Zhang CX, Hasi E, Dong ZB (2010) Has the Three Norths Forest Shelterbelt Program solved the desertification and dust storm problems in arid and semiarid China? Journal of Arid Environments 74(1): 13–22.
47. Hou QC, Wang YK, Yang G (1996) Several problems of vegetation construction in criss2cross belt of bind2water erosion. Bulletin of Soil and Water Conservation 16 (5): 36–40.
48. Li YY, Shao MA, Zheng JY, Li QF (2007) Impact of grassland recovery and reconstruction on soil organic carbon in the northern Loess Plateau. Acta Ecologica Sinica 27 (6): 1–9.

23

Enhanced Accumulation of Copper and Lead in Amaranth (*Amaranthus paniculatus*), Indian Mustard (*Brassica juncea*) and Sunflower (*Helianthus annuus*)

Motior M. Rahman*, Sofian M. Azirun, Amru N. Boyce

Institute of Biological Sciences, Faculty of Science, University of Malaya, Kuala Lumpur, Malaysia

Abstract

Background: Soil contamination by copper (Cu) and lead (Pb) is a widespread environmental problem. For phytoextraction to be successful and viable in environmental remediation, strategies that can improve plant uptake must be identified. In the present study we investigated the use of nitrogen (N) fertilizer as an efficient way to enhance accumulation of Cu and Pb from contaminated industrial soils into amaranth, Indian mustard and sunflower.

Methods/Principal Findings: Plants were grown in a greenhouse and fertilized with N fertilizer at rates of 0, 190 and 380 mg kg^{-1} soil. Shoots, roots and total accumulation of Cu and Pb, transfer factor (TF), translocation index were assessed to evaluate the transport and translocation ability of tested plants. Addition of N fertilizer acidified the industrial soil and caused the pH to decrease to 5.5 from an initial pH of 6.9. Industrial soil amended with N fertilizer resulted in the highest accumulation of Pb and Cu (for Pb 10.1–15.5 mg kg^{-1}, for Cu 11.6–16.8 mg kg^{-1}) in the shoots, which was two to four folds higher relative to the concentration in roots in all the three plants used. Sunflower removed significantly higher Pb (50–54%) and Cu (34–38%) followed by amaranth and Indian mustard from industrial soils with the application of N fertilizer. The TF was <1 while the shoot and root concentration (SC/RC) ratios of Pb and Cu were between 1.3–4.3 and 1.8–3.8, respectively, regardless of plant species.

Conclusions: Sunflower is the best plant species to carry out phytoextraction of Pb and Cu. In contrast, Pb and Cu removal by Indian mustard and amaranth shows great potential as quick and short duration vegetable crops. The results suggest that the application of N fertilizer in contaminated industrial soil is an effective amendment for the phytoextraction of Pb and Cu from contaminated industrial soils.

Editor: Malcolm Bennett, University of Nottingham, United Kingdom

Funding: This research work was financially supported by the University of Malaya, Kuala Lumpur, Malaysia. The funder had no role in study design, data collection and analysis, decision to publish, or preparation of the manuscript.

Competing Interests: The authors have declared that no competing interests exist.

* E-mail: mmotiorrahman@gmail.com

Introduction

A major environmental concern resulting from the dispersal of industrial and urban wastes generated by human activities is the contamination of soil and its environment [1]. Soils may become contaminated by the accumulation of toxic heavy metals and metalloids through emissions from rapidly expanding industrial areas, mine tailings, disposal of high metal wastes, leaded gasoline and paints, application of fertilizers and manures, sewage sludge, pesticides, wastewater irrigation, coal combustion residues, spillage of petrochemicals, and atmospheric deposition [2], [3]. Soils are the major sink for toxic metals released into the environment by the aforementioned anthropogenic activities and unlike organic contaminants which are oxidized to carbon (IV) oxide by microbial action, most metals do not undergo microbial or chemical degradation [4], and their total concentration in soils persists for a long time after their introduction [5], although changes in their chemical forms and bioavailability are possible. The presence of toxic metals in soil can severely inhibit the biodegradation of organic contaminants by microorganisms [6]. Among various types of soil pollutants, toxic metals pollution appears to be of great concern, especially in developing countries where it has caused soil quality deterioration either by aerial deposition or waste water discharge [7].

Regardless of the origin of the metals in the soil, high levels of many metals can result in the degradation of soil fertility and poor quality of agricultural products and poses a significant threat and hazard to human, animal and ecosystem well-being [8], [9]. Metals such as Cu, Pb and zinc (Zn) are particularly important since high quantities of these metals can decrease crop production due to the risk of biomagnifications and bioaccumulation in the food chain. There is also the risk of underground and surface water contamination [10], [11]. These cannot be removed biologically and under certain circumstances can transform from one oxidation state to another which in some metals makes them less toxic [12]. Toxic metal contamination of soil may pose risks and hazards to humans and the ecosystem through direct ingestion or contact with contaminated soil, the food chain, drinking of contaminated ground water, reduction in food quality via phytotoxicity, reduction in land usability for agricultural production causing food insecurity, and land tenure problems [13], [14].

Accumulation of toxic metals in crop plants is also of great concern because it can enter the food chain [15]. As plants acquire the necessary nutrients, such as N, phosphorus (P) and potassium (K), they also take in and accumulate toxic metals such as Pb and cadmium (Cd). Ingestion of vegetables grown in soils contaminated with toxic metals has been suggested as a possible risk to human and wildlife health. These health risks will depend on the chemical composition of the contaminated soil, its physical characteristics, the type of vegetables cultivated and the consumption rate [16]. The uptake of toxic metals by plants is often influenced by plant species, growth stage, soil type, metals and environmental factors. Toxic metal concentration in the soil solution plays a critical role in controlling metal availability to plants [17]. Many findings have shown that increasing the levels of toxic metals in the soil will bring about an increased uptake by the plants. However, the availability of the toxic metal ions are influenced by various factors including soil pH, the physical and chemical properties of soil, the clay content and manganese oxide concentration [18].

In developing countries with high population density and insufficient funds available for environmental restoration, low cost and ecologically sustainable remedial options are required to restore contaminated lands so as to reduce the associated risks, make the land resource available for agricultural production, enhance food security, and scale down land tenure problems [19]. Phytoextraction was developed as a result of efforts to find a more efficient and less hazardous technique to remediate contaminated soils [20]. It involved the removal of metals by plants through uptake and accumulation into biomass [1]. Interestingly, although phytoremediation was recognized and documented by humans more than 300 years ago, its scientific study and development was not conducted until the early 1980's [21]. However, progress in making phytoextraction a practical commercial technology has been hindered by the lack of strategies to optimize plant uptake of metals [22]. Contaminated soil can be amended by conventional ways such as physical, chemical, bioremediation and phytoremediation. Unfortunately, existing remediation methods for toxic metal removal from contaminated soils are expensive and disruptive [1]. In recent years, efforts have focused on the remediation strategies that are less expensive, less destructive and more sustainable [23−25]. Phytoremediation, defined as the use of plants to remove pollutants from the environment, is a promising technology for the remediation of contaminated soils and perhaps for the removal of metals from contaminated soil [26]. This technology can be applied to both organic and inorganic pollutants present in soil, water and air. In this respect, plants can be compared to solar driven pumps which can extract and concentrate certain elements from the environment [27]. Phytoremediation is aesthetically pleasant, soil-microorganisms friendly, enhances diversity, and derives energy from sunlight [28]. More essentially, it is able to retain the fertility status of the soil even after the removal of toxic metals. Many plants are capable of accumulating high levels of toxic metals in the soil and are potent tools to help clean up contaminated soil [29]. The success of phytoremediation depends on the solubility of heavy metals and the ability of plant to take in and partition toxic metals to the upper plant parts [30]. Several studies have been done with chelating agents for their ability to dissolve metals and enhance the uptake of metals by plants. However these chemicals have limitations due to their negative effects on the plant and soil properties [31−33]. Phytoremediation has been recommended as an alternative remediation technology for polluted soil with heavy metals but is generally perceived to be too slow. The enhance accumulation of trace pollutants in harvestable plant tissues is a prerequisite for such technology to be practical. Thus, an ideal plant selection for rehabilitation of polluted soil has to be considered based on faster and higher biomass producing plants that can tolerate and accumulate heavy metals in plant parts. Relationships between toxic metals in crops and pollution of toxic metals in soils have been well documented [34]. However, current information on the phytoremediation technology using tropical species grown with nitrogen fertilizer is limited. Taking this into account a study was undertaken to investigate the impact of N fertilizer on the accumulation of Cu and Pb in industrial soils using amaranth, Indian mustard and sunflower plants.

Materials and Methods

Soil Management and Experimental Design

Experiments were conducted in a greenhouse at the University of Malaya, Kuala Lumpur, Malaysia from September 2010 to December 2010. Soil samples were collected from the 0–30 cm depth randomly from 10 spots within an area of recycling electronic wastes and chemicals that contained heavy metals in Senai, Johor, (1° 28′ 0″ N, 103° 45′ 0″ E) Malaysia. No specific permits were required for the described field studies and no specific permissions were required for these activities. In addition the locations were not protected and the field studies did not involve endangered or protected species. The industrial soils were sandy loam in texture and thoroughly mixed and sieved through a 2-mm mesh to obtain homogenous soil composites before they were used as the growing medium in polyethylene pots (height 18-cm × diameter 14-cm = surface area 0.095 m^2). Each pot was filled with 2 kg of industrial soil. Amaranth, Indian mustard and sunflower were grown in industrial soil with N fertilizer at rates of 0, 190 and 380 mg kg^{-1} soil. Each plant species was tested in an individual experiment against the three levels of N fertilizer. The experiment was conducted under completely randomized design with six replications. The seeds of amaranth, Indian mustard and sunflower were sown as per treatment. Sufficient seeds were sown to ensure healthy germination. Plants were thinned after germination and five plants per pot were kept until harvesting. Prior to sowing 0.41 g and 0.82 g urea (46%N) were applied into the soils as per treatment schedule. Continuous monitoring was done to observe growth and all plants were harvested at the age of 56 d.

Plant and Soil Analysis

Prior to planting, the initial Pb and Cu concentrations in the industrial soil were analyzed using an atomic absorption spectrometer (AAS). In industrial soil, Cu (56.2±0.09 mg kg^{-1}) and Pb (40.2±0.07 mg kg^{-1}) were present. The initial status of soil N in industrial soil was 26.50±0.71 mg kg^{-1}. The soil pH was determined from the prepared soil suspension (1:2.5 soil water ratios) by using a combined pH meter model 900A (Thermo Orion, Ontario, Canada) at regular 14 d interval [35]. After harvesting all plant samples were oven-dried at 72°C for 48 h. Biomass accumulation and partitioning were determined for all the plants used. Shoot and root tissues of each plant species and soils were firstly ground by using mortar and pestle prior to toxic metal analysis. For each soil and plant parts, 800 mg of the ground particle was diluted with 1 L of distilled water to produce an 800 ppm mixture. Lead and Cu concentrations in each soil and plant parts were analyzed using a PE Analyst 400 AAS.

Calculation

The Cu and Pb transport from soil to the shoots was evaluated using the transfer factor (TF) = WC (mg kg^{-1})/TC (mg kg^{-1}) where, WC = elements concentration in whole plant; TC = elements soil concentration obtained with AAS [36]. The ability for Cu and Pb translocation from roots to shoots was calculated by the translocation index (TI) % = SQ (mg per pot)/WPQ (mg per pot)×100, where, SQ = element accumulation in the shoots; WPQ = element accumulation in the whole plant (shoots+roots). Percentage removal (%) was calculated from the total concentration (TC) of elements initially present in the soil [37].

Data Analysis

Statistical analysis was carried out by one-way ANOVA using general linear model and t-Test to evaluate significant differences between means of plant biomass, absorbed heavy metal concentrations in harvested plants at 95% level of confidence [38]. Further statistical validity of the differences among treatment means was estimated using the least significance difference (LSD) method.

Results and Discussion

Soil pH in Industrial Soil

Soil pH in the industrial soil was significantly affected by the application of N fertilizer regardless of plant species. From an initial pH of 6.9 the soil pH decreased significantly after the application of N fertilizer. Nitrogen fertilizer application at the rate of 380 mg kg^{-1} soil developed a more acidic soil pH during experimental period. The lowest soil pH was between 5.5–5.6, with the use of N fertilizer at rates of 190 mg kg^{-1} soil and 380 mg kg^{-1} soil at 14 d. Thereafter the soil pH increased slightly over time in both unfertilized and fertilized industrial soils regardless of plant species (Table 1). This could possibly be due to the stable saturated moisture in the soils and plants that will result in a dilution effect of the soil pH by the presence of hydroxyl ions (OH^-) in water, particularly in the unfertilized industrial soil experiments. The change in soil pH with N fertilizer treatment probably involved more complex reactions due to the application of varying levels of N fertilizer. The initial acidic condition of fertilized industrial soil at 14 d is possibly due to the application of N fertilizer and the subsequent microbial stimulated aerobic conversion of NH_4^+ to NO_3^- through nitrification, which will release H^+ ions into the soil. Enhanced nitrate uptake by the plants could potentially affect soil pH as a result of OH^- excretion and the intake of protons (H^+) during uptake of soil nitrate.

Soil pH greatly influences the availability of plant nutrients [39] and controls the solubility of heavy metals and other plant nutrients in the soil that subsequently influences the effectiveness of phytoremediation. Lead is relatively unavailable to plants when the soil pH is above 6.5 and even more so when soil phosphorus levels are high [40]. The decreased soil pH can bring about an increase in heavy metal sorption, due to the increased net negative surface charge on organic matter and soil oxides at high pH levels [41]. In the present study, adding N also acidified the industrial soil, which caused the pH decrease to 5.6 from an initial pH 6.9 within 56 d in the pot experiments. The optimal soil pH for plant growth has been documented to be around 6.4 even though; sunflower can tolerate soils ranging in from pH 5.7 to over 8.0. A pH range between 6.0 to 7.2 has been reported to be optimal for many soils [42] while optimal soil pH for amaranth and Indian mustard growth has been documented to be 6.4 [43] and 6.5 [44], respectively. Thus a balance between soil pH and heavy metal availability are required for enhance plant growth. In this study the soil pH increased from 5.8 to 6.5 after 56 d when the plants were grown with N fertilizer whilst the pH increased more than 6.5 when the plants were grown in unfertilized soil. The solubility of Cu is drastically increased at pH 5.5 [32], which are close to the ideal farmland pH of 6.0−6.5 [33]. Copper is an important essential element for plants, microorganisms, animals, and humans. The connection between soil and water contamination and metal uptake by plants is determined by many chemical and physical soil factors as well as the physiological properties of the crops.

Table 1. Soil pH influenced by N fertilizer with amaranth, Indian mustard and sunflower grown in industrial soils over time.

Plant species and N fertilizer (mg kg^{-1} soil)	Days after emergence				
	0	14	28	42	56
Amaranth:					
N 0	6.9 a	6.8 a	6.8 a	6.9 a	7.0 a
N 190	6.9 a	5.6 b	6.0 b	6.2 b	6.4 b
N 380	6.9 a	5.5 b	5.8 c	6.0 c	6.2 c
Indian Mustard:					
N 0	6.9 a	6.8 a	6.9 a	7.0 a	7.0 a
N 190	6.9 a	5.6 b	6.0 b	6.3 b	6.5 b
N 380	6.9 a	5.5 b	5.8 c	5.9 c	6.2 c
Sunflower:					
N 0	6.9 a	6.8 a	6.9 a	7.0 a	7.0 a
N 190	6.9 a	5.6 b	6.0 b	6.2 b	6.4 b
N 380	6.9 a	5.5 b	5.8 c	6.0 c	6.3 c

Means followed by the same letters are not significantly different for each treatment means ($P<0.05$).

Dry Matter Yield

The shoot, root and total dry matter yield was affected significantly ($P<0.001$) by application of N fertilizer in all tested plant species (Table 2). The shoot, root and total dry matter yield in the plants varied depending on the plant species. All the plant species grown with N fertilizer exhibited the highest shoot, root and total dry matter yield. Sunflower produced appreciably higher shoot, root and total dry matter than amaranth and Indian mustard. The shoot dry matter was significantly higher than roots in all tested species.

The shoot, root and whole plant accumulate content of Pb and Cu per pot was used to estimate translocation index and all the attributes were affected significantly by N fertilizer application in all tested plant species (Table 2). All the plant species grown in industrial soils with N fertilizer at the rate of 380 mg kg^{-1} soil exhibited the highest shoot, root and whole plant accumulate content per pot. Sunflower produced significantly higher shoot, root and whole plant accumulate content than amaranth and Indian mustard. The shoots accumulate content was significantly higher than roots accumulate content in all the tested plant species.

Higher dry matter accumulation was possibly due to the greater vegetative growth resulting from higher photosynthetic activities influenced by the addition of N fertilizer. Sunflower and amaranth responded well to varying levels of N fertilizer in the present study.

Table 2. Dry matter yield, lead and copper accumulation in plant tissues.

Plant species and N fertilizer (mg kg^{-1} soil)	Dry matter yield (g pot^{-1})			Roots accumulate content (RQ) (mg pot^{-1})		Shoots accumulate content (SQ) (mg pot^{-1})		Whole plant accumulate content (WPQ) (mg pot^{-1})	
	Root	Shoot	Total	Pb	Cu	Pb	Cu	Pb	Cu
Amaranth:									
N 0	1.6 b	10.0 b	11.6 b	0.007 b	0.005 b	0.06 c	0.07 c	0.07 c	0.07 c
N 190	1.9 a	12.1 a	14.0 a	0.010 a	0.009 a	0.12 b	0.14 b	0.13 b	0.15 a
N 380	1.9 a	12.7 a	14.6 a	0.011 a	0.009 a	0.14 a	0.16 a	0.15 a	0.16 a
Indian Mustard:									
N 0	1.6 b	9.8 b	11.4 b	0.004 b	0.005 a	0.05 c	0.06 c	0.06 c	0.07 c
N 190	1.8 a	11.3 a	13.1 a	0.005 a	0.006 a	0.11 b	0.14 b	0.12 b	0.15 b
N 380	1.9 a	12.0 a	13.9 a	0.005 a	0.007 a	0.14 a	0.17 a	0.15 a	0.18 a
Sunflower:									
N 0	1.6 b	11.0 b	12.6 b	0.008 c	0.008 b	0.09 c	0.10 c	0.10 c	0.11 c
N 190	2.4 a	13.1 a	15.5 a	0.013 b	0.011 a	0.19 b	0.19 b	0.20 b	0.20 b
N 380	2.5 a	13.6 a	16.1 a	0.015 a	0.012 a	0.21 a	0.23 a	0.23 a	0.24 a

Means followed by the same letters are not significantly different for each treatment means ($P<0.05$).

Indian mustard growth was slightly poorer compared to the other two species. The possible reason for this is the use of sandy loam soil in this study. Sunflower plants grown in industrial soils with N fertilizer exhibited significantly higher dry matter yield probably due to its broader leaves which would intercept more sunlight and enable higher rates of photosynthesis and as a result incorporate more carbon into the plants [45]. This was also observed in the case of Indian mustard grown in industrial soils with N fertilizer, where its broader leaves provide a greater surface area for photosynthesis compared to the same plant growing in unfertilized industrial soils. The results indicate that higher biomass accumulation is an important trait to enhance efficiency of phytoremediation. Soil amendments with the application of organic or inorganic fertilizer can promote higher biomass yield resulting in an increased heavy metal bioaccumulation capacity of the plants. Kenaf accumulated significantly higher biomass and absorbed higher amounts of Pb in fertilized soil compared to unfertilized soils [46]. Addition of fertilizers can significantly reduce the mobility of trace metals [47] but increases the amount of trace metals in plant shoots because of the higher total biomass [48]. The application of N and P enhanced the growth of shoot and root in Indian mustard and its subsequent ability to remove Cu from contaminated soil [49]. While N fertilizer combined with EDTA facilitated the phytoremediation efficiency of sunflower in Pb-polluted soil [50], high-dose of application of N fertilizer had an influence on the accumulation of Pb in edible parts of radish, carrot and potato. All these plants had higher concentrations of Pb when contaminated soils were treated with N fertilizer compared to untreated soils [51].

Nitrogen, Lead and Copper Accumulation

The N uptake in all the three plant species studied were significantly higher ($P<0.01$) in industrial soil amended with N fertilizer compared to unfertilized industrial soils (Table 3). All the plant species recorded higher N uptake (64.6−75.8 mg g^{-1}) in industrial soil with N application compared to unfertilized industrial soils (45.8−47.8 mg g^{-1}). The shoot and root N were also significantly higher in industrial soil with N fertilizer. The shoot N was significantly higher than root N irrespective of plant species.

In the present study, the total extracted Pb and Cu content varied depending on the plant species. Both Pb (20.0−21.6 mg kg^{-1}) and Cu (19.1−21.5 mg kg^{-1}) uptake by sunflower grown in industrial soil with N was significantly higher than in unfertilized industrial soil (Table 3). Indian mustard and amaranth also recorded significantly higher Pb and Cu uptake when grown with N fertilizer. Both Indian mustard and amaranth recorded higher Cu (15.7−18.1 mg kg^{-1} for Indian mustard; 18.7−21.4 mg kg^{-1} for amaranth) than Pb uptake (12.7−14.3 mg kg^{-1} for Indian mustard; 16.0−16.9 mg kg^{-1} for amaranth) from industrial soil with N fertilizer. Lead and Cu uptake by the different plant species studied were in the following order: sunflower>amaranth>Indian mustard for both unfertilized and N fertilized industrial soils. Lead and Cu concentration in shoots and roots of all the plant species studied were significantly higher ($P<0.05$) in industrial soil with N application than in unfertilized industrial soil (Table 3). Regardless of plant species and soils used, Pb and Cu concentration in the shoots were two to three folds higher than in the roots. All tested parameters showed identical results between N fertilizer levels.

In all the plant species, relatively more metal accumulated in the shoots than in the roots where it was observed that the shoot and root concentration ratio (SC/RC) in all the plant species studied for Pb was >1, suggesting that Pb translocation from root to shoot was remarkably higher and shoots acted as a sink for Pb (Table 4). Indian mustard recorded remarkably higher SC/RC ratio (>3) when grown with N fertilizer. The SC/RC ratio (>1) showed that the root translocation of Pb and its partitioning from root to shoot had a positive influence on the plant species.

Copper uptake in all the plant species resulted in a relatively higher concentration of Cu in the shoots compared to the roots (Table 3). Indian mustard exhibited appreciably higher SC/RC ratio (>3) followed by sunflower when industrial soil was fertilized with N. The SC/RC ratio of all plants species was >1, suggesting that Cu translocation from root to shoot was appreciable. Considering both Pb and Cu, the higher translocation index (TI) and transfer factor (TF) were obtained, in increasing order, in amaranth, mustard and sunflower with N amended soil (Table 4). For both Pb and Cu, the highest TI was obtained in Indian

Table 3. Nitrogen, lead and copper concentration in plant tissues.

Plant species and N fertilizer (mg kg^{-1} soil)	N (mg g^{-1})			Pb (mg kg^{-1})			Cu (mg kg^{-1})		
	Root (RC)	Shoot (SC)	Total	Root (RC)	Shoot (SC)	Total	Root (RC)	Shoot (SC)	Total
Amaranth:									
N 0	16.8 b	29.2 b	46.0 b	4.8 b	6.2 b	11.0 b	3.1 b	6.6 b	9.7 b
N 190	21.9 a	43.3 a	65.2 a	5.5 a	10.5 a	16.0 a	4.5 a	11.6 a	16.1 a
N 380	22.6 a	44.4 a	67.0 a	5.7 a	11.2 a	16.9 a	4.7 a	12.5 a	17.2 a
Indian Mustard:									
N 0	16.5 b	29.3 b	45.8 b	2.5 b	5.5 b	8.0 b	2.9 b	6.3 b	9.4 b
N 190	20.6 a	44.0 a	64.6 a	2.6 ab	10.1 a	12.7 a	3.5 a	12.2 a	15.7 a
N 380	21.1 a	45.3 a	66.4 a	2.7 a	11.6 a	14.3 a	3.8 a	14.3 a	18.1 a
Sunflower:									
N 0	17.0 b	30.8 b	47.8 b	5.2 b	8.5 b	13.7 b	5.1 a	9.2 b	14.3 b
N 190	23.7 a	50.7 a	74.4 a	5.8 a	14.2 a	20.0 a	4.6 b	14.5 a	19.1 a
N 380	24.3 a	51.5 a	75.8 a	6.1 a	15.5 a	21.6 a	4.7 b	16.8 a	21.5 a

Means followed by the same letters are not significantly different for each treatment means ($P<0.05$).

Table 4. Lead and copper removal, shoot and root concentration ratio, translocation index and transfer factor.

Plant species and N fertilizer (mg kg^{-1} soil)	Shoot and root concentration (SC/RC) ratio		Translocation index (TI)		Transfer Factor (TF)		Removal (%)	
	Pb	Cu	Pb	Cu	Pb	Cu	Pb	Cu
Amaranth:								
N 0	1.3 b	2.1 b	89.0 b	93.0 b	0.3 b	0.2 b	27.4 b	17.2 b
N 190	1.9 a	2.6 a	92.0 a	94.3 a	0.4 a	0.3 a	39.8 a	28.6 a
N 380	2.0 a	2.7 a	92.9 a	94.7 a	0.4 a	0.4 a	42.0 a	30.6 a
Indian Mustard:								
N 0	2.2 b	2.2 b	93.1 b	93.0 b	0.2 b	0.2 b	19.9 c	16.4 c
N 190	3.9 a	3.5 a	96.1 a	95.6 a	0.3 ab	0.3 a	31.6 b	27.9 b
N 380	4.3 a	3.8 a	96.4 a	96.0 a	0.4 a	0.3 a	35.6 a	32.2 a
Sunflower:								
N 0	1.6 b	1.8 b	91.8 b	92.5 b	0.3 b	0.3 b	34.1 c	25.4 c
N 190	2.4 a	3.2 a	93.0 a	94.5 a	0.5 a	0.3 b	49.8 b	34.0 b
N 380	2.5 a	3.6 a	93.3 a	95.1 a	0.5 a	0.4 a	53.7 a	38.3 a

Means followed by the same letters are not significantly different for each treatment means ($P<0.05$).

mustard and sunflower. Despite the high TI observed for Pb and Cu in all the species tested, the TF, which reflects the plant's capacity to transport metals from soil to shoots, was much higher for Pb than for Cu. Such behavior can probably be attributed to high Pb mobility in the soil due to the decreasing soil pH.

The results obtained showed that sunflower removed 49.8–53.7% of Pb 34–38.3% of Cu from industrial soil with N fertilizer compared to 34.1% of Pb and 25.4% of Cu from unfertilized industrial soil (Table 4). Amaranth removed 39.8–42% of Pb and 28.6–30.6% of Cu while Indian mustard removed 31.6–35.6% of Pb and 27.9–32.2% of Cu from

Table 5. Regression equation, correlation coefficient (r) and coefficients of determination (R^2) of different parameters.

Regression equation	Correlation coefficient (r)	Coefficients of determination (R^2)
Relationship between N fertilizer and biomass (B):		
Y_B (Amaranth) $= 0.011X_1+15.78$	0.97**	$R^2 = 0.939$
Y_B (Indian mustard) $= 0.013X_1+14.857$	0.95**	$R^2 = 0.909$
Y_B (Sunflower) $= 0.011X_1+17.044$	0.95**	$R^2 = 0.903$
Relationship between biomass and Pb uptake:		
Y_{Pb} (Amaranth) $= 1.519X_2-12.46$	0.99**	$R^2 = 0.9899$
Y_{Pb} (Indian mustard) $= 1.504X_2-13.849$	0.99**	$R^2 = 0.9962$
Y_{Pb} (Sunflower) $= 2.065X_2-20.798$	0.99**	$R^2 = 0.9996$
Relationship between biomass and Cu uptake:		
Y_{Cu} (Amaranth) Cu $= 1.935X_2-20.167$	0.99**	$R^2 = 0.9887$
Y_{Cu} (Indian mustard) $= 2.112X_2-21.504$	0.99**	$R^2 = 0.9938$
Y_{Cu} (Sunflower) $= 1.790X_2-15.705$	0.99**	$R^2 = 0.9748$
Relationship between N fertilizer and Pb uptake:		
Y_{Pb} (Amaranth) $= 0.017X_1+11.593$	0.94**	$R^2 = 0.883$
Y_{Pb} (Indian mustard) $= 0.018X_1+8.4324$	0.97**	$R^2 = 0.941$
Y_{Pb} (Sunflower) $= 0.022X_1+14.37$	0.96**	$R^2 = 0.913$
Relationship between N fertilizer and Cu uptake:		
Y_{Cu} (Amaranth) $= 0.021X_1+10.468$	0.94**	$R^2 = 0.879$
Y_{Cu} (Indian mustard) $= 0.025X_1+9.767$	0.97**	$R^2 = 0.949$
Y_{Cu} (Sunflower) $= 0.020X_1+14.614$	0.99**	$R^2 = 0.975$

X_1 = N fertilizer rate; X_2 = biomass;
**significant at 0.01 level of probability.

industrial soil with N fertilizer. The toxic metal removal efficiency by sunflower was 16−20% higher for Pb and 9−13% higher for Cu with the application of N fertilizer in industrial soils compared to unfertilized industrial soil. Similarly Pb removal efficiency by Indian mustard and amaranth was 12−16% and 13−15% higher, respectively. Copper removal efficiency by Indian mustard and amaranth was 11−15% and 11−14% higher in industrial soil with N fertilizer compared to unfertilized industrial soil.

A comparison of the metal content of the soil used in this study with others reported in the literature confirmed that it should be considered as a contaminated soil. The amounts of Cu (56.2±0.09 mg kg^{-1}) and Pb (40.2±0.07 mg kg^{-1}) recorded were above the maximum content commonly found in soils of industrial areas in Johor, Malaysia. Background concentrations of Pb that occur naturally in surface agricultural soils in the United States average 10 parts per million (ppm) with a range of 7 to 20 ppm [52]. Soils with Pb levels above this range are primarily the result of Pb contamination. The Canadian Council of Ministers of the Environment interim soil assessment criterion for Cu was set at 30 mg kg^{-1} [53]. However there is no reference value is available for Cu or Pb from the local environmental agency in Malaysia. The shoot and root ratios were calculated in order to evaluate the translocation of the elements inside the plant, from the roots to the shoots, and its potential accumulation in the biomass (Table 4). The SC/RC ratio values are important to estimate the potential of a plant for phytoremediation purposes. One of the selection criteria for hyperaccumulator's identification is the leaf's metal concentration. Accordingly a hyperaccumulator should concentrate more than 1,000 mg kg^{-1} of Ni, Pb or Cu and more than 100 mg kg^{-1} of Cd [54]. Sunflower exhibited the highest concentration of Pb in shoots followed by amaranth, while Indian mustard was the lowest for both the metals considered (Table 3). However, according to the criteria already presented, none of the plants species tested would be classified as hyperaccumulator for Cu or Pb. This contradicts with the hyperaccumulator characteristic of sunflower where the SC/RC ratio should be >1. In this study all the tested plant species achieved this target although they could not achieve a leaf metal's concentration of 1,000 mg kg^{-1} for Pb or Cu. A possible reason for this is the short experimental period (56 d) where sunflower did not reach maturity. As a result, the leaf metal's concentration in sunflower was <1,000 mg kg^{-1} compared to that reported in other studies. Another possible explanation might be the use of pot experiments, which could have limited plant development due to the small volume of soil used. Furthermore the immobilization of elements by root adsorption may have had some influence. This mechanism prevents element translocation from roots to shoots in some species working as a defense barrier, decreasing the phytoextraction potential of such species since only a small portion of ions associated with the roots are effectively absorbed [21].

Sunflower is able to secrete organic acids which acidify the rhizosphere and increases the solubility of toxic metals like Pb and Cu [55]. The N fertilized plants showed higher biomass and potential to facilitate phytoextraction process. Sunflower absorbed appreciably higher amounts of Pb than Indian mustard and amaranth in industrial soil with N fertilizer due to its hyperaccumulator characteristic [48]. In this study all the plant species contained relatively higher concentration of Pb in the shoots compared to roots as high metal concentrations are in available form in acidic soil due to the decreased soil pH. It is well known that Pb is an immobile metal in soil, since it readily forms a precipitate because of its low aqueous solubility within the soil matrix, and in many cases it is not readily bioavailable. In addition, many plants retain Pb in their roots via sorption and precipitation with only minimal transport to the above ground harvestable plant portions [56], [57]. The leafy vegetables like Indian mustard and amaranth are expected to absorb more Cu compared to Pb due to the physiological role of Cu in plants [17]. The SC/RC ratio for Cu uptake were >1 for all the plants which proved that Cu partitioning was performed from roots to shoots. The high values of SC/RC ratio (>1) implies that plant shoots acts as sink for the absorption of Cu. The result of this study suggests that application of inorganic N fertilizer could boost the biomass accumulation of plants and subsequently phytoremediation efficiency. However, careful measures have to be taken during fertilizer application to avoid anthropogenic source of heavy metal entering the soil and the human food chain. In the literature toxicity in plants is reported in literature to occur when Cu is found in the range of 20−100 mg kg^{-1}, and Pb in the range of 30−300 mg kg^{-1} [58], [59]. In this study the accumulations of Pb and Cu in all the tested plant species were below the toxic levels (Table 2).

The most serious source of exposure to soil Pb is through direct ingestion of contaminated soil or dust. In general, plants do not absorb or accumulate Pb. However, in soils with a high Pb content, it is possible for some Pb to be taken up. Studies have shown that Pb does not readily accumulate in the fruiting parts of vegetable and fruit crops (e.g., corn, beans, squash, tomatoes, strawberries, and apples). Higher concentrations are more likely to be found in leafy vegetables (e.g., lettuce) and on the surface of root crops (e.g., carrots). Generally, it has been considered safe to use garden produce grown in soils with total Pb levels less than 300 ppm. The risk of Pb poisoning through the food chain increases as the soil Pb level rises above this concentration. Even at soil levels above 300 ppm, most of the risk is from Pb contaminated soil or dust deposits on the plants rather than from uptake of Pb by the plant [40].

Correlation Analysis

There were significantly positive relationships between the accumulation of Pb in the three plant species grown in the contaminated industrial soil and the dosage of the N fertilizer applied to the industrial soil (Table 5). The slopes indicated that for each 1.0 mg N application in industrial soil, 0.017 to 0.022 mg of Pb was taken up into the plant tissues (Table 5). There was also a positive correlation between N fertilizer rate and Cu uptake in the three plant species. The slopes indicated that for each 1.0 mg N application in industrial soil, 0.020 to 0.025 mg of Cu was taken up into the plant tissues (Table 5). The regression equation showed that Pb uptake was higher in sunflower while Cu uptake was higher in Indian mustard among three plant species. A strong positive relationship was observed between the accumulation of Pb and biomass ($r = 0.99$), Cu and plant biomass accumulation ($r = 0.99$) in the three plant species growing in contaminated industrial soils (Table 5).

Lead and Cu concentration were positively correlated to plant biomass due to the increased plant tissue to accumulate more heavy metals. Furthermore the higher biomass indicates that plants were bigger and probably possessed higher absorption rate so that more heavy metal uptake took place [46]. The strong positive correlation between plant biomass and plant N content was due to the positive effect of N fertilizer application. As a result, the higher biomass of plants enhanced a greater absorption of more Pb and Cu and countered the effects of heavy metal immobilization [50]. The plant N content directly or indirectly affected the heavy metal concentrations in the plants, by boosting the biomass of the plants and subsequently their ability to accumulate more heavy metals. This positive correlation suggests that inorganic or other types of fertilizers, such as organic fertilizers, can be used to increase the biomass of plants and their capacity to accumulate heavy metals from the soil.

Conclusions

The Pb and Cu uptake ability of sunflower was appreciably greater than Indian mustard and amaranth. There was a positive relationship between N fertilizer application on plant growth and their ability to absorb Pb and Cu. Nitrogen fertilizer boosted growth parameters increasing root and shoot dry matter significantly, allowing a higher uptake of toxic metals by plants grown in fertilized industrial soil. Nevertheless, careful measures must be put in place to ensure that excessive N fertilizers are not applied to cause uncontrolled rapid above ground growth resulting in weaker plants. It is suggested that phosphorus be added to boost root growth to balance the high shoot growth in the presence of N fertilizer. In this study, sunflower was the best plant species to carry out phytoextraction of Pb and Cu due to its hyperaccumulator characteristic. Both Indian mustard and amaranth also showed Pb and Cu removal potential, as quick and short duration vegetable crops. This provides the opportunity to grow Indian mustard and amaranth several times on a piece of land to remove both Pb and Cu from contaminated soil. However, the results of the present study showed that N fertilizer application in industrial soil can lead to negative consequences on indigenous soil biota in acidic soils due to its impact on soil pH. It clearly shows the need to consider the soil characteristics of polluted soil before its remediation.

Acknowledgments

We are grateful for the logistic support provided by the University of Malaya, Kuala Lumpur, Malaysia. We would also like to thank two anonymous reviewers for thoughtful comments and students at the Science and Environmental Management, University of Malaya.

Author Contributions

Conceived and designed the experiments: MMR SMA ANB. Performed the experiments: MMR. Analyzed the data: MMR. Contributed reagents/materials/analysis tools: MMR SMA ANB. Wrote the paper: MMR ANB.

References

1. Ghosh M, Singh SP (2005) A review on phytoremediation of heavy metals and utilization of it's by products. Asian J Energy Environ 6: 214–231.
2. Khan SQ, Cao Y, Zheng M, Huang YZ, Zhu YG (2008) Health risks of heavy metals in contaminated soils and food crops irrigated with wastewater in Beijing, China. Environ Poll 152(3): 686–692.
3. Zhang MK, Liu ZY, Wang H (2010) Use of single extraction methods to predict bioavail-ability of heavy metals in polluted soils to rice. Com Soil Sci Plant Anal 41(7): 820–831.
4. Kirpichtchikova TA, Manceau A, Spadini L, Panfili F, Marcus MA, et al. (2006) Speciation and solubility of heavy metals in contaminated soil using X-ray microfluorescence, EXAFS spectroscopy, chemical extraction, and thermodynamic modeling. Geochim Cosmochim Acta 70(9): 2163–2190.
5. Adriano DC (2003) Trace elements in terrestrial environments: Biogeochemistry, Bioavailability and Risks of Metals 2nd ed Springer, New York, NY, USA.
6. Maslin P, Maier RM (2000) Rhamnolipid-enhanced mineralization of phenanthrene in organic-metal co-contaminated soils. Biorem J 4(4): 295–308.
7. Demirezen D, Aksoy A, Uruc K (2007) Effect of population density on growth, biomass and nickel accumulation capacity of *Lemna gibba* (Lemnaceae). Chemos 66: 553–557.
8. Blaylock MJ, Huang JW (2000) Phytoextraction of metals. *In* Phytoremediation of toxic metals: using plants to clean-up the environment. Raskin I, Ensley BD (eds) 53–70. Wiley, NY, USA.
9. Luo C, Shen Z, Li X, (2005) Enhanced phytoextraction of Cu, Pb, Zn and Cd with EDTA and EDDS. Chemos 59: 1–11.
10. Schmidt U (2003) Enhancing phytoextraction: the effect of chemical soil manipulation on mobility, plant accumulation and leaching of heavy metals. J Environ Qual 32: 1939–1954.
11. Nowack B, Schulin R, Robinson BH (2006) Critical assessment of chelant-enhanced metal phytoextraction. Environ Sci Tech 40: 5525–5532.
12. Garbisu C, Alkorta I (2001) Phytoextraction: a cost-effective plant-based technology for the removal of metals from the environment. Biores Tech 77: 229–236.
13. McLaughlin MJ, Zarcinas BA, Stevens DP, Cook N (2000) Soil testing for heavy metals. Com Soil Sci Plant Anal 31(11–14): 1661–1700.
14. Ling W, Shen Q, Gao Y, Gu X, Yang Z (2007) Use of bentonite to control the release of copper from contaminated soils. Aust J Soil Res 45(8): 618–623.
15. Cieslinski G, Van-Rees KCJ, Huang PM, Kozak LM, Rostad HPW, et al. (1996) Cadmium uptake and bioaccumulation in selected cultivars of durum wheat and flax as affected by soil type. Plant Soil 182: 115–124.
16. Cobb GP, Sands K, Waters M, Wixson BG, Dorward-King E (2000) Accumulation of heavy metals by vegetables grown in mine wastes. Environ Toxic Chem 19: 600–607.
17. Maimon A, Khairiah J, Ahmad MR, Aminah A, Ismail BS (2009) Comparative accumulation of heavy metals in selected vegetables, their availability and correlation in lithogenic and nonlithogenic fractions of soils from some agricultural areas in Malaysia. Adv Environ Biol 3: 314–321.
18. Xian X, Shokohiford G (1989) Effects of pH on chemical forms and plant availability of cadmium, zinc and lead in polluted soils. Water Air Soil Poll 47: 265–273.
19. Raymond AW, Felix EO (2011) Heavy metals in contaminated soils: a review of sources, chemistry, risks and best available strategies for remediation. International School Res Network Ecol 2011: 1–20.
20. Nascimento CWA, Xing B (2006) Phytoextraction: a review on enhanced metal availability and plant accumulation. Sci Agric 63: 299–311.
21. Lasat MM (2000) Phytoextration of metals from contaminated soil: a review of plant/soil/metal interaction and assessment of pertinent agronomic issues. J Hazard Sub Res 2:1–25.
22. Wang AS, Angle JS, Chaney RL, Delorme TA, Reeves RD (2006) Soil pH effects on uptake of Cd and Zn by *Thlaspi caerulescens*. Plant Soil 281: 325–337.
23. Gupta AK, Sinha S (2007a) Phytoextraction capacity of the plants growing on tannery sludge dumping sites. Biores Tech 98: 1788–1794.
24. Gupta AK, Sinha S (2007b) Phytoextraction capacity of the *Chenopodium album* L grown on soil amended with tannery sludge. Biores Tech 98: 442–446.
25. Xiaomei L, Qitang W, Banks MK (2005) Effect of simultaneous establishment of *Sedum alferdii* and *Zea mays* on heavy metal accumulation in plants. Inter J Phytorem 7: 43–53.
26. Dede G, Ozdemir S, Dede OH (2012) Effect of soil amendments on phytoextraction potential of *Brassica juncea* growing on sewage sludge. Inter J Environ Sci Tech 9: 559–564.
27. Salt DE, Smith RD, Raskin I (1998) Phytoremediation. Annual Rev Plant Physio Plant Mol Bio 49: 643–668.
28. Chaney RL, Angle JS, McIntosh MS, Reeves RD, Li YM, et al. (2005) Using hyperaccu-mulator plants to phytoextract soil Ni and Cd. Zeitschrift für Naturforschung C 60: 190–198.
29. Jarup L (2003) Hazards of heavy metal contamination. British Med Bull 68: 167–182.
30. Turgut C, Pepe MK, Cutright TJ (2004) The effect of EDTA and citric acid on phytoreme-diation of Cd, Cr, and Ni from soil using *Helianthus annuus*. Environ Pollu 131: 147–154.
31. Kaplan M, Orman S, Kadar I, Koncz J (2005) Heavy metal accumulation in calcareous soil and sorghum plants after addition of sulphur-containing waste as a soil amendment in Turkey. Agric Ecosys Environ 111: 41–46.
32. Martınez CE, Motto HL (2000) Solubility of lead, zinc and copper added to mineral soils. Environ Poll 107(1): 153–158.
33. Eriksson J, Andersson A, Andersson R (1997) The state of Swedish farmlands. Tech. Rep. 4778, Swedish Environmental Protection Agency, Stockholm, Sweden.
34. Syakalima MS, Choongo KC, Chilonda P, Ahmadu B, Mwase M, et al. (2001) Bioaccumulation of lead in wildlife dependent on the contaminated environment of the Kafue flats. Bull Environ Contam Toxicol 67: 438–445.
35. Thomas GW (1996) Soil pH and soil acidity. In: Sparks DL (ed) Methods of soil analysis, Part 3-Chemical methods, Soil Sci Soc Am Book Series # 5. 475–490. Madison, USA.
36. Lubben S, Sauerbeck D (1991) The uptake and distribution of heavy metals by spring wheat. Water Air Soil Poll 57–58: 239–247.
37. Paiva HN, Carvalho JG, Siqueira JO (2002) Translocation index of nutrients in cedro (*Cedrela fissilis* Vell.) and ipe-roxo (*Tabebuia impetiginosa* (Mart.) Standl.) seedlings submitted to increasing levels of cadmium, nickel and lead. Revista Arvore 26: 467–473.
38. SAS (2003) SAS Institute, SAS Version 9.1.2(c). SAS Institute, Inc., Cary, NC 449–453.
39. Motior MR, Abdou AS, Fareed HAID, Sofian AM (2011) Responses of sulfur, nitrogen and irrigation water to the uptake of nutrients by *Zea mays* grown in sandy calcareous soil. Aust J Crop Sci 5: 347–357.
40. Rosen CJ (2002) Lead in the Home Garden and Urban Soil Environment, College of Food, Agricultural and Natural Resources Sciences, University of Minnesota, Available: http://www.extension.umn.edu/distribution/horticulture/DG2543.html.Accessed 2012 Aug 4.

41. McBride MB (1994) Environmental Chemistry of Soils. Oxford University Press, New York.
42. Putnam DH, Oplinger ES, Hicks DR, Durgan BR, Noetzel DM, et al. (1990) Sunflower. Alternative Field Crops Manual. Oplinger ES Publication A3005, University of Wisconsin-Extension. Agric Bull Available: http://www.hort.purdue.edu/newcrop/afcm/sunflower.html. Accessed 2012 Jul 31.
43. Singh BP, Whitehead WF (1992) Response of vegetable amaranth to differing soil pH and moisture regimes. II International Symposium on Specialty and Exotic Vegetable Crops, Miami, USA.
44. Zaurov DE, Perdomo P, Raskin I (1999) Optimizing soil fertility and pH to maximize cadmium removed by Indian mustard from contaminated soils. J Plant Nutr 22: 977–986.
45. Martens SN, Breshears DD, Meyer CW (2000) Spatial distributions of understory light along the grassland/forest continuum: effects of cover, height, and spatial pattern of tree canopies. Ecolog Model 126: 79–93.
46. Ho WM, Ang LH, Lee DK (2008) Assessment of Pb uptake, translocation and immobilization in kenaf (*Hibiscus cannabilus* L.) for phytoremediation of sand tailings. J Environ Sci 20: 1341–1347.
47. Madejon E, de Mora AP, Felipe E, Burgos P, Cabrera F (2006) Soil amendments reduce trace element solubility in a contaminated soil and allow re-growth of natural vegetation. Environ Poll 139: 40–52.
48. Yang B, Lan CY, Shu WS (2005) Growth and heavy metal accumulation of *Vetiveria zizanioides* grown on lead/zinc mine tailings. Acta Ecologica Sinica 25: 45–50.
49. Wu LH, Li H, Luo YM, Christie P (2004) Nutrients can enhance phytoremediation of copper-polluted soil by Indian mustard. Environ Geochem Health 26: 331–335.
50. Lin C, Liu J, Liu L, Zhu T, Sheng L, et al. (2009) Soil amendment application frequency contributes to phytoextraction of lead by sunflower at different nutrient levels. Environ Exp Bot 65: 410–416.
51. Zhou Q (2003) Interaction between heavy metals and nitrogen fertilizers applied to soil-vegetable systems. Bull Environ Contam Toxicol 71: 338–344.
52. Holmgren CGS, Meyer MW, Chaney RL, Daniels RB (1993) Cadmium, lead, zinc, copper and nickel in agricultural soils in the United States. J Environ Qual 22: 335–348.
53. BC MOE (1989) Criteria for managing contaminated soils in British Columbia-Draft. Prepared by the Waste Management Program, British Columbia Ministry of the Environment, Victoria, British Columbia.
54. Brown SL, Chaney RL, Angle JS, Baker AJM (1995) Zinc and cadmium uptake by hyperac-cumulator *Thlaspi caerulescens* grow in the nutrient solution. Soil Sci Soc Am J 59: 125–133.
55. Cunningham SD, Berti WR (2000) Phytoextraction and phytostabilization: Technical, economical and regulatory considerations of the soil-lead issue. *In:* Phytoremediation of contaminated soil and water. Terry N, Banuelos G (eds) 359–376. Lewis Publication, Florida, USA.
56. Cui Y, Dong Y, Li H, Wang Q (2004a) Effect of elemental sulphur on solubility of soil heavy metals and their uptake by maize. Environ Intern 30: 323–328.
57. Cui Y, Wang Q, Christie P (2004b) Effect of elemental sulphur on uptake of cadmium, zinc, and sulphur by oilseed rape growing in soil contaminated with zinc and cadmium. Com Soil Sci Plant Anal 35: 2905–2916.
58. Kabata-Pendias A, Pendias H (2001) Trace Elements in Soil and Plants, 3rd ed. CRC Press, Boca Raton, FL, USA.
59. Pais I, Jones Jr (2000) The Handbook of Trace Elements. CRC Press, Boca Raton, FL, USA.

24

Dynamics of Soil Organic Carbon and Microbial Biomass Carbon in Relation to Water Erosion and Tillage Erosion

Nie Xiaojun[1,2], Zhang Jianhui[1]*, Su Zhengan[1]

1 Key Laboratory of Mountain Surface Processes and Ecological Regulation, Chinese Academy of Sciences, Institute of Mountain Hazards and Environment, Chinese Academy of Sciences and Ministry of Water Conservancy, Chengdu, China, 2 School of Surveying and Land Information Engineering, Henan Polytechnic University, Jiaozuo, China

Abstract

Dynamics of soil organic carbon (SOC) are associated with soil erosion, yet there is a shortage of research concerning the relationship between soil erosion, SOC, and especially microbial biomass carbon (MBC). In this paper, we selected two typical slope landscapes including gentle and steep slopes from the Sichuan Basin, China, and used the ^{137}Cs technique to determine the effects of water erosion and tillage erosion on the dynamics of SOC and MBC. Soil samples for the determination of ^{137}Cs, SOC, MBC and soil particle-size fractions were collected on two types of contrasting hillslopes. ^{137}Cs data revealed that soil loss occurred at upper slope positions of the two landscapes and soil accumulation at the lower slope positions. Soil erosion rates as well as distribution patterns of the <0.002-mm clay shows that water erosion is the major process of soil redistribution in the gentle slope landscape, while tillage erosion acts as the dominant process of soil redistribution in the steep slope landscape. In gentle slope landscapes, both SOC and MBC contents increased downslope and these distribution patterns were closely linked to soil redistribution rates. In steep slope landscapes, only SOC contents increased downslope, dependent on soil redistribution. It is noticeable that MBC/SOC ratios were significantly lower in gentle slope landscapes than in steep slope landscapes, implying that water erosion has a negative effect on the microbial biomass compared with tillage erosion. It is suggested that MBC dynamics are closely associated with soil redistribution by water erosion but independent of that by tillage erosion, while SOC dynamics are influenced by soil redistribution by both water erosion and tillage erosion.

Editor: David L. Kirchman, University of Delaware, United States of America

Funding: The funding for our study: the National Natural Science Foundation of China (41001157, 41271242), the 135 Strategic Program of the Institute of Mountain Hazards and Environment, CAS (SDS-135-1206), and the Doctoral Program Foundation of Henan Polytechnic University (B2010-47). The funders had no role in study design, data collection and analysis, decision to publish, or preparation of the manuscript.

Competing Interests: The authors have declared that no competing interests exist.

* E-mail: zjh@imde.ac.cn

Introduction

Soil erosion has a severe impact on soil organic carbon (SOC) pools, and as a consequence affects the soil carbon cycle. Historically, water erosion was assumed to be a major soil disturbance in agricultural slope landscapes. Numerous studies have emphasized the influences of water erosion on SOC removal and carbon sequestration in agricultural slope landscapes [1–6]. Since the widespread application of ^{137}Cs (i.e., Caesium-137) technique to assess soil erosion in the 1990s, field evidence from different regions of the world has documented that tillage erosion (i.e. the net movement of soil downslope in response to the action of farm implements) is another major soil disturbance in agricultural slope landscapes with intensive tillage [5,7–11]. The impacts of tillage erosion on within-field variability of SOC have recently been addressed [12–14]. For example, on steep hillslopes of the Sichuan Basin, China, there is a complete depletion of SOC at summit positions and remarkable accumulation at bottom positions after 15-yr tillage erosion [14].

Although the physical process of SOC transfer on the slope under water and tillage erosion has been well known, the SOC biochemical process (e.g. carbon mineralization), mostly linked with the carbon cycle, remains unclear in eroding landscapes. Polyakov and Lal (2004) reported that no significant differences in SOC mineralization exist between the eroding and control soils, but there is a 26% higher CO_2-efflux in the deposited soils compared with those without deposition [15]. In contrast, a recent study conducted by Hemelryck et al. (2011) shows that SOC mineralization in soils with deposition does not apparently differ from soils without deposition [16]. Van Oost et al. (2007) concluded that SOC mineralization associated with transport are relatively minor and that most deposited carbon is effectively preserved [17].

Indeed, it is difficult to make clear the relationship between erosion and SOC mineralization without considering the type of erosion event and soil microbial conditions. In fact, soil microbes participate in all of the biochemical process of SOC, thus microbial biomass affects the emission of CO_2 under soil erosion. Soil microbial biomass carbon (MBC) generally making up 1–5% of total SOC [18], acts as the most active component in the biochemical process of SOC turnover [19] and responds more rapidly to soil disturbance than does SOC [20]. Relatively, few studies have dealt with relationships between MBC dynamics and soil erosion [21]. In the process of water erosion, selective entrainment of fine soil particles is controlled by the transport capacity of the overland flow [22], and the labile SOC fraction adhered to fine soil particles is easily mineralized [23]. In contrast, tillage erosion unselectively moves the soil downslope [14,24], with no transport-related mineralization of SOC in the process of tillage erosion [25]. Given different mechanisms of soil transport, it would be expected that the two different processes of soil erosion may exert different impacts on MBC and the resultant SOC mineralization.

Caesium-137 (^{137}Cs) is a well-established tracer of soil erosion [26]. By comparing ^{137}Cs inventories of the study sites with those of reference sites, one can determine whether erosion (less ^{137}Cs in study sites than in reference sites) or deposition (more ^{137}Cs in study sites than in reference sites) has occurred. The relationships among patterns of soil erosion, SOC and some physio-chemical properties have been established for a number of fields and small agricultural watersheds in different climate zones [5,27–29], showing that this method is a valuable tool to investigate the erosion-carbon relationship. In this study, we use the same methodology to evaluate the relationship between erosion and carbon dynamics on cultivated slopes of the Sichuan Basin, China, where both water erosion and tillage erosion have occurred.

This study aims to (1) compare the difference in soil erosion between the two different slope landscapes (i.e., gentle and steep slopes), (2) examine spatial patterns of SOC and MBC distribution over eroded slopes, and (3) elucidate effects of water erosion and tillage erosion on the dynamics of SOC and MBC.

Materials and Methods

Study Area

The study area was located in Jianyang County of the Sichuan Basin, southwestern China (30°04′28″–30°39′00″N, 104°11′34″–104°53′36″E). This region is typical of the hilly areas of Sichuan (400–587 m above sea level) and has a humid subtropical climate (mean annual temperature 17°C; mean annual rainfall of 872 mm with 90% of precipitation between May and October). Most of the hillslopes have been dissected into slope segments with different slope gradients during previous long-term agricultural practices to minimize soil erosion and optimize farming operations. With the hoe, a predominant tillage implement in this area, one major tillage operation per year always starts at the bottom of the slope and moves upslope step by step, but at every step the tillage direction is always downslope (i.e., always pulling down). Due to long-term intensive hoeing, those in the upper parts of the slope are characterized by a thin soil layer, underlain by bedrock, whereas deeper soils are found in the lower parts. The soils in the study area, derived from purple mudstone and sandstone of Jurassic Age, are classified as Orthic Regosols in the FAO soil taxonomy. Soil texture is clay loam (27% clay, 29% silt, 44% sand), containing 1.0–2.5% organic matter and with a pH of 7.9–8.2. Crop rotation involves [wheat (*Triticum aestivum* L.), maize (*Zea mays* L.)/sweet potato (*Ipomoea batatas*)] and fertilizers (N: 330 kg ha^{-1} year^{-1} as urea and ammonium bicarbonate, P_2O_5:166 kg ha^{-1} year^{-1} as superphosphate) are uniformly applied on the cultivated slopes. Residue management is also uniformly on the slopes in the study area. For wheat and maize, the residues are cut by hand as low as possible and removed, retained residues with about 10-cm height above the ground are mixed into the tillage layer by manual hoeing. For sweet potato, all the residues are removed from the field.

Soil Sampling

Soil sampling was conducted on two types of contrasting hillslopes. Two neighboring cultivated slopes with a length of 44 and 53 m and a corresponding gradient of 7% and 4%, respectively, were considered as two replicates for gentle slope landscapes (Table 1). Another two neighboring slopes with a length of 17 and 22 m and a corresponding slope gradient of 26% and 19%, respectively, were considered as two replicates for steep slope landscapes (Table 1). Samples were collected using a soil corer (6.8-cm diameter) along the downslope transect at 10-m intervals on the two gentle slopes and at 5-m intervals on the two steep slopes. For steep slope landscapes, there were 4, 1 and 4 samples, respectively, in the upper, middle and lower positions. For gentle slope landscapes, 4, 3 and 4 samples were collected in the upper, middle and lower positions, respectively. The coordinates of each sampling point as well as elevation were measured using a survey-grade Differential Global Positioning System (DGPS).

For each sampling point, five soil cores were collected. Two cores for the determination of ^{137}Cs inventory were taken to the bedrock (depths range from 20 to 40 cm depending on thickness of total soil layer at different slope positions) and were then bulked to make a composite sample. In the context of this study, soil depth reached a maximum of 40 cm, and therefore sampling to this depth has all ^{137}Cs in soil profiles contained in samples. Another three core samples for the determination of SOC, MBC, and soil particle-size fractions were collected from the till layer (0–20 cm). One of major considerations for this is that soil redistribution by water and tillage happened mainly in the till layer. Another is that soil microbes are the most active in the till layer where a large portion of organic matter such as crop residues are concentrated, thus the effects of soil erosion on the biochemical process of SOC can be expected to be dominant. After the three soil cores were taken, they were mixed into a composite sample and then divided into two subsamples. One was sieved (2-mm) and immediately stored at 4°C in plastic bags loosely tied to ensure sufficient aeration and to prevent moisture loss until MBC analysis. The other was air-dried and sieved (2-mm) for analyses of SOC and particle-size fractions.

Laboratory Analysis

Soil samples for ^{137}Cs determination were air-dried, crushed and passed through a 2-mm mesh sieve to remove gravels. Samples of the <2 mm particle-size fraction were packed into plastic (PVC) beakers with a 320-cm^3 volume, and ^{137}Cs activity was measured using a hyperpure lithium-drifted germanium detector (HpC –40% efficiency) coupled with a Nuclear Data 6700 multichanel γ-ray spectrophotometer. Caesium-137 was detected at 662 KeV, and count time for each sample ranged from 40,000 to 60,000 s, providing a measurement precision of ±5%. The contents of ^{137}Cs were originally expressed as a unit mass basis (Bq kg^{-1}) and were then converted into an area basis (Bq m^{-2}) using the total weight of the bulked core sample and the cross-sectional area of the sampling device. Soil bulk densities (kg m^{-3}) were determined using oven-dried weight and sample volume.

SOC was determined using wet oxidation with $K_2Cr_2O_7$ and soil particle-size fractions were determined by pipette method following H_2O_2 treatment to destroy organic matter and dispersion of soil suspensions in Na-hexametaphosphate [30]. MBC was determined by fumigation with ethanol-free $CHCl_3$ and extraction with K_2SO_4 [31]. Measurement precision for SOC, MBC, and soil particle-size fractions is ±2%, ±4%, and ±2%, respectively.

Calculation of Soil Erosion Rates

The application of the ^{137}Cs technique and tillage erosion models provides a new perspective to assess the contribution of

Table 1. Total soil erosion rates, tillage erosion rates and water erosion rates.

	Gentle slope landscape		Steep slope landscape	
	4%-slope	**7%-slope**	**19%-slope**	**26%-slope**
Total length (m)	53	44	22	17
Sample number	6	5	5	4
Total soil erosion rate (t ha^{-1} yr^{-1})	24.41	28.01	39.77	44.83
Tillage erosion rate (t ha^{-1} yr^{-1})	6.96	9.29	26.16	39.67
Percentage of total erosion (%)	29	33	66	88
Water erosion rate (t ha^{-1} yr^{-1})	17.45	18.72	13.61	5.16
Percentage of total erosion (%)	71	67	34	12

water and tillage to total soil erosion [5,9–11]. For an eroding site where a total ^{137}Cs inventory A (Bq m^{-2}) is less than the reference inventory A_0 (Bq m^{-2}) at year t (date), soil erosion rates can be expressed as [32]:

$$Y=\frac{10dB}{P}\left[1-\left(1-\frac{A_0-A}{A_0}\right)^{1/(t-1963)}\right] \tag{1}$$

where Y is the soil erosion rate (t ha^{-1} yr^{-1}), d the sampling depth (m), B the bulk density of soil (kg m^{-3}) and P the particle size correction factor (assumed to be $P=1.0$). According to the previous study conducted in the study area [5], a local ^{137}Cs reference inventory was calculated as 1318 Bq m^{-2} with radioactivity decay corrected to the year 2008.

Based on a large number of experimental observations conducted previously in the study area, an empirical model of tillage erosion has been well established by Zhang et al. (2004) [33]. We adopted the model in this study because of similar terrain attributes such as the shape (linear), ranges of gradient and slope length, and even similar soil properties. In the empirical model, tillage erosion rates increases with increasing slope gradient and is inversely proportional to the slope length, expressed as:

$$R=(k_3+k_4S)10/L_d \tag{2}$$

where R is the tillage erosion rate (t ha^{-1} yr^{-1}), k_3 and k_4 are the soil transport coefficient (kg m^{-1} yr^{-1}), S is the slope gradient (m m^{-1}) and L_d is the downslope parcel length (m). Tillage transport coefficients in the study area reach 30.72 and 141.28 kg m^{-1} yr^{-1}, respectively, for k_3 and k_4 [33]. Water erosion rates were obtained by the differences between total soil erosion rates derived from ^{137}Cs data and tillage erosion rates [5,11].

Data Analysis

Statistical analyses were carried out using SPSS 11.0 for Windows version (SPSS Inc., US, 2002). Linear regression analysis was used to test the correlations between ^{137}Cs, SOC and MBC. One-way analysis of variance was used to test the significance of differences in MBC/SOC ratios between gentle and steep slope landscapes, and between upper and lower positions of the two slope landscapes.

Ethics Statement

In this study, soil sampling and sample determinations conducted were permitted by the local government (i.e. Jianyang County People's Government). We also obtained a permission from the local government for reporting research results to the public.

Results

Soil Redistribution

For steep slope landscapes, tillage erosion rates were 26.16 and 39.67 t ha^{-1} yr^{-1} and accounted for 66% and 88% of total erosion rates for the 19%- and 26%-slopes, respectively (Table 1). For gentle slope landscapes, tillage erosion rates were 9.29 and 6.96 t ha^{-1} yr^{-1} and contributed only 33% and 29% to total erosion rates for the 7%- and 4%-slopes, respectively (Table 1). The results indicate that tillage erosion dominates the process of soil redistribution in steep slope landscapes, whereas water erosion is a major process of soil redistribution in gentle slope landscapes. Differences in dominant processes of soil erosion between the two slope landscapes were also supported by the distribution of the <0.002-mm particle-size fraction on the eroded slopes. The fine particle fraction gradually increased downslope on each gentle slope, but such a trend of the fine particle was not observed on the two steep slopes (Figure 1). This result suggests that soil fine particles are preferentially transported downslope mainly by water in gentle slope landscapes, while the soils are unselectively or

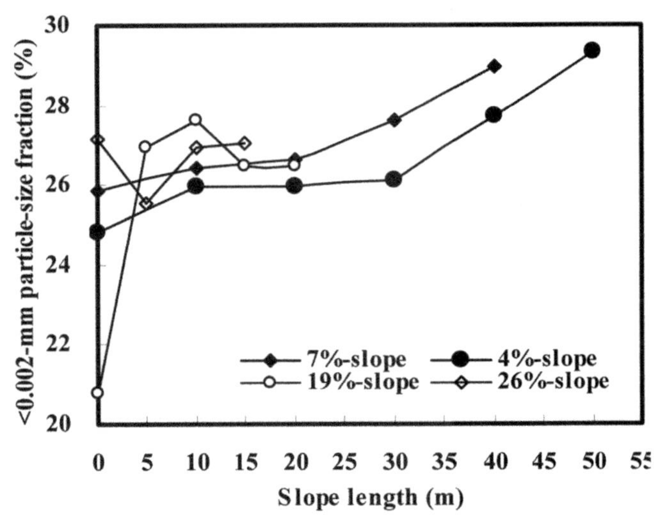

Figure 1. Distribution of the <0.002-mm particle-size fraction on eroded slopes.

entirely transported downslope due to the dominant process of tillage erosion in steep slope landscapes.

Figure 2 shows a similar ^{137}Cs distribution pattern in gentle and steep slope landscapes, i.e., ^{137}Cs inventory increased along the downslope transect. For gentle slope landscapes, ^{137}Cs inventory increased from 861 to 1336 Bq m^{-2} and 786 to 1289 Bq m^{-2} from the upper to lower positions of the 7%- and 4%-slopes, respectively (Table 2). For steep slope landscapes, ^{137}Cs inventory ranged from lows of 402 and 668 Bq m^{-2} at upper slope positions to highs of 1179 and 1002 Bq m^{-2} at lower slope positions of the 19%- and 26%-slopes, respectively (Table 2). The results suggest that soil loss occurs at upper slope positions and soil accumulation at lower slope positions.

SOC and MBC Dynamics

In gentle slope landscapes, both SOC and MBC contents increased downslope in a roughly consecutive increment (Figure 3a). SOC contents averaged 12.99 and 12.42 g kg^{-1} at lower slope positions of the 7%- and 4%-slopes with an increase of 44% and 31%, respectively, compared with those at respective upper slope positions (Table 2). From the upper to lower slope positions, MBC contents changed from 182.13 to 217.80 mg kg^{-1} with an increase of 20% on the 7%-slope, and from 168.78 to 221.13 mg kg^{-1} with an increase of 31% on the 4%-slope (Table 2). These patterns are in agreement with the distribution pattern in ^{137}Cs inventory that mirrors soil redistribution rates on the two gentle slopes. Correlation analysis also showed that inventories of SOC (kg m^{-2}) and MBC (g m^{-2}) were significantly correlated with ^{137}Cs inventory (Bq m^{-2}) in the landscape (r=0.82, P<0.01 and r=0.88, P<0.01, respectively, for SOC and MBC; see Figures 4, 5). As a consequence, it is clear that the dynamics of SOC and MBC are closely associated with soil redistribution in gentle slope landscapes.

Different distribution patterns between SOC and MBC contents were observed in steep slope landscapes, showing a rough increase in SOC content but a discrete pattern in MBC content along the downslope transect (Figure 3b). From the upper to lower slope positions, SOC contents changed from 7.33 to 10.15 g kg^{-1} with an increase of 38% on the 19%-slope, and 9.33 to 11.82 g kg^{-1} with an increase of 27% on the 26%-slope (Table 2). An increase in MBC content was observed only at middle positions of the 19%-slope, but little change in MBC between the upper and lower positions of each steep slope (Table 2). The distribution pattern in SOC content is consistent with ^{137}Cs inventory on each steep slope. A significant correlation (r=0.84, P<0.05) between SOC (kg m^{-2}) and ^{137}Cs (Bq m^{-2}) was also found in steep slope landscapes (Figure 4). However, MBC distribution could not be explained by ^{137}Cs data on each steep slope, as a result of no correlation (r=0.25, P>0.05) between MBC (g m^{-2}) and ^{137}Cs (Bq m^{-2}) in steep slope landscapes (Figure 5). Thus it is suggested that SOC dynamics are remarkably impacted by soil redistribution in steep slope landscapes, whereas MBC dynamics are independent of soil redistribution in the landscape. The close relationship between the <0.002-mm fine particle and MBC contents in gentle slope landscapes dominated by water erosion (r=0.86–0.93, P=0.02–0.03; data not shown) also showed that the microbial biomass decreased with decreasing soil fine particles under water erosion. However, no significant correlations between the two parameters (r= −0.22–0.79, P=0.11–0.78; data not shown) were found in steep slope landscapes, implying that the microbial biomass is scarcely impacted by tillage erosion which produces unselective soil particle movement on the slope.

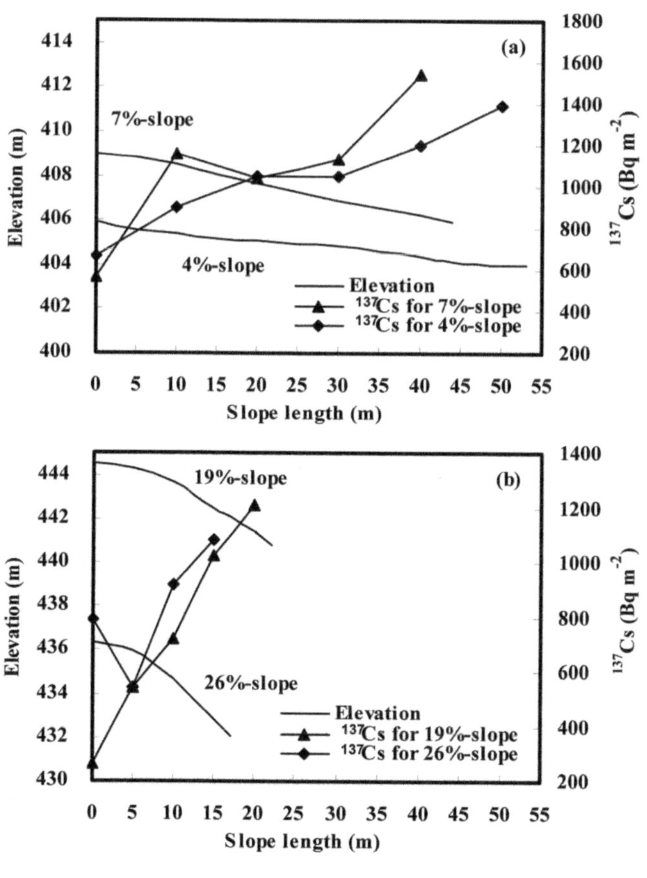

Figure 2. Distribution of ^{137}Cs inventories over eroded slopes. (a) gentle slope landscape; (b) steep slope landscape.

Relationship between MBC and SOC

A close correlation (r=0.91, P<0.01) between MBC and SOC contents (mass basis) was found in gentle slope landscapes, whereas the two were not correlated in steep slope landscapes (r=0.18, P>0.05) (Figure 6a). When the contents of SOC and MBC were expressed on a unit area basis (kg m^{-2} and g m^{-2}, respectively), a similar correlation (r=0.95, P<0.01) between the two variables was also found in gentle slope landscapes, but there was still no correlation (r=0.18, P>0.05) in steep slope landscapes (Figure 6b). In this study, additional information on the relationship between MBC and SOC was also obtained from changes in the ratio of MBC (g kg^{-1}) to SOC (g kg^{-1}). The MBC/SOC ratio was slightly greater at upper positions than at lower positions of each slope (Table 2). However, there were no statistically significant differences in MBC/SOC ratios between upper and lower positions of the two slope landscapes (P>0.05). The results indicate the difference in the MBC fraction per unit of SOC between the upper and lower positions was insignificant in the landscape. In terms of the whole slope, the MBC/SOC ratio in steep slope landscapes averaged 2.77×10^{-2}, and was significantly greater than that (mean 1.87×10^{-2}) in gentle slope landscapes (P<0.05). This showed a smaller MBC fraction per unit of SOC in gentle slope landscapes than in steep slope landscapes, suggesting that MBC is influenced by different processes of soil redistribution in the landscape.

Table 2. Descriptive statistics of ^{137}Cs inventories, SOC and MBC contents in different positions for gentle and steep slope landscapes.

Gentle slope landscape	**4%-slope**			**7%-slope**		
Slope position	Upper	Middle	Lower	Upper	Middle♀	Lower
Slope length	0–10	20–30	30–53	0–10	20	30–44
Sample number	2	2	2	2	1	2
Mean ^{137}Cs (Bq m^{-2})	786	1051	1296	861	1040	1336
Mean SOC (g kg^{-1})	9.33	9.84	12.42	9.20	9.39	12.99
Mean MBC (mg kg^{-1})	168.78	202.67	221.13	182.13	180.95	217.80
Mean MBC/SOC ratio	1.81×10^{-2}	2.06×10^{-2}	1.78×10^{-2}	1.99×10^{-2}	1.93×10^{-2}	1.69×10^{-2}
Steep slope landscape	**19%-slope**			**26%-slope**		
Slope position	Upper	Middle♀	Lower	Upper	Lower	
Slope length	0–5	10	10–22	0–10	10–17	
Sample number	2	1	2	2	2	
Mean ^{137}Cs (Bq m^{-2})	402	723	1117	668	1002	
Mean SOC (g kg^{-1})	7.33	8.49	10.15	9.33	11.82	
Mean MBC (mg kg^{-1})	242.45	308.44	234.41	257.09	273.62	
Mean MBC/SOC ratio	3.28×10^{-2}	3.63×10^{-2}	2.32×10^{-2}	2.83×10^{-2}	2.32×10^{-2}	

♀single value in the position.

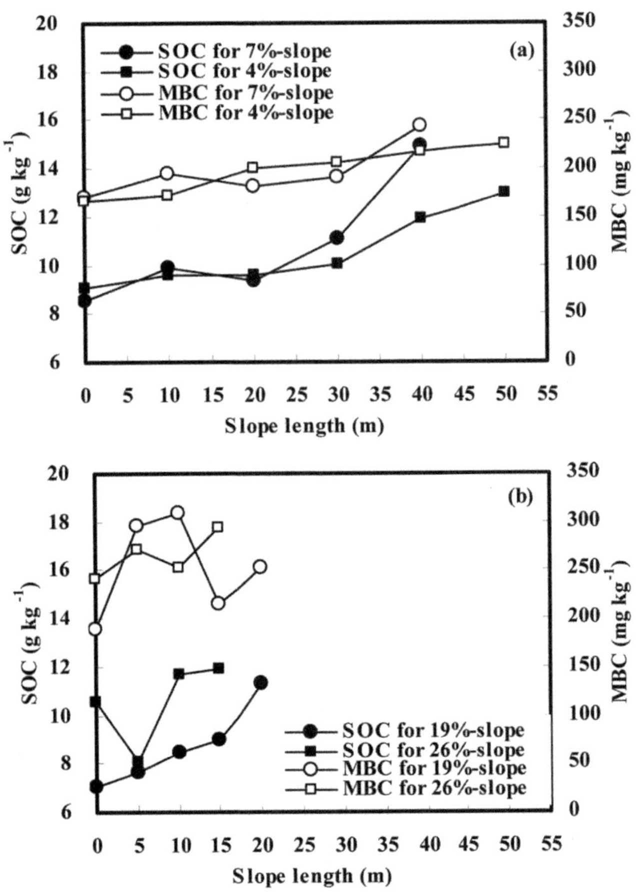

Figure 3. Distribution of SOC and MBC contents over eroded slopes. (a) gentle slope landscape; (b) steep slope landscape.

Discussion

The distribution pattern of SOC content was consistent with redistribution patterns of ^{137}Cs inventory in the two slope landscapes. The result agrees with previous studies [5,11,28], further confirming a close relationship between SOC translocation and soil redistribution by water and tillage. The MBC distribution pattern was in agreement with soil redistribution in gentle slope landscapes but independent of soil redistribution in steep slope landscapes. This is attributed to impacts of water-induced soil redistribution on SOC and MBC in gentle slope landscapes, and impacts of tillage-induced soil redistribution in steep slope landscapes. The difference in the relationship between MBC and SOC under the disturbances of water and tillage erosion

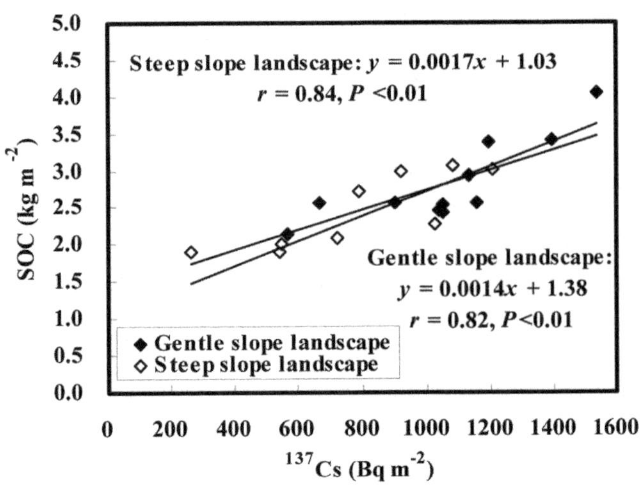

Figure 4. SOC contents vs. ^{137}Cs inventories in gentle and steep slope landscapes.

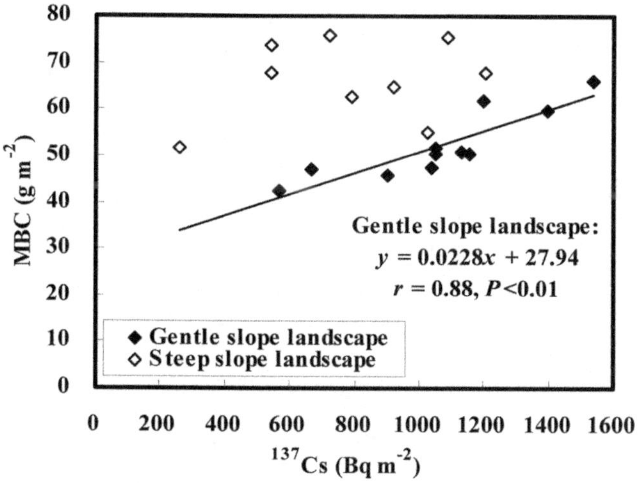

Figure 5. MBC contents vs. ^{137}Cs inventories in gentle and steep slope landscapes.

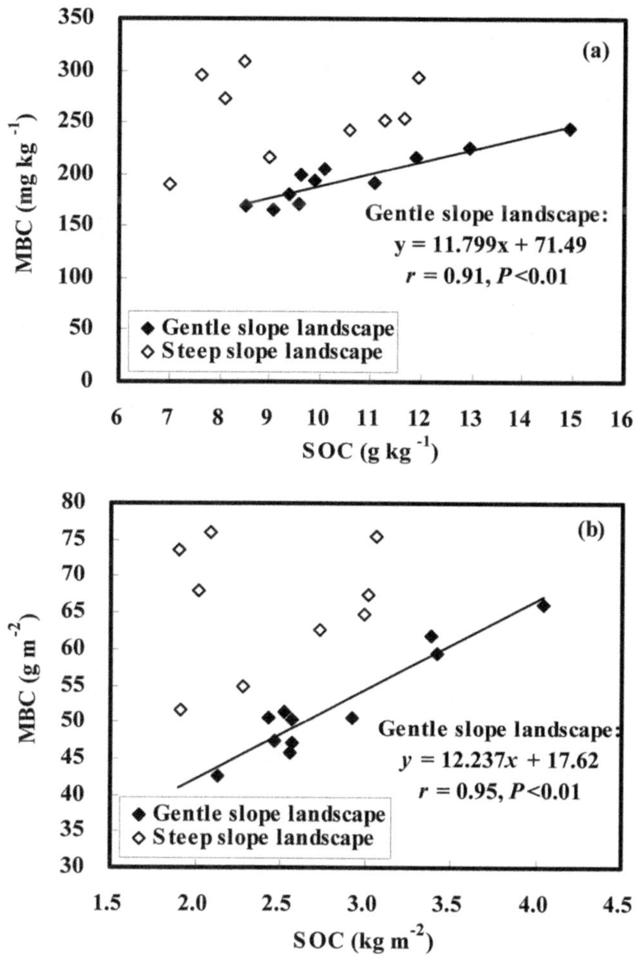

Figure 6. Relationship between MBC and SOC in gentle and steep slope landscapes. (a) mass basis; (b) area basis.

differed from the results of previous studies. A close correlation between MBC and SOC has been reported under various anthropic disturbances such as plant cover, fertilization, organic amendments, and tillage [34–37]. It is, therefore, that the dominant erosion process should be first considered when one assesses the dynamics of MBC and SOC in eroded slope landscapes.

In the two slope landscapes, lower SOC contents were found at upper slope positions with soil loss than at lower slope positions with soil accumulation. The result indicates the consistent physical process of SOC depletion in upper slope positions and SOC accumulation in lower slope positions regardless of water- or tillage-induced soil transport. However, changes in MBC contents were inconsistent between upper and lower positions in the two slope landscapes. On the two gentle slopes, MBC contents were apparently 20% and 31% higher in lower slope positions than upper slope positions, thus indicating that the physical transfer of MBC is related to soil redistribution. On the two steep slopes, however, there was little difference in MBC content between upper and lower positions, suggesting a negligible impact of soil redistribution on the physical dynamics of MBC. This similarity of MBC content between the two positions in steep slope landscapes could be attributed to unselective removal and the dilution effect due to the incorporation of the soil from deeper layers (e.g. subsoils) into the till layer by tillage (i.e. MBC contents in the newly formed surface soil decrease compared to the original surface soil, due to poor MBC in the subsoils) [14,17]. In the study area, the farmers have to till into subsoils, parent material, or even bedrocks to offset soil loss in the till layer. The soil translocated from the upslope moves entirely downslope and subsequently deposits at lower slope positions under consecutive tillage operations. Consequently, little impact on soil microbial biomass was present in the till layer in the line of slope, as a result of the redistribution of low fertility soils. No statistically significant difference ($P>0.05$) in MBC/SOC ratios between upper and lower positions in the two slope landscapes showed that the erosion-induced SOC depletion did not apparently decrease the ratio of MBC to SOC, and the deposition-induced SOC accumulation did not obviously increase this ratio either.

The MBC/SOC ratio represents the contribution of microbial biomass to organic carbon in soil [38], and is a more useful assessment index for soil health than either MBC or SOC [39]. The lower mean of MBC/SOC ratios in gentle slope landscapes than in steep slope landscapes ($P<0.05$) shows a smaller MBC contribution to SOC pools in gentle slope landscapes. This also implies that water erosion with selective removal of soil fine particles has a more severe disturbance to soil microbial biomass than tillage erosion with unselective soil movement. The difference in MBC/SOC between the two slope landscapes is probably associated with different responses of microbial biomass to distinctive changes in soil fine particles. Insam et al. (1989) reported that the clay content of soil has an important influence on the variance in MBC/SOC [40]. It is well known that soil clay (i.e. the <0.002-mm fine particle) has a strong adhesion to organic matter due to its large surface area, and thus is regarded as the mineral binding agents to form soil aggregates. In the process of water erosion, soil clay is preferentially removed by runoff, thus disrupting soil aggregates [5,41]. Accordingly, microbes incorporated into initial soil aggregates die due to their exposure to an unprotected circumstance [23], which results in a decline in soil microbial biomass. It should be noted that the same tillage operation including tillage intensity (one time per year) and tillage tool (hoe) is performed in the two slope landscapes, although tillage reduces aggregate stability by exposing the encapsulated SOC binding agents to mineralization [42]. As a result, impacts of aggregate breakdown by tillage would be similar between the two slope landscapes. In this sense, the difference in soil microbial biomass between the two landscapes resulted from different processes of soil redistribution (dominated by water erosion or by tillage erosion), excluding direct impacts of aggregate breakdown due to tillage operations.

Conclusion

In water-eroded slope landscapes, both SOC and MBC contents showed a significant correlation with soil redistribution rates. In tillage-eroded slope landscapes, SOC contents also exhibited a similar relationship, whereas MBC contents did not reveal any clear pattern. The MBC/SOC ratio was lower on the water-eroded slope than on the tillage-eroded slope, while the contribution of MBC to SOC pools was similar between the erosion-induced SOC depletion and SOC accumulation sites irrespective of landscape types. It is clear that water erosion has a severe disturbance to soil microbial biomass compared with tillage erosion. We conclude that SOC dynamics are associated closely with soil redistribution by water erosion and by tillage erosion as well, while MBC dynamics are significantly influenced by water-induced soil redistribution, but independent of tillage-induced soil redistribution.

Author Contributions

Conceived and designed the experiments: ZJH. Performed the experiments: NXJ SZA. Analyzed the data: NXJ ZJH. Contributed reagents/materials/analysis tools: NXJ. Wrote the paper: NXJ ZJH.

References

1. Verity GE, Anderson DW (1990) Soil erosion effects on soil quality and yield. Can J Soil Sci 70: 471–484.
2. Mabit L, Bernard C (1998) Relationship between soil ^{137}Cs inventories and chemical properties in a small intensively cropped watershed. Earth Planet Sci 327: 527–532.
3. Ritchie JC, McCarty GW (2003) ^{137}Caesium and soil carbon in a small agricultural watershed. Soil Till Res 69: 45–51.
4. Shukla MK, Lal R (2005) Erosional effects on soil organic carbon stock in an on-farm study on Alfisols in west central, Ohio. Soil Till Res 81: 173–181.
5. Zhang JH, Quine TA, Ni SJ, Ge FL (2006) Stocks and dynamics of SOC in relation to soil redistribution by water and tillage erosion. Global Change Biol 12: 1834–1841.
6. Nie XJ, Wang XD, Liu SZ, Liu HJ, Gu SX (2010) ^{137}Cs tracing dynamics of soil erosion, organic carbon and nitrogen in sloping farmland converted from original grassland in Tibetan plateau. Appl Radiat Isot 68: 1650–1655.
7. Lindstrom MJ, Nelson WW, Schumacher TE (1992) Quantifying tillage erosion rates due to moldboard plowing. Soil Till Res 24: 243–255.
8. Lobb DA, Kachanoski RG, Miller MH (1995) Tillage translocation and tillage erosion on shoulder slope landscape positions measured using ^{137}Cs as a tracer. Can J Soil Sci 75: 211–218.
9. Govers G, Quine TA, Desmet PJJ, Walling DE (1996) The relative contribution of soil tillage and overland flow erosion to soil redistribution on agricultural land. Earth Surf Proc Land 21: 929–946.
10. Quine TA, Walling DE, Chakela OK, Mandiringana OT, Zhang X (1999) Rates and patterns of tillage and water erosion on terraces and contour strips: Evidence from caesium-137 measurements. Catena 36: 115–142.
11. Li Y, Lindstrom MJ (2001) Evaluating soil quality–soil redistribution relationship on terraces and steep hillslope. Soil Sci Soc Am J 65: 1500–1508.
12. Heckrath G, Djurhuus J, Quine TA, Van Oost K, Govers G, et al. (2005) Tillage erosion and its effect on soil properties and crop yield in Denmark. J Environ Qual 34: 312–324.
13. Papiernik SK, Lindstrom MJ, Schumacher JA, Farenhorst A, Stephans KD, et al. (2005) Variation in soil properties and crop yield across an eroded prairie landscape. J Soil Water Conserv 60: 388–395.
14. Zhang JH, Nie XJ, Su ZA (2008) Soil profile properties in relation to soil redistribution by intense tillage on a steep hillslope. Soil Sci Soc Am J 72: 1767–1773.
15. Polyakov VO, Lal R (2004) Soil erosion and carbon dynamics under simulated rainfall. Soil Sci 169: 590–599.
16. Hemelryck HV, Govers G, Van Oost K, Merckx R (2011) Evaluating the impact of soil redistribution on the in situ mineralization of soil organic carbon. Earth Surf. Process. Landforms 36: 427–438.
17. Van Oost K, Quine TA, Govers G, Gryze SD, Six J, et al. (2007) The impact of agricultural soil erosion on the global carbon cycle. Science 318: 626–629.
18. Jenkinson DS, Ladd JN (1981) Microbial biomass in soil, measurement and turn over. In: Paul EA, Ladd JN, editor. Soil Biochemistry. New York: Marcel Dekker press, 415–471.
19. Marinari S, Mancinelli R, Campiglia E, Grego S (2006) Chemical and biological indicators of soil quality in organic and conventional farming systems in Central Italy. Ecol Indic 6: 701–711.
20. Bergstrom DW, Monreal CM, King DJ (1998) Sensitivity of soil enzyme activities to conservation practices. Soil Sci Soc Am J 62: 1286–1295.
21. Mabuhay JA, Nakagoshi N, Isagi Y (2004) Influence of erosion on soil microbial biomass, abundance and community diversity. Land Degrad Dev 15: 183–195.
22. Schiettecatte W, Gabriels D, Cornelis WM, Hofinan G (2008) Impact of deposition on the enrichment of organic carbon in eroded sediment. Catena 72: 340–347.
23. Schlesinger WH (1995) Soil respiration and changes in soil carbon stocks. In: Biotic Woodwell GM, Mackenzie FT, editor. Feedback in the Global Climatic System: Will the Warming Feed the Warming? New York: Oxford University Press, 159–168.
24. De Alba S, Lindstrom M, Schumacher TE, Malo DD (2007) Soil landscape evolution due to soil redistribution by tillage: a new conceptual model of soil catena evolution in agricultural landscapes. Catena 58: 77–100.
25. Van Oost K, Govers G, Quine TA, Heckrath G (2004) Comment on 'Managing soil carbon'. Science 305: 1567b.
26. Walling DE, Quine TA (1991) The use of caesium-137 measurements to investigate soil erosion on arable fields in the UK: potential applications and limitations. J Soil Sci 42: 147–165.
27. Quine TA, Zhang Y (2002) An investigation of spatial variation in soil erosion, soil properties, and crop production within an agricultural field in Devon, United Kingdom. J Soil Water Conserv 57: 55–65.
28. Ritchie JC, McCarty GW (2003) ^{137}Caesium and soil carbon in a small agricultural watershed. Soil Till Res 69: 45–51.
29. Mabit L, Bernard C, Makhlouf M, Laverdière MR (2008) Spatial variability of erosion and soil organic matter content estimated from ^{137}Cs measurements and geostatistics. Geoderma 145: 245–251.
30. Liu GS (1996) Soil Physical and Chemical Analysis and Description of Soil Profiles. Beijing: Chinese Standard Press, 123–166 (in Chinese).
31. Vance ED, Brookes PC, Jenkinson DS (1987) An extraction method for measuring soil microbial biomass-C. Soil Biol Biochem 19: 703–707.
32. Zhang XB, Higgitt DL, Walling DE (1990) A preliminary assessment of the potential for using caesium-137 to estimate rates of soil erosion in the Loess Plateau of China. Hydrolog Sci J 35: 267–276.
33. Zhang JH, Lobb DA, Li Y, Liu GC (2004) Assessment for tillage translocation and tillage erosion by hoeing on the steep land in hilly areas of Sichuan, China. Soil Till Res 75: 99–107.
34. García C, Hernández T, Roldán A, Martin A (2002) Effect of plant cover decline on chemical and microbiological parameters under Mediterranean climate. Soil Biol Biochem 34: 635–642.
35. Melero S, Porras JCR, Herencia JF, Madejon E (2006) Chemical and biochemical properties in a silty loam soil under conventional and organic management. Soil Till Res 90: 162–170.
36. Madejón E, Moreno F, Murillo JM, Pelegrín F (2007) Soil biochemical response to long-term conservation tillage under semi-arid Mediterranean conditions. Soil Till Res 94: 346–352.
37. Vineela C, Wani SP, Srinivasarao CH, Padmaja B, Vittal KPR (2008) Microbial properties of soils as affected by cropping and nutrient management practices in several long-term manurial experiments in the semi-arid tropics of India. Appl Soil Ecol 40: 165–173.
38. Sparling GP (1992) Ratio of microbial biomass to soil organic carbon as a sensitive indicator of changes in soil organic matter. Aust J Soil Res 30: 195–207.
39. Anderson TH, Domsch KH (1989) Ratios of microbial biomass carbon to total organic carbon in arable soils. Soil Bio Biochem 21: 471–479.
40. Insam H, Parkinson D, Domsch K (1989) Influence of macroclimate on the soil microbial biomass. Soil Bio Biochem 21: 211–221.
41. Lal R (2003) Soil erosion and the global carbon budget. Environ Int 29: 437–450.
42. Hajabbasi MA, Hemmat A (2000) Tillage impacts on aggregate stability and crop productivity in a clay-loam soil in central Iran. Soil Till Res 56: 205–212.

25

Warming Rather Than Increased Precipitation Increases Soil Recalcitrant Organic Carbon in a Semiarid Grassland after 6 Years of Treatments

Xiaoqi Zhou[1,2,3]*, Chengrong Chen[3], Yanfen Wang[1]*, Simeon Smaill[2], Peter Clinton[2]

1 Graduate University of Chinese Academy of Sciences, Beijing, China, **2** Scion (New Zealand Forest Research Institute), Christchurch, New Zealand, **3** Environmental Futures Centre, Griffith University, Nathan, Australia

Abstract

Improved understanding of changes in soil recalcitrant organic carbon (C) in response to global warming is critical for predicting changes in soil organic C (SOC) storage. Here, we took advantage of a long-term field experiment with increased temperature and precipitation to investigate the effects of warming, increased precipitation and their interactions on SOC fraction in a semiarid Inner Mongolian grassland of northern China since April 2005. We quantified labile SOC, recalcitrant SOC and stable SOC at 0–10 and 10–20 cm depths. Results showed that neither warming nor increased precipitation affected total SOC and stable SOC at either depth. Increased precipitation significantly increased labile SOC at the 0–10 cm depth. Warming decreased labile SOC ($P = 0.038$) and marginally but significantly increased recalcitrant SOC at the 10–20 cm depth ($P = 0.082$). In addition, there were significant interactive effects of warming and increased precipitation on labile SOC and recalcitrant SOC at the 0–10 cm depth (both $P < 0.05$), indicating that that results from single factor experiments should be treated with caution because of multi-factor interactions. Given that the absolute increase of SOC in the recalcitrant SOC pool was much greater than the decrease in labile SOC, and that the mean residence time of recalcitrant SOC is much greater, our results suggest that soil C storage at 10–20 cm depth may increase with increasing temperature in this semiarid grassland.

Editor: Ben Bond-Lamberty, DOE Pacific Northwest National Laboratory, United States of America

Funding: This study was conducted as part of a comprehensive research project (Global Change Multi-factor Experiment - Duolun) sponsored by Institute of Botany, Chinese Academy of Sciences. This research was supported jointly by Chinese Academy of Sciences (No. 2010-Biols-CAS-0103), Chinese Academy of Agriculture (No. 2012BAC19B04), National Natural Science Foundation of China (30925009), Griffith University and Postdoctoral Fellowship at Scion, New Zealand. The funders had no role in study design, data collection and analysis, decision to publish, or preparation of the manuscript.

Competing Interests: The authors have declared that no competing interests exist.

* E-mail: zhou318@yahoo.com.cn (XZ); yfwang@gucas.ac.cn (YW)

Introduction

The world's soils contain twice as much as carbon (C) as the atmosphere [1]. As a result, even minor soil organic carbon (SOC) losses from the soils can greatly enhance carbon dioxide concentrations in the atmosphere, which will have a positive feedback on climate [2]. It is predicted that global mean temperature will increase by 1.8–4.0°C at the end of this century [2]. Concurrently with rising temperature, global and regional precipitation regimes are predicted to change as well [2]. Changes in temperature and precipitation can greatly affect C and nitrogen (N) cycling in terrestrial ecosystems [3,4].

Recently, a meta-analysis including studies from around the world showed that warming had greatly increased soil respiration from the terrestrial ecosystems in the past 5 decades [5]. However, there is much debate whether increases in soil respiration mean that soils are releasing a substantial proportion of their stored C storage into the atmosphere [6]. In fact, the impact of climate change on the C balance of terrestrial ecosystems depends on changes in the plant-derived C entering soils as well as changes in the rate at which SOC is decomposed by heterotrophic organisms and released back to the atmosphere [7]. The SOC can be conceptually divided into several discrete fractions, i.e., labile SOC, recalcitrant SOC and stable SOC, characterized by distinct mean residence times (MRT) from days to years to millenia [7,8]. Labile SOC accounts for a small fraction of SOC and has a fast turnover with a MRT up to several months. By comparison, recalcitrant SOC is a larger pool and has a slower turnover rate with a MRT from dozens of years to hundreds of years [8]. Therefore, the ability to predict changes in recalcitrant SOC is of more importance than predicting the responses of labile SOC. Until now, studies have documented the decomposition and temperature sensitivity of recalcitrant SOC in response to warming using laboratory incubations [7–10] or modeling analysis [11,12]. However, these studies have focused on only one aspect of the C balance in soils. Available information is inadequate to confidently assess changes in recalcitrant SOC dynamics in response to warming [7]. Acquisition of new data describing recalcitrant SOC responses to warming will greatly advance our knowledge regarding the prediction of soil C storage, which may not be readily evident from gas exchange measurements [3–5].

Few studies have examined the effects of increased precipitation/drying on soil respiration in concert with warming [3,13], not to mention the interactive effects between these treatments [4]. In addition, few studies have examined the effects of increased precipitation on recalcitrant SOC, especially over the long term.

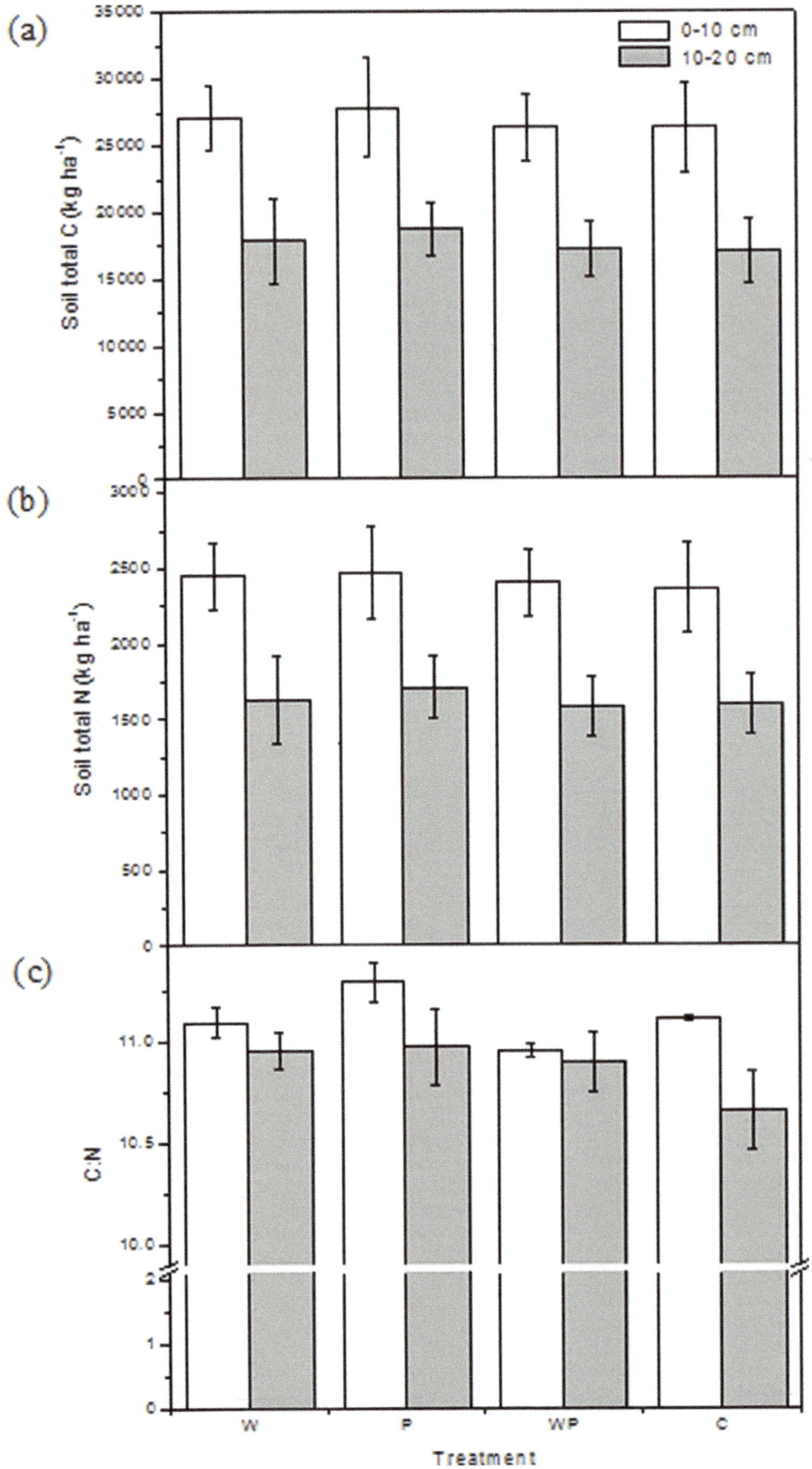
(a)
35000
30000
25000
20000
15000
10000
5000
0
Soil total C (kg ha-1)
0-10 cm
10-20 cm
(b)
3000
2500
2000
1500
1000
500
0
Soil total N (kg ha-1)
(c)
11.0
10.5
10.0
2
1
0
C:N
W
P
WP
C
Treatment

Figure 1. Soil Total C (kg ha^{-1}; a), N (kg ha^{-1}; b) and ratio of C to N (c) (mean ± SE) at two depths under warming, increased precipitation and their interactions. W, warming; P, increased precipitation; WP, warming plus increased precipitation; C, control.

Here, we used a long-term, multifactor climate change experiment artificially maintaining increased temperature and precipitation in a semiarid temperate grassland of northern China to investigate the responses of recalcitrant SOC to climate change. We measured labile SOC, recalcitrant SOC and stable SOC fractions at two depths of 0–10 and 10–20 cm in response to warming, increased precipitation and their interactions after 6 years of treatments.

Materials and Methods

Ethics Statement

All necessary permits were obtained for the described field studies and we thank Prof. Shiqiang Wan at the Institute of Botany, Chinese Academy of Sciences, for access to the experimental site and assistance in soil sampling.

Study Site

This study site was established in late April 2005 in a semiarid temperate steppe in Duolun County (42°02′N, 116°17′E, 1324 m a.s.l.), Inner Mongolia, China. The region is characterized as a moderate temperature zone with monsoon climate. Long-term mean annual precipitation and mean annual temperature is approximately 383 mm and 2.1°C, respectively. About 90% of the total precipitation falls during the period from May to October and monthly mean temperature ranges from −17.5°C in January to 18.9°C in July. The soil in this area is classified as chestnut according to Chinese classification or Haplic Calcisols according to the FAO classification with 62.8±0.1% sand, 20.3±0.1% silt, and 16.9±0.1% clay, respectively. Soil bulk density was 1.29±0.04 g cm^{-3} at the 0–10 cm depth and 1.35±0.03 g cm^{-3} at the 10–20 cm depth (Fig. S1). Soil pH was 6.84±0.07. Soil organic C and total N contents were 20.86±3.53 and 1.87±0.31 g kg^{-1}, respectively. The plant community at our experimental site was dominated by *Stipa krylovii*, *Artemisia frigida*, *Potentilla acaulis*, *Cleistogenes squarrosa*, *Allium bidentatum*, and *Agropyron cristatum* [14].

Experimental Design

This experiment used a paired and nested design with four treatments [4,14]. There were three blocks with an area of 44×28 m for each block. There was a pair of 10×15 m sub-blocks in each block, in which one plot was assigned as the increased precipitation treatment and the other one as the ambient precipitation treatment. Four 3×4 m plots were established in each 10×15 m sub-block with 1 m distance between the plots. The four plots were randomly assigned to warming and unwarmed control treatments with two replicates. Thus, there were 24 plots in total with six replicates for each treatment [control (C), warming (W), increased precipitation (P), and warming plus increased precipitation (WP)]. There were six sprinklers arranged in two rows in each of the precipitation treatment plot, with each sprinkler covering a circular area with a diameter of 3 m. A total amount of 120 mm precipitation (approximately 30% of the mean annual precipitation at this study site) was applied under the increased precipitation treatment in July and August with approx.15 mm week^{-1}. Each warmed plot was heated continuously by a 165×15 cm SR-2420 infrared radiators (Kalglo Electronics, Bethlehem, PA, USA) suspended 2.5 m aboveground since April 28, 2005. One 'dummy' heater with the same shape and size as the infrared radiator was used to simulate the shading effect of the infrared radiator in the unwarmed control plot. Long-term monitoring data showed that mean soil temperature at 10 cm depth were 1.39°C and 1.02°C higher in all warming plots than in the control and increased precipitation plots, respectively [15]. Increased precipitation enhanced soil moisture by 1.23% v/v ($P<0.01$) during the experimental period from 2005 to 2009 as compared to the control plots [14].

Soil Sampling and Total C and N Measurements

Soil samples were collected from all the 24 plots at two depths of 0–10 and 10–20 cm in early August, 2010. In each plot, one soil core (15 cm in depth and 8 cm in diameter) was taken to get a fresh sample. After passing through a 2 mm sieve, the soil samples

Table 1. Summary of results (P-value) of two-way factorial ANOVA on the effects of warming and increased precipitation and their interactions on total C, total N, ratio of C to N, microbial biomass C (MBC), labile organic C, recalcitrant organic C and stable organic C at two depths of 0–10 and 10–20 cm relative to control plots[a].

Treatment	Total C (kg ha^{-1})	Total N (kg ha^{-1})	C:N	MBC (kg ha^{-1})	Labile organic C (kg ha^{-1})[b]	Recalcitrant organic C (kg ha^{-1})	Stable organic C (kg ha^{-1})
0–10 cm							
W	0.909	0.970	**0.019**	0.535	0.933	0.446	0.886
P	0.909	0.929	0.757	**0.021**	**0.043**	0.902	0.850
W×P	0.709	0.799	**0.033**	0.645	**0.014**	**0.048**	0.574
10–20 cm							
W	0.887	0.847	0.547	0.735	**0.038**	*0.082*	0.146
P	0.844	0.879	0.450	0.736	0.905	0.543	0.754
W×P	0.659	0.731	0.293	0.474	0.539	0.906	0.202

W, warming; P, increased precipitation; WP, warming plus increased precipitation.
[a]Labile organic C was calculated from cumulative CO_2-C (mg kg^{-1} dry soil) evolved from 91-day incubation.
P values smaller than 0.05 are bold and *italics* indicates marginal significance ($P<0.1$).

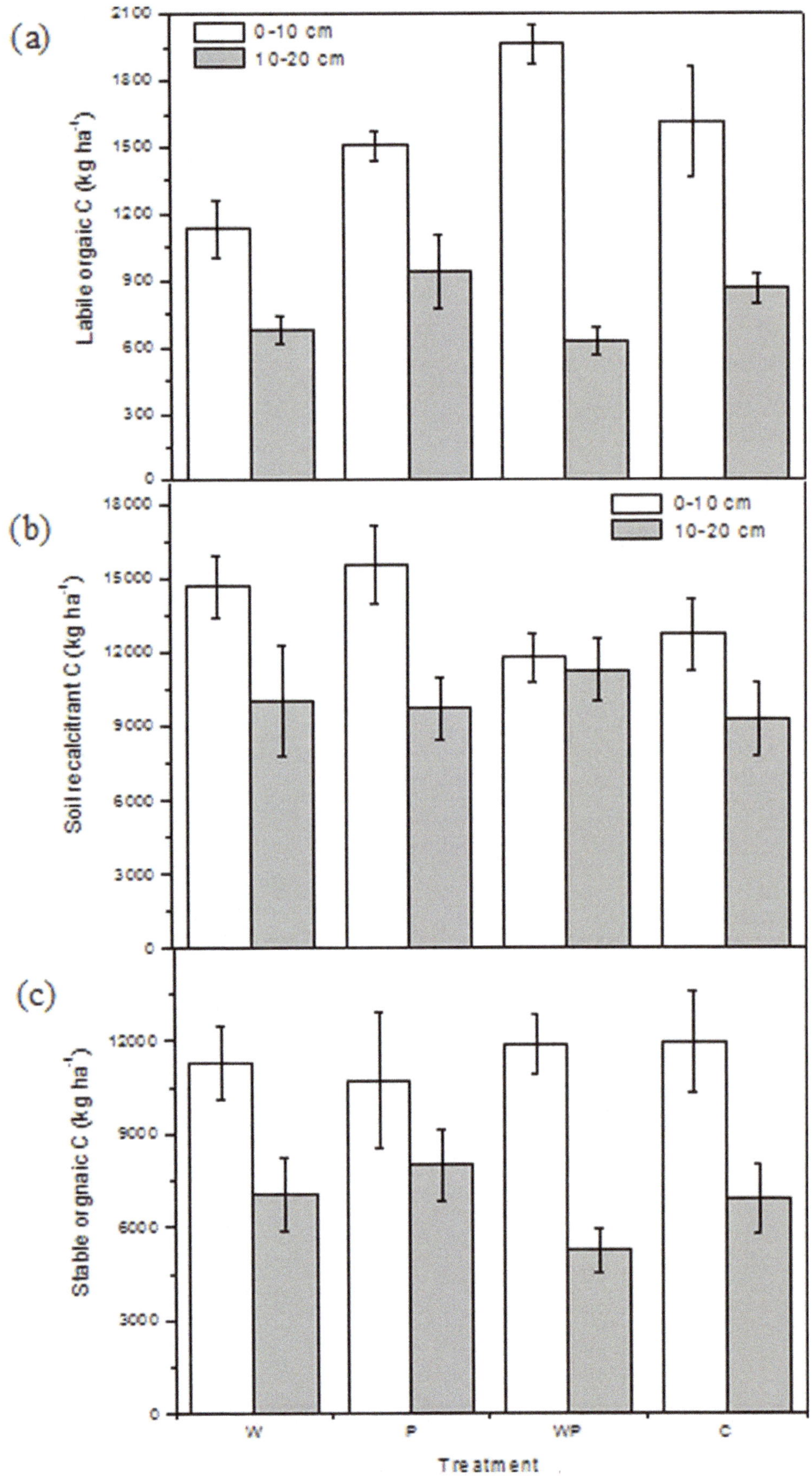
(a)
0-10 cm
10-20 cm
Labile orgaic C (kg ha^{-1})
2100
1800
1500
1200
900
600
300
0
(b)
0-10 cm
10-20 cm
Soil recalcitrant C (kg ha^{-1})
18000
15000
12000
9000
6000
3000
0
(c)
Stable orgnaic C (kg ha^{-1})
12000
9000
6000
3000
0
W
P
WP
C
Treatment

Figure 2. Soil labile organic C (kg ha^{-1}; a), recalcitrant organic C (kg ha^{-1}; b) and stable organic C (kg ha^{-1}; c) (mean ± SE) at two depths under warming, increased precipitation and their interactions. W, warming; P, increased precipitation; WP, warming plus increased precipitation; C, control.

were stored at 4°C prior to analysis. The sub-samples were air-dried and stored at room temperature for hot water extraction.

Soil moisture content was determined after being oven-dried at 105°C for 16 hours. Total C (TC) and total N (TN) of soil samples and crop C and N contents were determined using an Isoprime isotope ratio mass spectrometer with a Eurovector elemental analyzer (Isoprime-EuroEA 3000). Soil pH was measured at a 1:2.5 dry soil/water ratio.

Measurements of Soil Microbial Biomass C and Labile Organic C

Soil microbial biomass C (MBC) was measured by the chloroform fumigation-extraction method as described before [15]. Briefly, from one soil sample two 10 g (field moist) subsamples were collected and weighed out. One subsample was fumigated with chloroform for 24 h and extracted with 0.5 M K_2SO_4 in an end-to-end shaker for 1 h, then the supernatants were filtered through a Whatman no. 42 paper. The other subsample was directly extracted as above. The amounts of total C and N in the fumigated and un-fumigated soil extracts were determined using a SHIMADZU TOC-VCPH/CPN analyzer (fitted with a TN unit).

Cumulative CO_2-C using laboratory incubation was used to represent labile SOC [8–10]. Field soil samples (about 30 g) were adjusted to 60% of the water holding capacity and incubated aerobically in a sealed 1 L jar at 22°C. The CO_2 evolved from the soils over the 91-day incubation was trapped in 0.1 M NaOH and measured using 0.05 M HCl titration after the precipitation of carbonate with 1 ml of 1 M $BaCl_2$. Measurements occurred after 1, 3, 7, 14, 21, 28, 35, 42, 49, 56, 64, 70, 84 and 91 days.

Measurements of Recalcitrant Organic C and Stable Organic C

The stable SOC was measured using acid hydrolysis method [9]. Water-extractable SOC was removed by filtration through a 0.45 μm glass filter membrane using a pump-vacuum system, then the non-extracted soil material was captured on the membrane. This material was then mixed with a solution of 1 M NaCl to float and remove particulate matter. The remaining soil was then rinsed several times though a fiber membrane with deionized water and then dried at 60°C for 16 hours. After all identifiable fragments of plant material were carefully removed by hand, the soil was acidified in 6 N HCl for 12 h at 116°C to separate acid-soluble and acid-insoluble SOC fractions. After being rinsed with deionized water, the water and soil mixture were passed through a glass fiber membrane to isolate stable SOC. The residue was oven-dried at 60°C for 16 hours, then scraped off the fiber membrane and analyzed to determine the amount of stable SOC on a Eurovector elemental analyzer (Isoprime-EuroEA 3000). The amounts of recalcitrant SOC were determined by subtracting to labile SOC and stable SOC from total SOC for each plot. All SOC fractions were calculated on an area basis at the two depths.

$$SOC = \sum D_i * A * B_i * OC * 100,$$

where D_i, A, B_i, and OC represent thickness of the soil layer (cm), cross-sectional area (ha), bulk density (g cm^{-3}), organic C content (g kg^{-1}), respectively; i = 1 and 2. All SOC fractions were expressed as kg C ha^{-1}.

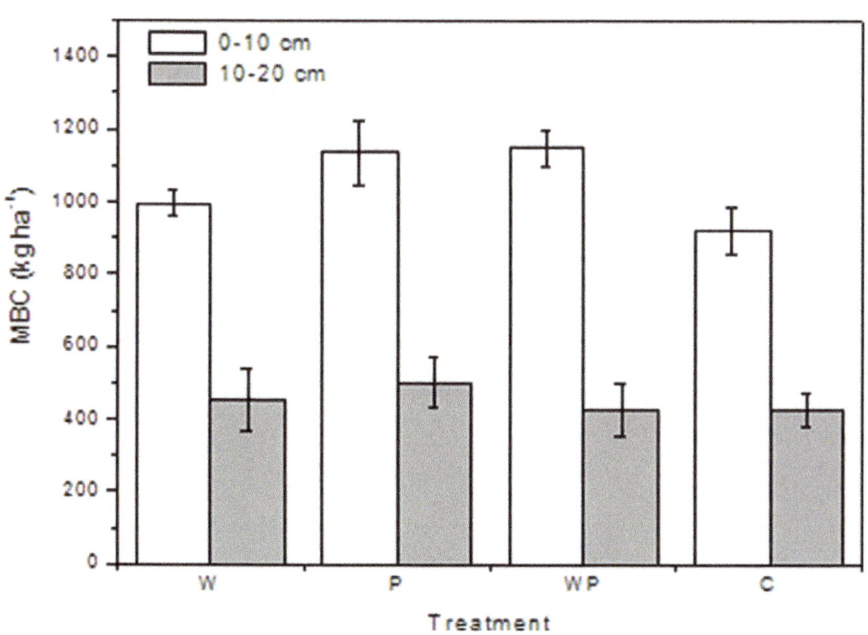

Figure 3. Soil microbial biomass C (MBC; kg ha^{-1}) at two depths under warming, increased precipitation and their interactions. W, warming; P, increased precipitation; WP, warming plus increased precipitation; C, control.

Statistical Analysis

For each soil depth, two-way analysis of variance (ANOVA) for a blocked split-plot design was used to determine the main and interactive effects of warming and increased precipitation on SOC fractions and biochemical properties. A correlation matrix of different properties was based on Pearson's correlation coefficients ($P<0.05$). All ANOVA and regression analyses were performed using SPSS 12.0 software (SPSS Inc., USA).

Results and Discussion

Warming significantly decreased C:N ratios at the 0–10 cm depth ($P<0.05$) relative to the control, but neither warming nor increased precipitation affected total C and N (Fig. 1) at either depth (Table 1). There were significant interactions between the warming and increased precipitation treatments on the ratios of C to N at the 0–10 cm depth (Table 1). It is very difficult to detect significant changes in soil total C and stable SOC over years as a result of climate change [16,17]. The lower C:N ratios under warming could be due to faster C losses via soil respiration [4], which was supported by lower labile SOC under warming (Fig. 2a).

As labile SOC is the most active fraction of total SOC, it is sensitive to management practices [18]. Labile SOC can serve as a short term C reservoir and is closely correlated with aboveground biomass [16] and ecosystem respiration [3]. We found that labile SOC accounted for 3.8–7.6% of total SOC across the plots, which agrees with previous research reporting labile SOC as a small fraction of total SOC [19,20]. The percentage of microbial biomass C in total SOC was 2.5–4.0% and was correlated well with microbial biomass C ($r^2=0.71$, $P<0.001$, $n=24$) across the treatments. Increased precipitation significantly increased labile SOC at the 0–10 cm depth (Fig. 2a), which was consistent with significantly greater microbial biomass C under increased precipitation (Fig. 3). Greater labile SOC and microbial biomass C under increased precipitation could be due to wetter soils as a result of the increased rainfall treatment, as there were significant correlations moisture contents with labile SOC ($r^2=0.72$, $P<0.001$, $n=24$) and microbial biomass C ($r^2=0.88$, $P<0.001$, $n=24$). Our results were supported by previous findings that increased precipitation increased net ecosystem productivity [21], plant biomass [14] and soil respiration [4]. The significant decrease in labile SOC at the 10–20 cm depth with warming was in contrast to a previous study that warming increased soil microbial biomass C and labile organic C in a tallgrass prairie [16]. However, our finding was supported by previous studies in which warming decreased microbial biomass C [22] and soil respiration [3,4]. The discrepancy between results from our grassland study and the tallgrass prairie study [16] could be due to complexity of ecosystems and various key controlling factors for C cycling in different ecosystems [23].

Neither increased precipitation nor its interaction with warming affected recalcitrant SOC (Fig. 2b) and stable SOC at either depth (Fig. 2c). Acid hydrolysis has been widely used to isolate and quantify stable SOC [24,25]. No significant changes have been found in stable SOC among the treatments, indicating that stable SOC is insensitive to warming, in contrast to labile SOC [26]. However, warming marginally but significantly increased recalcitrant SOC at the 10–20 cm depth ($P=0.082$; Fig. 2b). As recalcitrant SOC contains longer mean residence time compared with labile SOC, this result indicates that there is more C storage (albeit lower labile SOC) with warming compared with ambient temperatures. Warming has been reported to enhance soil respiration based on meta-analysis of gas exchange measurements [5]. However, these findings represented an increase in the rate of C cycling, not meaning that soils are losing a substantial amount of C into the atmosphere [4,6]. In addition, predictions of climate change largely depend on effects of warming on plant-C input into the soils and understanding of the temperature sensitivity of different SOC fractions [7,27]. Fang et al. [10] found similar temperature sensitivity for labile SOC and recalcitrant SOC. However, models based on the Arrhenius function show that it is theoretically possible to have increased temperature sensitivity in recalcitrant SOC compared with labile SOC [12,28,29], which implies lower recalcitrant SOC with increasing temperature. Many studies have documented one aspect of this process, i.e., temperature sensitivity of SOC decomposition [7,9,10]. Given plant-C input, it speculated that plant-derived C could compensate for the accelerated decomposition of recalcitrant SOC in boreal forest soils [30]. However, few studies have examined recalcitrant SOC in response to warming, especially over the long term, as is presented here.

Additionally, warming significantly interacted with increased precipitation in terms of their effects on labile and recalcitrant SOC at the 0–10 cm depth (both $P<0.05$, Table 1). Compared with the ambient temperature control, warming decreased labile SOC by 31.4% in the ambient precipitation treatment, but warming enhanced it by 24.7% under increased precipitation. Warming increased recalcitrant SOC by 15.6%, but decreased it by 2.1% with increased precipitation at the 0–10 cm depth as compared to the control plots (Fig. 2b). The reason for interaction of warming and increased precipitation is complex. One possible explanation is warming-induced soil drying, as moisture content plays a predominant role in ecosystem community [14], soil respiration [4] and microbial respiration [31] in this semiarid grassland and there was a significant correlation relationship between moisture content and labile SOC. Another explanation for the significant interactions is the potential effects of warming and increased precipitation on root productivity and other plant processes [32], given the close relationships between root turnover and ecosystem C cycling. Increased incidence of extreme rainfall events are a common factor in global warming predictions [2,33]. However, the responses of recalcitrant SOC processes to concurrent climate warming and changing precipitation are yet to be assessed in this region.

In summary, although warming decreased labile SOC at the 10–20 cm depth, recalcitrant SOC increased by 24.5% at the same depth. Given the longer mean residence time for recalcitrant SOC and the much greater size of this SOC pool, these results suggest that warming may substantially increase soil C storage at the 10–20 cm depth in this semiarid grassland.

Supporting Information

Figure S1 Soil moisture content (a) and bulk density (b) (mean ± SE) at two soil depths under warming, increased precipitation and their interactions. W, warming; P, increased precipitation; WP, warming plus increased precipitation; and C, control.

Acknowledgments

We greatly thank Editor-in-chief, Ben Bond-Lamberty, and two anonymous reviewers for careful revision and helpful comments on the early version of the manuscript.

Author Contributions

Revised the manuscript: SS PC. Conceived and designed the experiments: XZ CC YW. Performed the experiments: XZ. Analyzed the data: XZ. Contributed reagents/materials/analysis tools: XZ CC. Wrote the paper: XZ.

References

1. Batjes NH (1996) Total carbon and nitrogen in the soils of the world. Eur. J. Soil Sci. 47: 151–163.
2. IPCC Working Group 1 Climate Change 2007: The physical Science Basis (Cambridge Uni. Press, 2007).
3. Allison SD, Treseder KK (2008) Warming and drying suppress microbial activity and carbon cycling in boreal forest soils. Glob. Change Biol. 14: 2898–2909.
4. Liu WX, Zhang Z, Wan SQ (2009) Predominant role of water in regulating soil and microbial respiration and their responses to climate change in a semiarid grassland. Glob. Change Biol. 15: 184–195.
5. Bond-Lamberty B, Thomson A (2010) Temperature-associated increases in the global soil respiration record. Nature 464: 579–582.
6. Smith P, Fang CM (2010) A warm response by soils. Nature 464: 499–500.
7. Davidson EA, Janssens IA (2006) Temperature sensitivity of soil carbon decomposition and feedbacks to climate change. Nature 440: 165–173.
8. Fissore C, Giardina CP, Swanston CW, King GM, Kolka RK (2009) Variable temperature sensitivity of soil organic carbon in North American forests. Glob. Change Biol. 15: 2295–2310.
9. Conant RT, Drijber RA, Haddix ML, Parton WJ, Paul EA, Plante AF, et al. (2008) Sensitivity of organic matter decomposition to warming varies with its quality. Glob. Change Biol. 14: 1–10.
10. Fang C, Smith P, Moncrieff JB, Smith JU (2005) Similar response of labile and resistant soil organic matter pools to changes in temperature. Nature 433: 57–59.
11. Giardina CP, Ryan MG (2000) Evidence that decomposition rates of organic carbon in mineral soil do not vary with temperature. Nature 404: 858–861.
12. Hartley IP, Ineson P (2008) Substrate quality and the temperature sensitivity of soil organic matter decomposition. Soil Biol. Biochem. 40: 1567–1574.
13. Deng Q, Hui DF, Zhang DQ, Zhou GY, Liu JX, et al. (2012) Effects of precipitation increase on soil respiration: a three-year field experiment in subtropical forests in China. PloS One 7: e41493.
14. Yang HJ, Wu MY, Liu WX, Zhang Z, Zhang NL, et al. (2011) Community structure and composition in response to climate change in a temperate steppe. Glob. Change Biol. 17: 452–465.
15. Zhou XQ, Liu X, Rui YC, Chen CR, Wu HW, et al. (2011) Symbiotic nitrogen fixation and soil N availability under legume crops in an arid environment. J. Soils Sediments 11: 762–770.
16. Belay-Tedla A, Zhou XH, Su B, Wan SQ, Luo YQ (2009) Labile, recalcitrant, and microbial carbon and nitrogen pools of a tallgrass prairie soil in the US Great Plains subjected to experimental warming and clipping. Soil Biol. Biochem. 41: 110–116.
17. Song B, Niu SL, Zhang Z, Yang HJ, Li LH, et al. (2012) Light and heavy fractions of soil organic matter in response to climate warming and increased precipitation in a temperate steppe. Plos One 7: e33217.
18. Zhou XQ, Wu HW, Koetz E, Xu ZH, Chen CR (2012) Soil labile carbon and nitrogen pools and microbial metabolic diversity under winter crops in an arid environment. Appl. Soil Ecol. 53: 49–55.
19. Townsend AR, Vitousek PM, Desmarais DJ, Tharpe A (1997) Soil carbon pool structure and temperature sensitivity inferred using CO_2 and $^{13}CO_2$ incubation fluxes from five Hawaiian soils. Biogeochem. 38: 1–17.
20. Collins HP, Elliott ET, Paustian K, Bundy LG, Dick WA, et al. (2000) Soil carbon pools and fluxes in long-term corn belt agroecosystems. Soil Biol. Biochem. 32: 157–168.
21. Niu SL, Wu MY, Han Y, Xia JY, Li LH, Wan SQ (2008) Water-mediated responses of ecosystem carbon fluxes to climatic change in a temperate steppe. New Phytol. 177: 209–219.
22. Niinistö SM, Silvola J, Kellomäki S (2004) Soil CO_2 efflux in a boreal pine forest under atmospheric CO_2 enrichment and air warming. Glob. Change Biol. 10: 1–14.
23. Allison SD, Wallenstein MD, Bradford MA (2010) Soil-carbon response to warming dependent on microbial physiology. Nature Geosci 3: 336–340.
24. Fissore C, Giardina CP, Kolka RK, Trettin CC, King GM, Jurgensen MF, et al. (2008) Temperature and vegetation effects on soil organic carbon quality along a forested mean annual temperature gradient in North America. Glob. Change Biol. 14: 193–205.
25. Rovira P, Vallejo VR (2007) Labile, recalcitrant, and inert organic matter in Mediterranean forest soils. Soil Biol. Biochem. 39: 202–215.
26. Trumbore SE, Chadwick OA, Amundson R (1996) Rapid exchange between soil carbon and atmospheric carbon dioxide driven by temperature change. Science 272: 393–396.
27. Luo Y (2007) Terrestrial carbon cycle feedback to climate warming. Ann. Rev. Ecol. Evol. Syst. 38: 683–712.
28. Knorr W, Prentice IG, House JI, Holland EA (2005) Long-term sensitivity of soil carbon turnover to warming. Nature 433: 298–301.
29. Rey A, Jarvis P (2006) Modelling the effect of temperature on carbon mineralization rates across a network of European forest sites (FORCAST). Glob. Change Biol. 12: 1894–1908.
30. Karhu K, Fritze H, Hämäläinen K, Vanhala P, Jungner H, et al. (2010) Temperature sensitivity of soil carbon fractions in boreal forest soil. Ecology 91: 370–376.
31. Zhou XQ, Chen CR, Wang YF, Xu ZH, Hu ZY, et al. (2012) Effects of warming and increased precipitation on soil carbon mineralization in an Inner Mongolian grassland after 6 years of treatments. Biol. Fert. Soils 48: 859–866.
32. Bai WM, Wan SQ, Niu SL, Liu WX, Chen QS, Wang QB, et al. (2010) Increased temperature and precipitation interact to affect root production, mortality, and turnover in a temperate steppe: implications for ecosystem C cycling. Glob. Change Biol. 16: 1306–1316.
33. Allan RP, Soden BJ (2008) Atmospheric warming and the amplification of precipitation extremes. Science 321: 1481–1484.

26

Miscanthus Establishment and Overwintering in the Midwest USA: A Regional Modeling Study of Crop Residue Management on Critical Minimum Soil Temperatures

Christopher J. Kucharik[1,2,3]*, Andy VanLoocke[4], John D. Lenters[5], Melissa M. Motew[2]

1 Department of Agronomy, University of Wisconsin-Madison, Madison, Wisconsin, United States of America, **2** Nelson Institute Center for Sustainability and the Global Environment, University of Wisconsin-Madison, Madison, Wisconsin, United States of America, **3** Department of Energy Great Lakes Bioenergy Research Center, University of Wisconsin-Madison, Madison, Wisconsin, United States of America, **4** Department of Atmospheric Sciences, University of Illinois, Urbana, Illinois, United States of America, **5** School of Natural Resources, University of Nebraska-Lincoln, Lincoln, Nebraska, United States of America

Abstract

Miscanthus is an intriguing cellulosic bioenergy feedstock because its aboveground productivity is high for low amounts of agrochemical inputs, but soil temperatures below −3.5°C could threaten successful cultivation in temperate regions. We used a combination of observed soil temperatures and the Agro-IBIS model to investigate how strategic residue management could reduce the risk of rhizome threatening soil temperatures. This objective was addressed using a historical (1978–2007) reconstruction of extreme minimum 10 cm soil temperatures experienced across the Midwest US and model sensitivity studies that quantified the impact of crop residue on soil temperatures. At observation sites and for simulations that had bare soil, two critical soil temperature thresholds (50% rhizome winterkill at −3.5°C and −6.0°C for different Miscanthus genotypes) were reached at rhizome planting depth (10 cm) over large geographic areas. The coldest average annual extreme 10 cm soil temperatures were between −8°C to −11°C across North Dakota, South Dakota, and Minnesota. Large portions of the region experienced 10 cm soil temperatures below −3.5°C in 75% or greater for all years, and portions of North and South Dakota, Minnesota, and Wisconsin experienced soil temperatures below −6.0°C in 50–60% of all years. For simulated management options that established varied thicknesses (1–5 cm) of miscanthus straw following harvest, extreme minimum soil temperatures increased by 2.5°C to 6°C compared to bare soil, with the greatest warming associated with thicker residue layers. While the likelihood of 10 cm soil temperatures reaching −3.5°C was greatly reduced with 2–5 cm of surface residue, portions of the Dakotas, Nebraska, Minnesota, and Wisconsin still experienced temperatures colder than −3.5°C in 50–80% of all years. Nonetheless, strategic residue management could help increase the likelihood of overwintering of miscanthus rhizomes in the first few years after establishment, although low productivity and biomass availability during these early stages could hamper such efforts.

Editor: Stephen S. Fong, Virginia Commonwealth University, United States of America

Funding: This research was partially supported by the U.S. Department of Energy's Office of Science through the Midwestern Regional Center for the National Institute for Climatic Change Research at Michigan Technological University, under Award Number DE-FC02-06ER64158. This work was also funded in part by the U.S. Department of Energy Great Lakes Bioenergy Research Center (Department of Energy Biological and Environmental Research Office of Science DE-FC02-07ER64494) and U.S. Department of Energy Biomass Program Office of Energy Efficiency and Renewable Energy (DE-AC05-76RL01830). The funders had no role in study design, data collection and analysis, decision to publish, or preparation of the manuscript.

Competing Interests: The authors have declared that no competing interests exist.

* E-mail: kucharik@wisc.edu

Introduction

The recent push towards developing new bioenergy cropping systems has focused on identifying highly productive plants – other than *Zea mays* (maize) – to provide biomass for lignocellulosic biorefineries [1]. Ideal bioenergy cropping systems should lead to improved soil, water, and air quality across agricultural regions, as well as reducing emissions of greenhouse gases without competing with food sources [2]. One of the plants of interest, miscanthus, is a highly productive C_4 perennial rhizomatous grass, which is not native to many temperate regions, but its bioenergy potential is now being studied extensively in Europe, the US, and Canada [3–5]. Specifically in the Midwest US, *Miscanthus* × *giganteus* is being studied as a model cellulosic feedstock because for low amounts of agrochemical inputs, its productivity is extremely high, ~60% higher than maize total aboveground biomass [5], and double that of another C_4 grass contender, switchgrass (*Panicum virgatum*), regardless of climate and nitrogen fertilizer applied [6].

Before cellulosic feedstocks can supplant maize grain as the dominant source there are significant technical obstacles to overcome [7]. Key barriers include developing an economically viable process to break down cellulose and establishing highly productive plants in environmental conditions that are more harsh than their native regions [8]. In the case of miscanthus, cultivation in the US has largely focused on *Miscanthus* × *giganteus*, which is a natural sterile triploid hybrid of *Miscanthus* × *sinensis* and *Miscanthus* × *sacchariflorus* [9]. Because the triploid *M.* × *giganteus* clones are sterile, establishment results from the planting of rhizomes at a typical depth of 5 to 10 cm [10], and is therefore more costly to establish on the basis of time and money [11].

A key factor that may affect the viability of rhizome propagation for establishing miscanthus in non-native regions is the susceptibility to damage at cold temperatures [11]. Temperatures below freezing can lead to significant miscanthus production losses in two forms. First, soil temperatures that fall below a critical threshold can damage rhizomes, where vulnerability appears elevated in the first winter after establishment [3]. Second, air temperatures below zero after the emergence of new leaves can damage young vegetation, and rhizomes may not sprout again [1,3]. There appears to be a wide variation in frost tolerance among different genotypes because miscanthus has been found to exist naturally from warm subtropical regions to more northern locations of the subarctic [12,13]. Previous research conducted as part of the European Union (EU) Miscanthus Productivity Network [3,14,15] suggested that at some sites in northern Europe including Denmark, Ireland, and Germany, rhizomes in newly established stands did not survive the first year (winter). This has led to subsequent research on the cold tolerance of different genotypes, including *M. sacchariflorus* and *M. sinesnsis* [13]. Results suggested that the lethal temperature at which 50% of shoots were killed was −8°C for *M. giganteus*, −7.5°C for *M. sacchariflorus*, and between −6 and −9°C for two hybrids of *M. sinensis* [1,16]. For rhizomes, the lethal temperature at which 50% of rhizomes were killed was −3.5°C for *M. giganteus* and *M. sacchariflorus*, and −4.5°C and −6°C, respectively for two different hybrids of *M. sinensis* [1,13].

In the US, research has focused predominantly on *M.* × *giganteus* in the Midwest, a region that typically sees extended periods of cold Arctic air outbreaks during the late fall and winter, and correspondingly is at risk to experience near-surface soil temperatures below 0°C as well as frequent freeze-thaw cycles. These cold temperature dynamics create a wide range of uncertainty concerning overwintering of miscanthus in the Midwest US [7]. Recent research has documented the impacts of rhizome size, planting depth, and cold storage on the success of establishment, and arrived at the conclusion that *M.* × *giganteus* rhizomes are best suited to be planted at a depth of 10 cm [8]. Heaton et al. [7] suggested that at this 10 cm depth in Illinois, mature stands of *M.* × *giganteus* have been able to consistently survive winter air temperatures as cold as −20°C and soil temperatures below −6°C. At the University of Wisconsin-Madison's agricultural research station near Arlington, Wisconsin, multiple plots of *M.* × *giganteus* that originated from hand planted rhizomes in spring of 2008 experienced 90% winterkill during the winter of 2008–09, which was representative of other plantings in the Midwest US that year [7].

Crop residue, the unharvested portion of above-ground biomass, can have a significant impact on the surface energy balance and soil properties including temperature. It has been well documented that surface residue management in agricultural systems can aide in conserving soil moisture by reducing maximum soil temperatures by up to 10°C and decreasing evapotranspiration due to a higher surface albedo [17–21]. Residue can also act as an effective insulating barrier to the mineral soil during winter, increasing soil temperatures by 5 to 8°C in the central US [21,22]. Therefore, improved or adapted agronomic management of *M.* × *giganteus* residue in the first several years may be key to increasing the likelihood of successful establishment.

Our primary objective was to investigate how the risk of *M.* × *giganteus* rhizome threatening temperatures could be reduced with strategic residue management. This objective was based on using observations and an agroecosystem model to examine historical spatial and temporal patterns of extreme minimum 10 cm soil temperatures across the Midwest US. Given previous research results, we focus on rhizome losses because this type of damage appears to be more devastating because frost damage to leaves appears to be survivable. Specifically, we created a simulated reconstruction of daily wintertime soil temperatures at high spatial resolution (0.08333°, 5 minute, or ~10 km) across the Midwest US (a region bounded by 36°N to 50°N lat and −79°W and −105°W lon) from 1948–2007, and quantified the frequency that lethal soil temperature thresholds, previously suggested for two miscanthus genotypes (−3.5°C and −6.0°C), were reached at 10 cm depth. Through model sensitivity studies, we investigate how varied thicknesses of prostrate layers of miscanthus straw and corn residue impact wintertime soil temperatures, and how management post harvest could reduce the risk of miscanthus establishment failure. We hypothesize that a prostrate thatch layer of miscanthus straw or corn residue of 1 cm or greater will increase wintertime soil temperatures compared to removal of the residue post harvest, and reduce the risk of soil temperatures at 10 cm reaching critical rhizome kill thresholds. To address our objectives, we employ a dynamic agroecosystem model, Agro-IBIS [21,23] that has been recently modified to include representation of miscanthus and switchgrass [24,25]. We conduct a validation of Agro-IBIS simulated snow depth and 10 cm soil temperatures using several Midwest observational datasets. We conclude with an analysis of trends in the coldest annual soil temperatures to determine whether climate change has led to decreased risk of winterkill of *M.* × *giganteus.*

Methods

Model description

Agro-IBIS is a process-based ecosystem model capable of simulating managed and natural ecosystem dynamics of North America, with coupled carbon, water, and energy cycles. Agro-IBIS was developed by adapting a Global Dynamic Vegetation Model (DGVM), called the Integrated Biosphere Simulator (IBIS) [26,27], to simulate corn, soybean, and wheat cropping systems across the continental US [28], and most recently miscanthus and switchgrass [24,25]. Agro-IBIS simulates the energy, water, carbon, and momentum balance of the soil-plant-atmosphere system at a 60-min time step. The model includes two vegetation layers with eight potential forest plant functional types (PFTs) in the upper canopy, and two grasses (cool and warm season) and two shrub PFTs in the lower canopy. Row crops and miscanthus are simulated as part of the lower canopy layer. The model version used in this study includes 11 soil layers of varying thicknesses to a 250 cm depth, which are parameterized with one of eleven soil textural categories and corresponding physical attributes [29]. A three-layer thermodynamic snow model simulates the energy balance of the snow surface and changes in snow cover in terms of temperature, fractional coverage, and total snow thickness [30]. Physiologically-based formulations of leaf-level photosynthesis, stomatal conductance [31–33] and respiration control canopy exchange processes, and parameters vary according to generalized vegetation categories (e.g., trees, shrubs, C_3 and C_4 grasses or crops). The reader is referred to Li et al. [34] for more details about root water uptake and hydrology and Soylu et al. [35] for description of one dimensional water movement through the soil profile.

Agro-IBIS simulates crop growth transitions through phenological stages of development using an accumulated thermal time approach, and characterizes seasonal changes in carbon (C) allocation to specific crop C pools (i.e. leaf, stem, root, and grain). Leaf area index (LAI) is calculated each timestep using the accumulated leaf tissue C multiplied by a crop specific leaf area value. Canopy and land surface processes in Agro-IBIS are based on the key differences in C_3 and C_4 physiology, daily phenology, and carbon allocation so that coupled carbon-water exchange is responsive to agricultural management and environmental stresses. For a complete description of the modeling approach, the reader is referred to several other publications [23,27,28,36]. IBIS and Agro-IBIS have been validated extensively from the individual farm scale to the global scale to improve model formulations and parameterizations [21,23–25,27,28,30,37–41].

Other modeling approaches and the selection of Agro-IBIS

We briefly review two other well validated models (HYDRUS and SHAW) that are often used to study the impacts of agricultural management on soil heat and water flow and offer reasoning why Agro-IBIS was selected to carry out the study objectives. HYDRUS is a variably saturated soil water flow model that numerically solves the mixed-based Richards' equation for saturated - unsaturated water flow and heat transport [42,43]. HYDRUS simulates water and heat movement in one-dimensional, variably saturated homogeneous media and represents infiltration, evaporation, root water uptake and transpiration, soil water storage, deep drainage, and groundwater recharge. The Simultaneous Heat And Water (SHAW) model is a one-dimensional physically based model of water and heat transport in soils [44], and is capable of simulating infiltration, evapotranspiration (ET), interception, and other hydrologic processes. SHAW has been used to address soil tillage and residue effects on soil freezing and soil water conservation [45] and the interaction between vegetation, soil properties, and other land surface characteristics on frozen soil processes [46]. The Agro-IBIS soil physics module uses Richard's equation to calculate the time rate of change of liquid soil moisture, and the vertical flux of water is modeled according to Darcy's Law. The water budget of soil is controlled by the rate of infiltration, evaporation of water from the soil surface, the transpiration stream originating from plants, and redistribution of water in the profile [27]. Each soil layer in Agro-IBIS is described in terms of soil temperature, volumetric water content, and ice content for any time step [27].

Therefore, while the three models discussed have similar capabilities in simulating soil heat and water flow, HYDRUS and SHAW have some limitations that make them difficult to apply to this study. First, they are designed as point models, and therefore operate at a limited spatial scale compared to Agro-IBIS, which makes it difficult to answer questions across broad regional scales. Second, HYDRUS and SHAW require temporal changes in leaf area index (LAI) and some other vegetation characteristics to be input by the user; therefore, they cannot explicitly simulate phenological development, growth (i.e. photosynthesis), or differences in carbon allocation among different plant species such as miscanthus and corn. Lastly, HYDRUS and SHAW do not explicitly represent carbon and nutrient cycling coupled to water and energy exchanges in the soil-plant-atmosphere system. Therefore, in the context of studying the impact of variable plant growth and development, litter decomposition, and residue management from 1978–2007 on the water cycle and the magnitude of minimum wintertime soil temperatures, HYDRUS and SHAW have limited abilities to generate a historical record for the entire Midwest USA. While SHAW and HYDRUS are exceptional modeling tools in their own right, the expanded capabilities of Agro-IBIS made it a better choice for this particular study.

Agro-IBIS inputs

Climate inputs required at each model time step for each grid cell include solar radiation, temperature, precipitation, relative humidity, and wind speed. ZedX Inc. (Bellefonte, PA) developed a daily gridded weather dataset at 10 km resolution (0.08333°) for a sixty-year period from 1948 through 2007, which included all six variables needed as model input. ZedX Inc. (Bellefonte, PA) generated the gridded weather data using statistical interpolation of observational data that was subject to a rigorous quality control procedure. Before the data were input into the interpolation algorithm, quality control checks were performed which included assessments of plausibility, checks against observational extremes, and checks against neighboring stations using a quality control threshold based on standard deviations. The spatial extent of this dataset spanned from 24°N to 52°N latitude, and 50°W to 130°W longitude. Three data sets were used to generate the gridded maximum and minimum daily temperatures and precipitation. Input station data for Canada and Mexico were obtained from the Global Historical Daily Climatology (GHCND) database, and the National Climatic Data Center (NCDC) TD3200 and TD3210 station data were used for the United States. Relative humidity and wind speed were generated using the Global Summary of the Day (GSOD) daily gridded data. The 10-km gridded data of solar radiation were produced using coarser resolution NCEP/NCAR reanalysis 1 data [47] and the NCEP/DOE AMIP 2 reanalysis data [48]. Hourly variations in climatic variables are simulated through the use of empirical formulations of temperature, specific humidity, precipitation, and radiation variability [29].

Land surface inputs at model initialization include soil textural class at each soil layer to a depth of 250 cm. The dominant soil texture for each soil layer in each 5-minute grid cell was derived from the USDA State Soil Geographic Database (STATSGO) 1 km resolution dataset [49]. The standard thicknesses of the 11 soil layers are 5 cm (layers 1 and 2), 10 cm (layers 3–5), 20 cm (layers 6–8), and 50 cm (layers 9–11), coinciding with the CONUS-Soil dataset. From the assignment of a textural category in each grid cell and each soil layer, the porosity, field capacity, wilting point, saturated air-entry potential and hydraulic conductivity, and moisture release curve "b" (Campbell) coefficient are obtained from a look-up table [29]. Soil moisture is used in combination with snow and vegetative properties to determine the land surface albedo in the absence of surface residues.

Implementation

The geographic region delineated for this study was between −79W and −105W longitude, and 35N to 50N latitude, excluding portions of Canada. While we carry out simulations of miscanthus growing everywhere across this region, this is done purely as a scientific exercise and not as a specific recommendation. We performed multiple simulations to (1) validate simulations of snow depth and 10 cm soil temperature against historical observations, and (2) examine the effects of changing land cover and management on annual minimum wintertime soil temperatures deemed critical to miscanthus rhizome overwintering, and how those temperatures are impacted by differing soil surface residue thicknesses, laying prostrate and evenly distributed (Table 1). Simulations represented potential (natural) vegetation (POTVEG); maize managed with conventional tillage (MAIZE+TILL) leaving no surface residue or stubble post harvest; maize managed with no-tillage (MAIZE+NOTILL) that left a 5 cm thick surface residue layer post harvest, but no standing stubble; *M.* × *giganteus* with an annual fall harvest leaving varying thicknesses of surface residue (1.0, 2.5, and 5.0 cm), with the intent to have a thatch layer present during each subsequent winter and spring, (MISCAN+R; Table 1). The prescribed range of residue thicknesses for miscanthus used in regional simulations are consistent with observations [50].

Given the focus of this study on regional soil temperatures and the important connection between snow depth and wintertime minimum soil temperatures [22,51], we reassessed Agro-IBIS simulations of soil temperature and monthly and seasonal snow depth across the Midwest US. Numerous studies have discussed the insulating properties of snowpack, thereby decreasing the depth of frost penetration, as well as the difficulty in modeling the transition from snowpack to a existence of a bare soil surface during spring in temperate latitudes [52,53]. We first used snow depth and soil temperature datasets previously constructed by Lenters et al. [30] and Hu and Feng [54], respectively. Iowa soil temperatures from the Hu and Feng [54] dataset were selected for model validation because of the high number of soil temperature observations available from 1982–2002. The majority of stations in the Hu and Feng [54] dataset had soil temperature readings made beneath soils void of vegetative or residue cover but could not always be confirmed.

We also used additional sources of soil temperature data at agricultural research sites to further validate Agro-IBIS. First, simulated 10 cm soil temperatures for the 2009–2011 period were compared to 10 cm soil temperature data (chromel-constantan thermocouples) collected in fields managed for conventionally tilled maize and switchgrass with a sparse thatch layer at the University of Wisconsin-Madison Arlington Agricultural Research Station, near Arlington, WI (43.31°N Lat, −89.38°W Lon). Next, we used observed 10 cm soil temperatures (Hydra Probe II, Stevens, Portland, OR USA) for the 2009–2011 period collected in five *M.* × *giganteus* plots at the University of Illinois Energy Farm (40.06°N Lat, −88.20°W Lon; for full site and plot description see Zeri et al. [55] and Smith et al. [56]). For this comparison, the model was driven with a climate data set developed specifically for the University of Illinois Energy Farm [24,25]. Lastly, we compiled 10 cm soil temperature data from 125 observation sites in the Midwest US – across eleven states – that have collected continuous 10 cm daily soil temperature data during a portion of the years 1981 through 2011 (SI, Table S1). Given our inability to exactly replicate specific site management history, soil profiles, and hourly meteorological conditions at each of the 125 observation sites in this comparison, we used a more conservative 3-day running mean of 10 cm soil temperatures in Agro-IBIS to estimate the simulated extreme minimum soil temperatures, and to compare with the daily extreme minimums measured at each site. These station data were used to assess the ability of Agro-IBIS to simulate the average annual extreme minimum 10 cm soil temperatures across a large geographic extent, and the frequency that the 10 cm soil temperature falls below −3.5°C and −6°C, respectively. The majority of stations had soil temperature readings made beneath soils void of vegetative or residue cover with the exception of a subset of the 125 sites located in Illinois, which had grass cover (Carl Bernacchi, personal communication).

The POTVEG, MAIZE+TILL, and MAIZE+NOTILL model simulations were used to validate monthly and seasonal (Nov-Apr) averages of snow depth for the 1963–1995 time period [30] and monthly 10 cm soil temperatures from 1980–2002 across Iowa [54]. For validation of simulated monthly and seasonal snow depth, we used the same observational snow depth dataset from the National Weather Service Summary of the Day that was used in a previous validation of the IBIS-2 model by Lenters et al. [30]. Agro-IBIS spatial averages were formed for all grid cells in northern quadrants (grid cells within 43.5° to 47.5°N lat and −94.0 and −83.0°W lon) and southern quadrants (from 39.5° to 43.49°N lat and −94.0° and −83.0°W lon) to compare with averages for 34 station observations. The simulation of extreme minimum soil temperatures was validated using the daily average 10 cm soil temperature output from the MAIZE+TILL simulation was compared with minimum temperature data obtained from the 125 sites (SI, Table S1).

Regional simulations were conducted from 1940 to 2007. However, because of changes in climate across the region [57,58], we limited our historical analyses of minimum 10 cm soil temperatures and frequency of occurrences to a shortened 30-year period from 1978–2007. The first eight years of the 1940–2007 simulations were discarded as spin-up, needed to bring the soil water balance into equilibrium. We drove those eight years with randomly selected years of climate data. Simulations assumed that nitrogen (N) was not a limiting factor to plant growth. For simulations of maize, yearly changes in optimal planting dates and cultivar selection (total growing degree days required to physiological maturity) were simulated. All simulations were performed with a static atmospheric CO_2 concentration of 370 ppm.

The Midwest US has experienced significant warming temperatures during the past several decades [57,58], particularly during winter and springtime. Given these changes, we also investigated whether soil temperatures have experienced similar warming,

Table 1. Description of Agro-IBIS model runs.

Model Simulation	Description	Residue Layer?
POTVEG	Potential vegetation representing natural vegetation types that could grow in each grid cell based on bioclimatic limits of each plant functional type; dynamic vegetation modeling	No
MAIZE+TILL	Continuous maize managed with conventional tillage; fall harvest removes all aboveground vegetation, leaving a bare soil surface	No
MAIZE+NOTILL	Continuous maize managed with no-tillage; fall harvest management leaves dry plant matter on field with an assumed thickness of 5 cm.	Yes
MISCAN+$R_{1\ cm}$	Miscanthus grown each year; fall harvest management leaves dry plant matter on field with an average thickness of 1 cm.	Yes
MISCAN+$R_{2.5\ cm}$	Miscanthus grown each year; fall harvest management leaves dry plant matter on field with an average thickness of 2.5 cm.	Yes
MISCAN+$R_{5\ cm}$	Miscanthus grown each year; fall harvest management leaves dry plant matter on field with an average thickness of 5 cm.	Yes

thereby lowering the risk of unsuccessful overwintering of miscanthus. We used Agro-IBIS and our daily climate dataset from 1948–2007 to analyze trends in annual average soil temperatures, as well as changes in the extreme minimum values at a 10 cm depth. We also analyzed trends in the annual average soil temperatures at three depths in the MAIZE+TILL simulation from 1967–2002 to coincide with the Hu and Feng [54] study for comparison. All statistics were performed off-line using a commercial software package [59]. To be considered statistically significant, trends had to differ from zero at $P<0.05$.

Simulating the effects of residue in Agro-IBIS

For the MAIZE+NOTILL and the three MISCAN+R regional simulations, the top model soil layer (0–5 cm for MAIZE+NOTILL, and varying thicknesses for MISCAN+R; Table 1) was modified to represent an organic residue layer, lying prostrate, that persisted throughout the year. The variables modified to represent a thatch layer and the values for key maize and miscanthus residue properties are presented in Table 2. We also used several additional sensitivity analyses at eight geographic locations that experience a wide range in average wintertime soil temperatures and snowfall. These additional model runs were used to further investigate the impact of (1) a wider range of ten different miscanthus straw residue thicknesses (from 1 cm to 20 cm), (2) varying residue albedo (from 0.15 to 0.50), and (3) porosity (bulk density) of residue material (from 0.5 to 0.99) on annual average extreme minimum soil temperatures. The ten additional miscanthus residue thickness simulations were performed to develop more easily interpreted response curves (e.g., soil temperature warming response to residue thickness), as well as to investigate how changing residue thicknesses impacts interannual variability in minimum soil temperatures. These types of responses would be difficult to illustrate succinctly with a series of spatial maps. We note that crop residue thicknesses greater than about 10 cm should be considered extreme scenarios that were used solely for the purpose of building response curves, and are not an easily implemented or recommended residue management option. In the case of miscanthus residue albedo and porosity, there is currently very little published data concerning values for these variables. In order to understand whether our simulation results could be biased due to choosing a mean value for miscanthus residue albedo and bulk density, we further investigated whether large changes in these quantities can have a significant impact on soil temperature responses associated with residue management.

Results

Validation of simulated snow depth

Observed monthly mean snow depth in the northern region illustrated gradual increases from November through February, with an average maximum of approximately 30.3 cm occurring in February, declining to 19.6 cm in March and 3.4 cm in April (Fig. 1a). Compared with a previous version of IBIS that was executed over the Midwest with different climate and soils datasets at coarser spatial resolution [30], Agro-IBIS simulations exhibited increased snow depth in all months from December through March for the POTVEG scenario, which is also how land cover in Lenters et al. [30] was parameterized. In the POTVEG scenario, simulated snow depths were within ±5–10% of observations in all months from November through April, and the model captured the timing of the observed seasonal maximum snow depth in February. Model simulations for MAIZE+TILL scenario also showed higher simulated monthly mean snow depth from December through February compared to previous IBIS-2 simulations, but were approximately 15% and 75% lower than observed averages for February and March, respectively, and simulated maximum values occurred in January (Fig. 1a).

Observations of mean snow depth in the southern region illustrated gradual increases in monthly values from November through January, with an average maximum of approximately 8.0 cm occurring in January, declining to 7.4 cm in February and 2.0 cm in March (Fig. 1b). Compared to the previous IBIS-2 model results for a POTVEG scenario [30], simulations exhibited increased snow depth in all months for a similar vegetation parameterization and were 10–50% (i.e. 1–4 cm) higher than observed values in all months but matched the observed January maximum (Fig. 1b). Model simulations for MAIZE+TILL scenario also suggested improved simulated monthly mean snow depth from December through February compared to previous IBIS-2 simulations, were within 2 cm of observed values from December through March, and correctly simulated the timing of the observed snow depth maximum in January (Fig. 1b).

In the current study, all three scenarios showed significant improvement over the previous IBIS-2 model validation (slope = 0.61; $r^2 = 0.78$) across the northern region (Fig. 2a). Both cropping system scenarios resulted in a negative bias and underestimated mean annual snow depth (slope = 0.73; $r^2 = 0.81$ for MAIZE+TILL), while the POTVEG scenario closely captured the observed mean annual snow depth for each year (slope = 1.02, $r^2 = 0.82$). In the southern region (Fig. 2b), all three scenarios

Table 2. Plant residue biophysical values used to modify Agro-IBIS to simulate the effects of crop residue on soil surface energy balance and heat transfer.

Quantity	Maize residue	Reference	Miscanthus straw	Reference
Residue layer thickness (m)	0.05	[18]	0.022–0.042	[50]
Roughness length (m)	0.012	[76]	0.0065	[77]
Bulk density (kg m^{-3})	36.4	[78]	22.0	[50]
Cellulose density (kg m^{-3})	1450	[18]	1350	[79]
Thermal conductivity (W m^{-1} K^{-1})	0.126	[80]	0.08	[81]
Specific heat (J kg^{-1} K^{-1})	1900	[82]	1335	[81]
Porosity	0.975	[78]	0.96	[83]
Albedo	0.25	[84]	0.32	[85]
Fractional cover	0.95	[18]	0.90	Kucharik (unpublished data)

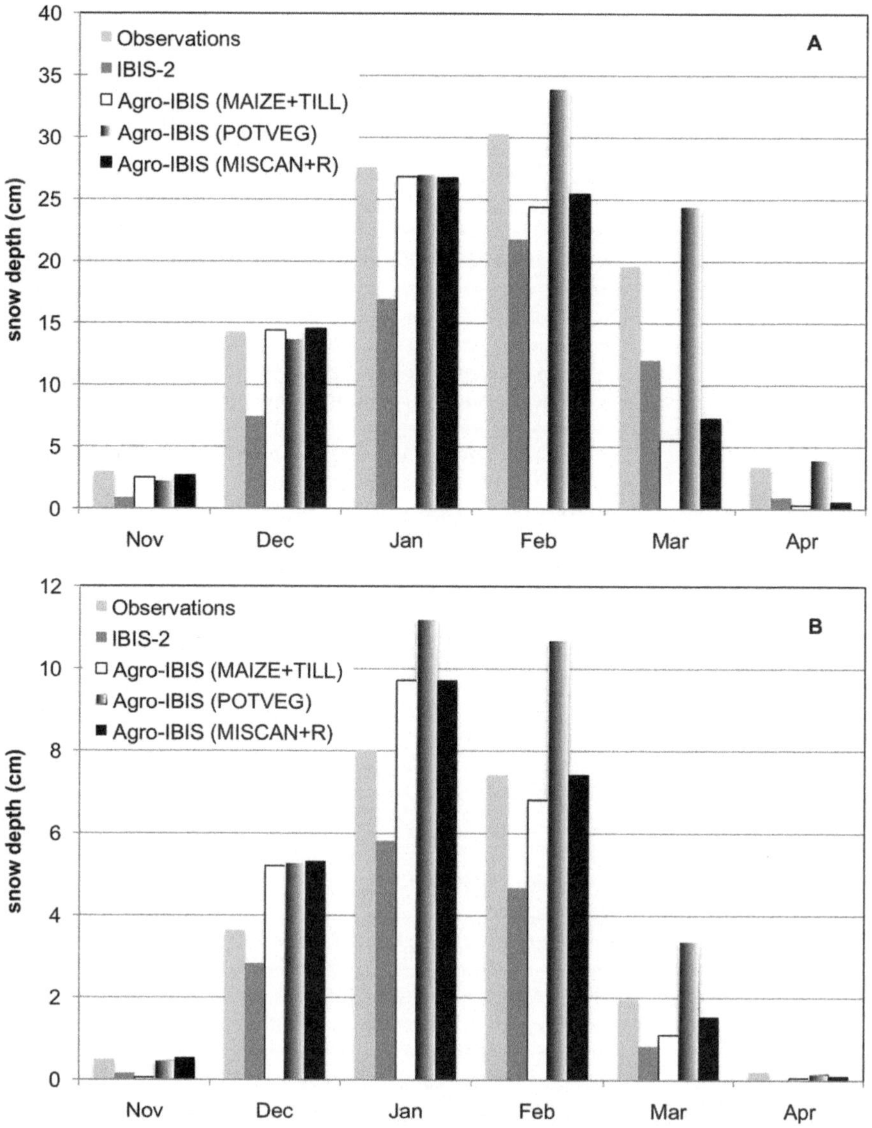

Figure 1. Comparison of observed and simulated monthly mean snow depths. Long-term monthly mean snow depths (1963–1995) for 34 Midwest station observations compared to previous IBIS and current Agro-IBIS simulations in (a) northern (43.5° to 47.5°N lat; −94.0 and −83.0°W lon) and (b) southern (39.5° to 43.49°N lat; −94.0 and −83.0°W lon) areas where observation stations were located. Observations and IBIS-2 results were obtained directly from previous statistical analyses for comparison here (Lenters et al., 2000).

plotted showed improvement over the previous IBIS-2 model validation (slope = 0.61; $r^2 = 0.79$). Both cropping system scenarios resulted in close approximations to mean annual snow depth and simulated interannual variability well (slope = 0.93; $r^2 = 0.88$ for MAIZE+TILL and slope = 0.99), while the POTVEG scenario generally overestimated the observed mean annual snow depth (slope = 1.23), but captured interannual variability in snow depth as well as the cropping system simulations ($r^2 = 0.88$).

Validation of simulated monthly soil temperatures

The MAIZE+TILL simulations had a warm bias compared to the Iowa observations of about 3 to 8°C from March to June (Fig. 3a), which corresponds with the bias of an early spring snowmelt across the northern regions of the study area (Figs. 1, 2). The MAIZE+NOTILL and POTVEG simulations were in much better agreement with observed values across Iowa, although the MAIZE+NOTILL model runs showed a warm bias from September to January, and January is the month when the coldest monthly average temperatures occur (Fig. 3a). The MAIZE+TILL simulation exhibited a small cold bias in the December-February time period of about 0.25 to 3°C.

Simulated soil temperatures at the University of Wisconsin-Madison site showed a similar pattern to the findings in the state of Iowa for the two maize simulations (Fig. 3b). There was a general model warm bias in early spring to early summer of 2–7°C for the MAIZE+TILL simulation (bare soil), very similar to observations from June through December, and a slight cold bias of about 1°C in January. However, the presence of the 5 cm thick residue layer in both the MAIZE+NOTILL and MISCAN+$R_{5\ cm}$ simulations caused soil temperatures to have a cool bias of about 1–6°C during the peak of the summer and then remain warmer than typical observed values from October through February (Fig. 3b).

Simulated 3-day running mean 10 cm soil temperatures at the University of Illinois Energy Farm site for miscanthus without a

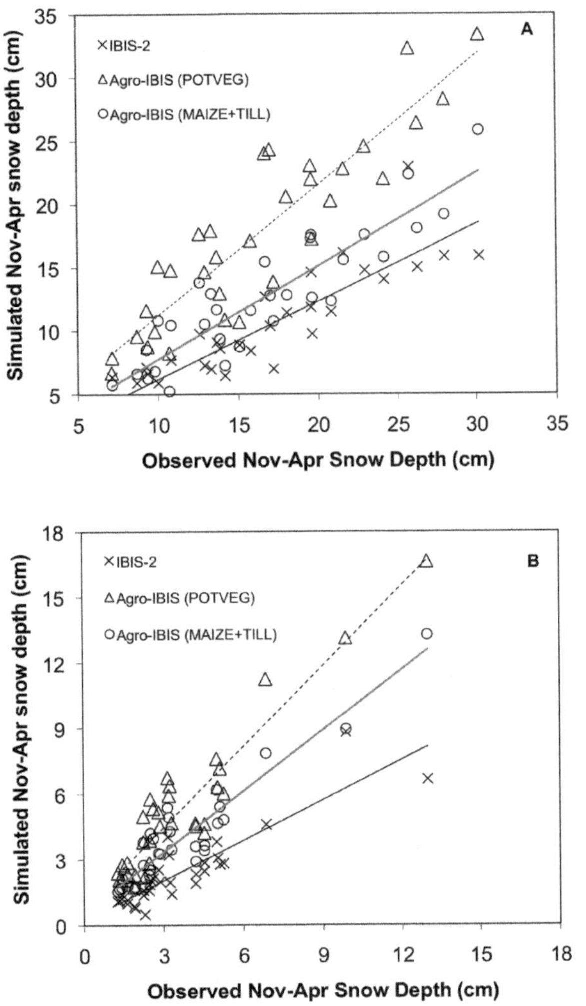

Figure 2. Observed versus simulated interannual variability of mean Nov-Apr snow depth. Comparison of observed versus Agro-IBIS model simulated mean Nov-Apr snow depth for each year from 1963–1995 in (a) northern (43.5° to 47.5°N lat; −94.0 and −83.0°W lon) and (b) southern (39.5° to 43.49°N lat; −94.0 and −83.0°W lon) areas where observation stations were located. Previous IBIS-2 model results (Lenters et al., 2000) are plotted for comparison. Linear regression fits between observations and model results are denoted by the following lines: thin black (IBIS-2); gray solid (MAIZE+TILL); black dashed (POTVEG).

residue layer present compare well with observations (Fig. 4a; $r^2 = 0.92$, P<0.0001) with a slope of 0.99 (S.E. 0.008) and intercept of 0.66 (S.E. 0.126). There were relatively large model errors at the freezing point, however, simulated values are typically within 2.5°C when observed temperatures were less than −1°C. At the monthly time scale, simulated values were ±1.7°C of the observed values for all months except April, and within 1.6°C during the winter months (December, January, February; Fig. 4b). Overall simulated values showed no consistent bias relative to observations, however, there was a slight underestimate (<0.8°C) of monthly mean 10 cm soil temperature during winter months.

Validation of simulated annual extreme soil temperatures

Overall, Agro-IBIS performed exceedingly well for extreme minimum temperatures greater than approximately −6°C, but there was an increasing cold bias in model simulations as average annual observed extreme minimum temperatures decreased from about −6°C to −12°C (Fig. 5a; $r^2 = 0.78$, P<0.0001). For example, the simulated cold bias was about −0.5°C at −6°C (observed temperature), and −2°C at −8°C. The validation exercise suggested that a 2nd degree polynomial model fit best approximated the relationship between simulated and observed average annual extreme minimum 10 cm temperatures (Fig. 5a).

Before comparing simulations to the observed fraction of years below the −3.5°C and −6°C thresholds, simulated annual extreme minimum soil temperatures were adjusted based on the regression analysis from the validation exercise (Fig. 5a). The simulated annual average minimum 10 cm soil temperatures had the mean bias removed by adjusting simulated values using the mean response relationship (2nd degree polynomial) between observed and simulated quantities and its numerical deviation from a linear relationship with a slope = 1.0. The model corrected data, also shown in Fig. 5a, exhibited greatly improved agreement, particularly at the coldest soil temperatures ($r^2 = 0.77$, P<0.0001, slope = 0.947). This statistical adjustment process did not significantly affect our ability to capture the soil temperature variability among observational sites, denoted by the similar r^2 values (0.78 vs. 0.77) for the original and corrected model fits (Fig. 5a). The linear model fit through the observed data and model simulated output for frequency of occurrence of −3.5°C temperatures (Fig. 5b) resulted in an $r^2 = 0.76$ (P<0.0001), with a slope of 0.86 (S.E. 0.043) and intercept of 0.046 (S.E. 0.029). For the frequency of occurrence of −6.0°C temperatures across the region (Fig. 5c), the linear model fit through the observed data and model simulated output resulted in an $r^2 = 0.76$ (P<0.0001), with a slope of 0.72 (S.E. 0.036) and intercept of 0.067 (S.E. 0.018).

Average annual extreme minimum 10 cm soil temperatures: 1978–2007

Analysis of the average annual extreme minimum soil temperatures at a 10 cm depth for the control bare soil case (MAIZE+TILL) model simulation suggest that an absence of any residue layer after fall crop harvest would result in the majority of the Midwest region commonly experiencing temperatures below 0.0°C each year (Fig. 6a). The −3.5°C and −6°C thresholds are generally reached over smaller regions of the Upper Midwest, but are still present in core areas of the Corn Belt. The coldest average annual extreme soil temperatures at 10 cm are in the −8°C to −11°C range confined to large portions of North Dakota and Minnesota (Fig. 6a). However, the MAIZE+NOTILL simulations (Fig. 6b) suggest that a 5 cm thick, continuous cover of maize residue helps to provide a widespread insulating effect on minimum soil temperatures. Most of the region experienced a warming of the extreme minimum temperatures from 2.5°C to 6°C for MAIZE+NOTILL compared to the bare soil (MAIZE+–TILL) simulation. However, this analysis suggests regions that typically experience greater snowfall in the far northern portions (e.g., North Dakota and Minnesota), the upper peninsula of Michigan, and on the eastern side of lake Michigan, would not warm as much during the winter with a persistent 5 cm thick residue layer. The areas that saw the greatest warming impact of residue were located in the central portions of the Midwest, across northern Iowa, eastern South Dakota, southern and central Minnesota, and much of Wisconsin.

Analysis of the average annual extreme minimum soil temperatures at a 10 cm depth for the series of miscanthus simulations with varied residue thicknesses (1, 2.5, and 5 cm) suggest that as residue thicknesses increase, the magnitude of the insulating effect on annual minimum soil temperatures also increases (Figs. 6c–6e). The residue simulations for the 5 cm thick residue layer for

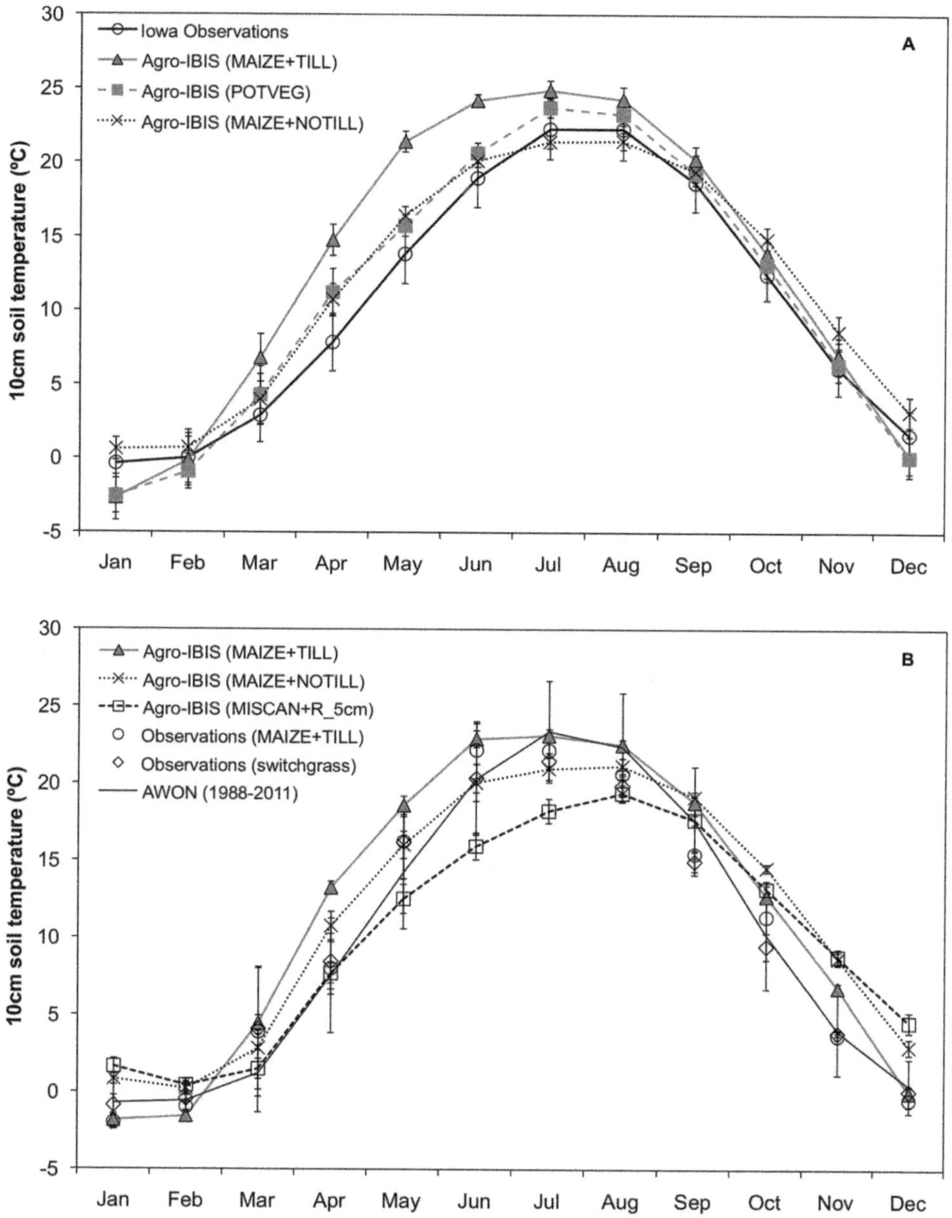

Figure 3. Comparison of observed and simulated soil temperatures in Iowa and Wisconsin. (a) Comparison of monthly average 10 cm soil temperatures for three Agro-IBIS model simulations with observational station data from Hu and Feng (2003) averaged over the state of Iowa for 1982–2002. Error bars are ±1 S.E. for both simulated and observed values. (b) Comparison of monthly average 10 cm soil temperatures for three Agro-IBIS models simulations compared with observational data at the University of Wisconsin-Madison Arlington Agricultural Research Station in tilled maize and switchgrass study plots for July 2009- June 2012. Observed data (maize and switchgrass) represent the monthly mean among 3 replicate plots (n = 3). Long-term averages (1988–2011) of monthly mean 10 cm soil temperature collected at the UW-Madison Automated Weather Observing Network (AWON) site at Arlington are plotted for comparison. Error bars are ±1 S.E. for all simulated and observed values.

miscanthus (MISCAN+$R_{5\ cm}$; Fig. 6e) suggest that miscanthus straw has a slightly increased insulating effect than maize stover, given different biophysical properties (Table 2). Regionally, the largest difference between miscanthus straw and maize stover was in southern portions, with a ~2°C increase in the extreme minimum temperatures while differences in the far northwest portion were typically less than 0.5°C (Fig. 6f). The impacts of a 1 cm thick miscanthus straw layer, compared to the bare soil scenario, were minimal over northwest portions of the region, where no warming of minimum soil temperatures occurred (Fig. 7a). More widespread warming of minimum soil temperatures of 0.5°C to as much as 6°C occurred with miscanthus residue thicknesses of 2.5 cm (Fig. 7b) and 5 cm (Fig. 7c), respectively, compared to the bare soil case. However, even in the 2.5 cm scenario, the magnitude of warming was minimal across far northwestern portions of the region (0°C to only 0.5°C), as well as areas in central Wisconsin and northcentral Nebraska that had soils with higher sand content (Fig. 7b). Thus, many areas still had annual average minimum 10 cm soil temperatures that were below −3.5°C and −6.0°C when miscanthus residue was less than or equal 2.5 cm thick (Figs. 6c,d). The largest magnitude of warming associated with both 1 cm and 2.5 cm miscanthus

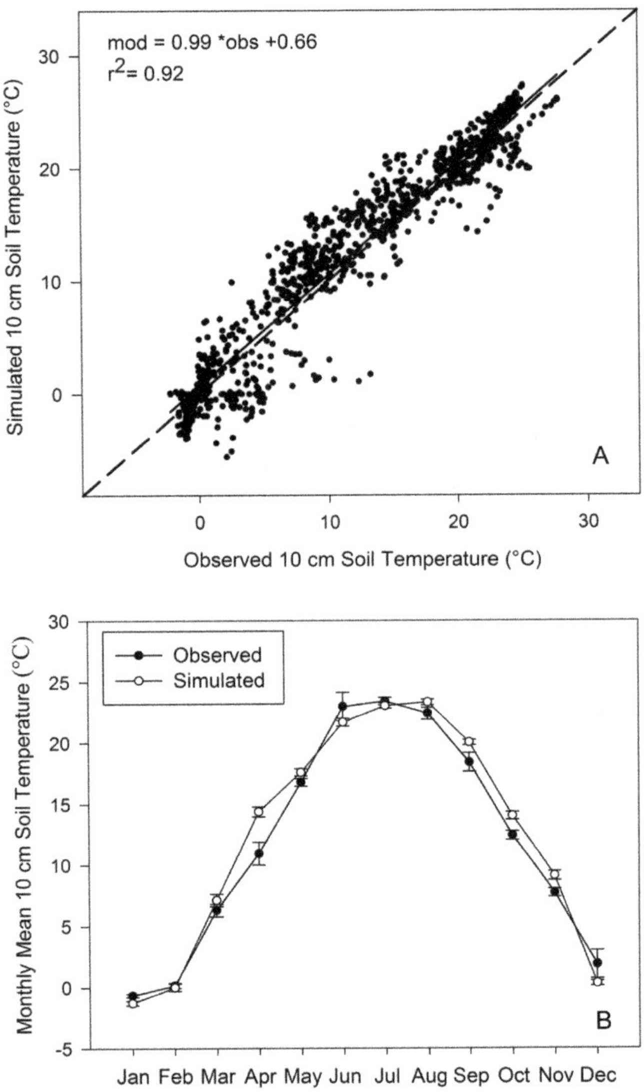

Figure 4. Comparison of observed and simulated soil temperatures in Illinois. (a) Comparison of 3-day running mean 10 cm soil temperatures for Agro-IBIS model simulations of miscanthus with no residue layer with observed values at the University of Illinois Energy Farm from 2009 to 2011. Observed values are the daily mean (n = 1 to 5) of observation for the miscanthus plots, n was less than five for periods when sensors were damaged, with n at least 3 for 80% of the days. (b) Comparison of monthly mean 10 cm soil temperatures for the same site and simulation. Data points are the mean (n = 3) of the 2009–20011 monthly values, error bars are ±1 S.E. for both simulated and observed values.

residue thicknesses was focused in a corridor from eastern Nebraska, through Iowa, southern Minnesota, northern Illinois and southern Wisconsin.

Expanded investigation of varied miscanthus residue depth on minimum soil temperatures

The warming of annual average minimum temperatures with 20 cm of miscanthus straw residue could be as little as approximately 4°C, to as great as 12°C depending on geographic location (Fig. 8). Although the soil warming effects were maximized for a residue thickness of around 20 cm, the shape of the response curves suggests that even greater warming could occur with greater thicknesses regardless of location throughout the Midwest US (Fig. 8). However, we reiterate that residue thicknesses greater than approximately 10 cm are extreme scenarios and not realistic management options in the field. The 5 cm and 10 cm thick residue layers for miscanthus produced about 60% and 84%, respectively, of the warming benefit associated with the thickest residue cover simulated. Additionally, the interannual variability of the coldest 10 cm soil temperatures was reduced as residue thickness increased (Fig. 8).

Likelihood of reaching critical minimum soil temperatures: 1978–2007

In the bare soil control simulation (MAIZE+TILL), large portions of the upper Midwest would have experienced 10 cm soil temperatures below −3.5°C in 75–95% of the years, and the risk of those temperatures being reached is not completely eliminated unless fields are located in far southern regions (Fig. 9a). The risk for these cold temperatures is reduced for the three scenarios of varied miscanthus straw thicknesses, with a corresponding relationship between the magnitude of reduced probabilities and residue thickness (Figs. 9b–d). In these simulations, the likelihood of reaching −3.5°C was considerably lower over southern and eastern portions of the Midwest US, but across the Dakotas as well as portions of Nebraska, Minnesota, and Wisconsin, about 40–80% of years had 10 cm soil temperatures reaching −3.5°C even with a 5 cm residue layer (Fig. 8d). While the effectiveness of a 1 cm thick miscanthus residue layer was much lower, even this small amount greatly reduced the probability of reaching −3.5°C across large portions of Kansas, Missouri, Illinois, Indiana, and Ohio (Fig. 9b).

In the bare soil control simulation (MAIZE+TILL), smaller portions of the upper Midwest would have experienced 10 cm soil temperatures below −6.0°C in about 75% of the years and large sections of the Dakotas, Minnesota, and Wisconsin, reached this threshold in 50 to 60% of the years (Fig. 10a). The risk for these very cold temperatures was greatly reduced, but not completely eliminated, in the three miscanthus simulations (Figs. 10b–d). Similar to the results presented in Figure 9, the likelihood of reaching the −6.0°C threshold is reduced the most with a 5 cm residue layer (Fig. 10c), and the least in the 1 cm thickness scenario (Fig. 10b). However, even in the 5 cm simulation, the −6.0°C threshold was reached in 60–90% of all years in far northern regions of North Dakota and Minnesota (Fig. 10d).

Impact of varied miscanthus residue albedo and bulk density on minimum soil temperatures

In general, variations in residue albedo or bulk density had small impacts on the simulated annual minimum soil temperatures. For residue thicknesses of 1 cm and 5 cm, a change in residue albedo from 0.15 to 0.5 contributed to minimum soil temperatures at 10 cm that were 0.3°C to 0.6°C colder. Over a more realistic range of likely albedo values (0.2 to 0.4), the contribution was only 0.1 to 0.2°C, or about 5–10% of the mean annual minimum 10 cm soil temperatures for the two residue thicknesses simulated in this sensitivity study. Bulk density values were varied, with particle (cellulose) density held fixed, to generate a range in porosity of the 1 cm and 5 cm layers from 0.5 to 0.99. Here, the net effect on minimum soil temperatures was even less than for albedo, contributing to a net change of 0.01–0.05°C.

Soil temperature trends

Our regionally averaged results for soil temperature trends (e.g., a spatial average for the entire study region) produced values equal

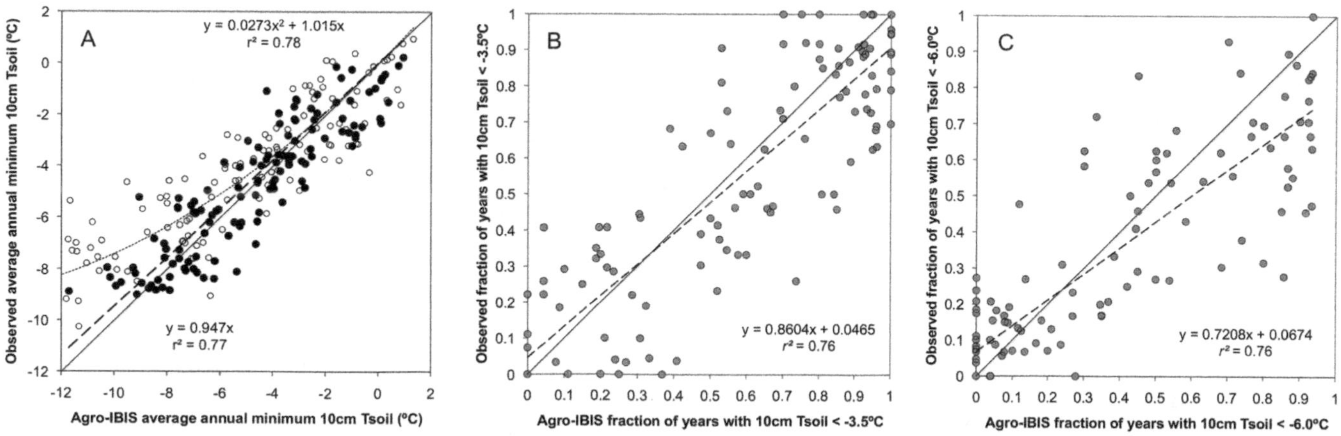

Figure 5. Comparison of observed and simulated annual extreme minimum 10 cm soil temperatures. (a) Agro-IBIS average annual extreme minimum 10 cm soil temperature (based on 3-day running mean temperatures) compared with observations from 125 observation sites (SI, Table S1) from the Midwest USA for original model values (open circles and dotted regression line), and statistically adjusted model values (filled circles and dashed regression line); (b) Agro-IBIS simulated frequency of occurrence of 10 cm soil temperatures (3-day running mean) reaching −3.5°C and (c) −6.0°C, respectively, compared with results from 125 observation sites in the Midwest USA. In these comparisons, Agro-IBIS was only simulated for a period from the beginning year that data was available for each observation station through 2007, which denotes the last year that gridded daily climate data was available.

to 0.250°C $(10\ \text{yr})^{-1}$ at 10 cm, 0.252°C $(10\ \text{yr})^{-1}$ at 60 cm, and 0.253°C $(10\ \text{yr})^{-1}$ at 100 cm during 1967–2002 for the MAIZE+TILL simulations (data not shown). We also compared observed trends in the annual extreme minimum 10 cm soil temperatures for station observations in our study (SI, Table S1) with the average simulated response at a subset of those sites that had continuous records of at least 27 years. The observed trend in annual extreme minimum 10 cm soil temperature, averaged across all 36 stations, was 0.88°C $(10\ \text{yr})^{-1}$ (S.D. 0.097) compared to 0.78°C $(10\ \text{yr})^{-1}$ (S.D. 0.030) for Agro-IBIS simulations.

The overall trend (°C per 60 years) in 10 cm annual average soil temperature varied widely from 1948–2007 (Fig. 11a). Some regions of the Midwest experienced significant cooling in contrast to warming of about 1–3°C. For the entire 1948–2007 period our analysis produced regionally averaged trends that were much lower than those for the 1967–2002 period; 0.059°C $(10\ \text{yr})^{-1}$ at 10 cm; 0.052°C $(10\ \text{yr})^{-1}$ at 60 cm, and 0.049°C $(10\ \text{yr})^{-1}$ at 100 cm. There are similar spatial patterns in the trends in the annual extreme minimum 10 cm soil temperatures from 1948–2007, with the most significant warming of about 2–3°C having taken place in the northern Plains (Fig 11b). However, a dramatically different spatial pattern emerges when the analysis of trends is limited to the 1981–2007 time period (Fig 11c). Here, southern regions appear to have experienced the most significant warming up to 3–4°C of change in the annual extreme minimum temperatures, with a smaller magnitude of change across more northern states. The regionally averaged annual average 10 cm soil temperature trend was 0.33°C $(10\ \text{yr})^{-1}$ for 1981–2007, and the spatial patterns of change across the Midwest were similar to trends in extreme minimum temperatures.

Discussion

This study has described a dataset for annual extreme minimum temperatures across the Midwest US for both soils that are managed without leaving residue on the soil surface after harvest of crops as well as for soils that have a full cover, 5 cm thick prostrate residue layer for maize in place, and for varying thicknesses of miscanthus straw. Our results indicated that strategic residue management in the region has potential to help increase extreme minimum soil temperatures that can threaten overwintering of miscanthus rhizomes in the first year of establishment and potentially in later years. Based on previous research that has investigated the ability of winter wheat to survive extremely cold winters in the central US [60], it appears that a combination of leaving behind standing stubble after harvest to preferentially trap snow, coupled with a prostate residue layer, offers the highest likelihood of insulating soils and to increase the odds that miscanthus rhizomes can survive the first winter after establishment. However, as our results have shown, the thickness of that residue layer has significant bearing on the insulating effect.

Influence of residue, soils, snowpack, and management on minimum extreme soil temperatures

The Midwest US is subjected each year to rapid and extreme temperature changes. However, due to the large presence of human management of agricultural lands, as well as differences in the timing of snowfall and the buildup of a consistent snow cover each year, air temperatures should not be perceived as a guide for estimating extreme minimum soil temperatures near the surface. For example, while air temperatures across regions of the Midwest can drop to −25°C to −40°C during the winter for extended periods of time, the absolute coldest soil temperatures at 10 cm were about −12° to −16°C, based on both observed data and simulated results for individual years in our study. According to observations in the region over the past 30 years, the average 10 cm minimum soil temperatures are in the range of −8°C to −10°C (Fig. 5a), which is much warmer than the typical average low air temperatures. This is attributed to the insulating effect of snow cover. The timing and duration of a consistent snow pack is highly influential in determining minimum soil temperatures in mid-January to early March, which is historically when soils reach their lowest temperatures [22,51,61].

Agro-IBIS simulations suggested that miscanthus straw could function better as an insulator than maize residue for a comparable 5 cm thick layer on the soil surface, which is largely attributed to differences in biophysical properties (Table 2). However, both plant residue types keep soils warmer during

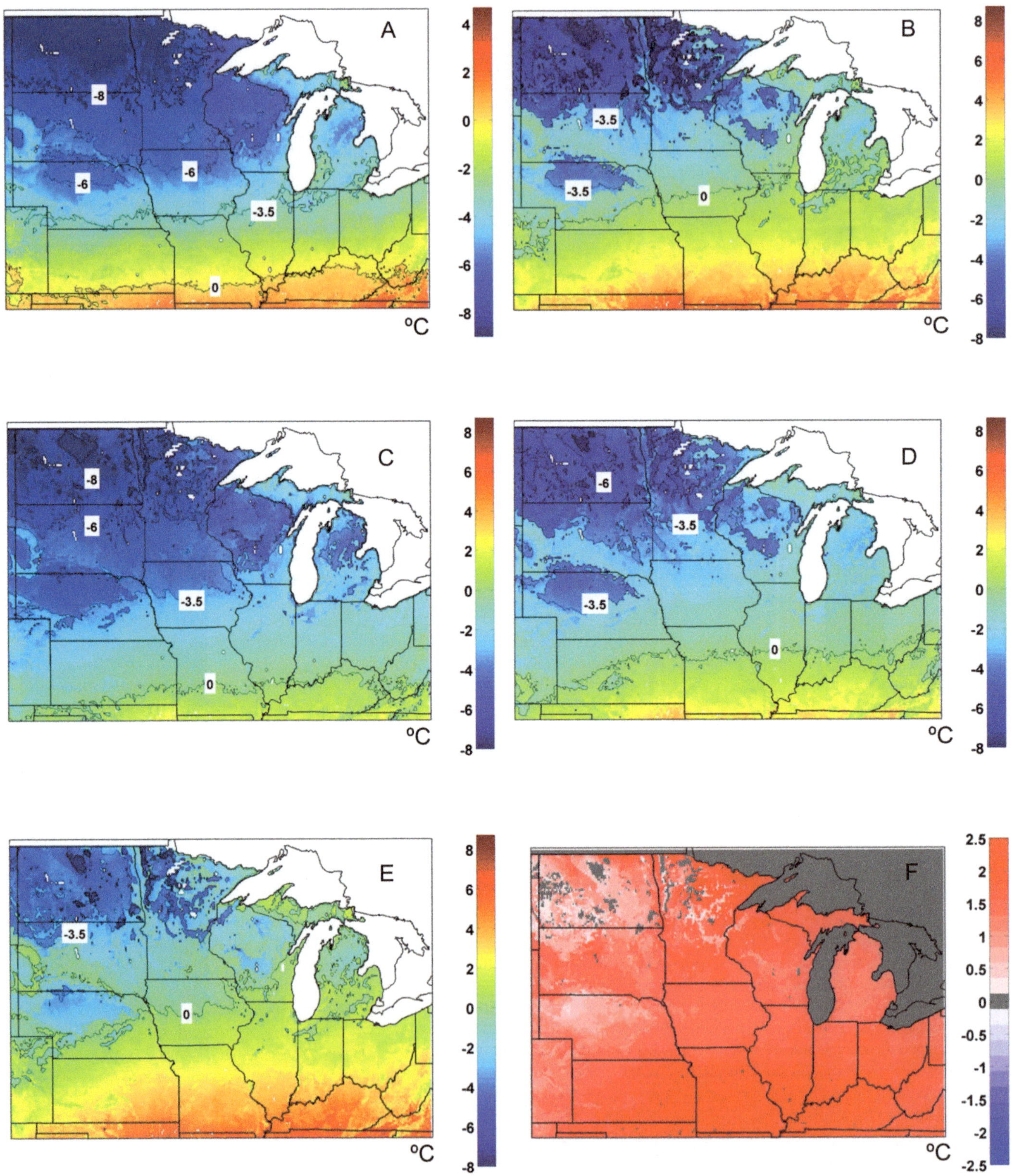

Figure 6. Impacts of soil surface residue management on annual average extreme minimum 10 cm soil temperatures. Average annual (1978–2007) extreme minimum 10 cm soil temperatures (based on 3-day running mean) for (a) MAIZE+TILL (bare soil post harvest), (b) MAIZE+NOTILL, (c) MISCAN+$R_{1\ cm}$, (d) MISCAN+$R_{2.5\ cm}$, and (e) MISCAN+$R_{5\ cm}$ simulations. The position of the 0.0°C, −3.5°C, −6.0°C, and −8.0°C 10 cm soil temperature isopleths are highlighted by labels on solid black lines; (f) differences in average annual extreme minimum 10 cm soil temperatures for MISCAN+$R_{5\ cm}$ – MAIZE+NOTILL (Fig. 6e–Fig. 6b results).

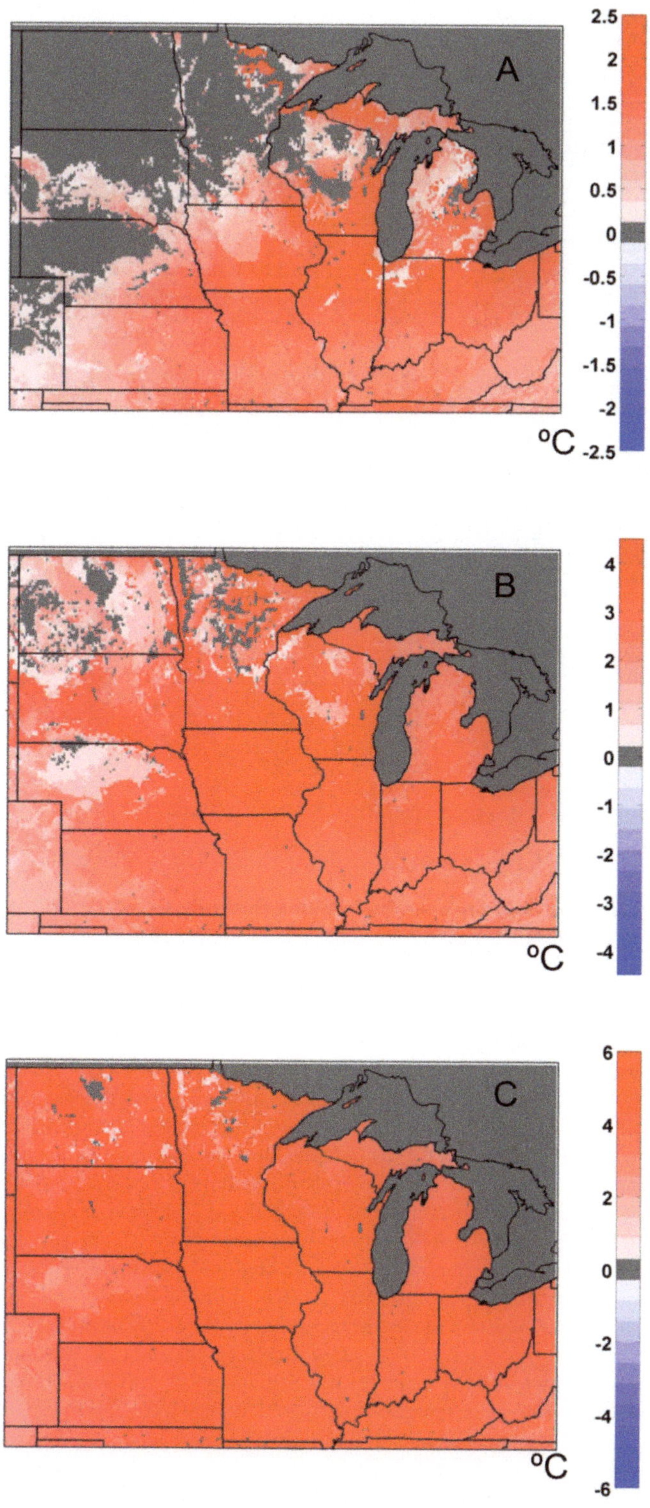

Figure 7. Soil temperature differences between different residue layer thicknesses. Average annual (1978–2007) extreme minimum 10 cm soil temperature differences, based on a 3-day running mean, for the following paired simulations: (a) MISCAN+$R_{1\ cm}$ minus MAIZE+TILL (bare soil post harvest), (b) MISCAN+$R_{2.5\ cm}$ minus MAIZE+TILL, (c) MISCAN+$R_{5\ cm}$ minus MAIZE+TILL.

winter, thereby decreasing the risk of 10 cm soil temperatures going below −3.5°C over a large part of the Midwest. However, for decreased thicknesses of miscanthus residue (1 and 2.5 cm), the

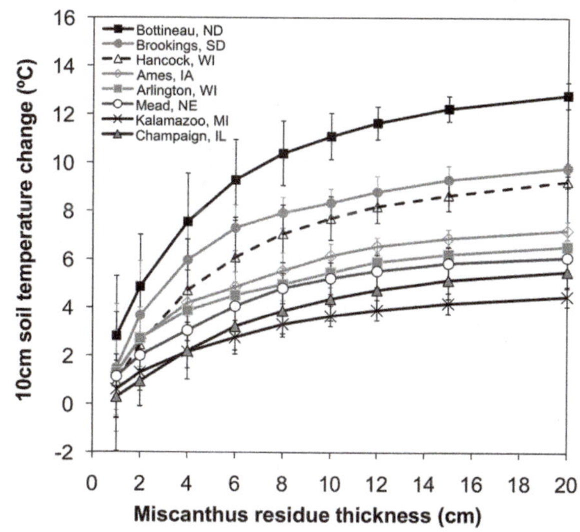

Figure 8. Impacts of miscanthus residue thickness on annual average extreme 10 cm soil temperatures. Average annual (1978–2007) extreme minimum 10 cm soil temperature changes at eight locations in the Midwest, based on a 3-day running mean, for MISCAN+R simulations with varied residue thicknesses relative to the MAIZE+TILL (bare soil) simulation.

warming effect is considerably less, and in some locations across the Midwest, negligible. Our expanded sensitivity tests at eight locations concerning the impacts of a wide range of residue thickness on the magnitude of warming of the minimum soil temperatures suggested that 60% of the maximum warming benefit occurs with a 5 cm thick layer, and 84% with a 10 cm thick layer (Fig. 8). For 1 cm and 2.5 cm thick layers, only 15% and 40% of the maximum warming benefit occurs. While thicker residue layers clearly increase the warming benefit, data from field experiments (Table 2) suggest that a nominal thickness of only a few centimeters is likely easier to sustain across large fields given issues related to wind blowing loose residue around. We emphasize that the 20 cm residue thickness is an extreme scenario that was created to understand system behavior, and is not suggested as a recommended management practice.

The simulated soil warming attributed to residue in this study is in agreement with previous studies for an experiment in Minnesota that examined the effect of different maize residue management options on soil temperatures in the 0.05 to 0.3 m soil layer. Sharratt et al. [22] reported that wintertime minimum soil temperatures at a 1 cm depth were 5 to 8°C warmer over a three year period for a management scenario that left 60 cm of stubble standing in combination with residue laying prostrate on the soil surface compared to a treatment that had 0 cm stubble with all residue removed from the soil surface. While the magnitude of the warming effect associated with residue was greater in Sharratt et al. [22] then in our study, their temperatures were reported for a 1 cm depth. They also found that snow depth was influenced by the residue treatments, whereby fields that had stalks cut closer to the ground (e.g., 30 cm compared to 60 cm) had a lower average snow depth during the winter.

However, not all Midwest regions may see a significant soil warming from straw or maize residue due to the confounding influence of snow cover on winter soil temperatures. Model results illustrated that even with the addition of a 5 cm thick residue layer of either miscanthus straw or maize residue, this had minimal effects on the extreme minimum temperatures in small portions of

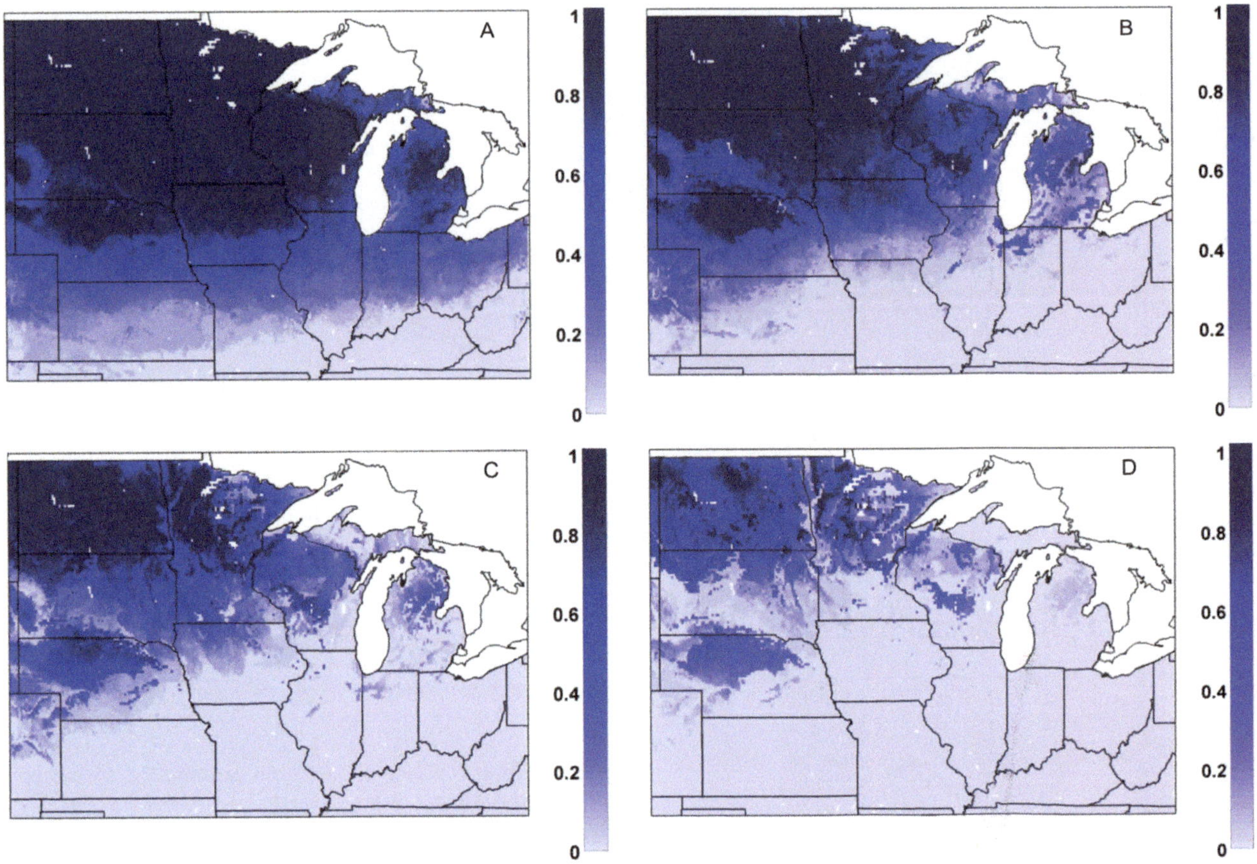

Figure 9. Frequency of 10 cm soil temperatures reaching −3.5°C or colder for varied miscanthus residue thicknesses. Fraction of total years during the 1978–2007 time period that simulated annual 10 cm soil temperatures were at or below a −3.5°C threshold (based on a 3-day running mean) for (a) MAIZE+TILL (bare soil), (b) MISCAN+$R_{1\ cm}$, (c) MISCAN+$R_{2.5\ cm}$, and (d) MISCAN+$R_{5\ cm}$ simulations.

North Dakota and Minnesota, and to a lesser extent the sandier soils of north and central Nebraska and the central sands region of Wisconsin (Figs. 6–7). This may be attributed to two reasons: first, across the most northern regions of the Midwest, snowfall comes earlier, and due to colder temperatures, snowpack depth and duration is generally greater and longer, respectively, than in more southern locations [62]. Sharratt et al. [63] suggested that approximately 15–42 cm of snow cover is needed to insulate the top portion of the soil profile from cold wintertime temperatures, resulting in near steady-state soil temperatures. Thus, the potential impact of a residue layer may be minimized in regions that historically have deeper and more consistent snow cover (e.g., North Dakota and northern Minnesota, as well as lake effect snow regions in Michigan), which might explain the simulated spatial patterns and results in our study. However, we note that in northern locations that typically have a significant, consistent snowpack during winter (e.g., >15 cm), observed and simulated 10 cm soil temperatures still reached well below 0.0°C; thus the timing of a building and retreating snowpack is also crucial, besides the thickness, in determining extreme minimum soil temperatures. Second, some of the spatial patterns of soil temperature change attributed to residue management suggest that soil texture plays an important role also in determining the magnitude of extreme minimum soil temperatures. In sandy soil regions, we hypothesize that a thicker residue layer may add a more prominent insulating effect on these soils that lose heat more rapidly due to their lower average volumetric water content in fall and inherent mineral properties [29].

These results suggest that in the first years of establishment of miscanthus, a soil surface residue layer could increase the probability of successful overwintering of the plant rhizomes, but the thickness of that layer is highly deterministic to the overall soil warming. This management option, coupled with leaving standing stubble that could preferentially trap snow, would likely provide the greatest likelihood of maximizing soil warming [22,51]. However, during the first year or two of establishment when *M.* × *giganteus* might be the most susceptible to winterkill or damage to rhizomes, the amount of biomass produced may not be sufficient to support a residue layer thickness that significantly reduces the risk of lethal soil temperatures. Several studies from the literature suggest that *M.* × *giganteus* will take at least three years to reach the expected yield ceilings, and during the first year, annual productivity can typically be in the 1–4 Mg ha^{-1} range [4,15,64–66]. With a typical residue bulk density of 22 kg m^{-3} (Table 2), 2.2 Mg ha^{-1} of aboveground biomass is needed for each 1 cm of thatch depth. Thus, to achieve a 5 cm thatch thickness, approximately 11 Mg ha^{-1} of aboveground biomass would be required; these values may not be observed until year three and beyond [50,55,67]. Given these results, a producer might be faced with a new dilemma in the context of residue management. For example, a farmer may not have enough miscanthus biomass to sell to make a profit after the first year, so they would probably mow the crop. In the next two years, they will have to hedge the

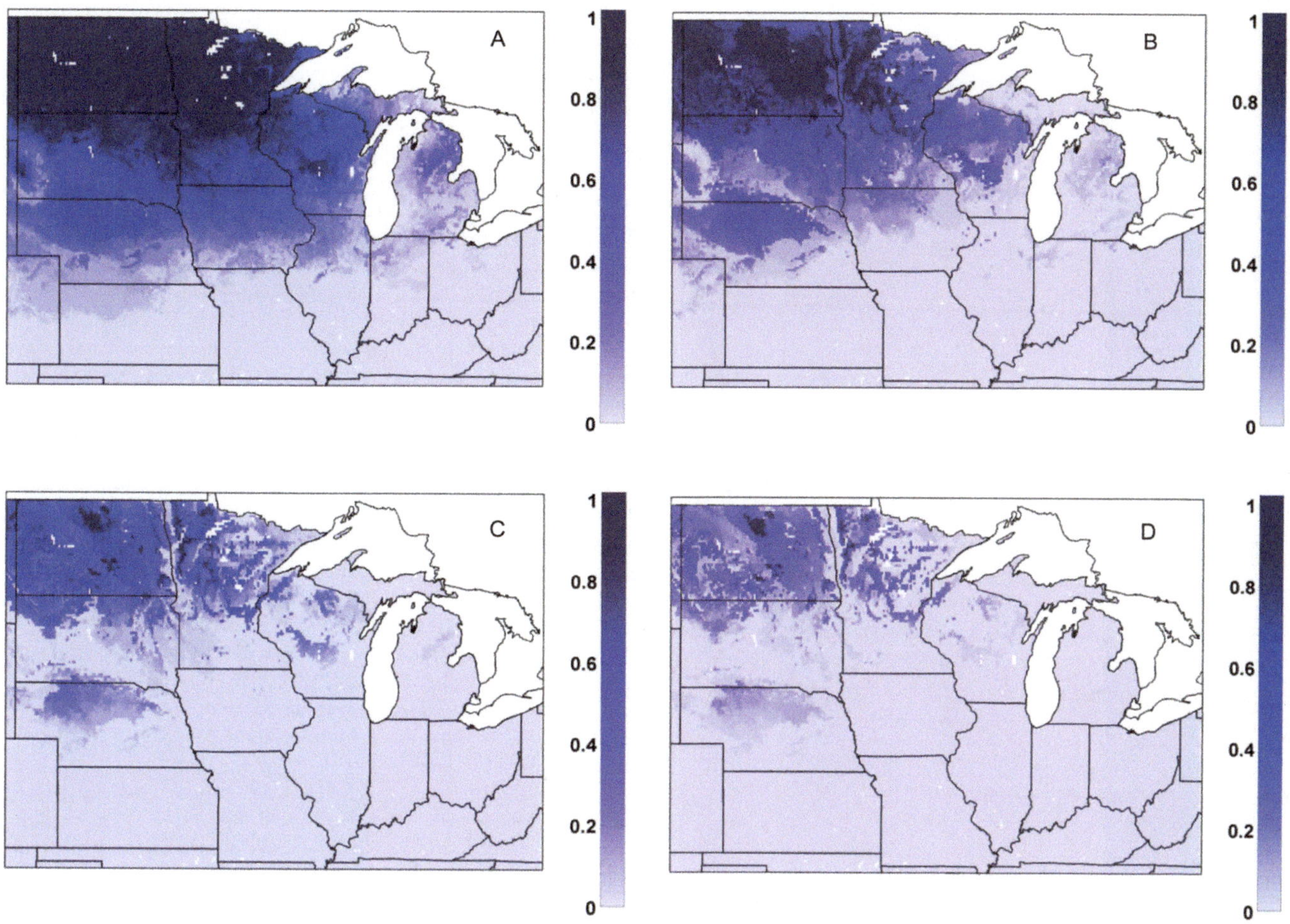

Figure 10. Frequency of 10 cm soil temperatures reaching −6.0°C or colder for varied miscanthus residue thicknesses. Fraction of total years during the 1978–2007 time period that simulated annual 10 cm soil temperatures were at or below a −6.0°C threshold (based on a 3-day running mean) for (a) MAIZE+TILL (bare soil), (b) MISCAN+$R_{1\ cm}$, (c) MISCAN+$R_{2.5\ cm}$, and (d) MISCAN+$R_{5\ cm}$ simulations.

risk of maximizing profits vs. minimizing the cold temperature threat. Furthermore, they may face a more complicated decision on what to do with available residue from nearby corn fields; should it be used to build a solid residue layer to protect rhizomes?

Additional studies that quantify the relationship between miscanthus productivity, residue bulk density, and thatch thickness during early establishment years will help determine whether additional sources of crop residue from nearby fields (e.g., maize) are needed to build a more substantial protective layer to reduce the odds of winterkill. The research presented here showed that very small differences in thatch thickness lead to significant differences in minimum winter soil temperatures. Therefore, field trials should continue to be established that strategically manage residue in varied amounts on fields as well as investigate how stubble height and snow depth variations impact soil temperatures over several years to account for interannual variability. These data would also prove useful to help identify an optimal growing region in the US and Canada for miscanthus. Currently, there is also a lack of biophysical information on the properties of miscanthus straw (e.g., bulk density, albedo, heat capacity), which is needed to better constrain the parameterization of agroecosystem models.

Soil temperature trends: are soils warming or cooling?

This study, as well as three other studies, have documented long-term changes in soil temperatures but arrived at different conclusions [54,68,69]. Using a limited number (38) of observation stations in the US that had a period of record from 1967–2002, Hu and Feng [54] reported that annual average 10 cm and 100 cm soil temperatures across these sites were increasing at a rate of 0.3°C $(10\ yr)^{-1}$, and the sites that had the greatest rate of warming were across northern regions. Sinha et al. [69] also reported warming soil temperatures at 10 cm during 1967–2006 in regions of Minnesota, Illinois, and Indiana, as well as a reduction in the number of days with soil frost in the Midwest. In contrast, Isard et al. [68] used a biophysical modeling approach and reported that even though wintertime air temperatures from 1951–2000 were increasing, wintertime soil temperatures at 50 cm depth were decreasing across the Great Lakes region, likely due to thinning and more variable snowpacks. However, a study by Dyer and Mote [62] suggested that minimal changes in North American snow depth has occurred in the November through January period from 1960–2000, but noted an earlier onset and acceleration of spring snowmelt in the March and April timeframe. While the long-term observational data presented in Hu and Feng [54] do not corroborate a reported trend of decreasing soil temperatures across Wisconsin and Michigan by Isard et al. [68], there were no

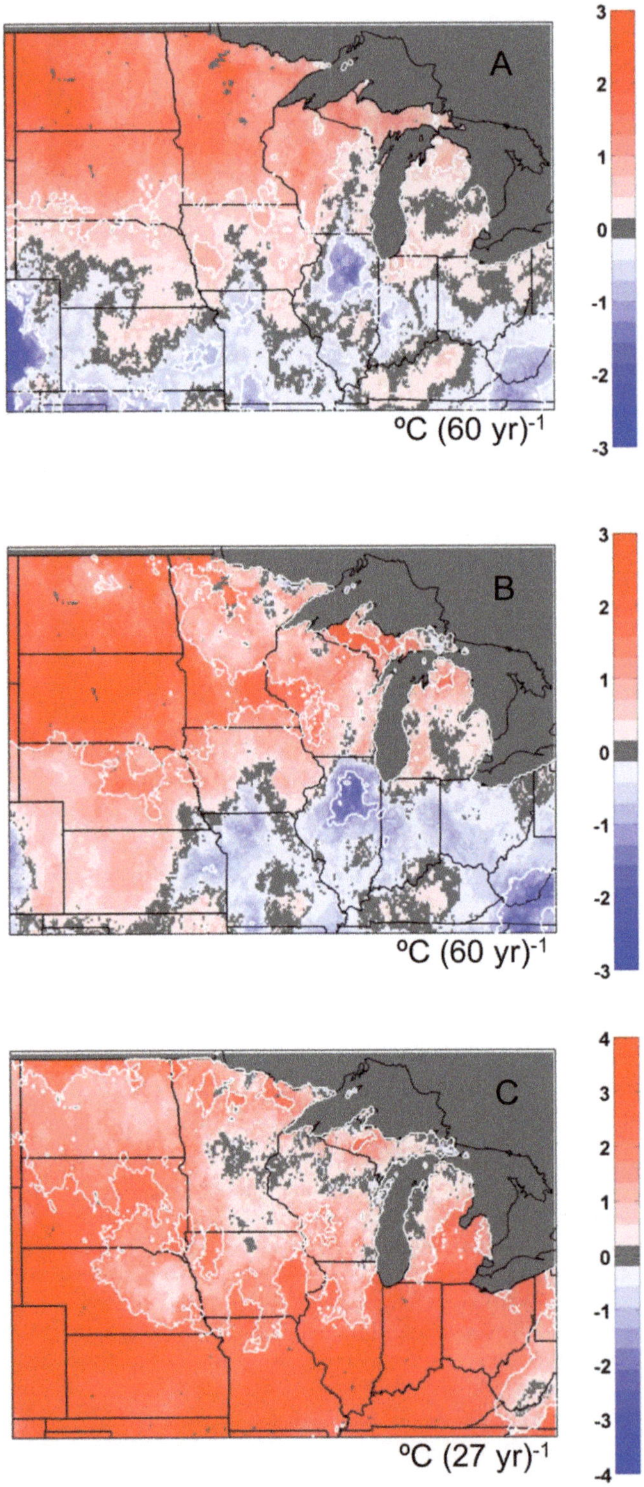

Figure 11. Simulated soil temperature trends across the Midwest US. (a) Total change (from linear regression) in annual average 10 cm soil temperatures for 1948–2007 for the MAIZE+TILL (bare soil) simulation; (b) total change (from linear regression) in annual extreme minimum 10 cm soil temperature for the MAIZE+TILL (bare soil) simulation from 1948–2007; (c) total change (from linear regression) in annual extreme minimum 10 cm soil temperature for the MAIZE+TILL (bare soil) simulation from 1981–2007. Regions bounded by solid white lines indicate trends with $P<0.05$.

observation stations across Wisconsin and Michigan included in the aforementioned analysis.

In our study, depending on the initial year for the calculation of linear trends (e.g., Fig. 11b compared to Fig. 11c), the magnitude of 10 cm soil temperature trends varied significantly across the Midwest. As with any type of linear trend analysis over time, the time period of choice can have a significant influence on the results. Agro-IBIS results for the 1948–2007 time period (Fig. 11b) illustrated a reduced warming signal, or no change in extreme minimum soil temperatures, across eastern Wisconsin and central lower Michigan, but are not necessarily indicative of a widespread cooling trend as suggested by Isard et al. [68]. We also analyzed Agro-IBIS extreme minimum soil temperature trends for the 50–60 cm soil depth, and did not find a significant difference from trends occurring at 10 cm (Fig. 11b). However, Agro-IBIS depicted several large areas of Missouri, Illinois, Indiana, and Ohio that exhibiting a cooling trend of annual average 10 cm soil temperatures (Fig. 11a), as well as the extreme minimum values (Fig. 11b) over a longer timeframe. Based on Agro-IBIS and soil temperature observation trends that agree, on average, for the 1981–2007 period for 36 station locations (SI, Table S1), we conclude that the coldest wintertime soil temperatures at 10 cm have been warming at a rate of approximately 0.8°C $(10\ \text{yr})^{-1}$ to 0.9°C $(10\ \text{yr})^{-1}$ over the last several decades. There is evidence to suggest that significant warming of soils and a reduction in soil frost would continue in the future, based on the modeling study of Sinha et al. [70]. They used future climate scenarios for the mid (2040–2069) and late (2070–2099) 21st century to drive a macroscale land surface model and found that increased wintertime soil temperatures, increased frequency of freeze-thaw cycles, and a reduction in soil frost days across the region would occur with continued climate change.

Challenges for ecosystem models

The model showed an overall good ability to simulate the dynamics of 10 cm soil temperatures observed at a number of locations in the Midwest (Figures 3–4), supporting the accuracy of our model estimates. The differences between observed soil temperatures in the tilled maize and switchgrass plots at the Arlington, WI site were attributed to a lack of an established residue layer in switchgrass experimental plots attributed to harvest of the majority of aboveground vegetation for three straight years. The simulation of interannual variability of annual extreme minimum temperatures at 10 cm proved more difficult than reproducing the average annual extreme soil temperatures at 10 cm. When we compared the frequency of occurrence of −3.5°C and −6.0°C soil temperatures with observations across the region (Figs. 5b,c), the model fit was not as good as we found when comparing the average annual extreme minimum temperatures (Fig. 5a). This result may be attributed to difficulties in simulating snow cover dynamics and the ability to capture the physical properties (i.e. density, compaction, water content) of snowpack on any particular day. Many ecosystem models currently do not have complicated dynamics such as the ability of standing vegetation or crop stubble to preferentially capture snow and lead to a deeper and longer duration of snow depth, which could lead to an accentuated insulating effect, higher soil temperatures, and decreased frost penetration [51,71] and alter the surface albedo [72]. However, this is an area of great potential for future model improvement if more data are collected in a variety of land management settings. We also understand that physical processes in soils such as freeze-thaw cycles, and soil ice and frost formation can influence soil structure and infiltration [52,53,73], which ultimately affect heat transfer, and make soil temperature prediction particularly challenging at short timescales. Incorporation of these factors or refinement of current modeling approaches as more data become available should also improve ecosystem model realism.

While simulated snow depth agreed quite well with observations from November through March for the POTVEG simulation in the northern portion of the study region (Figs. 1a, 2a), snowmelt occurred more rapidly in March compared to observations when the model was parameterized with maize (row crops) or miscanthus (grasses). Additionally, the seasonal timing of the maximum snow depth occurred one month earlier than observations in those scenarios. Across the southern region (Fig. 1b, 2b), simulated snow depth agreed quite well with observations across all months and years for maize as prescribed vegetation, whereas the POTVEG yielded the poorest comparison with observations.

The connection between snowpack, the timing of snowmelt, and simulated soil temperatures (Fig. 3) was obvious. When comparing simulated soil temperatures to observations, the scenarios that yielded lower than observed snow depth and an earlier occurrence of snowmelt in March-April produced soil temperatures that were 5–10°C warmer than observed values during the late winter and spring (Fig. 3a). Thus, we conclude that accurate simulation of snowpack and snowmelt in ecosystem models is a crucial, but potentially overlooked, step towards simulating an accurate portrayal of changing energy balance at the soil surface when transitioning from winter to spring. This transition period typically coincides with rapid changes in ecosystem processes (e.g., net ecosystem exchange, ET), and the timing of warming and drying of soils impacts farmer management across the Corn Belt [74]. Subsequently, land cover and management choices further influence plant phenology, leaf area index, and canopy architecture, which all play important roles on changing radiation interception and energy balance [21,38,52,61].

However, not all of the simulated error should probably be attributed to the model. When taking a closer examination of where and when the model performed well, we note that in the northern regions simulations of snow depth compared best with observations for the POTVEG simulations (Fig. 1a, 2a). Across the south, the opposite was true; the best comparisons with observations were found with simulations of maize (Fig. 1b, 2b). This is potentially a result of snow depth observations across the northern study having a higher likelihood of being collected in land cover/land use settings that are more reflective of natural vegetation (trees, shrubs, grasses), and therefore the snow depth in POTVEG simulations would likely agree better in those areas given closer agreement in plant phenology and LAI. Across the south, a higher proportion of land cover is in crops and might be the more likely land cover type where observations are collected, and therefore it might not be surprising that snow depth from the crop simulations across the southern region (Fig. 1b, 2b) compares the best across the south. Furthermore, accurate measurements of snow depth are known to be difficult to attain, attributed to a variety of factors, but recent improvements in technology may improve accuracy [75].

As discussed, there are several factors, concerning both modeling and observations, that contribute to perceived simulation error, and could call into question the statistical significance of the results. While using observational data across a wide range of research sites in the Midwest to validate the model would be considered a positive attribute of this study, we do not have a clear picture of the land-use history at those sites that could influence soil temperatures and we are focusing on the ability of the model to simulate the coldest (extreme) soil temperatures each year, which may not be sustained for more than a few days. We also know how difficult it is to simulate snow depth and density at any given site without having hourly weather data, and how influential hourly air temperature and other atmospheric conditions are in determining the ratio of liquid precipitation to accumulated snow depth. In Agro-IBIS, an air temperature threshold of 1.1°C is used to determine whether precipitation falls in liquid (rain) or frozen (snow) form, and a constant snow density is assumed. While the gridded climate data that we employed in this study are of extremely high quality and 5 min spatial resolution, some of the model bias for extremely cold soil temperatures (<-6.0°C) is likely induced by uncertainty in simulation of snowfall, snowmelt, and snow density, which is attributed to differences between the gridded climate data used to drive the model and what actually occurred at each site. For example, when additional model validation was performed at two specific locations (experimental stations in Wisconsin and Illinois) and implemented site-specific meteorological (hourly) and management data as drivers, the model agreed with observational data quite well, especially at the monthly time scale. Therefore, when the model had the best available land management and meteorological data to drive simulations, the comparisons were very strong.

The statistical significance of the results should be perceived as strong across large spatial regions like the Midwest US, and a good representation of how varied residue management impacts the typical average minimum soil temperatures. However, subtle changes in land management that effect soil structure, surface roughness, and the ability of the landscape to preferentially trap snow, as well as the short timescale temporal weather patterns, particularly those producing rapid changes in air temperature without snowpack, can lead to widely varying results. These temperature responses may be even greater than the magnitude of changes associated with varied residue management. The results presented here are robust as a broad generalization of what could happen at any particular Midwest US location in the context of residue management on soil temperatures based on the mean climate, but there are a range of other factors in any single winter season that could lead to a significant departure from the simulated mean responses.

Conclusions

There are numerous factors that will influence the ability of miscanthus to overwinter in the first year or two after establishment, including rhizome size, end of season harvest management, planting depth, soil water content, rhizome moisture content at the end of the season, and soil characteristics [4,7,10]. However, there is no denying that residue management plays a significant role in soil thermal dynamics, particularly on wintertime soil temperatures, and therefore on miscanthus rhizome survival during the early stages of establishment. The results presented here illustrated that very small differences in thatch thickness, on the order of a few centimeters, lead to significant differences in minimum winter soil temperatures. Continued field research across a wide climate gradient, and assessing varied management scenarios, will help to fill the gaps in our understanding of rhizome winter survival. A potential wild card might be how long-term climate change impacts snowpack variability and annual minimum soil temperatures, the latter of which have been shown to be warming over the last several decades. In the case of miscanthus, producers could be presented with a new cropping system to support the production of renewable fuels in the future. However, they may be faced with dilemmas on how to best manage miscanthus in its establishment phase to ensure long-term survival while simultaneously remaining profitable.

Supporting Information

Table S1 Soil temperature observation sites used in model validation and minimum soil temperature assessment. List of observation sites that were used to validate Agro-IBIS and for further assessment of annual minimum extreme 10 cm soil temperatures across the Midwest US. Soil surface refers to ground cover present; initial year is the beginning of the observation record used (starting January 1) and end year denotes the last year of the observation record used in this study (last day of record is December 31). Italicized and bolded station names denote the 36 locations that were used in an assessment of soil temperature trends simulated by Agro-IBIS.

Acknowledgments

We thank Tom Voigt, Emily Heaton, Daryl Herzmann, Kent Berns, and Gregg Sanford for helpful discussions. We extend gratitude to Nathan Healy, Michael Cruse, and James Tesmer for collecting and processing soil temperature data.

Author Contributions

Conceived and designed the experiments: CJK AV. Performed the experiments: CJK AV. Analyzed the data: CJK AV JDL MMM. Wrote the paper: CJK AV.

References

1. Zub HW, Brancourt-Hulmel M (2010) Agronomic and physiological performances of different species of Miscanthus, a major energy crop. A review. Agron Sustain Dev 30: 201–214.
2. Robertson GP, Dale VH, Doering OC, Hamburg SP, Melillo JM, et al. (2008) Agriculture - Sustainable biofuels Redux. Science 322: 49–50.
3. Lewandowski I, Clifton-Brown JC, Scurlock JMO, Huisman W (2000) Miscanthus: European experience with a novel energy crop. Biomass Bioenerg 19: 209–227.
4. Heaton EA, Dohleman FG, Long SP (2008) Meeting US biofuel goals with less land: the potential of Miscanthus. Global Change Biol 14: 2000–2014.
5. Dohleman FG, Long SP (2009) More Productive Than Maize in the Midwest: How Does Miscanthus Do It? Plant Physiol 150: 2104–2115.
6. Heaton E, Voigt T, Long SP (2004) A quantitative review comparing the yields of two candidate C-4 perennial biomass crops in relation to nitrogen, temperature and water. Biomass Bioenerg 27: 21–30.
7. Heaton EA, Dohleman FG, Miguez AF, Juvik JA, Lozovaya V, et al. (2010) Miscanthus: A Promising Biomass Crop. Adv Bot Res 56: 75–137.
8. Pyter RJ, Dohleman FG, Voigt TB (2010) Effects of rhizome size, depth of planting and cold storage on Miscanthus × giganteus establishment in the Midwestern USA. Biomass Bioenerg 34: 1466–1470.
9. Greef JM, Deuter M (1993) Syntaxonomy of Miscanthus-X-Giganteus Greef-Et-Deu. Angew Bot 67: 87–90.
10. Pyter R, Heaton E, Dohleman F, Voigt T, Long S (2009) Agronomic Experiences with Miscanthus × giganteus in Illinois, USA. Biofuels: Methods and Protocols 581: 41–52.
11. Khanna M, Dhungana B, Clifton-Brown J (2008) Costs of producing miscanthus and switchgrass for bioenergy in Illinois. Biomass Bioenerg 32: 482–493.
12. Numata M (1974) Grassland vegetation. In: Numata M, The flora and vegetation of Japan. Tokyo: Elsevier. 125–147.
13. Clifton-Brown JC, Lewandowski I (2000) Overwintering problems of newly established Miscanthus plantations can be overcome by identifying genotypes with improved rhizome cold tolerance. New Phytol 148: 287–294.
14. Clifton-Brown JC, Lewandowski I (2000) Water use efficiency and biomass partitioning of three different Miscanthus genotypes with limited and unlimited water supply. Ann Bot 86: 191–200.
15. Clifton-Brown JC, Lewandowski I, Andersson B, Basch G, Christian DG, et al. (2001) Performance of 15 Miscanthus genotypes at five sites in Europe. Agron J 93: 1013–1019.
16. Farrell AD, Clifton-Brown JC, Lewandowski I, Jones MB (2006) Genotypic variation in cold tolerance influences the yield of Miscanthus. Ann Appl Biol 149: 337–345.
17. Horton R, Bristow KL, Kluitenberg GJ, Sauer TJ (1996) Crop residue effects on surface radiation and energy balance - review. Theor Appl Climatol 54: 27–37.
18. Sauer TJ, Hatfield JL, Prueger JH, Norman JM (1998) Surface energy balance of a corn residue-covered field. Agr Forest Meteorol 89: 155–168.
19. Steiner JL, Schomberg HH (1996) Impacts of crop residue at the Earth-atmosphere interface: introduction. Theor Appl Climatol 54: 1–4.
20. Wilhelm WW, Johnson JMF, Hatfield JL, Voorhees WB, Linden DR (2004) Crop and soil productivity response to corn residue removal: A literature review. Agron J 96: 1–17.
21. Kucharik CJ, Twine TE (2007) Residue, respiration, and residuals: Evaluation of a dynamic agroecosystem model using eddy flux measurements and biometric data. Agr Forest Meteorol 146: 134–158.
22. Sharratt BS (2002) Corn stubble height and residue placement in the northern US Corn Belt Part I. Soil physical environment during winter. Soil Till Res 64: 243–252.
23. Kucharik CJ (2003) Evaluation of a Process-Based Agro-Ecosystem Model (Agro-IBIS) across the US Corn Belt: Simulations of the Interannual Variability in Maize Yield. Earth Interact 7.
24. Vanloocke A, Bernacchi CJ, Twine TE (2010) The impacts of Miscanthus × giganteus production on the Midwest US hydrologic cycle. GCB Bioenergy 2: 180–191.
25. VanLoocke A, Twine TE, Zeri M, Bernacchi CJ (2012) A regional comparison of water use efficiency for miscanthus, switchgrass and maize. Agr Forest Meteorol 164: 82–95.
26. Foley JA, Prentice IC, Ramankutty N, Levis S, Pollard D, et al. (1996) An integrated biosphere model of land surface processes, terrestrial carbon balance, and vegetation dynamics. Global Biogeochem Cycles 10: 603–628.
27. Kucharik CJ, Foley JA, Delire C, Fisher VA, Coe MT, et al. (2000) Testing the performance of a Dynamic Global Ecosystem Model: Water balance, carbon balance, and vegetation structure. Global Biogeochem Cycles 14: 795–825.
28. Kucharik CJ, Brye KR (2003) Integrated BIosphere Simulator (IBIS) yield and nitrate loss predictions for Wisconsin maize receiving varied amounts of Nitrogen fertilizer. J Environ Qual 32: 247–268.
29. Campbell GS, Norman JM (1998) An introduction to Environmental biophysics. New York, NY: Springer-Verlag.
30. Lenters JD, Coe MT, Foley JA (2000) Surface water balance of the continental United States, 1963–1995: Regional evaluation of a terrestrial biosphere model and the NCEP/NCAR reanalysis. J Geophys Res-Atmos 105: 22393–22425.
31. Collatz JG, Ball JT, Grivet C, Berry JA (1991) Physiological and environmental regulation of stomatal conductance, photosynthesis and transpiration: a model that includes a laminar boundary layer. Agr Forest Meteorol 53: 107–136.
32. Collatz GJ, Ribas-Carbo M, Berry JA (1992) Coupled photosynthesis-stomatal conductance model for leaves of C_4 plants. Aust J Plant Physiol 19: 519–538.
33. Farquhar GD, von Caemmerer S, Berry JA (1980) A biochemical model of photosynthetic CO_2 assimilation in leaves of C_3 species. Planta 149: 78–90.
34. Li KY, Coe MT, Ramankutty N (2005) Investigation of hydrological variability in west Africa using land surface models. J Clim 18: 3173–3188.
35. Soylu ME, Istanbulluoglu E, Lenters JD, Wang T (2011) Quantifying the impact of groundwater depth on evapotranspiration in a semi-arid grassland region. Hydrol Earth Syst Sci 15: 787–806.
36. Donner SD, Kucharik CJ (2003) Evaluating the impacts of land management and climate variability on crop production and nitrate export across the Upper Mississippi Basin. Global Biogeochem Cycles Doi 10.1029/2001gb001808.
37. Delire C, Foley JA (1999) Evaluating the performance of a land Surface/ecosystem model with biophysical measurements from contrasting environments. J Geophys Res-Atmos 104: 16895–16909.
38. Kucharik CJ, Barford CC, El Maayar M, Wofsy SC, Monson RK, et al. (2006) A multiyear evaluation of a Dynamic Global Vegetation Model at three AmeriFlux forest sites: Vegetation structure, phenology, soil temperature, and CO2 and H2O vapor exchange. Ecol Model 196: 1–31.
39. Twine TE, Kucharik CJ (2008) Evaluating a terrestrial ecosystem model with satellite information of greenness. J Geophys Res-Biogeo 113.
40. El Maayar M, Price DT, Delire C, Foley JA, Back AT, et al. (2001) Validation of the Integrated BIosphere Simulator over Canadian deciduous and coniferous boreal forest stands. J Geophys Res-Atmos 106: 14339–14355.
41. Molling CC, Strikwerda JC, Norman JM, Rodgers CA, Wayne R, et al. (2005) Distributed runoff formulation designed for a precision agricultural landscape modeling system. J Am Water Resour Assoc 41: 1289–1313.
42. Simunek J, Sejna H, Saito H, Sakai M, Van Genuchthen Mt (2008) The HYDRUS-1D Software Package for Simulating the Movement of Water, Heat, and Multiple Solutes in Variably Saturated Media, Version 4.08, HYDRUS Software Series 3. Riverside, CA: University of California Riverside. 330.
43. Simunek J, van Genuchten MT (2008) Modeling nonequilibrium flow and transport processes using HYDRUS. Vadose Zone J 7: 782–797.
44. Flerchinger GN (1991) Sensitivity of Soil Freezing Simulated by the Shaw Model. Trans ASAE 34: 2381–2389.
45. Flerchinger GN, Sauer TJ, Aiken RA (2003) Effects of crop residue cover and architecture on heat and water transfer at the soil surface. Geoderma 116: 217–233.
46. Flerchinger GN, Caldwell TG, Cho J, Hardegree SP (2012) Simultaneous Heat and Water (Shaw) Model: Model Use, Calibration, and Validation. Trans ASAE 55: 1395–1411.
47. Kalnay E, Kanamitsu M, Kistler R, Collins W, Deaven D, et al. (1996) The NCEP/NCAR 40-year reanalysis project. Bull Am Meteorol Soc 77: 437–471.

48. Kanamitsu M, Ebisuzaki W, Woollen J, Yang SK, Hnilo JJ, et al. (2002) NCEP-DOE AMIP-II Reanalysis (R-2). Bull Am Meteorol Soc 83: 1631–1643.
49. Miller DA, White RA (1998) A conterminous United States multilayer soil characteristics dataset for regional climate and hydrology modeling. Earth Interact 5: 42.
50. Amougou N, Bertrand I, Cadoux S, Reous S (2012) Miscanthus × giganteus leaf senescence, decomposition and C and N inputs to soil. GCB Bioenergy 10:1111/j.1757-1707.2012.01192.x.
51. Sharratt BS, Benoit GR, Voorhees WB (1998) Winter soil microclimate altered by corn residue management in the northern Corn Belt of the USA. Soil Till Res 49: 243–248.
52. Cherkauer KA, Lettenmaier DP (2003) Simulation of spatial variability in snow and frozen soil. J Geophys Res-Atmos 108.
53. Sinha T, Cherkauer KA (2008) Time Series Analysis of Soil Freeze and Thaw Processes in Indiana. J Hydrometeorol 9: 936–950.
54. Hu Q, Feng S (2003) A daily soil temperature dataset and soil temperature climatology of the contiguous United States. J Appl Meteorol 42: 1139–1156.
55. Zeri M, Anderson-Teixeira K, Hickman G, Masters M, DeLucia E, et al. (2011) Carbon exchange by establishing biofuel crops in Central Illinois. Agric Ecosyst Environ 144: 319–329.
56. Smith CM, David MB, Mitchell CA, Masters MD, Anderson-Teixeira KJ, et al. (2012) Reduced nitrogen losses following conversion of row crop agriculture to perennial biofuel crops. J Environ Qual 10.2134/jeq2012.0210.
57. Portmann RW, Solomon S, Hegerl GC (2009) Spatial and seasonal patterns in climate change, temperatures, and precipitation across the United States. Proc Nat Acad Sci USA 106: 7324–7329.
58. Kucharik CJ, Serbin SP, Vavrus S, Hopkins EJ, Motew MM (2010) Patterns of Climate Change across Wisconsin from 1950 to 2006. Phys Geogr 31: 1–28.
59. MathWorks (2010) MATLAB 7.11. Natick, MA.
60. Aase JK, Siddoway FH (1979) Crown-Depth Soil Temperatures and Winter Protection for Winter-Wheat Survival. Soil Sci Soc Am J 43: 1229–1233.
61. Zheng D, Hunt Jr., R.H Running, S.W. (1993) A daily soil temperature model based on air temperature and precipitation for continental applications. Climate Res 2.
62. Dyer JL, Mote TL (2006) Spatial variability and trends in observed snow depth over North America. Geophys Res Lett 33.
63. Sharratt BS, Baker DG, Wall DB, Skaggs RH, Ruschy DL (1992) Snow Depth Required for near Steady-State Soil Temperatures. Agr Forest Meteorol 57: 243–251.
64. Jørgensen U, Muhs H.J. (2001) Micanthus breeding and improvement. In: Jones MB, Walsh, M., Miscanthus for energy and fibre. London: Earthscan. 68–85.
65. Price L, Bullard M, Lyons H, Anthony S, Nixon P (2004) Identifying the yield potential of Miscanthus × giganteus: an assessment of the spatial and temporal variability of M-x giganteus biomass productivity across England and Wales. Biomass Bioenerg 26: 3–13.
66. Gauder M, Graeff-Honninger S, Lewandowski I, Claupein W (2012) Long-term yield and performance of 15 different Miscanthus genotypes in southwest Germany. Ann Appl Biol 160: 126–136.
67. Heaton EA, Flavell RB, Mascia PN, Thomas SR, Dohleman FG, et al. (2008) Herbaceous energy crop development: recent progress and future prospects. Curr Opin Biotechnol 19: 202–209.
68. Isard SA, Schaetzl RJ, Andresen JA (2007) Soils cool as climate warms in the great lakes region: 1951–2000. Ann Assoc Am Geogr 97: 467–476.
69. Sinha T, Cherkauer KA, Mishra V (2010) Impacts of Historic Climate Variability on Seasonal Soil Frost in the Midwestern United States. J Hydrometeorol 11: 229–252.
70. Sinha T, Cherkauer KA (2010) Impacts of future climate change on soil frost in the midwestern United States. J Geophys Res-Atmos 115.
71. Cherkauer KA, Lettenmaier DP (1999) Hydrologic effects of frozen soils in the upper Mississippi River basin. J Geophys Res-Atmos 104: 19599–19610.
72. Kunkel KE, Isard SA, Hollinger SE, Gleason H, Belding M (1999) Spatial heterogeneity of albedo over a snow-covered agricultural landscape. J Geophys Res-Atmos 104: 19551–19557.
73. Hillel D (2004) An introduction to environmental soil physics. San Diego: Academic Press.
74. Kucharik CJ (2006) A multidecadal trend of earlier corn planting in the central USA. Agron J 98: 1544–1550.
75. Kunkel KE, Palecki MA, Hubbard KG, Robinson DA, Redmond KT, et al. (2007) Trend identification in twentieth-century US snowfall: The challenges. J Atmos Oceanic Technol 24: 64–73.
76. Sauer TJ, Hatfield JL, Prueger JH (1996) Aerodynamic characteristics of standing corn stubble. Agron J 88: 733–739.
77. Bristow KL, Campbell GS, Papendick RI, Elliott LF (1986) Simulation of Heat and Moisture Transfer through a Surface Residue Soil System. Agr Forest Meteorol 36: 193–214.
78. Shen Y, Tanner CB (1990) Radiative and Conductive Transport of Heat through Flail-Chopped Corn Residue. Soil Sci Soc Am J 54: 653–658.
79. Miao Z, Grift TE, Hansen AC, Ting KC (2011) Energy requirement for comminution of biomass in relation to particle physical properties. Ind Crops Prod 33: 504–513.
80. Chung SO, Horton R (1987) Soil Heat and Water-Flow with a Partial Surface Mulch. Water Resour Res 23: 2175–2186.
81. Wagenaar BM, VandenHeuvel EJMT (1997) Co-combustion of Miscanthus in a pulverised coal combustor: Experiments in a droptube furnace. Biomass Bioenerg 12: 185–197.
82. Van Wijk WR, DeVries DA (1963) Periodic temperature variations in a homogeneous soil. In: Van Wijk WR, Physics of the plant environment. Amsterdam: North-Holland. 102–143.
83. Clemmensen AW (2004) Physical characteristics of Miscanthus composts compared to peat and wood fiber growth substrates. Compost Sci Util 12: 219–224.
84. Tanner CB, Shen Y (1990) Water-Vapor Transport through a Flail-Chopped Corn Residue. Soil Sci Soc Am J 54: 945–951.
85. Sui HJ, Zeng DC, Chen FZ (1992) A Numerical-Model for Simulating the Temperature and Moisture Regimes of Soil under Various Mulches. Agr Forest Meteorol 61: 281–299.

PERMISSIONS

All chapters in this book were first published in PLOS ONE, by The Public Library of Science; hereby published with permission under the Creative Commons Attribution License or equivalent. Every chapter published in this book has been scrutinized by our experts. Their significance has been extensively debated. The topics covered herein carry significant findings which will fuel the growth of the discipline. They may even be implemented as practical applications or may be referred to as a beginning point for another development.

The contributors of this book come from diverse backgrounds, making this book a truly international effort. This book will bring forth new frontiers with its revolutionizing research information and detailed analysis of the nascent developments around the world.

We would like to thank all the contributing authors for lending their expertise to make the book truly unique. They have played a crucial role in the development of this book. Without their invaluable contributions this book wouldn't have been possible. They have made vital efforts to compile up to date information on the varied aspects of this subject to make this book a valuable addition to the collection of many professionals and students.

This book was conceptualized with the vision of imparting up-to-date information and advanced data in this field. To ensure the same, a matchless editorial board was set up. Every individual on the board went through rigorous rounds of assessment to prove their worth. After which they invested a large part of their time researching and compiling the most relevant data for our readers.

The editorial board has been involved in producing this book since its inception. They have spent rigorous hours researching and exploring the diverse topics which have resulted in the successful publishing of this book. They have passed on their knowledge of decades through this book. To expedite this challenging task, the publisher supported the team at every step. A small team of assistant editors was also appointed to further simplify the editing procedure and attain best results for the readers.

Apart from the editorial board, the designing team has also invested a significant amount of their time in understanding the subject and creating the most relevant covers. They scrutinized every image to scout for the most suitable representation of the subject and create an appropriate cover for the book.

The publishing team has been an ardent support to the editorial, designing and production team. Their endless efforts to recruit the best for this project, has resulted in the accomplishment of this book. They are a veteran in the field of academics and their pool of knowledge is as vast as their experience in printing. Their expertise and guidance has proved useful at every step. Their uncompromising quality standards have made this book an exceptional effort. Their encouragement from time to time has been an inspiration for everyone.

The publisher and the editorial board hope that this book will prove to be a valuable piece of knowledge for researchers, students, practitioners and scholars across the globe.

LIST OF CONTRIBUTORS

Hu Baolan, Liu Shuai, Shen Lidong, Zheng Ping, Xu Xiangyang and Lou Liping
Department of Environmental Engineering, Zhejiang University, Hangzhou, Zhejiang, China

W. H. Gera Hol and Freddy ten Hooven
Netherlands Institute of Ecology, Department of Terrestrial Ecology, Wageningen, The Netherlands

Wietse de Boer
Netherlands Institute of Ecology, Department of Microbial Ecology, Wageningen The Netherlands
Wageningen University, Department of Soil Quality, Wageningen, The Netherlands

Wim H. van der Putten
Netherlands Institute of Ecology, Department of Terrestrial Ecology, Wageningen, The Netherlands
Wageningen University, Laboratory of Nematology, Wageningen, The Netherlands

Qingfeng Meng, Shanchao Yue, Xinping Chen, Zhenling Cui and Fusuo Zhang
Center for Resources, Environment and Food Security, China Agricultural University, Beijing, China

Youliang Ye
College of Resources and Environmental Sciences, Henan Agricultural University, Zhengzhou, China

Wenqi Ma
College of Resources and Environmental Sciences, Hebei Agricultural University, Baoding, China

Yanan Tong
College of Resources and Environmental Sciences, Northwest Sci-Tech University of Agriculture and Forestry, Yangling, China

Xia Zhu
Chengdu Institute of Biology, Chinese Academy of Sciences, Chengdu, China
University of Chinese Academy of Sciences, Beijing, China
Department of Land, Air, and Water Resources, University of California Davis, Davis, California, United States of America

Lucas C. R. Silva, Timothy A. Doane and William R. Horwath
Department of Land, Air, and Water Resources, University of California Davis, Davis, California, United States of America

Xia Li, Hang Wang, ShaoHua Gan and GuangMing Tian
College of Environmental and Resource Science, Zhejiang University, Hangzhou, China

DaQian Jiang
Department of Earth and Environmental Engineering, Columbia University,
New York, New York, United States of America

ZhiJian Zhang
College of Environmental and Resource Science, Zhejiang University, Hangzhou, China
Department of Earth and Environmental Engineering, Columbia University, New York, New York, United States of America
China Academy of West Region Development, Zhejiang University, Hangzhou, China

Shaohua Chen, Jianjun Luo, Meiying Hu, Peng Geng and Yanbo Zhang
Laboratory of Insect Toxicology, Key Laboratory of Pesticide and Chemical Biology, Ministry of Education, South China Agricultural University, Guangzhou, People's Republic of China

Anzhou Ma, Xuliang Zhuang, Junmei Wu, Mengmeng Cui and Guoqiang Zhuang
Research Center for Eco-Environmental Sciences, Chinese Academy of Sciences, Beijing, P.R. China

Di Lv
Insitute of Microbiology, Chinese Academy of Sciences, Beijing, P.R. China

Chunzhao Liu
National Key Laboratory of Biochemical Engineering, Institute of Process Engineering, Chinese Academy of Sciences, Beijing, P.R. China

Yi Wang
National Key Laboratory of Crop Genetic Improvement, and Microelement Research Center, Huazhong Agricultural University, Wuhan, China
College of Resources and Environmental Sciences, Henan Agricultural University, Zhengzhou, China

Xiangsheng Ye, Guangda Ding and Fangsen Xu
National Key Laboratory of Crop Genetic Improvement, and Microelement Research Center, Huazhong Agricultural University, Wuhan, China

Bin Zhang
State Key Laboratory of Forest and Soil Ecology, Institute of Applied Ecology, Chinese Academy of Sciences, Shenyang, P. R. China
Graduate University of Chinese Academy of Sciences, Beijing, P. R. China

Chao Liang
Department of Soil Science, University of Wisconsin-Madison, Madison, Wisconsin, United States of America

Hongbo He
State Key Laboratory of Forest and Soil Ecology, Institute of Applied Ecology, Chinese Academy of Sciences, Shenyang, P. R. China

Xudong Zhang
State Key Laboratory of Forest and Soil Ecology, Institute of Applied Ecology, Chinese Academy of Sciences, Shenyang, P. R. China
National Field Observation and Research Station of Shenyang Agroecosystems, Shenyang, P. R. China

Shuping Qin and Chunsheng Hu
Key Laboratory of Agricultural Water Resources, Center for Agricultural Resources Research, Institute of Genetic and Developmental Biology, The Chinese Academy of
Sciences, Shijiazhuang, Hebei, China

Oene Oenema
Wageningen University and Research, Alterra, Wageningen, The Netherlands

Jake E. Simpson
Centre for Ecology and Hydrology, Natural Environment Research Council, Wallingford, Oxfordshire, United Kingdom
Department of Biology, Imperial College London, London, United Kingdom

Eleanor Slade
Wildlife Conservation Research Unit, Department of Zoology, University of Oxford, Oxford, Oxfordshire, United Kingdom

Terhi Riutta
Environmental Change Institute, School of Geography and the Environment, University of Oxford, Oxford, Oxfordshire, United Kingdom

Michele E. Taylor
Centre for Ecology and Hydrology, Natural Environment Research Council, Wallingford, Oxfordshire, United Kingdom

Stefano Casalegno, Richard Inger, Caitlin DeSilvey and Kevin J. Gaston
Environment and Sustainability Institute, University of Exeter, Penryn, Cornwall, United Kingdom

Caio T. C. C. Rachid, Adriana L. Santos, Alexandre S. Rosado and Raquel S. Peixoto
Institute of Microbiology Paulo de Góes, Federal University of Rio de Janeiro, Rio de Janeiro, Rio de Janeiro, Brazil

Marisa C. Piccolo
Center for Nuclear Energy in Agriculture, Universidade de São Paulo, Piracicaba, São Paulo, Brazil

Fabiano C. Balieiro and Heitor L. C. Coutinho
Embrapa Solos, Rua Jardim Botânico, Rio de Janeiro, Rio de Janeiro, Brazil

James M. Tiedje
Center for Microbial Ecology, Michigan State University, East Lansing, Michigan, United States of America

Xiaorong Wei, Liping Qiu, Mingan Shao and Xingchang Zhang
State Key Laboratory of Soil Erosion and Dryland Farming in the Loess Plateau, Northwest A & F University, Yangling, China

William J. Gale
College of Resources and Environment, Northwest A & F University, Yangling, China

Xuemei Xiao, Zhihui Cheng, Huanwen Meng, Lihong Liu, Hezi Li and Yinxin Dong
College of Horticulture, Northwest A&F University, Yangling, Shaanxi, China

Adam J. Heathcote, Christopher T. Filstrup and John A. Downing
Department of Ecology, Evolution, and Organismal Biology, Iowa State University, Ames, Iowa, United States of America

Fasheng Zhang
Institute of Applied Ecology, Chinese Academy of Sciences, Shenyang, China
Eastern Cereal and Oilseed Research Centre, Agriculture and Agri-Food Canada, Ottawa, Canada

Guanghua Yin, Zhenying Wang and Zuoxin Liu
Institute of Applied Ecology, Chinese Academy of Sciences, Shenyang, China

Neil McLaughlin and Xiaoyuan Geng
Eastern Cereal and Oilseed Research Centre, Agriculture and Agri-Food Canada, Ottawa, Canada

Ramona Marasco, Eleonora Rolli, Daniele Daffonchio, Claudia Sorlini, Francesca Mapelli and Sara Borin
Dipartimento di Scienze per gli Alimenti, la Nutrizione e l'Ambiente, Università degli Studi di Milano, Milan, Italy

Besma Ettoumi and Ameur Cherif
Laboratoire Microorganismes et Biomolécules Actives, Université Tunis El Manar, Tunis, Tunisia and Laboratoire Biotechnologie et Valorisation des Bio-Géo Ressources, Institut Supérieur de Biotechnologie, Université de La Manouba, Sidi Thabet, Ariana, Tunisia

Gianpiero Vigani and Graziano Zocchi
Dipartimento di Scienze Agrarie e Alimentari-Produzione, Territorio, Agroenergia; Università degli Studi di Milano, Milan, Italy

Ayman F. Abou-Hadid and Usama A. El-Behairy
Department of Horticulture, Ain Shams University, Cairo, Egypt

Dejun Li
Department of Microbiology and Plant Biology, University of Oklahoma, Norman, Oklahoma, United States of America

Marijn van der Velde, Linda See, Juraj Balkovič, Steffen Fritz, Nikolay Khabarov and Michael Obersteiner
International Institute for Applied Systems Analysis (IIASA), Ecosystem Services and Management Program, Laxenburg, Austria

Liangzhi You
International Food Policy Research Institute (IFPRI), Washington D.C., United States of America
College of Economics and Management, Huazhong Agricultural University, Wuhan, China

Stanley Wood
Global Development Program, Bill & Melinda Gates Foundation, Seattle, Washington, United States of America

Silja Emilia Hannula and Wietse de Boer
Netherlands Institute of Ecology (NIOO-KNAW), Department of Microbial Ecology, Wageningen, The Netherlands

Johannes van Veen
Netherlands Institute of Ecology (NIOO-KNAW), Department of Microbial Ecology, Wageningen, The Netherlands
Insititute of Biology, Leiden University, Leiden, The Netherlands

Xiaobo Yi and Li Wang
College of Resources and Environment, Northwest A&F University, Yangling, Shaanxi, China
State Key Laboratory of Soil Erosion and Dryland Farming on the Loess Plateau, Northwest A&F University, Yangling, Shannxi, China

Motior M. Rahman, Sofian M. Azirun and Amru N. Boyce
Institute of Biological Sciences, Faculty of Science, University of Malaya, Kuala Lumpur, Malaysia

Nie Xiaojun
Key Laboratory of Mountain Surface Processes and Ecological Regulation, Chinese Academy of Sciences, Institute of Mountain Hazards and Environment, Chinese
Academy of Sciences and Ministry of Water Conservancy, Chengdu, China
School of Surveying and Land Information Engineering, Henan Polytechnic University, Jiaozuo, China

Zhang Jianhui and Su Zhengan
Key Laboratory of Mountain Surface Processes and Ecological Regulation, Chinese Academy of Sciences, Institute of Mountain Hazards and Environment, Chinese
Academy of Sciences and Ministry of Water Conservancy, Chengdu, China

Xiaoqi Zhou
Graduate University of Chinese Academy of Sciences, Beijing, China
Scion (New Zealand Forest Research Institute), Christchurch, New Zealand
Environmental Futures Centre, Griffith University, Nathan, Australia

Chengrong Chen
Environmental Futures Centre, Griffith University, Nathan, Australia

Yanfen Wang
Graduate University of Chinese Academy of Sciences, Beijing, China

Simeon Smaill and Peter Clinton
Scion (New Zealand Forest Research Institute), Christchurch, New Zealand

Christopher J. Kucharik
Department of Agronomy, University of Wisconsin-Madison, Madison, Wisconsin, United States of America
Nelson Institute Center for Sustainability and the Global Environment, University of Wisconsin-Madison, Madison, Wisconsin, United States of America

Department of Energy Great Lakes Bioenergy Research Center, University of Wisconsin-Madison, Madison, Wisconsin, United States of America

Andy VanLoocke
Department of Atmospheric Sciences, University of Illinois, Urbana, Illinois, United States of America

John D. Lenters
School of Natural Resources, University of Nebraska-Lincoln, Lincoln, Nebraska, United States of America

Melissa M. Motew
Nelson Institute Center for Sustainability and the Global Environment, University of Wisconsin-Madison, Madison, Wisconsin, United States of America

Index